S0-EQO-301

The electromagnetic

interaction

R. L. ARMSTRONG / J. D. KING
UNIVERSITY OF TORONTO

The electromagnetic

interaction

PRENTICE-HALL, INC.
ENGLEWOOD CLIFFS, NEW JERSEY

Library of Congress Cataloging in Publication Data

ARMSTRONG, ROBIN L.
  The electromagnetic interaction.

  Includes bibliographical references.
  1. Electromagnetic interactions. I. King,
James Douglas, joint author.
II. Title.
QC794.A72   537   72-13889
ISBN 0-13-249110-9

The electromagnetic interaction

R. L. Armstrong / J. D. King

© *1973 by Prentice-Hall, Inc., Englewood Cliffs, N.J.*
All rights reserved. No part of this book may be reproduced in any
form or by any means without permission in writing from the publisher.

10 9 8 7 6 5 4 3 2 1

Printed in the United States of America

PRENTICE-HALL INTERNATIONAL, INC.	*London*
PRENTICE-HALL OF AUSTRALIA, PTY. LTD.	*Sydney*
PRENTICE-HALL OF CANADA, LTD.	*Toronto*
PRENTICE-HALL OF INDIA PRIVATE LIMITED	*New Delhi*
PRENTICE-HALL OF JAPAN, INC.	*Tokyo*

Contents

Preface, ix

1 An introduction to the electromagnetic interaction, 1

1.1 The importance of the electromagnetic interaction, 2
1.2 The historical development of electromagnetism, 2
1.3 Scope and organization of this text, 5

2 Charge and current, 7

2.1 Introduction, 8
2.2 Coulomb's law, 9
2.3 Coulomb scattering, 10
2.4 Forces between currents, 13
2.5 Systems of units, 14

3 The electric field, 19

3.1 Introduction, 20
3.2 Definition of the electric field, 20
3.3 The elementary electric charge, 23
3.4 The electric dipole, 25
3.5 Molecular dipole moments, 28
3.6 Electric potential, 32
3.7 The electric dipole revisited, 36
3.8 The electric quadrupole, 37
3.9 Electric moments of elementary particles and nuclei, 40
3.10 The moments of a charge distribution, 43

4 Gauss' law, 48

4.1 Introduction, 49
4.2 Proof of Gauss' law, 49
4.3 Some applications of Gauss' law, 51
4.4 Gauss' law for gravitational interactions, 55

5 Capacitance, 59

5.1 Introduction, 60
5.2 The parallel-plate capacitor, 60
5.3 The oscilloscope, 61
5.4 The cylindrical capacitor, 64
5.5 Combinations of capacitors, 65
5.6 Energy stored in a capacitor, 67

6 The magnetic field, 73

6.1 Introduction, 74
6.2 Magnetic force on a moving charge, 74

v

CONTENTS

6.3 Cosmic Rays, 77
6.4 Motion of a charge in a magnetic field, 79
6.5 Mass spectrometers, 82
6.6 Van Allen radiation zones, 84
6.7 Magnetic force on a current, 86

7 Ampère's law, 91

7.1 Introduction, 92
7.2 The force between current elements, 92
7.3 Ampère's law, 95
7.4 The relation between electric and magnetic fields, 97
7.5 The field of a point charge in uniform motion, 102
7.6 Fusion power reactors, 104
7.7 Current loops; the magnetic dipole moment, 106
7.8 Electron, nuclear, and atomic magnetic dipole moments, 108
7.9 Magnetic dipole moment in an inhomogeneous magnetic field, 111

8 Electromagnetic induction, 117

8.1 Introduction, 118
8.2 Electromotive force (emf), 118
8.3 Induced emf, 119
8.4 The betatron, 123
8.5 The magnetic field of the earth, 125
8.6 Inductance, 127
8.7 Energy stored in a magnetic field, 129
8.8 Oscillations in LC circuits, 131

9 Maxwell's equations: electromagnetic waves, 136

9.1 Introduction, 137
9.2 Some useful vector mathematics, 137
9.3 Gauss' law: Maxwell's first and fourth equations, 138
9.4 Ampère's law: Maxwell's second equation, 139
9.5 Faraday's law: Maxwell's third equation, 143
9.6 Summary of Maxwell's equations, 144
9.7 Electromagnetic waves, 145
9.8 The invariance of Maxwell's equations, 150
9.9 Electromagnetic radiation, 154
9.10 Electromagnetic waves from space, 162

10 Dielectric materials, 172

10.1 Introduction, 173
10.2 Polarization and susceptibility, 173
10.3 Electric displacement, 175
10.4 Static dielectric constants of polar and nonpolar media, 182
10.5 Ferroelectric crystals, 187

11 Current, resistance, and the free electron theory of metals, 192

11.1 Introduction, 193
11.2 Ohm's law, 193
11.3 The separation of living cells according to their volumes, 196
11.4 The Fermi electron gas, 199
11.5 Temperature variation of electrical conductivity, 203
11.6 Electromotive force, 204
11.7 The flow of current in circuits containing resistance, capacitance, and inductance, 208

12 Band theory of solids: conductors, semiconductors, and insulators, 218

12.1 Introduction, 219
12.2 Metallic crystals, 219
12.3 Band theory, 221
12.4 Conductors and insulators, 222
12.5 Semiconductors, 224
12.6 The Hall effect, 228
12.7 Effective mass, 230
12.8 p-n junctions: transistors, 231
12.9 Superconductors, 234

13 Magnetic properties of matter, 241

13.1 Introduction, 242
13.2 Magnetic parameters, 242
13.3 Diamagnetism, 245
13.4 Paramagnetism, 247
13.5 Ferromagnetism, 252
13.6 Antiferromagnetism and ferrimagnetism, 255
13.7 Rock magnetism and continental drift, 256

14 Electromagnetic waves in matter: classical effects, 261

14.1 Introduction, 262
14.2 The general form of Maxwell's equations, 262
14.3 Plane waves using complex notation, 263
14.4 Plane waves in isotropic dielectrics, 264
14.5 The Fizeau experiment, 270
14.6 Plane waves in conducting media, 271
14.7 Dispersion, 272
14.8 Cerenkov radiation, 276
14.9 Polarization, 280
14.10 Unpolarized radiation, 284
14.11 Double refraction, 286

15 Electromagnetic waves in matter: quantum effects, 293

15.1 Introduction, 294
15.2 The interaction of photons with matter, 294
15.3 Optical absorption, 295
15.4 Absorption of ultraviolet radiation, x-rays, and γ-rays, 300
15.5 Biological interactions, 306
15.6 The Einstein coefficients, 310

16 Magnetic resonance, 316

16.1 Introduction, 317
16.2 The resonance phenomenon, 318
16.3 Energy absorption and spin lattice relaxation, 321
16.4 Bloch's equation, 326
16.5 Chemical exchange, 330
16.6 Chemical shifts and electron-coupled spin interactions, 333
16.7 Nuclear magnetic dipolar interactions in solids, 336
16.8 Nuclear spin-lattice relaxation, 338
16.9 Electron paramagnetic resonance in solids, 340
16.10 Nuclear quadrupole resonance, 344
16.11 The proton Magnetometer, 348

17 Double beam interference, 356

17.1 Introduction, 357
17.2 Interference of coherent radiation, 357
17.3 Temporal coherence, 360
17.4 Spatial coherence, 365
17.5 Interferometers in radio astronomy, 368
17.6 Intensity interferometry, 370
17.7 The Michelson interferometer, 371
17.8 The Michelson-Morley experiment, 374

18 Multiple beam interference, 380

18.1 Introduction, 381
18.2 Multiple reflections from a thin plate, 381
18.3 The Fabry-Perot interferometer, 383
18.4 Multilayer films, 387

19 Diffraction, 393

19.1 Introduction, 394
19.2 Fraunhofer and Fresnel diffraction, 395
19.3 Fraunhofer diffraction by a single slit (qualitative), 396
19.4 The Fresnel-Kirchhoff formula for Fraunhofer diffraction, 399
19.5 Diffraction gratings, 404
19.6 Fresnel diffraction, 406

19.7 *Diffraction of x-rays and particles, 409*
19.8 *Spatial filtering, 413*
19.9 *Holography, 417*

20 Nonlinear optics and lasers, 424

20.1 *Introduction, 425*
20.2 *Nonlinear optical effects, 425*
20.3 *The principle of the laser, 429*
20.4 *Amplification in a medium and the threshold condition for laser action, 433*
20.5 *Some basic kinds of lasers, 434*
20.6 *Nonlinear optical effects observed with lasers, 442*

APPENDICES

A Some useful vector relations, 451

A.1 *Gradient and divergence in curvilinear coordinates, 452*
A.2 *The divergence theorem, 453*
A.3 *Stokes' theorem, 454*

B Rotating coordinate axes, 456

C Complex numbers, 459

D Matrix algebra, 463

E Periodic table of the elements, 467

F Physical constants and conversion factors, 470

F.1 *Physical constants, 471*
F.2 *Conversion factors, 471*

ANSWERS TO ODD-NUMBERED PROBLEMS, 473

INDEX, 481

Preface

Although the course of study presented in this book follows logically a first course in physics based on our earlier text *Mechanics, Waves, and Thermal Physics*, any course in which mechanics and wave motion are discussed will serve as an adequate prerequisite. This text presents a development of basic electromagnetism with applications to a wide variety of problems in physics, astronomy, geophysics, chemistry, and biology. That is, through the examples discussed we have attempted to provide a significant interdisciplinary flavor. Many references to articles in *Scientific American, Contemporary Physics, Physics Today*, and other scientific journals are included to encourage the reader to seek additional information in areas in which he is particularly interested. Rather more material is presented than can normally be covered in any single course, thereby allowing an instructor a considerable amount of freedom in the selection of material for his course.

The first chapter gives a brief historical resumé of the development and statement of the importance of electromagnetism.

Chapters two through nine discuss the basic experimental laws of electricity and magnetism culminating in Maxwell's free space equations. The effects of matter on electric and magnetic fields are deliberately not discussed in order to emphasize that the existence of these fields does not require the presence of matter.

The electric and magnetic properties of matter are treated in chapters ten through thirteen.

The interaction of electromagnetic waves and matter is considered in chapters fourteen through twenty. Some well-known classical and quantum effects are treated in chapters fourteen and fifteen. Visible light is introduced as a particular example of electromagnetic waves. Chapter sixteen is devoted to a discussion of magnetic resonance, the interaction of microwave and radio-frequency radiation with matter. Chapters seventeen through nineteen deal with interference and diffraction phenomena; chapter twenty contains a discussion of nonlinear optics and lasers.

Appendices containing information on vectors, complex numbers, matrix algebra, and rotating coordinate axes are also included.

The questions and problems at the end of each chapter are arranged in the order of the sections to which they refer. The more difficult problems are designated by the symbol *.

Photographs of some scientists who have made important contributions to the development of electromagnetism appear on the title pages of each chapter. The set of photographs, although intended to be representative of the personalities involved, is certainly not all-inclusive.

Finally, we wish to express particular thanks to two talented University of Toronto physics students: to Peter Johannsonn for his constructive criticism of the manuscript and for the many questions and problems that he contributed and to Henry Van Driel for the large number of solutions to the problems that he provided.

R. L. ARMSTRONG
J. D. KING

The electomagnetic interaction

RICHARD P. FEYNMAN

1
An introduction to the electromagnetic interaction

1.1 THE IMPORTANCE OF THE ELECTROMAGNETIC INTERACTION

There are just four basic **forces** or **interactions** that have been observed to occur in natural phenomena. The **gravitational interaction** is an attractive force between masses; the **electromagnetic interaction** is associated with charges and may be either attractive or repulsive; the **nuclear interaction** is responsible for the binding together of the protons and neutrons in the atomic nucleus; and the **weak interaction** gives rise to the radioactive decay of atomic nuclei.

The electromagnetic interaction is particularly important since it largely determines the physical and chemical properties of matter over the whole range from atoms to living cells. Giant biological molecules contain tens to hundreds of thousands of atoms assembled into complex geometrical configurations. The shape and stability of such molecules are dependent on the delicate balancing of electromagnetic interactions between neighboring atoms. The atoms of which solids are composed are, in general, arranged in regular patterns with the regularity of order extending over distances that are enormous compared to atomic dimensions. It is the electromagnetic interaction that is responsible for this macroscopic ordering. We could continue with many more examples, but we shall be content to say that most of chemistry, biology and physics is concerned with the electromagnetic interaction. It follows that an understanding of this interaction is essential to all scientists.

1.2 THE HISTORICAL DEVELOPMENT OF ELECTROMAGNETISM[1,2,3]

During the latter half of the eighteenth century Charles Augustin Coulomb investigated the forces between electrically charged objects. He found that the force between two such objects acted along the line joining them, that it was proportional to the product of their charges, and that it was inversely proportional to the square of their separation. This result shows a striking similarity to the law of gravitation.

The discovery of the magnetic effects associated with electric currents came early in the nineteenth century. Hans Christian Oersted observed that a current flowing in a loop of wire exerted magnetic forces on permanent magnets. Subsequently, André Marie Ampère showed that the

---

[1] R. Taton [ed.], *History of Science* (3 volumes) (New York: Basic Books, Inc., 1963).

[2] W. F. Magie, *A Source Book in Physics* (Cambridge, Mass.: Harvard University Press, 1963).

[3] M. H. Shamos [ed.], *Great Experiments in Physics* (New York: Holt, Rinehart & Winston, Inc., 1959).

magnetic force resulting from the flow of current in a circuit can be considered as a sum of contributions from infinitesimal segments of the circuit. The force due to each segment falls off inversely as the square of the distance.

Up to this point in the development of electromagnetic theory scientists thought of the electric and magnetic forces in terms of the "action-at-a-distance" concept. The beginning of the change toward a "field theory" view of the interactions came in the first half of the nineteenth century. Michael Faraday introduced **lines of force** which, at an arbitrary point in space, point in the direction of the force that would be exerted on a test charge located at that point. The number of lines of force per unit area was made proportional to the magnitude of the force. His fundamental idea was that charged objects establish a field in the region of space around them and that the interaction between objects occurs via this field. Faraday's discovery of **electromagnetic induction**, in which a time-varying magnetic field through a circuit induces voltage in the circuit, established the importance of the concept of a magnetic field.

Karl Frederick Gauss, a contemporary of Faraday, was able to relate the number of lines of force out of a region to the total charge within the region. This was the first mathematical formulation of field theory.

James Clerk Maxwell accepted the idea that the electric and magnetic fields were the fundamental quantities and was able to devise a set of partial differential equations obeyed by these fields. He immediately noted that the equations predicted that an electromagnetic disturbance originating at one charged object would travel out as a wave with a speed that could be deduced from electrical and magnetic measurements. The predicted speed, within the small experimental error of measurement, agreed with the speed of light. Maxwell's contribution to physics was very significant. He had provided convincing evidence of the superiority of a field theory to an action-at-a-distance theory and also established the link between electromagnetism and optics. Neither the revolution of quantum mechanics nor the advent of the special theory of relativity has in any way reduced the significance of the electromagnetic field equations Maxwell proposed over one hundred years ago.

During the early period of the development of optics there were two diametrically opposed theories of light. Isaac Newton on the one hand believed in a corpuscular theory in which light consisted of a stream of infinitesimal particles; Christian Huygens held the view that light was a type of wave motion. Newton objected to the wave theory on the basis that it could not explain the sharp shadows cast by objects illuminated with visible light. He failed to realize that, if light waves were of sufficiently short wavelength, the observation of sharp shadows presented no contradiction.

Early in the nineteenth century the work of Thomas Young and

Augustin Fresnel provided the explanation of the **interference** and **diffraction** of light in terms of a wave theory. Fresnel also deduced laws of reflection and refraction which were subsequently shown to predict correctly the fraction of incident light in the reflected and refracted beams as well as the directions of these beams.

Physicists of the nineteenth century tried to understand a wave model of light in terms of the transverse vibrations of a hypothetical elastic medium which they named the **ether**.[4] However, the properties required of the ether were unrealistic. This was the situation when Maxwell formulated his equations. In terms of these equations the electric and magnetic fields are the fundamental entities concerned with optics and electromagnetism and there is no need to introduce the ether at all. Furthermore, Maxwell's equations yield only transverse vibrations whereas an elastic-solid medium transmits longitudinal as well as transverse waves.

A few years later Heinrich Hertz produced electromagnetic waves of several centimeters wavelength in the laboratory. From this time Maxwell's equations have received universal acceptance, and optics has been treated as a branch of electromagnetism.

Early in the twentieth century Max Planck introduced the quantum concept and shortly afterward Albert Einstein predicted the existence of the photon and used Planck's quantum hypothesis to deduce the law of the photoelectric effect. Whereas a wave theory predicts a continuous distribution of energy in a wave, the existence of photons requires that the energy in an electromagnetic wave be quantized in discrete packets as for a corpuscular theory. For some time this apparent contradiction provided a serious impediment to the development of quantum theory. It is now recognized that both the corpuscular and wave points of view simultaneously have their merits—the wave picture correctly describes the average behavior of the photons but the energy is carried by the photons in discrete amounts. For long wavelength radiation, such as radio-frequency radiation, the photon energy is small and ordinary sources provide such enormous numbers of photons that their discreteness is not of practical importance. On the other hand, for short wavelength radiation, such as gamma-ray radiation, the photon energy is large and the discreteness is of striking significance. The incorporation of wave-particle duality into a single theory of electromagnetism falls into the difficult realm of **quantumelectrodynamics**, invented by Paul Dirac. Later significant developments of the theory were pioneered by Richard Feynman, Julian Schwinger, and Shin-Ichiro Tomonaga.

A second important development of electromagnetic theory in the twentieth century has been its role in the quantum theory of atomic and molecular structure. Niels Bohr proposed that the hydrogen atom consists

---

[4] *Encyclopaedia Britannica* 9th. Ed. (1875–89), VIII, 568.

of a proton and an electron which interact with each other through the electrostatic interaction but whose motion is governed by quantum mechanics. Louis de Broglie suggested that a wave phenomenon was associated with the motion of electrons. Erwin Schroedinger described this wave by a wave equation. The solutions of his equation provide accurate descriptions of the structure of atoms, molecules, and bulk matter. The forces that enter Schroedinger's theory are well described by classical electromagnetic theory; it is in the laws of mechanics that this theory differs from the classical theory. Other important contributions to quantum mechanics and the atomic theory have been made by Werner Heisenberg, Paul Dirac, Wolfgang Pauli, Max Born, and Eugene Wigner.

In the 1940's and 1950's the development of **magnetic resonance** took place following the early work of Isidor Rabi, Felix Bloch, and Edward Purcell. This remarkable technique has revolutionized the study of organic chemistry.

The development of **optical masers** or **lasers** in the 1960's gave new life to the field of optics. The invention of the maser and the formulation of the theory of coherent atomic radiation was carried through by Charles Townes, Nikolai Basov, and Aleksandr Prokhorov.

The above brief account is included to provide some insight into the historical development of the subject of electromagnetism. It by no means constitutes a complete history of events but does make reference to some of the more important contributions.

1.3 SCOPE AND ORGANIZATION OF THIS TEXT

It is assumed that the reader of this text has completed a first course in physics which included the study of mechanics, waves, and the thermal properties of matter. In addition, it is assumed that the reader is familiar with the basic concepts of the special theory of relativity and the quantum and statistical character of natural phenomena. This background material may be found, for example, in our previous text,[5] *Mechanics, Waves, and Thermal Physics*. Frequent references to this text will be made and indicated by the designation *MWTP* followed by an appropriate section number. The present approach further implies that the reader has attained some proficiency in the application of mathematical techniques to physics, especially those of integral and differential calculus, and has some acquaintance with the elementary facts of electricity.

In the first half of the book, the basic laws of electricity and magnetism leading up to Maxwell's equations in vacuum are discussed. Emphasis is

---

[5]R. L. Armstrong and J. D. King, *Mechanics, Waves, and Thermal Physics* (Englewood Cliffs, N.J.: Prentice-Hall, Inc., 1970).

placed upon the field approach to the description of the interaction between charged particles. The remainder of the book is devoted to the study of the interaction of electromagnetic fields with matter. Throughout the book many sections are included which show the relevance of the electromagnetic interaction to scientific problems of current interest in physics, chemistry, and biology, and which illustrate the fundamental importance of the electromagnetic interaction in all areas of science.

CHARLES A. DE COULOMB

Courtesy of The New York Public Library, Astor, Lenox and Tilden Foundations.

Charge and current 2

2.1 INTRODUCTION

Charge is one of the four "fundamental" physical quantities in terms of which all other physical quantities are ultimately defined. The charge associated with an object is a measure of its electrical content. Charge is found in nature in two forms called **positive charge** and **negative charge**. It is found by experiment that objects possessing similar charges repel one another, whereas objects possessing dissimilar charges attract one another. Charge, like mass, serves as a label to characterize the elementary particles of physics. All **electrons** carry a negative charge of definite magnitude; all **protons** carry a positive charge equal in magnitude to the electronic charge. Matter is made up of atoms that are composed of a small positively charged nucleus surrounded by a "cloud" of electrons. In its ground state an atom has no net charge. The atomic model that best accounts for this observation and explains, as well, a multitude of other properties of the atom, pictures the atom to consist of a nucleus containing Z protons and an equal number of extranuclear electrons; Z is the **atomic number** of the nucleus. The remainder of the nuclear mass is contributed by **neutrons** that carry no net charge. A short-range attractive nuclear force acts between all the particles (**nucleons**) in the nucleus. This force produces stability in the nucleus which would otherwise tend to break up because of the repulsive electric force between protons. According to the Bohr model of the atom the force of attraction between the positively charged nucleus and the extranuclear electrons provides the centripetal force required for the electrons to orbit the nucleus.

In the atomic model just described, the electrons are only very weakly bound to the atom and one, or even more, may be removed very easily. Macroscopic objects may therefore acquire states of nonzero charge. Very often this is accomplished merely by bringing two dissimilar objects into contact. Anyone who has walked across a rug on a dry winter day and reached out for a doorknob has witnessed the build-up of a charge upon his own person. Separation of charged particles from atoms can occur in diverse ways. For example, charge separation occurs in chemical reactions, in radioactive decay, during nuclear reactions, and even when atoms are bound together into a regular crystalline structure to form a metal.

Current is the term introduced to describe the movement of charge. Quantitatively, the current is defined in terms of the amount of charge that moves past a point or through an area in one second; that is

$$I = nqAv_d$$

where n is the number density of particles of charge q traveling with speed v_d perpendicular to cross-sectional area A. Currents were known and studied long before the discovery of electrons and protons. Having no criterion by which to determine which type of charge moved, scientists arbitrarily chose to talk about the movement of positive charge when discussing currents. This caused no difficulty until it was discovered that, in fact, it is the electrons that usually move. Meanwhile, the concept of "positive" current flow in the discussion of electrical circuits and devices had become firmly established, indeed so firmly established that the positive current flow convention is still often encountered in modern works on electricity and magnetism. In this text, when discussing currents, we shall be referring to the flow of negative charge unless we explicitly state otherwise.

2.2 COULOMB'S LAW

By carrying out experiments involving the interaction between charges at rest, we can arrive at several properties that are characteristic of **electrostatic** forces.

1. There are two kinds of electrical charge. Charges of the same kind repel each other while charges of the opposite kind attract each other.
2. The force acts along the line joining the charges.
3. The force between two charges is proportional to the magnitude of one charge multiplied by the magnitude of the other charge.
4. The force between two charges is inversely proportional to the square of the distance separating them.

From statement (3) we can write that the magnitude F of the force is given by

$$F \propto q_1 q_2$$

where q_1 and q_2 are the magnitudes of the two charges. From statement (4) we can write

$$F \propto \frac{1}{r^2}$$

where r is the separation of the charges. These two equations can be combined to yield the single equation

$$F \propto \frac{q_1 q_2}{r^2}$$

or

$$F = \frac{k_e q_1 q_2}{r^2}$$

where k_e is a constant. This is known as **Coulomb's law** after Charles Augustin Coulomb.

If q_1 and q_2 are both positive or both negative, F will be a positive quantity. However, if one of q_1 and q_2 is positive while the other is negative, F will be negative. Therefore, a positive value for F indicates repul-

sion, according to statement (1), while a negative value indicates attraction.
If we define $\hat{\mathbf{r}}_{12}$ to be a unit vector directed from q_1 toward q_2 as shown in Fig. 2.1, the force $\mathbf{F}_{12}$ exerted on charge q_1 by charge q_2 is given by

$$\mathbf{F}_{12} = -\frac{k_e q_1 q_2}{r_{12}^2}\hat{\mathbf{r}}_{12}$$

while the force $\mathbf{F}_{21}$ exerted on charge q_2 by charge q_1 is

$$\mathbf{F}_{21} = \frac{k_e q_1 q_2}{r_{12}^2}\hat{\mathbf{r}}_{12}.$$

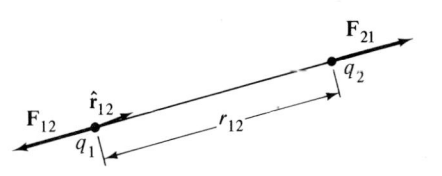

Fig. 2.1. The unit vector $\hat{\mathbf{r}}_{12}$ is directed from q_1 toward q_2.

This is the vector form of Coulomb's law. Note that $\mathbf{F}_{12} = -\mathbf{F}_{21}$, which is just Newton's third law.[1]

Suppose that we have a collection of N charges as shown in Fig. 2.2. The force between each pair is given by Coulomb's law. The force $\mathbf{F}_1$ experienced by charge q_1 because of its interaction with all the remaining charges is

$$\mathbf{F}_1 = \mathbf{F}_{12} + \mathbf{F}_{13} + \cdots + \mathbf{F}_{1N}$$

$$= -\sum_{i=2}^{N}\frac{k_e q_1 q_i}{r_{1i}^2}\hat{\mathbf{r}}_{1i}$$

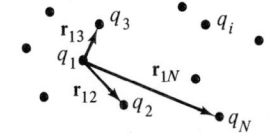

Fig. 2.2. A collection of N charges $q_1, q_2, \ldots, q_i, \ldots, q_N$.

where $\mathbf{F}_{1i}$ is the force on q_1 due to q_i, r_{1i} is the distance between q_1 and q_i, and $\hat{\mathbf{r}}_{1i}$ is a unit vector directed from q_1 toward q_i.

Example. Two charges $+q_1$ and $q_2 = +3q_1$ are fixed in space. Where must a third charge $+q_3$ be placed in order that it experience no net force?

Solution. Since the forces exerted by the two fixed charges on q_3 must be equal and opposite for no net force, q_3 must lie somewhere along the line between q_1 and q_2. Therefore,

$$F_{31} = \frac{k_e q_1 q_3}{r_{13}^2} = F_{32} = \frac{k_e q_2 q_3}{r_{23}^2} = \frac{3k_e q_1 q_3}{r_{23}^2}$$

or

$$r_{23}^2 = 3r_{13}^2$$
$$r_{23} = 1.73 r_{13}.$$

The equilibrium configuration is shown in Fig. 2.3.

Fig. 2.3. The charge q_3 experiences no net force.

2.3 COULOMB SCATTERING

Early in this century Ernest Rutherford discovered the nucleus of the atom[2] by bombarding thin gold targets with alpha particles and study-

---
[1] *MWTP*, Section 7.6.
[2] E. N. da C. Andrade, "The Birth of the Nuclear Atom," Scientific American, (November 1966).

COULOMB SCATTERING

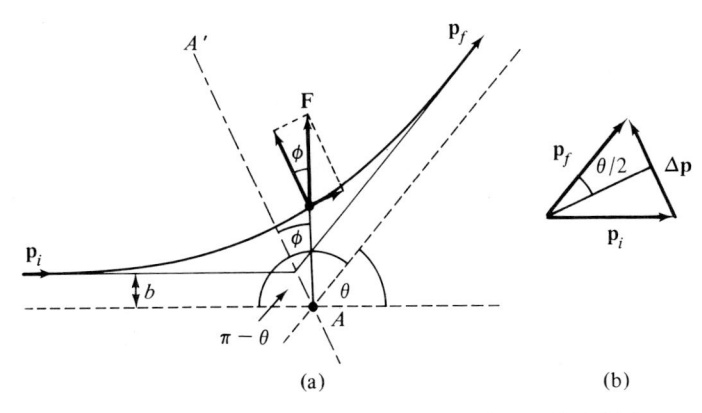

Fig. 2.4. Coulomb scattering. (a) A particle with initial momentum $\mathbf{p}_i$ is deflected by its Coulomb interaction with a scattering center at A. The final momentum of the particle is $\mathbf{p}_f$. (b) The relation between the initial momentum $\mathbf{p}_i$, the change in momentum $\Delta \mathbf{p}$, and the final momentum $\mathbf{p}_f$ of the scattered particle.

ing the deflections suffered by the alpha particles as a result of the electromagnetic interaction. This is a special case of the general problem of the scattering of positively charged particles by the massive nuclei of atoms. The problem is illustrated in Fig. 2.4. A particle of initial momentum $\mathbf{p}_i$, charge q_1, interacts via the Coulomb force with a fixed scattering center, charge q_2, located at A. As a result of this interaction the momentum of the particle is changed to $\mathbf{p}_f$. From conservation of energy it follows that the overall effect of the interaction is to change the direction of the particle's momentum but not its magnitude. Since the Coulomb force falls as the square of the separation r, the path of the particle is nearly linear far from the scattering center. We will deduce the relation between the angle defined by the directions of the initial and final momenta of the particle (the **scattering angle** θ) and the distance of closest approach of the particle to the scattering center in the absence of the Coulomb force (the **impact parameter** b).

Since the Coulomb force depends only on the separation r, the path of the particle will be symmetrical with respect to the line AA' bisecting the angle $(\pi - \theta)$. The momentum change $\Delta \mathbf{p}$ is shown in Fig. 2.4(b); its magnitude Δp is

$$\Delta p = 2p \sin\left(\frac{\theta}{2}\right)$$

where $p = |\mathbf{p}_i|$, and its direction is that of AA'. Now

$$\Delta \mathbf{p} = \int \mathbf{F}\, dt$$

where the right-hand side is the impulse[3] of the Coulomb force **F**. From this equation we conclude that only the component $F \cos \phi$ of the Coulomb force parallel to the direction of AA' contributes to $\Delta \mathbf{p}$. The component of the Coulomb force perpendicular to AA' causes the particle to slow down as it approaches the scattering center and to speed up as it recedes from the scattering center. From symmetry it follows that the overall effect of this component of the force is zero. We may therefore write

$$\Delta p = \int_{-\infty}^{\infty} F \cos \phi \, dt$$
$$= \int_{-\infty}^{\infty} \frac{k_e q_1 q_2}{r^2} \cos \phi \, dt.$$

In order to evaluate this integral we make use of the fact that the angular momentum of the system is a constant of the motion.[4] Taking the scattering center as the origin, we can write

$$pb = mr^2 \left(\frac{d\phi}{dt}\right).$$

The term on the left is the asymptotic magnitude of the angular momentum when the particle is far from the scattering center; the term on the right is a general expression for the magnitude of the angular momentum when the particle is a distance r from the scattering center. Therefore,

$$\frac{1}{r^2} = \frac{m}{pb}\left(\frac{d\phi}{dt}\right)$$

and

$$\Delta p = \frac{k_e q_1 q_2 m}{pb} \int_{-\infty}^{\infty} \cos \phi \left(\frac{d\phi}{dt}\right) dt$$
$$= \frac{k_e q_1 q_2 m}{pb} \int_{-(\pi-\theta)/2}^{(\pi-\theta)/2} \cos \phi \, d\phi.$$

The limits of integration are obtained from a consideration of Fig. 2.4(a). Carrying out the integration gives

$$\Delta p = \frac{2k_e q_1 q_2 m}{pb} \cos \left(\frac{\theta}{2}\right).$$

But, as we saw earlier,

$$\Delta p = 2p \sin \left(\frac{\theta}{2}\right)$$

so that

$$2p \sin \left(\frac{\theta}{2}\right) = \frac{2k_e q_1 q_2 m}{pb} \cos \left(\frac{\theta}{2}\right)$$

[3] *MWTP*, Section 7.4.
[4] *MWTP*, Section 8.2.

and
$$b = \frac{k_e q_1 q_2}{2(p^2/2m)} \cot\left(\frac{\theta}{2}\right)$$
$$= \frac{k_e q_1 q_2}{2K} \cot\left(\frac{\theta}{2}\right)$$
where
$$K = \frac{p^2}{2m}$$
is the initial kinetic energy of the particle. This is the fundamental relation between impact parameter and scattering angle for Coulomb scattering.

2.4 FORCES BETWEEN CURRENTS

The effect of an electric current on a permanent magnet in the form of a compass needle was first noted by Hans Christian Oersted. Subsequently, André Marie Ampère investigated the interaction of one electric current upon another and was able to establish experimentally the law of force between two currents. For the present we shall confine our discussion to the case of two very long parallel currents. A more general discussion is presented in Section 7.2.

For the two very long parallel currents I_1 and I_2 shown in Fig. 2.5, it is found experimentally that the **magnetic force** between them has the following properties:

1. The force per unit length varies directly as the magnitude of each current.
2. The force acts in the direction perpendicular to that of the currents.
3. The force per unit length varies inversely as the distance between the two currents.
4. The force is attractive if the currents are in the same direction and repulsive if they are in oppsite directions.

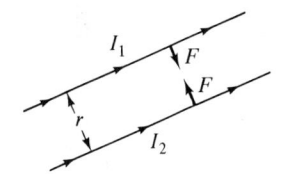

Fig. 2.5. Two parallel currents experience a force F between them.

This force is associated with moving charges only. It is not the same as the electrostatic force that acts between stationary charges.

According to statement (1) we can write for the magnitude F of the force per unit length between the currents
$$F \propto I_1 I_2,$$
while from statement (3) we can write
$$F \propto \frac{1}{r}.$$
Combining these two expressions gives the equation
$$F = \frac{k_m I_1 I_2}{r}$$

where k_m is a constant. It is left as an exerise for the reader (Problem 19) to write a vector equation for the force.

2.5 SYSTEMS OF UNITS

In the MKS system of units, the fundamental units of length, mass, and time are the meter, kilogram, and second, respectively. In order to discuss electromagnetic phenomena quantitatively it is necessary to introduce a fourth fundamental unit. Either of the force laws introduced in the two preceding sections could be used to define the fourth fundamental unit; by international agreement, the **ampere** has been chosen. The ampere (A) is defined as that current which, flowing in each of two parallel wires separated by one meter, results in a force of 2×10^{-7} N per meter of length on each conductor. Since the force law is

$$F = \frac{k_m I_1 I_2}{r}$$

we have

$$F = 2 \times 10^{-7} \text{ N} \cdot \text{m}^{-1} = k_m \frac{(1)(1)}{(1)} \text{A}^2 \cdot \text{m}^{-1}$$

or

$$k_m = 2 \times 10^{-7} \text{ N} \cdot \text{A}^{-2}.$$

The unit of charge is the **coulomb** (C). It is defined as the quantity of charge that passes through a given area per second due to a current flow of one ampere through that area. The constant k_e in Coulomb's law must then be determined experimentally. Experiments show that if a point charge of 1 C were to be placed 1 m from a similar point charge in a vacuum, the force of repulsion between the charges would be 8.9880×10^9 N. Since

$$F = \frac{k_e q_1 q_2}{r^2},$$

the value of k_e is

$$k_e = 8.9880 \times 10^9 \text{ N} \cdot \text{m}^2 \cdot \text{C}^{-2}$$
$$= \frac{1}{4\pi(8.8538 \times 10^{-12})}$$
$$= \frac{1}{4\pi\epsilon_0}$$

where $\epsilon_0 = 8.8538 \times 10^{-12}$ $\text{C}^2 \cdot \text{N}^{-1} \cdot \text{m}^{-2}$ is known as the **permittivity constant** or as the **vacuum permittivity**. The constant ϵ_0 and the factor 4π are inserted into Coulomb's law for practical reasons; several important

SEC. 2.5   SYSTEMS OF UNITS

equations which are derived from Coulomb's law will appear in a simpler form. With the introduction of the factor 4π the system of units is referred to as the **rationalized MKS system**. This is the system of units that will be used throughout this text.

Two CGS systems of units (based on the centimeter, gram, and second as fundamental units) are also often used in textbooks and articles on electromagnetism. These systems lead to somewhat more complicated units in electricity and magnetism, but have the advantage of showing more clearly the fundamental symmetry between electric and magnetic effects. In the CGS electrostatic system, the constant k_e in Coulomb's law is set equal to unity to define the CGS electrostatic unit (esu) of charge, the **statcoulomb**. The relation between the coulomb and the statcoulomb is

$$1 \text{ coulomb} = 3 \times 10^9 \text{ statcoulomb}.$$

The CGS electromagnetic system is frequently used to discuss magnetic effects. The CGS electromagnetic unit (emu) of current, the **abampere**, is defined in terms of the effect produced at the center of a circular loop by a current in the loop. The corresponding emu of charge, the **abcoulomb**, is the quantity of charge passing a point in the loop per second when the current in the loop is 1 abampere. The relation between the coulomb and the abcoulomb is

$$1 \text{ coulomb} = 0.1 \text{ abcoulomb}.$$

Example. Compare the electrical force of attraction between the electron and proton in a hydrogen atom to the gravitational force of attraction between the two particles.

Solution. The Bohr model[5] for the hydrogen atom pictures the electron as traveling about the proton in a circular orbit of radius $r = 5.3 \times 10^{-11}$ m when the atom is in its ground state. In this state, the electrical force is constant and its magnitude is given by

$$F_e = \frac{1}{4\pi\epsilon_0} \frac{q_1 q_2}{r^2}$$
$$= \frac{1}{4\pi\epsilon_0} \frac{e^2}{r^2}$$

where $e = 1.6 \times 10^{-19}$ C is the magnitude of the charge on both the electron and the proton. The gravitational force is also constant and is given by

$$F_g = G \frac{m_e m_p}{r^2}$$

where m_e and m_p are the masses of the electron and proton, and are equal to 9.1×10^{-31} kg and 1.7×10^{-27} kg, respectively. Taking the ratio of the forces yields

[5] *MWTP*, Section 18.4.

$$\frac{F_e}{F_g} = \frac{e^2}{4\pi\epsilon_0 G m_e m_p}$$

$$= \frac{(1.6 \times 10^{-19})^2}{4\pi(8.9 \times 10^{-12})(6.7 \times 10^{-11})(9.1 \times 10^{-31})(1.7 \times 10^{-27})}$$

$$= 2.2 \times 10^{39}.$$

That is, the electrical force of attraction is 2.2×10^{39} times greater than the gravitational force of attraction. Gravitational forces are negligible on the atomic scale of sizes. Note also that the ratio of the two forces is independent of the distance separating the charges since both forces are of the "inverse square" type.

Example. A 12 eV proton beam carrying 5.0×10^{12} protons·m^{-3} travels from right to left. The cross-sectional area of the beam is 2.0×10^{-6} m². Deduce the current.

Solution. The kinetic energy of the protons in the beam is

$$K = 12 \times 1.6 \times 10^{-19} \text{ J}.$$

The speed of the protons is

$$v = \left(\frac{2K}{m_p}\right)^{1/2}$$

$$= \left(\frac{2 \times 12 \times 1.6 \times 10^{-19}}{1.7 \times 10^{-27}}\right)^{1/2}$$

$$= 4.8 \times 10^4 \text{ m·sec}^{-1}.$$

The number of protons passing a fixed point each second is

$$4.8 \times 10^4 \times 2.0 \times 10^{-6} \times 5.0 \times 10^{12} = 4.8 \times 10^{11}.$$

Therefore, the charge per second passing a fixed point, or the current, is

$$4.8 \times 10^{11} \times 1.6 \times 10^{-19} = 7.7 \times 10^{-8} \text{ A}.$$

Since the conventional direction of a current is that associated with the motion of negative charges, the current is from left to right if the proton beam moves from right to left. Note that if an electron beam rather than a proton beam were considered, the direction of the current would have been from right to left, that is, in the direction of the beam.

QUESTIONS AND PROBLEMS

1. What evidence could be given that would support the belief that the total charge in the universe is very nearly, if not exactly, zero?
2. If at some time all the positively charged particles in the universe became negatively charged and vice versa, would we be able to tell?
3. The **faraday** (F) is the net charge on one g-mole of singly ionized atoms or molecules. Show that

$$1 \text{ F} = 96{,}490 \text{ C}.$$

What difficulties would you face if you tried to isolate 1 g-mole of Cl⁻ ions from NaCl in a test tube?

4. Three charges, each of magnitude $+q$, are situated at the corners of an equilateral triangle of side l. Find the electric force on each of them.

5. Two identical spherical conductors are each charged with unlike charges. At a separation r, the force on each is F. Later they are brought into contact and again separated by a distance r. Again the force between them has magnitude F, only this time it is repulsive. What is the ratio of the two charges initially on the conductors?

6. Four charges q_1, q_2, q_3 and q_4 are placed at each of the four corners of a square of side a. Find the force experienced by a test charge q placed at the center if:
   (a) $q_1 = q_2 = q_3 = q_4$;
   (b) $q_1 = q_2 = q_3$, $\quad q_4 = -q_1$;
   (c) $q_1 = 2q_2 = 3q_3 = 4q_4$.

7. A **positronium** atom consists of an electron and a **positron** separated by 0.10 nm and moving in circular orbits about their common center of mass. Calculate the strength of the Coulomb interaction and the speed of each particle. A **positron** is identical to an electron except for the sign of its charge.

*8. Two fixed negative charges $-q$ are separated by a distance R. A positive charge $+q$ located midway between them is displaced along the right bisector of the line joining the charges and is released. Show that for small displacements the positive charge will execute simple harmonic motion of frequency

$$f = \frac{2q}{\pi}\left(\frac{k_e}{mR^3}\right)^{1/2}.$$

where m is the mass of the positive charge.

*9. A charge q_1 is a distance d from an infinite line of point charges q each separated from the other by a distance d. One of the point charges is directly opposite the charge q_1. Show that the resultant force on q_1 is given by

$$F = \frac{k_e q_1 q}{d^2} \sum_{n=-\infty}^{+\infty} \frac{1}{(n^2 + 1)^{3/2}}.$$

Show that in the limit of a continuous charge distribution ($\lambda = q/d$ C·m⁻¹) that this expression for the force becomes

$$F = \frac{2k_e q_1 \lambda}{d}.$$

*10. A particle of mass m and charge q is located midway between two fixed charges, each of magnitude q and separated by a distance l. Find the frequency of oscillation when the middle charge is displaced a small distance along the line joining the charges and is released.

11. An alpha particle with kinetic energy 8.0 MeV is incident on a gold foil. Deduce the scattering angle for an alpha particle scattered by a gold nucleus for impact parameters 1.0×10^{-x} m with $x = 12, 13, 14$.

*12. The scattering of a charged particle by a scattering center having a charge of the same sign was discussed in Section 2.3. Adapt the analysis to the case when the charges are of unlike sign.

*13. A 6.0 MeV alpha particle is scattered 10° by a collision with a gold nucleus. What is the corresponding impact parameter? If the gold foil is 2.0×10^{-5} cm thick, what fraction of the incident alpha particles would be scattered through an angle equal to or greater than 10°?

14. What is the minimum distance a 2 MeV alpha particle can approach a fixed gold nucleus? At the point of closest approach what is the potential energy?

15. A **cyclotron** is a machine for accelerating charged particles to quite high kinetic energies. A certain machine produces a beam of 5.0 MeV protons. The cross-sectional area of the beam is 2.0 mm². If the current is 10 μA, calculate the number of protons per mm³ in the beam.

16. A copper conductor of cross-sectional area 2.0 mm² carries a current of 5.0 A. Assuming one conduction electron per atom, calculate the average speed of these electrons. Comment in some detail on your answer.

17. In the Bohr model for the ground state of the hydrogen atom, the electron travels in a circular orbit of radius 5.3×10^{-11} m. Calculate the current associated with the orbiting electron.

18. Two parallel straight wires each carry a current of 5.0 A. At what separation is the attractive force between them 5.0×10^{-5} N·m^{-1}?

19. Write a vector equation for the force between two very long parallel currents.

20. Two very long trains are passing each other on parallel tracks separated by 6 m. One is going east at 80 m·sec^{-1} and the other west at the same speed. Each train is made up of box cars 20 m long and each boxcar has accumulated +1 C of charge due to friction with the rails. What is the force per unit length exerted on each train due to magnetic (parallel current) effects of the other?

ROBERT A. MILLIKAN

The electric field 3

3.1 INTRODUCTION

Just as the gravitational interaction between objects possessing mass[1] acts through a vacuum, electric and magnetic forces between charges do not require an intervening medium. It is convenient for the discussion of this **action-at-a-distance** phenomenon to introduce the concept of a **field**. Accordingly, the force which is experienced by an object in some region of space is ascribed to the existence of an appropriate field in that region of space. The field is viewed as some sort of tension or stress in empty space that reveals itself by producing forces on objects in the field. The field, of course, owes its existence to suitable sources; the gravitational field has mass as its source, while the electromagnetic field is produced by charged particles. It is often very difficult, however, to identify the sources of a given field; this is true particularly for the electromagnetic field. For this reason we tend to speak of a field as if it has an existence of its own and to ignore the sources of the field. It is quite feasible experimentally to catalog the properties of a field in a given region of space by moving a small test particle throughout that region of space and noting the force it experiences as a function of both position and time. It is essential that the test particle be small enough that its presence does not cause the field to depart appreciably from what it would be if the test particle were not there.

Theoretically, both gravitational and electric fields extend to infinity because of the inverse square dependence on distance in both Newton's law of gravitation and Coulomb's law. For practical purposes the fields are effectively zero in regions of space sufficiently far removed from any sources. Moreover, since electrical charges come in two kinds, cancellations usually occur resulting in very rapid reduction of the field with distance from the charges.

Because of the very great importance of the field concept in electromagnetism, we shall examine the characteristics of fields carefully in this chapter by referring to the electric field. Most of what we say will be applicable to the magnetic field also. Because of the similarity of Coulomb's law and Newton's law of gravitation, we should expect many ideas that we have developed in discussing gravity to carry over to the discussion of the electric field. For example, we have shown that the gravitational force is a conservative force.[2] This is so because it is a power law force; it follows immediately that the electric force is also conservative.

3.2 DEFINITION OF THE ELECTRIC FIELD

Let us suppose that a very small test charge q is brought into a region of space where at a certain point it experiences an electric force **F**. The

---
[1] *MWTP*, Sections 7.6 and 7.11.
[2] *MWTP*, Section 7.9.

electric field **E** (sometimes called the **electric intensity**) at that point is defined to be

$$\mathbf{E} = \frac{\mathbf{F}}{q}.$$

The electric field is a **vector field** and **E** has the direction of the force experienced by the charge q when q is positive. The unit of **E** is $N \cdot C^{-1}$.

Suppose that we bring the test charge q into the region near an **isolated charge** q_1 (see Fig. 3.1). From Coulomb's law we have for the force **F** experienced by q

$$\mathbf{F} = \frac{1}{4\pi\epsilon_0} \frac{qq_1}{r^2} \hat{\mathbf{r}}$$

Fig. 3.1. A test charge q at a distance r from an isolated point charge q_1.

where $\hat{\mathbf{r}}$ is a unit vector in the direction of **r**. The electric field **E** at a distance r due to the charge q_1 is then

$$\mathbf{E} = \frac{\mathbf{F}}{q} = \frac{1}{4\pi\epsilon_0} \frac{q_1}{r^2} \hat{\mathbf{r}}.$$

Since every finite charge can be viewed as a collection of point charges (normally electrons and protons), the total electric field due to a finite charge may be written as a sum of contributions from a collection of point charges q_i by the vector equation

$$\mathbf{E} = \sum_i \mathbf{E}_i = \frac{1}{4\pi\epsilon_0} \sum_i \frac{q_i}{r_i^2} \hat{\mathbf{r}}_i.$$

We should realize that the presence of q produces forces on all the charges q_i which tends to alter the configuration of the charges q_i and hence to alter the electric field produced by them. To be more accurate we should define the field by the equation

$$\mathbf{E} = \lim_{\Delta q \to 0} \frac{\Delta \mathbf{F}}{\Delta q} = \frac{d\mathbf{F}}{dq} = \sum_i \frac{d\mathbf{F}_i}{dq}$$

wherein a test charge Δq and the force $\Delta \mathbf{F}$ that it experiences become vanishingly small and are represented by the infinitesimals dq and $d\mathbf{F}$.

Michael Faraday, a pioneer in the study of electricity and magnetism, introduced the concept of **lines of force** as a geometrical construct to represent the magnitudes and directions of electric fields. The rules for drawing lines of force are as follows:

1. The tangent to a line of force at any point gives the direction of **E** at that point.
2. The density of the lines of force (that is, the number per unit cross-sectional area perpendicular to the direction of the lines) is proportional to the magnitude of **E**.

Each line of force originates on a positive charge and ends on a negative charge. The lines of force about isolated positive and negative charges are shown in Fig. 3.2.

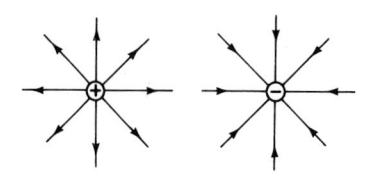

Fig. 3.2. Lines of force about isolated positive and negative charges.

Example. Three positive charges $(2)^{1/2}q_1$, $(2)^{1/2}q_1$, and $2q_1$ are situated at three corners of a square of side d as shown in Fig. 3.3(a). Find the electric field at the fourth corner of the square.

Solution. We first compute the electric field arising from each of the charges. The magnitudes of the fields E_1, E_2, E_3 are given by

$$E_1 = \frac{1}{4\pi\epsilon_0} \frac{(2)^{1/2}q_1}{d^2} = E_2$$

$$E_3 = \frac{1}{4\pi\epsilon_0} \frac{2q_1}{[(2)^{1/2}d]^2} = \frac{1}{4\pi\epsilon_0} \frac{q_1}{q^2}.$$

The directions of the fields are shown in Fig. 3.3(b). To obtain the resultant field **E** we perform the vector addition

$$\mathbf{E} = \mathbf{E}_1 + \mathbf{E}_2 + \mathbf{E}_3.$$

The fields $\mathbf{E}_1$ and $\mathbf{E}_2$ may be resolved into components parallel and perpendicular to $\mathbf{E}_3$. The components perpendicular to $\mathbf{E}_3$ cancel; the components parallel to $\mathbf{E}_3$ each have magnitude

$$E_1 \cos 45° = \frac{1}{4\pi\epsilon_0} \frac{(2)^{1/2}q_1}{d^2} \frac{1}{(2)^{1/2}} = \frac{1}{4\pi\epsilon_0} \frac{q_1}{q^2}.$$

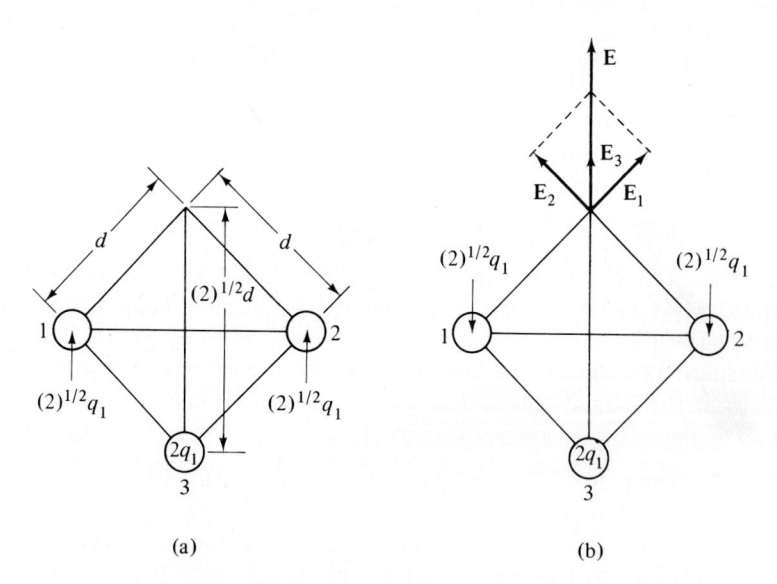

Fig. 3.3. (a) Three positive charges located at three corners of a square. (b) The electric field **E** at the fourth corner arising from the sum of the fields $\mathbf{E}_1$, $\mathbf{E}_2$, and $\mathbf{E}_3$ due to the individual charges.

The resultant field has magnitude

$$E = \frac{1}{4\pi\epsilon_0} \frac{3q_1}{d^2}$$

and points along the diagonal of the square away from the charge $2q_1$, as indicated in Fig. 3.3(b).

3.3 THE ELEMENTARY ELECTRIC CHARGE

Robert A. Millikan devised a method of studying the motion of charged single droplets of water vapor under the combined action of electric and gravitational fields. The importance of this work was that it provided a convincing demonstration of the discreteness of electric charge and thereby helped to establish the atomic theory of matter. To overcome the problem of evaporation he later substituted oil droplets for the water droplets and the experiment has come to be known as the **Millikan oil-drop experiment**.

The experiment is extremely simple in principle. Oil droplets are sprayed into a region of constant electric field. A microscope is used to observe a particular droplet. If the droplet has picked up a charge from an ion in the air or from the spraying process, then the electric field can be adjusted so that the droplet remains motionless under the combined action of the electric and gravitational fields. That is, for a balance condition

$$\mathbf{F}_e = -\mathbf{F}_g$$

with $\mathbf{F}_e$ the electric force and $\mathbf{F}_g$ the gravitational force experienced by the droplet. Therefore

$$Eq = mg$$

where E is the magnitude of the electric field, q the charge on the droplet, m its mass and g the acceleration due to gravity. In practice, it was the determination of the mass of the droplet that provided the difficulty. Since no estimate of the size of the droplet could be obtained directly, the following indirect procedure was adopted.

The droplet under observation was allowed to fall through the air with no electric field present. As the speed of the droplet increased, the **viscous force** $\mathbf{F}_v$ due to the surrounding air increased until a balance between the viscous force and the net gravitational force was achieved. Thereafter, the droplet fell with constant velocity known as the **terminal velocity** $\mathbf{v}_t$. According to **Stokes' law** for a sphere falling in a viscous medium,

$$\mathbf{F}_v = -6\pi\eta r \mathbf{v}_t$$

where η is the **viscosity** for air and r is the radius of the droplet. The net gravitational force experienced by the droplet is

$$\mathbf{F}_g = \tfrac{4}{3}\pi r^3 (\rho_0 - \rho_a)\mathbf{g}$$

where ρ_0 is the density of the droplet and ρ_a the density of air. When the two forces are balanced

$$6\pi\eta r v_t = \tfrac{4}{3}\pi r^3 (\rho_0 - \rho_a)g$$

or

$$r^2 = \frac{9}{2}\frac{\eta v_t}{(\rho_0 - \rho_a)g}.$$

By measuring v_t, the radius of the droplet could be calculated and therefore its mass. In fact, because of the small size of the oil droplets, the atomicity of air caused small deviations from Stokes' law which had to be taken into account. Another small correction is necessary to take account of the deviation from sphericity of the droplets due to the viscous retarding force of the air.

In practice, it was difficult to precisely balance the gravitational and electric forces. The following approach circumvented that difficulty. The droplet was first allowed to fall freely and its terminal velocity was measured. Then a sufficiently strong electric field was applied so that the droplet moved upward due to a net force F_u given by

$$F_u = Eq - mg.$$

The terminal velocity was again measured. From these measurements the charge on the droplet could be determined (see Problem 13).

Using this technique Millikan and his colleagues measured the charge on thousands of droplets. The results showed that the charge was always an integral multiple of 1.61×10^{-19} C. In some instances droplets carrying a single unit of charge were found; in others, the charge on a droplet was observed to change by one unit due to the gain or loss of a single unit of charge. That is, the experiment provided a direct measurement of the unit of electric charge with the result $e = 1.61 \times 10^{-19}$ C. (The charge carried by the electron is $-e$.)

Today physicists are searching for fractionally charged particles called **quarks**.[3] A symmetry-based classification scheme for the elementary particles of high-energy physics allows for the possibility of quarks carrying charges of $+2e/3$ and $-e/3$. At the time of this writing no conclusive evidence for their existence has been presented. According to the quark model the proton which possesses a charge $+e$ consists of two quarks carrying charge $+2e/3$ and one quark carrying charge $-e/3$; the uncharged neutron consists of one quark with charge $+2e/3$ and two with charge $-e/3$.

[3] V. F. Weisskopf, "The Three Spectroscopies," *Scientific American*, May, 1968.

3.4 THE ELECTRIC DIPOLE

An important charge arrangement is the **electric dipole**. In its simplest form it consists of two equal but opposite charges ($+q$ and $-q$) separated by a distance d (see Fig. 3.4). An electric dipole has associated with it an **electric dipole moment p** defined by $\mathbf{p} = q\mathbf{d}$ where $\mathbf{d}$ is the displacement of the positive charge from the negative charge. The electric field at the point P is given by the vector sum of the fields $\mathbf{E}_1$ and $\mathbf{E}_2$ due to the charges $+q$ and $-q$. Applying the law of cosines to the triangles PAB and PDC in Fig. 3.4, we have the following relations:

$$E^2 = E_1^2 + E_2^2 + 2\mathbf{E}_1 \cdot \mathbf{E}_2$$
$$= E_1^2 + E_2^2 - 2E_1 E_2 \cos\theta$$
$$d^2 = r_1^2 + r_2^2 - 2\mathbf{r}_1 \cdot \mathbf{r}_2$$
$$= r_1^2 + r_2^2 - 2r_1 r_2 \cos\theta$$

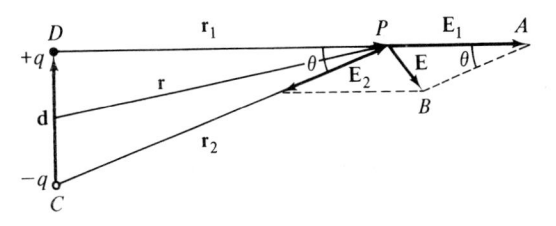

Fig. 3.4. Two charges $+q$ and $-q$ separated by a small distance d constitute an electric dipole.

where

$$E_1 = \frac{1}{4\pi\epsilon_0}\frac{q}{r_1^2}, \quad E_2 = \frac{1}{4\pi\epsilon_0}\frac{q}{r_2^2}$$

are the magnitudes of $\mathbf{E}_1$ and $\mathbf{E}_2$. Eliminating $\cos\theta$ and substituting for E_1 and E_2 gives

$$E^2 = E_1^2 + E_2^2 + \frac{E_1 E_2}{r_1 r_2}[d^2 - (r_1^2 + r_2^2)]$$
$$= \frac{q^2}{(4\pi\epsilon_0)^2}\left\{\frac{1}{r_1^4} + \frac{1}{r_2^4} + \frac{1}{r_1^3 r_2^3}[d^2 - (r_1^2 + r_2^2)]\right\}$$

and

$$E = \frac{q}{4\pi\epsilon_0}\left\{\frac{1}{r_1^4} + \frac{1}{r_2^4} + \frac{1}{r_1^3 r_2^3}[d^2 - (r_1^2 + r_2^2)]\right\}^{1/2}.$$

This expression is exact but rather difficult to evaluate generally. The field is easy to calculate, however, for the special points P_1 and P_2 shown in Fig. 3.5. Letting r be the distance from the center of the dipole to the point at which the field is to be evaluated, we see from Fig. 3.5 that for the point P_1 on the axis of the dipole

$$r_1 = r - \frac{d}{2} \quad \text{and} \quad r_2 = r + \frac{d}{2}.$$

Clearly,

$$\mathbf{E}(P_1) = \mathbf{E}_1(P_1) + \mathbf{E}_2(P_1)$$

becomes

$$E(P_1) = E_1(P_1) - E_2(P_1)$$

$$= \frac{q}{4\pi\epsilon_0\left(r - \frac{d}{2}\right)^2} - \frac{q}{4\pi\epsilon_0\left(r + \frac{d}{2}\right)^2}$$

$$= \frac{q}{4\pi\epsilon_0}\left[\frac{2rd}{\left(r - \frac{d}{2}\right)^2\left(r + \frac{d}{2}\right)^2}\right].$$

For points P_1 sufficiently far removed from the dipole, $r \gg d$ and

$$E(P_1) = \frac{2qd}{4\pi\epsilon_0 r^3}$$

$$= \frac{2p}{4\pi\epsilon_0 r^3}.$$

For the point P_2 we have

$$r_1^2 = r^2 + \frac{d^2}{4} = r_2^2$$

and

$$E_1 = \frac{q}{4\pi\epsilon_0\left(r^2 + \frac{d^2}{4}\right)} = E_2.$$

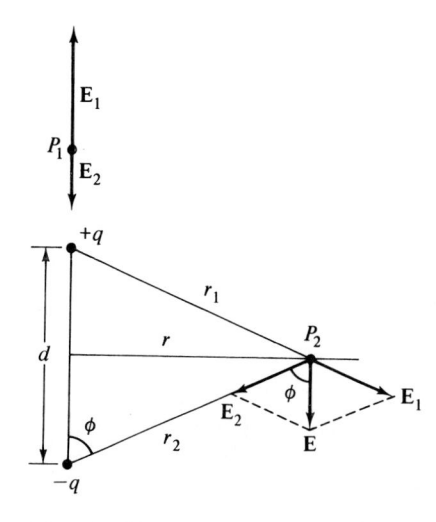

Fig. 3.5. The dipole field at points P_1 and P_2.

The components of $\mathbf{E}_1$ and $\mathbf{E}_2$ perpendicular to the dipole cancel while the component parallel to the dipole is

$$E(P_2) = \frac{2q}{4\pi\epsilon_0\left(r^2 + \frac{d^2}{4}\right)} \cos\phi$$

where

$$\cos\phi = \frac{d}{2r_2} = \frac{d}{2\left(r^2 + \frac{d^2}{4}\right)^{1/2}}.$$

Therefore,

$$E(P_2) = \frac{2q}{4\pi\epsilon_0\left(r^2 + \frac{d^2}{4}\right)} \cdot \frac{d}{2\left(r^2 + \frac{d^2}{4}\right)^{1/2}}$$

$$= \frac{qd}{4\pi\epsilon_0\left(r^2 + \frac{d^4}{4}\right)^{3/2}}.$$

For points P_2 sufficiently far removed from the dipole, this reduces to

$$E(P_2) = \frac{p}{4\pi\epsilon_0 r^3}.$$

SEC. 3.4 THE ELECTRIC DIPOLE

The magnitude of the dipole field at the points P_1 and P_2 is proportional to the dipole moment p and inversely proportional to the cube of the distance from the dipole at distances large compared to the charge separation d. This is true in general for all points in space (provided $r \gg d$) as we shall see in Section 3.7. If d were zero, the field would be zero as would be expected for two equal and opposite charges. However, for a dipole the charges are separated by the distance d so that the fields $\mathbf{E}_1$ and $\mathbf{E}_2$ do not quite cancel but produce a field that decreases as r^{-3} at large distances. That is, the field of a dipole falls off more rapidly than does the field of a single charge. The lines of force about an electric dipole are as shown in Fig. 3.6.

Example. An electric dipole of moment $\mathbf{p}$ is placed in a uniform external electric field E. The dipole makes an angle θ with the field. (a) Deduce an expression for the torque experienced by the dipole. (b) Deduce an expression for the potential energy of the dipole.

Solution. (a) Two equal and opposite forces $\mathbf{F}$ and $-\mathbf{F}$ act on the charges as indicated in Fig. 3.7. As a result the dipole experiences a torque $\boldsymbol{\tau}$ which tends to align it in the direction of the field. The magnitude of the torque is

$$\tau = 2F\left(\frac{d}{2}\right)\sin\theta$$
$$= qd\, E \sin\theta$$
$$= pE \sin\theta.$$

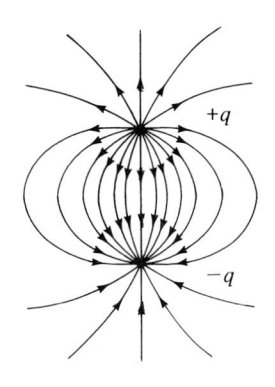

Fig. 3.6. The lines of force about an electric dipole.

In vector notation

$$\boldsymbol{\tau} = \mathbf{p} \times \mathbf{E}.$$

(b) The change in potential energy of the dipole is equal to the work that must be done on the dipole to reorient it in the field. The work required to change the orientation of the dipole from θ_1 to θ_2 is

$$W = \int dW = \int_{\theta_1}^{\theta_2} \tau\, d\theta$$
$$= \int_{\theta_1}^{\theta_2} pE \sin\theta\, d\theta$$
$$= -pE \cos\theta_2 + pE \cos\theta_1.$$

The potential energy U of the dipole may therefore be taken as

$$U = -pE \cos\theta.$$

In vector notation

$$U = -\mathbf{p}\cdot\mathbf{E}.$$

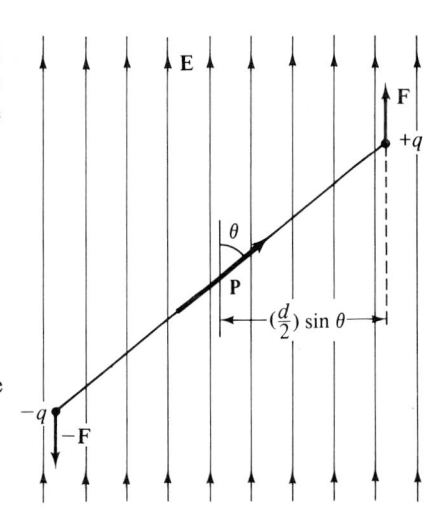

Fig. 3.7. An electric dipole in a uniform external electric field.

Note that the potential energy is minimum when the dipole is aligned parallel to the external field ($\theta = 0°$) and maximum when the dipole is aligned antiparallel to the field ($\theta = 180°$). The torque due to the external field acts to move the dipole into alignment with the field.

3.5 MOLECULAR DIPOLE MOMENTS

Homonuclear diatomic molecules such as hydrogen (H_2) and nitrogen (N_2) have a charge distribution that is symmetric with respect to their centers of mass. Such molecules do not possess permanent electric dipole moments and they are called **nonpolar molecules**. Heteronuclear diatomic molecules such as potassium chloride (KCl) and hydrogen chloride (HCl) have an asymmetrical charge distribution as the result of a net transfer of electronic charge from one atom to the other. Such molecules exhibit permanent electric dipole moments and are known as **polar molecules**. A simple model for the KCl molecule is that of two ions, K^+ and Cl^-, separated by a distance r. For this model we predict a dipole moment equal to the product of the electronic charge and the internuclear separation. Since the internuclear separation in KCl is 0.2667 nm, the assumption of **ionic binding**[4] leads to a predicted dipole moment of 12.8 Debye units (D).[5] The observed dipole moment is 10.5 D. The discrepancy is due to the fact that the electronic charge tends to concentrate more in the region between the two atoms than the model of two ions suggests. The ionic binding approximation is much worse for HCl where the predicted value is 6.1 D as compared to an experimental value of 1.0 D. In this case most of the electronic charge occupies the region between the two atoms and the assumption of covalent binding in which the valence electrons are shared between the two atoms is more realistic.

Polyatomic molecules may or may not possess permanent dipole moments depending on their structure. In fact, the absence of a dipole moment, or its magnitude if it exists, often provides an important clue to the structure of a polyatomic molecule. For example, the carbon dioxide (CO_2) molecule is a linear symmetric molecule O—C—O and is nonpolar; the nitrous oxide (N_2O) molecule is a linear asymmetric molecule N—N—O and has a small dipole moment of 0.17 D; and the water (H_2O) molecule is triangular with the oxygen atom at the apex and has a large dipole moment of 1.84 D.

When considering polyatomic molecules, we often assume that the

---

[4]*MWTP*, Section 21.5.

[5]The **Debye unit** named after Peter Debye is the common unit for quoting molecular dipole moments:
$$1\ D = 3.336 \times 10^{-30}\ C \cdot m = 10^{-18}\ esu \cdot cm.$$

resultant dipole moment of the molecule is the vector sum of the dipole moments associated with each of the chemical bonds. For example, a hydrogen—nitrogen bond (H—N) is assumed to contribute a dipole moment **p**(H → N) of magnitude 1.3 D to any molecule of which it forms a part. The ammonia (NH$_3$) molecule is a pyramidal structure in which each H—N bond makes an angle of 68° with the symmetry axis of the molecule. Therefore, assuming **bond moment additivity,** we predict a dipole moment of magnitude $3p$(H → N) cos 68° = 1.5 D for the NH$_3$ molecule. This compares well with the experimental value of 1.47 D, although this close agreement is actually fortuitous. In general, the assumption of bond moment additivity does not hold to within 10%.

Dipole moments of thymine—thymine dimers

The union of biology and physics called **biophysics** provides one of today's active frontiers of research. This section contains the first of several examples of the application of physics to a problem in biology. For a good, modern introduction to biology, we recommend *Molecular Biology of the Gene* by James D. Watson.[6]

Deoxyribonucleic acid, commonly known as **DNA,** is the genetic material of all cells. The most important feature of DNA is that it usually consists of two very long, thin chains twisted about each other in the form of a regular **double helix.** The diameter of the helix is 1.8 nm and each chain makes a complete turn every 3.4 nm. DNA is a **macromolecule** formed by the union of smaller molecules called **deoxynucleotides.** There are 10 nucleotides on each chain for every turn of the helix. Each nucleotide contains a phosphate group, a sugar group, and a base. The phosphate group of one nucleotide is linked to the sugar group of the adjacent nucleotide. This arrangement produces a very regular backbone for the chain. Four bases are present: **thymine** (T), **cytosine** (C), **adenine** (A), and **guanine** (G). It is the order of the bases that carries the genetic information. This order is highly irregular and varies from one molecule to another. The two chains are joined together by hydrogen bonds between pairs of bases. A is always paired with T and G with C. A schematic illustration of the double helix is shown in Fig. 3.8.

An entertaining account of the discovery of the structure of DNA by Francis Crick, James Watson, and Maurice Wilkins is contained in *The Double Helix.*[7]

The thymine molecule has approximately the planar structure shown

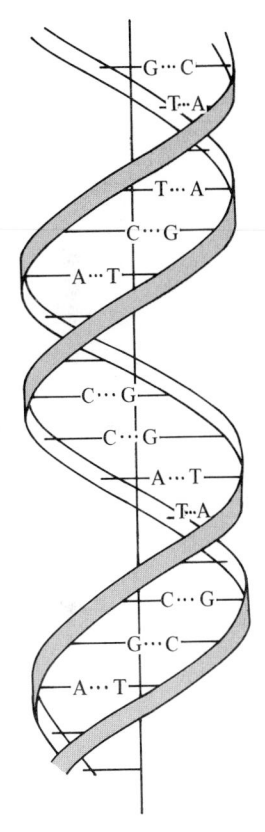

Fig. 3.8. Schematic illustration of the DNA double helix. The two sugar-phosphate backbones twist about on the outside. The hydrogen-bonded base pairs form the core.

---

[6]J. D. Watson, *Molecular Biology of the Gene,* 2nd ed. (New York: W. A. Benjamin, Inc., 1970).

[7]J. D. Watson, *The Double Helix* (New York: Atheneum Publishers, 1968).

in Fig. 3.9. The directions of the bond dipole moments are indicated. Their magnitudes are as follows:[8]

$$H \to C = 0.4 \text{ D}$$
$$CH_3 \to C = 0.4 \text{ D}$$
$$H \to N = 1.3 \text{ D}$$
$$C \Rightarrow O = 3.0 \text{ D}$$
$$C \to N = 0.5 \text{ D}$$
$$C \Rightarrow C = 0 \text{ D}.$$

We now calculate the molecular dipole moment of thymine assuming bond moment additivity. The x component $p_x(T)$ of the dipole moment is

$$\begin{aligned} p_x(T) &= [p(H \to N) - p(C \Rightarrow O) - p(C \to N) \\ &\quad - p(H \to C) - p(CH_3 \to C)] \cos 30° \\ &= (1.3 - 3.0 - 0.5 - 0.4 - 0.4)(0.866) \\ &= -2.6 \text{ D.} \end{aligned}$$

The y component $p_y(T)$ is

$$\begin{aligned} p_y(T) &= [p(C \to N) + p(H \to N) + p(C \Rightarrow O)] \\ &\quad + [-p(H \to N) - p(C \Rightarrow O) - 3p(C \to N) \\ &\quad - p(CH_3 \to C) + p(H \to C)] \cos 60° \\ &= [0.5 + 1.3 + 3.0] + [-1.3 - 3.0 \\ &\quad - 3(0.5) - 0.4 + 0.4](0.5) \\ &= 1.9 \text{ D.} \end{aligned}$$

Therefore, the magnitude $p(T)$ of the thymine dipole moment $\mathbf{p}(T)$ is

$$\begin{aligned} p(T) &= [p_x^2(T) + p_y^2(T)]^{1/2} \\ &= [(-2.6)^2 + (1.9)^2]^{1/2} \\ &= 3.2 \text{ D.} \end{aligned}$$

The direction of $\mathbf{p}(T)$ is such that it makes an angle

$$\begin{aligned} \alpha &= \tan^{-1} \frac{1.9}{2.6} \\ &= 36° \end{aligned}$$

with the negative x axis.

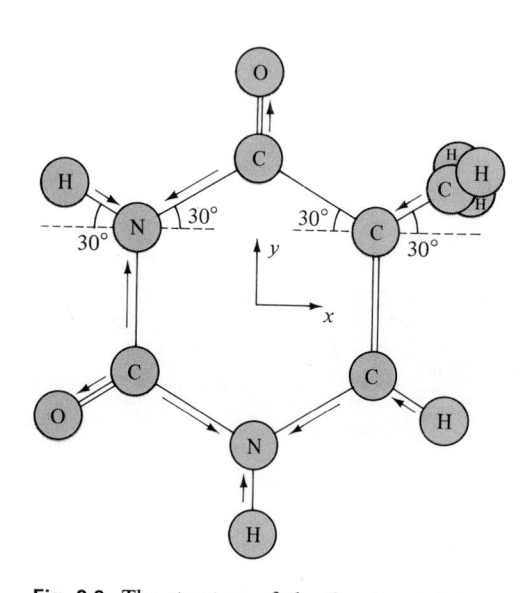

Fig. 3.9. The structure of the thymine molecule. The arrows indicate the directions of the bond dipole moments.

[8] L. Pauling, *The Nature of the Chemical Bond*, 2nd ed. (Ithaca, N.Y.: Cornell University Press, 1948).

MOLECULAR DIPOLE MOMENTS

If ultraviolet radiation incident on DNA is absorbed by adjacent thymine molecules, they fuse together to form a **thymine—thymine (T—T) dimer**. This **radiation damage**, if unrepaired, is normally lethal since the fused thymine molecules do not act as faithful templates for the production of progeny strains. Several distinct types of repair processes are now known to exist.[9] In normal cells much radiation damage is usually repaired before it has time to express itself. It is of interest to relate the manner in which the geometric configuration of the T—T dimers formed in DNA was identified.

Samples of thymine were irradiated by ultraviolet light. Following irradiation four photoproducts which we label A, B, C, D, were separated out. The dipole moments were measured to be 6.04 D, 2.79 D, 5.75 D, and 0 D, respectively. Molecular weight determinations revealed that each photoproduct consisted of T—T dimers. The binding between thymine molecules was shown to occur as indicated in Fig. 3.10. Four possible geometric configurations for T—T dimers were suggested. They are illustrated in Fig. 3.11. Resultant dipole moment magnitudes p_I, p_{II}, p_{III}, and p_{IV} for forms I, II, III, and IV were calculated from the value of $p(T)$ by vector addition. They may be shown to be $p_I = 6.0$ D, $p_{II} = 4.7$ D, $p_{III} = 3.8$ D, and $p_{IV} = 0$ (see Problem 23). The identification of the four photoproducts A, B, C, and D with the four proposed forms I, II, III, and IV is then as follows:

$$A(p_A = 6.04 \text{ D}) \equiv I(p_I = 6.0 \text{ D})$$

$$B(p_B = 2.79 \text{ D}) \equiv III(p_{III} = 3.8 \text{ D})$$

$$C(p_C = 5.75 \text{ D}) \equiv II(p_{II} = 4.7 \text{ D})$$

$$D(p_D = 0 \text{ D}) \equiv IV(p_{IV} = 0 \text{ D}).$$

The identification of photoproduct A with form I has subsequently been substantiated by x-ray diffraction measurements on single crystals produced from this photoproduct. Of the four photoproducts only A is also a photoproduct of the ultraviolet irradiation of DNA. In fact, if we con-

Fig. 3.10. The binding between thymine molecules to form T—T dimers: (a) "parallel" case; (b) "antiparallel" case.

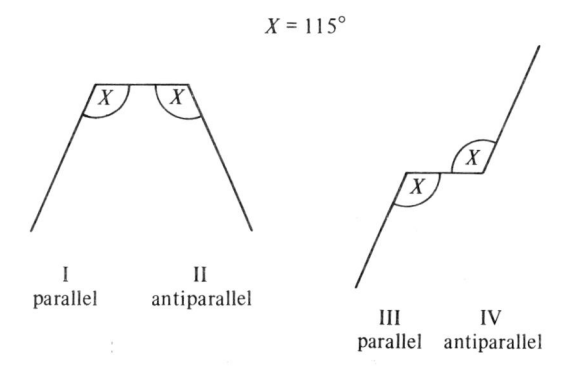

Fig. 3.11. Four possible geometric configurations or conformations for T—T dimers (cross sections perpendicular to the planes of the thymine molecules).

---

[9]P. C. Hanawalt and R. H. Haynes, "The Repair of DNA," *Scientific American*, February, 1967. Available as *Scientific American Offprint 1061* (San Francisco: W. H. Freeman and Co., Publishers).

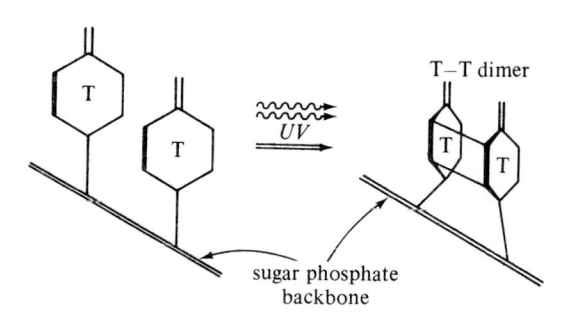

Fig. 3.12. Formation of T—T dimers in DNA by ultraviolet irradiation.

sider the structure of DNA and look at adjacent thymines, form I is the obvious choice for T—T dimers, as indicated in Fig. 3.12.

3.6 ELECTRIC POTENTIAL

The electric field produced by a charge or collection of charges can be described by a scalar quantity, the **electric potential**, as well as by the vector **E**. A charged particle placed in an electric field has potential energy due to its interaction with the field and the electric potential V at a point is defined as the potential energy per unit charge placed at that point. The **(electric) potential energy** U of a charge q at some point in space is defined as the amount of work W required to be done on the charge in order to bring it from infinity to the specified point. Thus

$$V = \frac{U}{q} = \frac{W}{q}.$$

We note that the potential V and the potential energy U are both zero at infinity. This need not worry us since only changes in potential energy can be observed. The unit of potential is the joule·coulomb^{-1} (J·C^{-1}), which is also known as the **volt** (V).

A test charge q in an electric field **E** experiences a force **E**q, by the definition of **E**. In order to move the charge q in the direction d**s** shown in Fig. 3.13 requires a force $-$**E**q to balance the force **E**q plus an additional force in the direction d**s**. If we consider the charge to be moved slowly, this additional force is very small and, in the limit of infinitesimally small velocity, approaches zero. In this limit the work done by the extermal force in moving the charge q a distance d**s** is

$$dW = \mathbf{F} \cdot d\mathbf{s} = (-q\mathbf{E}) \cdot d\mathbf{s}.$$

The change in potential energy is $dU = dW$ so that

$$dU = dW = -q\mathbf{E} \cdot d\mathbf{s}.$$

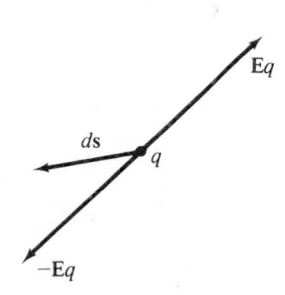

Fig. 3.13. To move a test charge q a distance ds in a field **E** requires the application of an external force $-$**E**q to balance the force **E**q.

SEC. 3.6   ELECTRIC POTENTIAL   33

The difference in potential energy between two points in space is then

$$U_B - U_A = \int_A^B dU = -\int_A^B q\mathbf{E}\cdot d\mathbf{s}.$$

If the point A is chosen to be infinity, we have (since $U_\infty = 0$ by convention)

$$U_B - U_\infty = U_B = -q\int_\infty^B \mathbf{E}\cdot d\mathbf{s}.$$

The potential at point B is then

$$V = \frac{U_B}{q} = -\int_\infty^B \mathbf{E}\cdot d\mathbf{s}.$$

Isolated point charge

For an isolated point charge $+q_1$,

$$\mathbf{E} = \frac{q_1}{4\pi\epsilon_0}\frac{\hat{\mathbf{r}}}{r^2}.$$

The difference in potential between the ends of the vector $d\mathbf{s}$ is (see Fig. 3.14)

$$dV = \frac{dW}{q} = -\mathbf{E}\cdot d\mathbf{s}$$

$$= -\frac{q_1}{4\pi\epsilon_0 r^2}\hat{\mathbf{r}}\cdot d\mathbf{s}$$

$$= -\frac{q_1}{4\pi\epsilon_0}\frac{dr}{r^2}$$

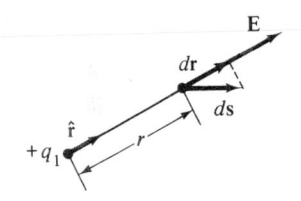

Fig. 3.14. Calculation of the potential near a point charge q_1.

where $\hat{\mathbf{r}}\cdot d\mathbf{s} = dr$. A possible ambiguity arises here since $\hat{\mathbf{r}}\cdot d\mathbf{s}$ is a scalar quantity that may be either positive or negative depending on whether the angle between the directions of $\hat{\mathbf{r}}$ and $d\mathbf{s}$ is less than or greater than 90°. By convention $\hat{\mathbf{r}}\cdot d\mathbf{s}$ is taken always to be positive with the result that the potential due to a positive charge is zero at infinity and positive at all other points. The potential at a distance r from the point charge is then

$$V(r) = -\frac{q_1}{4\pi\epsilon_0}\int_\infty^r \frac{dr}{r^2}$$

$$= -\frac{q_1}{4\pi\epsilon_0}\left(-\frac{1}{r}\right)\Big|_\infty^r$$

$$= \frac{q_1}{4\pi\epsilon_0 r}.$$

System of point charges

For several (positive) point charges the field $\mathbf{E}$ is

$$\mathbf{E} = \mathbf{E}_1 + \mathbf{E}_2 + \ldots + \mathbf{E}_N$$
$$= \sum_{i=1}^{N} \mathbf{E}_i.$$

The difference in potential between the ends of the vector $d\mathbf{s}$ is now

$$dV = -\mathbf{E}\cdot d\mathbf{s} = -\sum_{i=1}^{N} \mathbf{E}_i \cdot d\mathbf{s}$$
$$= -\mathbf{E}_1\cdot d\mathbf{s} - \mathbf{E}_2\cdot d\mathbf{s} - \mathbf{E}_3\cdot d\mathbf{s} - \ldots - \mathbf{E}_N\cdot d\mathbf{s}$$
$$= -\frac{1}{4\pi\epsilon_0}\left(q_1\frac{dr_1}{r_1^2} + q_2\frac{dr_2}{r_2^2} + \ldots + q_N\frac{dr_N}{r_N^2}\right)$$

and the potential $V(P)$ at some point P is

$$V(P) = \frac{1}{4\pi\epsilon_0}\left(\frac{q_1}{r_1} + \frac{q_2}{r_2} + \ldots + \frac{q_N}{r_N}\right)$$
$$= \frac{1}{4\pi\epsilon_0}\sum_{i=1}^{N}\frac{q_i}{r_i}$$
$$= \sum_{i=1}^{N} V(r_i).$$

That is, the potential at point P due to a collection of point charges is the sum of the potentials at P due to the individual charges. This behavior follows from the fact that the potential is a scalar quantity, and vector addition is therefore not required.

For a continuous charge distribution, the total potential at point P is

$$V(P) = \frac{1}{4\pi\epsilon_0}\int_v \frac{\rho\,dv}{r}$$

where ρ is the charge in a volume element dv (that is, ρ is the **charge density** in the distribution) while r is the distance from the volume element to P. The integration is carried out over the total volume of the charge distribution.

The relation between potential and field

Since the potential is a function only of position coordinates, its total differential dV is, by definition, (in Cartesian coordinates)

$$dV = \frac{\partial V}{\partial x}dx + \frac{\partial V}{\partial y}dy + \frac{\partial V}{\partial z}dz.$$

We have already seen that

$$dV = -\mathbf{E}\cdot d\mathbf{s}$$

where, in Cartesian coordinates, $d\mathbf{s} = \mathbf{i}\,dx + \mathbf{j}\,dy + \mathbf{k}\,dz$. Therefore,

$$dV = -E_x\,dx - E_y\,dy - E_z\,dz$$

where E_x, E_y, and E_z are the components of $\mathbf{E}$ in the x, y, and z directions,

SEC. 3.6　　　　　　　　　　　　　ELECTRIC POTENTIAL

respectively. Comparing the two expressions for dV we see that

$$E_x = -\frac{\partial V}{\partial x}, \quad E_y = -\frac{\partial V}{\partial y}, \quad \text{and} \quad E_z = -\frac{\partial V}{\partial z}.$$

Analogous expressions can be written for other coordinate systems (see Appendix A).

At this point it is convenient to introduce the vector differential operator $\mathbf{\nabla}$ (called **del**) which is defined as

$$\mathbf{\nabla} = \mathbf{i}\frac{\partial}{\partial x} + \mathbf{j}\frac{\partial}{\partial y} + \mathbf{k}\frac{\partial}{\partial z}$$

(in Cartesian coordinates).

When it operates directly on a scalar quantity, such as the potential V, it produces a vector quantity $\mathbf{\nabla} V$ or grad V called the **gradient** of V where

$$\mathbf{\nabla} V = \text{grad } V = \mathbf{i}\frac{\partial V}{\partial x} + \mathbf{j}\frac{\partial V}{\partial y} + \mathbf{k}\frac{\partial V}{\partial z}.$$

The three components of grad V give the rates at which the scalar quantity V changes in each of the coordinate directions. That is $\partial V/\partial x$, $\partial V/\partial y$, and $\partial V/\partial z$ are each potential gradients. Now, the electric field $\mathbf{E}$ can be written

$$\mathbf{E} = \mathbf{i} E_x + \mathbf{j} E_y + \mathbf{k} E_z$$
$$= -\mathbf{i}\frac{\partial V}{\partial x} - \mathbf{j}\frac{\partial V}{\partial y} - \mathbf{k}\frac{\partial V}{\partial z}$$

or

$$\mathbf{E} = -\text{grad } V = -\mathbf{\nabla} V.$$

Therefore, the electric field can be derived from the scalar potential V. **This is a general property of conservative fields.**

Since the potential V is a scalar quantity, it is often simpler to determine the electric field by first calculating the potential and then taking the gradient of the potential to determine $\mathbf{E}$.

Example. Three charges are located in the xy plane: a charge $+q_1$ at the point $(\frac{1}{2}, 0)$, a charge $+q_1$ at the point $(-1, 0)$, and a charge $-3q_1$ at the point $(0, 1)$.
(a) Calculate directly the electrostatic force on a charge $+q$ at the origin.
(b) By taking the gradient of the electric potential, determine the electrostatic force on a charge $+q$ at the origin.

Solution. The distribution of charges is shown in Fig. 3.15.
(a) The electrostatic force $\mathbf{F}$ experienced by a charge $+q$ at the origin is

$$\mathbf{F} = \frac{1}{4\pi\epsilon_0}(qq_1 - 4qq_1)\mathbf{i} + \frac{1}{4\pi\epsilon_0}3qq_1\mathbf{j}$$

$$= \frac{3qq_1}{4\pi\epsilon_0}(-\mathbf{i} + \mathbf{j}).$$

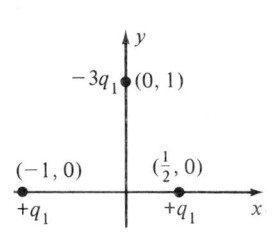

Fig. 3.15. Three fixed charges in the xy plane.

(b) The electric potential at a point (x, y) near the origin is

$$V(x, y) = \frac{q_1}{4\pi\epsilon_0}\left\{\frac{1}{[(\frac{1}{2} - x)^2 + y^2]^{1/2}} + \frac{1}{[(1 + x)^2 + y^2]^{1/2}} - \frac{3}{[x^2 + (1 - y)^2]^{1/2}}\right\} \simeq \frac{q_1}{4\pi\epsilon_0}\left[\frac{1}{(\frac{1}{2} - x)} + \frac{1}{(1 + x)} - \frac{3}{(1 - y)}\right].$$

The electric field at the point (x, y) is

$$\mathbf{E}(x, y) = -\frac{\partial V(x, y)}{\partial x}\mathbf{i} - \frac{\partial V(x, y)}{\partial y}\mathbf{j}$$

$$\simeq \frac{q_1}{4\pi\epsilon_0}\left\{\left[\frac{-1}{(\frac{1}{2} - x)^2} + \frac{1}{(1 + x)^2}\right]\mathbf{i} + \frac{3}{(1 - y)^2}\mathbf{j}\right\}.$$

The electric field $\mathbf{E}$ at the origin $(x = y = 0)$ is

$$\mathbf{E} = \frac{3q_1}{4\pi\epsilon_0}[-\mathbf{i} + \mathbf{j}].$$

Therefore, the force $\mathbf{F}$ experienced by a charge $+q$ at the origin is

$$\mathbf{F} = \frac{3qq_1}{4\pi\epsilon_0}[-\mathbf{i} + \mathbf{j}]$$

in agreement with the result obtained in (a).

3.7 THE ELECTRIC DIPOLE REVISITED

The potential at the point P due to the charges $+q$ and $-q$ shown in Fig. 3.16 is

$$V = \frac{1}{4\pi\epsilon_0}\left(\frac{q}{r_1} - \frac{q}{r_2}\right)$$

$$= \frac{q}{4\pi\epsilon_0}\left(\frac{r_2 - r_1}{r_1 r_2}\right).$$

For $d \ll r$, we can set

$$r_1 r_2 = r^2 \quad \text{and} \quad r_2 - r_1 = d\cos\theta.$$

Therefore,

$$V = \frac{qd\cos\theta}{4\pi\epsilon_0 r^2} = \frac{p\cos\theta}{4\pi\epsilon_0 r^2}.$$

Let us take the point O at the center of the dipole as the origin of a spherical polar coordinate system. In this coordinate system

$$E_r = -\frac{\partial V}{\partial r}, \quad E_\theta = -\frac{1}{r}\frac{\partial V}{\partial \theta},$$

$$E_\phi = -\frac{1}{r\sin\theta}\frac{\partial V}{\partial \phi}$$

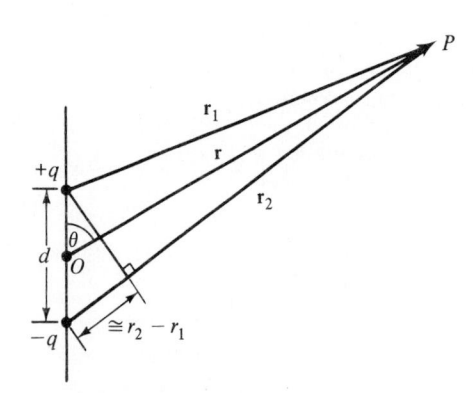

Fig. 3.16. The electric dipole.

THE ELECTRIC QUADRUPOLE

(see Appendix A). Since V is independent of ϕ, $E_\phi = 0$. The other components of **E** are

$$E_r = -\frac{\partial V}{\partial r} = \frac{2p \cos \theta}{4\pi\epsilon_0 r^3}$$

and

$$E_\theta = -\frac{1}{r}\frac{\partial V}{\partial \theta} = \frac{p \sin \theta}{4\pi\epsilon_0 r^3}.$$

The magnitude E of the field at the general point $P(r, \theta)$ is

$$\begin{aligned} E &= (E_r^2 + E_\theta^2)^{1/2} \\ &= \frac{p}{4\pi\epsilon_0 r^3}(4\cos^2\theta + \sin^2\theta)^{1/2} \\ &= \frac{p}{4\pi\epsilon_0 r^3}(3\cos^2\theta + 1)^{1/2}. \end{aligned}$$

The field is proportional to p and inversely proportional to r^3 as was stated in Section 3.4. For points on the axis of the dipole ($\theta = 0°$ or $180°$), $E_\theta = 0$ and the field, of magnitude

$$E = \frac{2p}{4\pi\epsilon_0 r^3},$$

is along the axis. For points on a plane through the center of the dipole and perpendicular to its axis ($\theta = 90°$) $E_r = 0$ and the field, of magnitude

$$E = \frac{p}{4\pi\epsilon_0 r^3},$$

is parallel to the dipole axis. The magnitude and direction at other points may be readily determined.

3.8 THE ELECTRIC QUADRUPOLE

Charge distributions such as those shown in Fig. 3.17 are known as **electric quadrupoles**. They can be thought of as consisting of two equal and opposite dipoles that do not coincide in space so that their electric effects at distant points do not quite cancel. The **linear electric quadrupole** shown in Fig. 3.17(a) is more amenable to calculation for it possesses cylindrical symmetry with respect to the z axis. We will determine the field due to the linear quadrupole in the same manner as we did for the electric dipole. We leave the calculation of the field for configuration (b) to Problem 29.

The electric potential at point P due to the

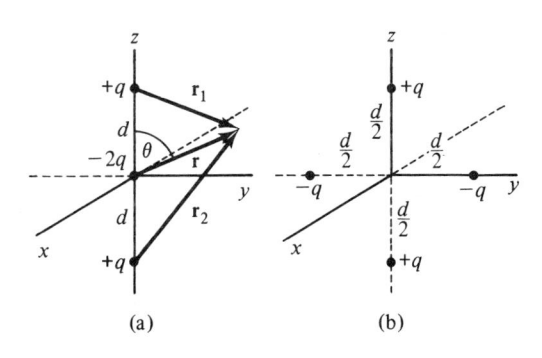

Fig. 3.17. Electric quadrupoles.

linear quadrupole is

$$V = \frac{1}{4\pi\epsilon_0}\left(\frac{q}{r_1} - \frac{2q}{r} + \frac{q}{r_2}\right).$$

Now,

$$r_1^2 = r^2 + d^2 - 2dr\cos\theta$$
$$= r^2\left(1 - \frac{2d\cos\theta}{r} + \frac{d^2}{r^2}\right)$$
$$r_1 = r\left(1 - \frac{2d\cos\theta}{r} + \frac{d^2}{r^2}\right)^{1/2}$$
$$\frac{1}{r_1} = \frac{1}{r}\left[1 - \left(\frac{2d\cos\theta}{r} - \frac{d^2}{r^2}\right)\right]^{-1/2}.$$

When $d \ll r$, the term in parentheses is $\ll 1$ and we can expand the quantity in brackets using the binomial series

$$(1 - x)^{-1/2} = 1 + \tfrac{1}{2}x + \tfrac{3}{8}x^2 + \cdots.$$

Therefore,

$$\frac{1}{r_1} = \frac{1}{r}\left[1 + \frac{1}{2}\left(\frac{2d\cos\theta}{r} - \frac{d^2}{r^2}\right) + \frac{3}{8}\left(\frac{2d\cos\theta}{r} - \frac{d^2}{r^2}\right)^2 + \cdots\right]$$
$$= \frac{1}{r}\left[1 + \frac{d\cos\theta}{r} - \frac{d^2}{2r^2} + \frac{3}{2}\frac{d^2\cos^2\theta}{r^2} - \frac{3}{2}\frac{d^3\cos\theta}{r^3} + \frac{3}{8}\frac{d^4}{r^4} + \cdots\right]$$
$$= \frac{1}{r} + \frac{d\cos\theta}{r^2} - \frac{d^2}{2r^3} + \frac{3d^2\cos^2\theta}{2r^3} - \cdots.$$

Neglecting terms in r^{-4} and higher, we have

$$\frac{1}{r_1} = \frac{1}{r} + \frac{d\cos\theta}{r^2} + \frac{d^2}{2r^3}(3\cos^2\theta - 1).$$

Similarly,

$$\frac{1}{r_2} = \frac{1}{r} - \frac{d\cos\theta}{r^2} + \frac{d^2}{2r^3}(3\cos^2\theta - 1).$$

Upon substitution into the expression for the electric potential we find

$$V = \frac{qd^2(3\cos^2\theta - 1)}{4\pi\epsilon_0 r^3}.$$

We define the **scalar quadrupole moment** eQ of a set of point charges q_i relative to the z axis (where the z axis is a symmetry axis) as (see Fig. 3.18)

$$eQ = \sum_i q_i(3z_i^2 - r_i^2).$$

Therefore, for the linear quadrupole (see Fig. 3.17)

$$eQ = q(3d^2 - d^2) - 2q(0 - 0) + q(3d^2 - d^2) = 2q(3d^2 - d^2) = 4qd^2$$

so that

$$V = \frac{eQ}{4} \cdot \frac{(3\cos^2\theta - 1)}{4\pi\epsilon_0 r^3}.$$

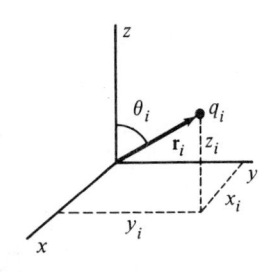

Fig. 3.18. Coordinates of a charge q_i.

SEC. 3.8 THE ELECTRIC QUADRUPOLE

It follows that
$$E_r = \frac{3eQ}{4} \cdot \frac{(3\cos^2\theta - 1)}{4\pi\epsilon_0 r^4}$$
and
$$E_\theta = \frac{3eQ}{4} \frac{\sin 2\theta}{4\pi\epsilon_0 r^4}.$$

Note that the electric field produced by an electric quadrupole is proportional to the electric quadrupole moment and inversely proportional to the fourth power of the distance at large distances.

Example. A quadrupole experiences no torque in a uniform electric field; however, a torque may be exerted on a quadrupole by a nonuniform field. A linear quadrupole is placed in an external electric field having a constant electric field gradient. The quadrupole makes an angle θ with the field gradient. (a) Deduce an expression for the torque experienced by the quadrupole. (b) Derive an expression for the potential energy of the quadrupole.

Solution.
(a) The electric field **E** at a point with coordinate dz is given by the expression
$$\mathbf{E}(dz) = \mathbf{E}(z=0) + \left(\frac{\partial E}{\partial z}\right)_{z=0} dz.$$
However,
$$\boldsymbol{\nabla} E = \mathbf{k}\left(\frac{\partial E}{\partial z}\right) = \mathbf{k}\left(\frac{\partial E}{\partial z}\right)_{z=0}$$
since $\boldsymbol{\nabla} E$ is a constant vector in the z direction. Therefore,
$$\mathbf{E}(dz) = \mathbf{E}(z=0) + |\boldsymbol{\nabla} E|\, dz.$$
Since we are concerned only with the gradient of **E** and not its absolute value, we can put $\mathbf{E}(z=0) = 0$ for convenience. Integrating gives the field **E** at a point with coordinate z as
$$\mathbf{E} = \int_0^z |\boldsymbol{\nabla} E|\, dz$$
$$= |\boldsymbol{\nabla}(E)| \int_0^z dz$$
$$= |\boldsymbol{\nabla}(E)|\, \mathbf{z}.$$

The resulting forces on the charges constituting the quadrupole are shown in Fig. 3.19(a). Note that the net force experienced by the quadrupole is zero; however, it experiences a net torque. This torque τ is equivalent to that produced by two equal and opposite forces **F** and $-\mathbf{F}$ as shown in Fig. 3.19(b). The torque tends to align the quadrupole in the direction of the field gradient. The magnitude of the torque is
$$\tau = 2Fd\sin\theta = 2q|\boldsymbol{\nabla} E|\, zd\sin\theta$$
$$= 2qd^2|\boldsymbol{\nabla} E|\sin\theta\cos\theta = \frac{eQ}{2}|\boldsymbol{\nabla} E|\sin\theta\cos\theta$$
$$= \frac{eQ}{4}|\boldsymbol{\nabla} E|\sin 2\theta.$$

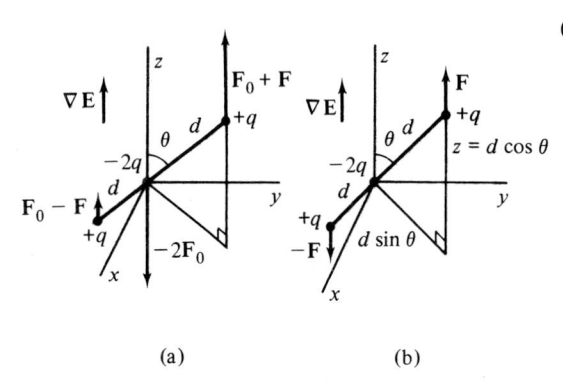

Fig. 3.19. (a) An electric quadrupole in an external electric field with a constant electric field gradient in the z direction. (b) The resultant torque experienced by the quadrupole.

(b) The change in potential energy of the quadrupole is equal to the work that must be done on the quadrupole to reorient it in the presence of the field gradient. The work required to change the orientation of the quadrupole from θ_1 to θ_2 is

$$W = \int dW = \int_{\theta_1}^{\theta_2} \tau \, d\theta$$
$$= \int_{\theta_1}^{\theta_2} \frac{eQ}{2} |\nabla E| \sin \theta \cos \theta \, d\theta$$
$$= \frac{eQ}{4} |\nabla E| \sin^2 \theta_2 - \frac{eQ}{4} |\nabla E| \sin^2 \theta_1.$$

The potential energy U of the quadrupole may therefore be taken as

$$U = \frac{eQ}{4} |\nabla E| \sin^2 \theta.$$

3.9 ELECTRIC MOMENTS OF ELEMENTARY PARTICLES AND NUCLEI

Elementary particles as well as the nuclei of atoms possess charge and might be expected to have electric dipole and quadrupole moments. (The neutron, although possessing no net charge, does have an extended electric charge distribution in its internal structure, wherein equal amounts of positive and negative charge are found). On the assumption that the charge within a nucleus or elementary particle is distributed in order to minimize the total energy of the system, we might expect nuclei and elementary particles to exhibit spherical symmetry or, at worst, ellipsoidal symmetry (see Fig. 3.20). An ellipsoid of revolution is a three-dimensional object formed by rotating an ellipse about one of its axes. If the rotation is about the major axis, as in Fig. 3.20, the ellipsoid is **prolate** while rotation about the minor axis produces an **oblate** ellipsoid.

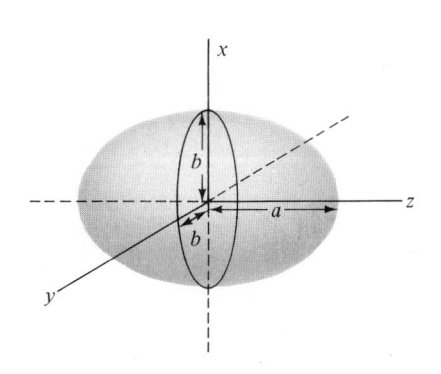

We note first that symmetry considerations rule out the possibility of the existence of an electric dipole moment for spherical or ellipsoidal systems with axial symmetry. This can be shown by an application of quantum mechanics which is, unfortunately, beyond the scope of this text. We do, however, expect no electric dipole moment for particles and nuclei and none has yet been measured. Experiments designed to measure such moments, if they should exist, cannot tell us that the electric dipole moment of the particle or nucleus concerned is identically zero, but rather give limits on the magnitude of the moment. Two recent measurements on the neutron[10] yielded values for the electric dipole moment of $(-2 \pm 3) \times 10^{-24}\ e$ meters and $(+2.4 \pm 3.9) \times 10^{-24}\ e$ meters. For comparison, we note that two charges $+e$ and $-e$ separated by a distance d have an electric dipole moment ed. Since the neutron has a radius of about 10^{-15} m, we see that the separation of the centers of positive and negative charge in a neutron can be no greater than about 10^{-9} of the neutron diameter. Experiments have been performed on other particles and nuclei with similar results.

Nuclear electric quadrupole moments can be estimated if we assume that the nuclear charge of the protons is distributed uniformly throughout the nuclear volume. The volume of an ellipsoid with semiaxes a and b as shown in Fig. 3.20 is $4\pi ab^2/3$. The charge density ρ for a nucleus of atomic number Z is then

$$\rho = \frac{3Ze}{4\pi ab^2}.$$

The scalar quadrupole moment for a continuous charge distribution is defined to be

$$eQ = \int_v \rho(3z^2 - r^2)\, dv$$

which, upon evaluation, yields

$$Q = \tfrac{2}{5} Z(a^2 - b^2).$$

Evaluation of the integral for the quadrupole moment

We wish to evaluate the integral

$$eQ = \int_v \rho(3z^2 - r^2)\, dv$$

for a prolate ellipsoidal nucleus of constant charge density

$$\rho = \frac{3Ze}{4\pi ab^2}.$$

Substituting for ρ we obtain

$$Q = \frac{3Z}{4\pi ab^2} \int_v (3z^2 - r^2)\, dv.$$

The integration can be carried out most conveniently using cylindrical coor-

---

[10] P. D. Miller, W. B. Dress, J. K. Baird, and N. F. Ramsey, "Limit to the Electric Dipole Moment of the Neutron," *Physical Review Letters*, **19** (August, 1967), 381. C. G. Shull and R. Nathans, "Search for a Neutron Electric Dipole Moment by a Scattering Experiment," *Physical Review Letters*, **19** (August, 1967), 384.

dinates (ρ, z, ϕ) as shown in Fig. 3.21. A point $P(x, y, z)$ on the surface of the ellipsoid has cylindrical coordinates (ρ, z, ϕ) where

$$\rho^2 = x^2 + y^2$$
$$x = \rho \cos \phi$$

and

$$y = \rho \sin \phi.$$

The equation for the ellipsoid in Cartesian coordinates is

$$\frac{x^2}{b^2} + \frac{y^2}{b^2} + \frac{z^2}{a^2} = 1$$

or

$$\frac{\rho^2}{b^2} + \frac{z^2}{a^2} = 1$$

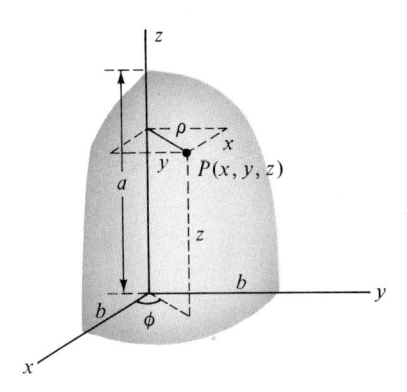

Fig. 3.21. A point P with Cartesian coordinates (x, y, z) has cylindrical coordinates (ρ, z, ϕ) where $\rho^2 = x^2 + y^2$, $x = \rho \cos \phi$, and $y = \rho \sin \phi$.

in cylindrical coordinates. The volume element in cylindrical coordinates is

$$dv = \rho \, d\rho \, dz \, d\phi.$$

Since

$$r^2 = x^2 + y^2 + z^2$$
$$= \rho^2 + z^2,$$

the integral becomes

$$Q = \frac{3Z}{4\pi ab^2} \int_v (3z^2 - \rho^2 - z^2) \rho \, d\rho \, dz \, d\phi$$

$$= \frac{3Z}{4\pi ab^2} \int_0^{2\pi} d\phi \int_{-a}^{a} dz \int_0^{\rho=b[1-(z^2/a^2)]^{1/2}} \rho(2z^2 - \rho^2) \, d\rho$$

$$= \frac{3Z}{2ab^2} \int_{-a}^{a} dz \left[z^2 b^2 \left(1 - \frac{z^2}{a^2}\right) - \frac{b^4}{4}\left(1 - \frac{z^2}{a^2}\right)^2 \right]$$

$$= \frac{3Z}{a} \int_0^{a} dz \left[z^2 - \frac{z^4}{a^2} - \frac{b^2}{4} + \frac{b^2 z^2}{2a^2} - \frac{b^2 z^4}{4a^4} \right]$$

$$= \frac{3Z}{a} \int_0^{a} dz \left[z^2 \left(1 + \frac{b^2}{2a^2}\right) - \frac{z^4}{a^2}\left(1 + \frac{b^2}{4a^2}\right) - \frac{b^2}{4} \right]$$

$$= \frac{3Z}{a} \left[\frac{a^3}{6a^2}(2a^2 + b^2) - \frac{a^5}{20a^4}(4a^2 + b^2) - \frac{b^2 a}{4} \right]$$

$$= \frac{3Z}{a} \left(\frac{2a^3}{15} - \frac{2ab^2}{15} \right)$$

$$Q = \tfrac{2}{5} Z(a^2 - b^2).$$

We note first that the quadrupole moment is zero for spherical nuclei ($a = b$), is positive for prolate nuclei ($a > b$), and is negative for oblate nuclei ($a < b$). Since for most nuclei a and b are nearly equal, it is convenient to define the nuclear radius R as their mean value

$$R = \frac{a+b}{2}$$

and to measure the departure from sphericity by a parameter η defined by

$$\eta = \frac{b-a}{R} = 2\left(\frac{b-a}{b+a}\right).$$

The quadrupole moment is now written as

$$Q = \tfrac{4}{5}\eta Z R^2.$$

The quantity ηZ gives a rough measure of the number of protons whose cooperation is required in order to produce the observed quadrupole moment. The values of Q, η, and ηZ for a few representative nuclei are given in Table 3.1.

Table 3.1 Quadrupole Moments and Ellipticities of Some Nuclei

Nucleus (Z)	Q 10^{-30} m²	η	$|\eta Z|$
Hydrogen (1)	0	0	0
Deuterium (1)	+0.273	+0.095	0.095
Aluminium-27 (13)	+15.6	+0.074	0.96
Sulphur-33 (16)	−8	−0.027	0.43
Sulphur-35 (16)	+6	+0.019	0.30
Gallium-71 (31)	+14.6	+0.015	0.47
Germanium-73 (32)	−20	−0.02	0.64
Indium-113 (49)	+114	+0.055	2.70
Antimony-123 (51)	−120	−0.053	2.70
Lutetium-175 (71)	+590	+0.148	10.5
Bismuth-209 (83)	−40	−0.008	0.7

3.10 THE MOMENTS OF A CHARGE DISTRIBUTION

It is often necessary to deal with the electric field produced by a collection or distribution of charges. The study of the electrical behavior of atoms and molecules is an important example. In general, we are concerned with some sort of charge distribution centered about, or at least in the region of, the origin of a convenient coordinate system such as that shown in Fig. 3.22. Denoting $\rho(r')$ as the density of charge at position $\mathbf{r}'$, we have for the potential at point P due to the charge in a volume element dv

$$V(P) = \frac{1}{4\pi\epsilon_0} \int_v \frac{\rho(r')\,dv}{R}.$$

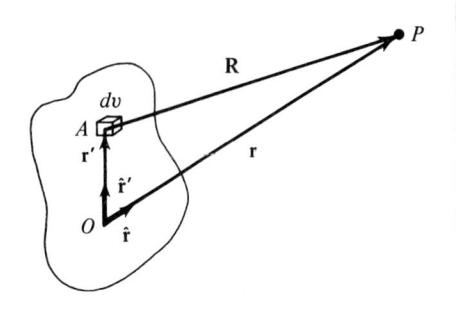

Fig. 3.22. A general charge distribution.

Applying the law of cosines to triangle OAP we have

$$R^2 = r^2 + r'^2 - 2\mathbf{r} \cdot \mathbf{r}'$$
$$= r^2 + r'^2 - 2rr'(\hat{\mathbf{r}} \cdot \hat{\mathbf{r}}')$$

so that

$$\frac{1}{R} = (r^2 - 2rr'\hat{\mathbf{r}} \cdot \hat{\mathbf{r}}' + r'^2)^{-1/2}.$$

Since $r' \ll r$ for points far from the charge distribution,

$$\frac{1}{R} = \frac{1}{r}\left[1 - \left(\frac{2rr'\hat{\mathbf{r}} \cdot \hat{\mathbf{r}}'}{r^2} - \frac{r'^2}{r^2}\right)\right]^{-1/2}$$
$$= \frac{1}{r}\left\{1 + \frac{r'\hat{\mathbf{r}} \cdot \hat{\mathbf{r}}'}{r} + \left[\frac{3r'^2(\hat{\mathbf{r}} \cdot \hat{\mathbf{r}}')^2}{r^2} - \frac{r'^2}{r^2}\right] + \cdots\right\}.$$

Therefore,

$$V(P) = \frac{1}{4\pi\epsilon_0} \int_v \frac{\rho(r')\,dv}{r} + \frac{1}{4\pi\epsilon_0} \int_v \frac{r'\hat{\mathbf{r}} \cdot \hat{\mathbf{r}}'}{r^2} \rho(r')\,dv$$
$$+ \frac{1}{4\pi\epsilon_0} \int_v [3(\hat{\mathbf{r}} \cdot \hat{\mathbf{r}}')^2 - 1]\frac{r'^2}{r^3} \rho(r')\,dv + \cdots.$$

Now r is a constant and may be taken from under the integral sign. The integrals have constant values characteristic of the charge distribution in question so that we may write

$$V(P) = \frac{A_0}{4\pi\epsilon_0 r} + \frac{A_1}{4\pi\epsilon_0 r^2} + \frac{A_2}{4\pi\epsilon_0 r^3} + \cdots.$$

The integral A_0 is obviously the total charge of the distribution and is called the **monopole moment**. It indicates the potential that would result if the total charge were concentrated at the origin. The second integral A_1 is the dipole moment and is a generalization of the definition given for two equal and opposite point charges. It is left as an exercise for the reader (Problem 31) to show that the two definitions give the same result for two equal and opposite point charges. The third integral A_2 has nine components and is related to the **quadrupole moment**. The nine components form the **quadrupole moment tensor** (a tensor is a generalization of a vector). Higher-order terms in the expansion for $V(P)$ generate higher-order moments, such as octupole moments and hexadecapole moments, which are of particular importance in the description of the behavior of the atomic nucleus.

QUESTIONS AND PROBLEMS

1. Can lines of force ever intersect? Explain your answer.
2. Sketch the lines of force for the following charge configurations: (a) A charge $-q$ a short distance from a charge $+3q$. (b) A charge $+q$ near a

uniformly charged negative surface. (c) A charge $+q$ near a uniformly charged positive surface. (d) A positively charged wire parallel to a negatively charged wire. (e) An electric quadrupole (Fig. 3.17a). (f) Three charges $+q$, $+q$, and $-q$ on the corners of an equilateral triangle.

3. An electron is subjected to a force of 2×10^{-4} N directed to the north. What are the magnitude and direction of the electric field?

4. Calculate the electric field produced at the point $(-3, 0)$ by point charges $+2q$ located at $(2, 0)$, $-q$ at $(0, -1)$, and $+3q$ at $(1, 1)$.

5. Two point charges of 3.0×10^{-5} C and 6.0×10^{-5} C are separated by 5.0 cm. Find three points on the line joining them where the electric field is zero.

6. What is the magnitude of the electric field at a distance 10^{-12} m from the center of a cobalt nucleus? A cobalt nucleus contains 27 protons.

7. Deduce the most general path traced out by a charged particle traveling in a uniform electric field.

*8. Show that the magnitude of the electric field due to a uniform distribution of electric charge along an infinitely long wire is given by

$$E = \frac{2k_e \lambda}{r}$$

where λ is the density of charge per meter along the wire and r is the perpendicular distance to the wire. What is the direction of the field?

9. At what speed must an electron be projected so that it will move in a circular path about a positively and uniformly charged wire?

*10. Find an expression for the electric field due to a uniform distribution of charge on an infinite plane surface.

11. Write an essay on the life and work of Robert A. Millikan.

*12. A droplet of oil of radius r, density ρ_0 and charge $+q$ is in a container that is filled with a substance whose density increases with depth due to compression. The density at the top is $\rho_m (\rho_m > \rho_0)$ and increases by K g·cm^{-3} for each additional 1.0 cm of depth. An electric field **E** is present and is directed downward. At what depth below the surface will the drop experience no net force? Assuming no viscous damping, calculate the frequency of vertical oscillations about this point.

*13. Deduce the expression for the charge q on an oil droplet in a Millikan experiment in terms of the terminal velocity of free fall v_d and the terminal upward velocity v_u in an electric field **E** of sufficient strength to overcome gravity.

14. An electric dipole makes an angle of 30° with respect to a uniform electric field of magnitude 2×10^3 N·C^{-1}. A torque of 5×10^{-2} N·m acts on it. Calculate the potential energy and the dipole moment.

15. At what frequency will an electric dipole **p** with moment of inertia I oscillate in an electric field **E**? Assume small angle displacements about the electric field lines.

16. A dipole aligned originally parallel to a uniform electric field is brought to an antiparallel alignment with the expenditure of 0.1 J of energy. What was the torque exerted on the dipole when it made an angle of 45° with the field?

*17. An electric dipole in a nonuniform electric field experiences a translational force. Derive an expression for this force.

18. Describe one mechanism for the repair of radiation damage in DNA.

19. Illustrate the molecular structures of adenine and guanine.

*20. Amino acids are a class of biologically important molecules from which protein is synthesized. Glycine (see Fig. 3.23) is the smallest molecule of this class. Estimate the dipole moment you would expect this molecule to have assuming bond moment additivity and neglecting any moment in the COOH structure which is not parallel to the C—C bond. Take the H—O, C—O and C—C bond moments to be

$$H \rightarrow O = 1.51D$$
$$C \rightarrow O = 0.8D$$
$$C \rightarrow C = 0D.$$

Fig. 3.23. The structure of the tetrahedral glycine molecule. Bond angles are 109° about the C–C bond and 120° otherwise.

21. The internuclear separations in the hydrogen fluoride (HF), hydrogen chloride (HCl), hydrogen bromide (HBr), and hydrogen iodide (HI) molecules are 0.0917, 0.127, 0.142, and 0.162 nm, respectively. Assuming an ionic model, calculate the dipole moment of each molecule. The observed dipole moments are 1.91, 1.03, 0.78, and 0.38D. Comment on the comparison between calculated and experimental values. What generalization might one make about the halogen series of atoms on the basis of the above results?

22. The structure of the cytosine molecule is shown in Fig. 3.24. Assuming the molecule to be planar and assuming bond moment additivity, deduce the molecular dipole moment. Take the C ⇒ N bond of magnitude ≃ 1.6 D to be directed from the carbon to the nitrogen atom.

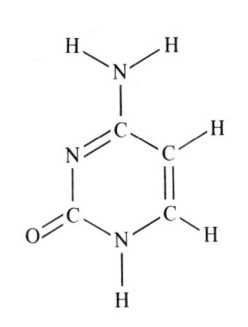

Fig. 3.24. The structure of the cytosine molecule.

23. Show that the values for the dipole moments of the four configurations of T—T dimers indicated in Fig. 3.11 are:

$$p_\text{I} = 6.0 \text{ D} \qquad p_\text{II} = 4.7 \text{ D} \qquad p_\text{III} = 3.8 \text{ D} \qquad p_\text{IV} = 0.0 \text{ D}$$

24. Six charges $+q$ are located on the corners of a regular hexagon whose sides are of length ℓ. Calculate the electric potential at the center of the hexagon. How much energy would be required to move a charge $-q$ from infinity to the center of the hexagon?

*25. Calculate, by direct integration, the electric potential as a function of radial distance from an infinitely long and uniformly charged wire. Explain the physical meaning of your result.

26. Calculate the work necessary to charge a spherical shell of radius R and a total charge $+q$. (Hint: The electric field for $r > R$ is the same as if all the charge on the sphere were located at a point at the center.)

27. Three charges are located as follows: $+3q$ at $(1, 0)$, $-2q$ at $(0, -1)$, and

$+q$ at (1, 1). Write an expression for the electric potential and from it calculate the x component of the electric field.

*28. Work out an expression for the potential energy of two coplanar dipoles separated by a distance R. Assume that R is large compared to the size of the dipoles and that the dipoles make angles θ_1 and θ_2 with the y axis.

*29. Calculate the electric field due to the quadrupole shown in Fig. 3.17(b).

30. Derive an expression for the torque exerted on a linear electric quadrupole by a point charge q a distance r from the center of the quadrupole. Assume $r \gg a$, the dimension of the quadrupole.

31. Show that the general definition of the dipole moment (Section 3.10) reduces to the definition given in terms of two equal and opposite point charges.

*32. Show that the dipole moment of an arbitrary charge distribution is independent of the choice of origin providing that the net charge is zero.

*33. Calculate the dipole moment of a spherical shell of radius R with charge density $\sigma = \sigma_0 \cos\theta$, where θ is the polar angle.

34. Molecules are examples of charge distributions. Give examples of molecules for which the first nonvanishing moment of the charge distribution allowed by molecular symmetry is: (a) the quadrupole moment, (b) the octupole moment and (c) the hexadecapole moment.

KARL F. GAUSS

4 Gauss' law

4.1 INTRODUCTION

Karl Friedrich Gauss, among his many outstanding contributions to mathematics and the sciences, developed a general relation between the surface integral of the electric field vector and the net charge enclosed within the surface. A simple mathematical statement of the law is as follows:

$$\int_S E_n \, dS = \frac{q}{\epsilon_0}$$

where E_n is the component of the electric field normal to the element of surface area dS at the position of dS. The symbol $\int_S$ refers to integration over any closed surface surrounding the charge q and a double integration is implied. Because of its general nature, Gauss' law can be useful in the solution of many problems, especially those involving continuous charge distributions and having a high degree of symmetry.

4.2 PROOF OF GAUSS' LAW

Before proceeding to a proof of Gauss' law for surfaces of arbitrary shape we must discuss the surface integral a bit more and introduce the concept of **solid angle**. Let us consider the element of surface area dS of some arbitrary surface enclosing the single charge $+q$ (see Fig. 4.1). The vector d**S** is defined to be a vector normal to the surface element dS with magnitude dS equal to the area of the surface element. The scalar product $\mathbf{E} \cdot d\mathbf{S} = E \, dS \cos \theta = E_n \, dS$ is then the product of the component of the electric field normal to the surface and the area of the surface element. Now dS' is the projection of dS in a direction perpendicular to **E** and has magnitude $dS \cos \theta$ (that is, if the vector d**S** is rotated into the direction of **E**, dS and dS' would be parallel, which indicates that the angle between dS and dS' is the same as the angle between d**S** and **E**). Therefore, $E_n \, dS = E \, dS'$ and, since the magnitude of **E** is measured by the density of electric lines, we may interpret $\mathbf{E} \cdot d\mathbf{S}$ as the total number of lines of force going through the surface area dS. The surface integral

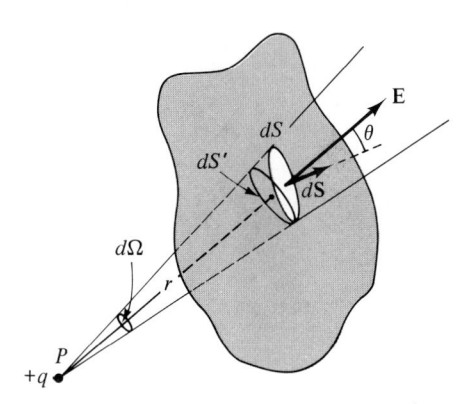

Fig. 4.1. Graphical representation of $\mathbf{E} \cdot d\mathbf{S} = E_n \, dS$ and of the solid angle $d\Omega$.

$$\int_S E_n \, dS = \int_S \mathbf{E} \cdot d\mathbf{S} = \Phi_E$$

is equal to the total number of lines of force passing through the surface and is called the **electric flux** Φ_E through the surface.

The element of **ordinary angle** $d\theta$ is defined as

$$d\theta = \frac{dl}{r} \text{ rad}$$

where dl is an element of arc of a circle of radius r. The **total** planar angle is

$$\theta = \oint d\theta = \frac{1}{r} \oint dl = \frac{2\pi r}{r}$$
$$= 2\pi \text{ rad}.$$

The symbol $\oint$ means integration about a closed curve, in this case a circle. Notice that the quantity θ is dimensionless.

The general definition of angle $d\theta$ follows from a consideration of Fig. 4.2. The angle $d\theta$ subtended at point P by an element dl of the general curve C is

$$d\theta = \frac{dl'}{r} = \frac{dl \cos \phi}{r}$$

where dl' is an element of arc of a circle of radius r about the point P and ϕ is the angle between dl and dl'. The total planar angle about point P is

$$\theta = \oint d\theta = \frac{1}{r} \oint dl' = \frac{2\pi r}{r} = 2\pi \text{ rad}$$

for any contour lying in a plane and enclosing the point.

An element of **solid angle** $d\Omega$ can be defined in a similar manner to be the angle subtended at the center of a sphere by an element of the spherical surface. The surface element dS' in Fig. 4.1 is a portion of a spherical surface S' of radius r centered at P. The solid angle subtended at P by the element dS' is by definition

$$d\Omega = \frac{dS'}{r^2} = \frac{dS \cos \theta}{r^2} \text{ steradians (sterad)}.$$

Solid angles are also dimensionless. Note that dS and dS' both subtend the same solid angle. The total solid angle about any point is

$$\Omega = \int_S d\Omega = \frac{1}{r^2} \int_{S'} dS' = \frac{4\pi r^2}{r^2} = 4\pi \text{ sterad}.$$

The definition of solid angle is quite general and any surface element dS subtends a solid angle

$$d\Omega = \frac{dS \cos \theta}{r^2}$$

at a point a distance r away. The total solid angle about a point is 4π sterad for any surface en-

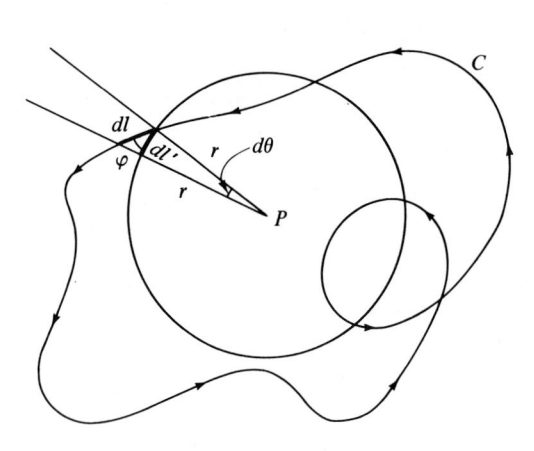

Fig. 4.2. The total planar angle about point P is 2π rad for any contour lying in a plane and enclosing the point.

closing the point. This may be shown by generalizing Fig. 4.2 to three dimensions.

Let us now reconsider the surface integral

$$\int_S E_n \, dS = \int_S \mathbf{E} \cdot d\mathbf{S} = \int_S E \, dS \cos\theta = \Phi_E.$$

Since for a single charge q we have

$$E = \frac{q}{4\pi\epsilon_0 r^2},$$

the surface integral becomes

$$\int_S E \, dS \cos\theta = \int_S \left(\frac{q}{4\pi\epsilon_0 r^2}\right) dS \cos\theta = \frac{q}{4\pi\epsilon_0} \int_S d\Omega = \frac{q}{4\pi\epsilon_0}(4\pi) = \frac{q}{\epsilon_0}$$

where we have used the fact that the total solid angle about a point is 4π sterad regardless of the shape of the surface. Therefore, Gauss' law is valid for a point charge.

It will be valid also for any amount of charge distributed in any way within the surface. Since

$$\mathbf{E} = \sum_i \mathbf{E}_i,$$

$$\int_S \mathbf{E} \cdot d\mathbf{S} = \int_S \sum_i E_i \, dS \cos\theta_i = \int_S \sum_i \frac{q_i}{4\pi\epsilon_0 r_i^2} dS \cos\theta_i$$

$$= \sum_i \left[\frac{q_i}{4\pi\epsilon_0} \int_S d\Omega_i\right] = \sum_i \frac{q_i}{\epsilon_0} = \frac{q}{\epsilon_0}.$$

For a continuous charge distribution Gauss' law takes the form

$$\int_S \mathbf{E} \cdot d\mathbf{S} = \frac{1}{\epsilon_0} \int_v \rho \, dv = \frac{q}{\epsilon_0}$$

where ρ is the charge density inside the closed surface and dv is a volume element.

Gauss' law predicts a zero surface integral when there is no charge inside the surface. Such a surface is pictured in Fig. 4.3 along with a nearby charge q. Since electric lines only terminate on charges, the number of electric lines entering the closed surface is equal to the number of electric lines leaving the closed surface and the net number of lines emerging is zero. In other words, the total electric flux through the surface is zero.

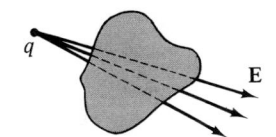

Fig. 4.3. A charge q near a closed surface.

4.3 SOME APPLICATIONS OF GAUSS' LAW

Electric field due to a spherical charge distribution

Let us consider a positive charge q distributed uniformly throughout a sphere of radius R. To determine the field at points external to the spher-

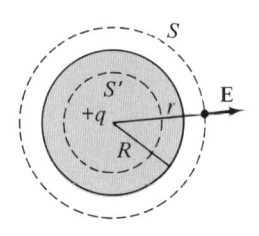

Fig. 4.4. A spherical charge distribution with charge density ρ.

ical charge distribution we consider a spherical Gaussian surface S of radius r concentric with the spherical charge distribution (see Fig. 4.4). The electric field **E** will be normal to the Gaussian surface because of the spherical symmetry. From Gauss' law

$$\int_S \mathbf{E} \cdot d\mathbf{S} = E \int_S dS = E(4\pi r^2)$$

$$= \frac{q}{\epsilon_0}$$

or

$$E = \frac{q}{4\pi\epsilon_0 r^2}.$$

This is identical with the expression for a point charge q. Thus, the external field due to a spherically symmetric charge distribution is just that field which would be produced if the charge were all located at the center of the distribution.

To calculate the field at points $r < R$ we consider the spherical Gaussian surface S' shown in Fig. 4.4. As before, the field at the surface of S' must be normal to S' due to the symmetry. From Gauss' law we have

$$E \int_{S'} dS = E(4\pi r^2) = \frac{1}{\epsilon_0} \int_v \rho \, dv.$$

The charge density ρ is

$$\rho = \frac{q}{\frac{4}{3}\pi R^3} = \frac{3q}{4\pi R^3}$$

so that

$$\frac{1}{\epsilon_0} \int_v \rho \, dv = \frac{3q}{4\pi\epsilon_0 R^3} \int_v dv = \frac{3q}{4\pi\epsilon_0 R^3} \cdot \frac{4}{3}\pi r^3 = \frac{q}{\epsilon_0}\left(\frac{r}{R}\right)^3.$$

Therefore,

$$E(4\pi r^2) = \frac{q}{\epsilon_0}\left(\frac{r}{R}\right)^3$$

or

$$E = \frac{q}{4\pi\epsilon_0}\frac{r}{R^3}, \qquad r \leq R.$$

The variation of the magnitude of the electric field as a function of distance from the center of the distribution is shown in Fig. 4.5.

Electric field due to a charged conducting spherical shell

The charge will be distributed uniformly over the shell since it is a conductor. If this were not so, portions of the charge distribution would experience unbalanced forces that would result in a further redistribution of charge since charge carriers are free to move within a conductor. If the total charge on the shell is $+q$ (see Fig. 4.6), the electric field **E** will be directed radially outward for all $r \geq R$ due to the symmetry and

$$E = \frac{q}{4\pi\epsilon_0 r^2}$$

as in the calculation for the uniform spherical distribution.

The spherical Gaussian surface S' for $r < R$, however, has no charge enclosed within it and it follows immediately from Gauss' law that

$$\int_S \mathbf{E} \cdot d\mathbf{S} = 0.$$

Fig. 4.5. The magnitude of the electric field of a uniform spherical charge distribution of radius R as a function of distance from the center of the distribution.

Since $d\mathbf{S}$ is finite, either $\mathbf{E} = 0$, or $\mathbf{E} \perp d\mathbf{S}$. However, because of the spherical symmetry, the electric field at any point, if not zero, must be directed radially either toward or away from the center of the sphere. Since $d\mathbf{S}$ is also a radial vector, we may therefore conclude that $\mathbf{E} = 0$ everywhere inside the spherical conducting shell. It is possible to show this is so if, and only if, the force between point charges varies inversely as the square of the distance between them (see Problem 4). The accuracy of Coulomb's law has been tested experimentally. This was first carried out by Robert Cavendish who showed that, if the force varies as $r^{-(2+x)}$, then $x \leq 0.02$, and later by Clerk Maxwell who showed that $x \leq 0.000046$. More recent experiments have placed an upper limit of 10^{-9} on the value of x.

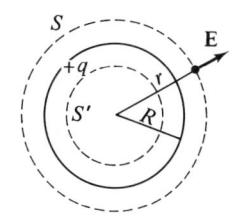

Fig. 4.6. A spherical conducting sphere with total charge $+q$.

The field inside a charged conductor is zero in general regardless of its shape. If we consider a **solid** conductor in which charge (in the form of electrons) is free to move, we see that the mutual repulsion between like charges will separate them and they will flee from each other until they all reside on the surface and are spread out uniformly over it. Therefore, in general, **the charge on a conductor resides on its surface and the field below the surface is zero whether the conductor is solid or hollow.**

Electric field due to a large plane sheet of uniform charge density

Let us consider a section of a two-dimensional charged sheet as shown in Fig. 4.7. If the sheet is very large, it follows from symmetry that, at all points not too close to the edge, the electric field should be normal to the plane of the sheet. (This would be true at all points if the sheet were infinite.) A suitable Gaussian surface is a cylinder whose axis is perpendicular to the sheet. If the cylinder has cross-sectional area A, the charge enclosed

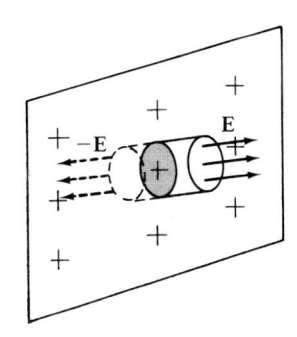

Fig. 4.7. Calculation of the electric field due to a sheet of charge.

within it will be σA where σ is the charge density of the sheet in $C \cdot m^{-2}$. From symmetry, we see that no electric flux will emerge through the curved surface of the cylinder while E will be constant over the ends. From Gauss' law

$$\int_S \mathbf{E} \cdot d\mathbf{S} = E \int_S dS = E(2A) = \frac{\sigma A}{\epsilon_0}$$

and

$$E = \frac{\sigma}{2\epsilon_0}$$

is the field for the large charged sheet. Note that it is independent of distance from the sheet (for distances much less than the dimensions of the sheet).

Electric field due to a long linear uniform charge distribution

For a very long linear uniform charge distribution, the field at points distant from the ends will be directed radially outward for a positive charge distribution due to the symmetry. We consider again a cylindrical Gaussian surface with its axis coincident with the linear charge distribution (see Fig. 4.8). If μ is the linear charge density of the distribution (in $C \cdot m^{-1}$), the total charge enclosed within the surface is μl where l is the length of the cylinder. There will be no electric flux through the ends of the cylinder and the electric field will be normal to the curved surface. From Gauss' law we have

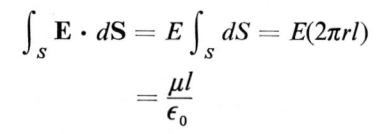

$$\int_S \mathbf{E} \cdot d\mathbf{S} = E \int_S dS = E(2\pi r l)$$
$$= \frac{\mu l}{\epsilon_0}$$

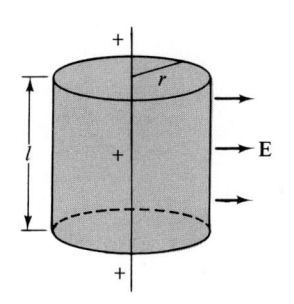

Fig. 4.8. Calculation of the electric field due to a linear charge distribution.

and

$$E = \frac{\mu}{2\pi\epsilon_0 r}$$

is the field due to the very long linear uniform charge distribution.

Example. A thin conducting spherical shell of radius 0.50 m carries a charge of 9 μC. Deduce the magnitude of the electric field 0.25 m, 0.50 m, and 1.00 m from the center of the sphere.

Solution. The electric field within a conducting surface is everywhere zero. Therefore,

$$E(0.25 \text{ m}) = 0.$$

The magnitude of the electric field at the inner surface of the sphere is zero. Therefore,

$$E(0.50 \text{ m})_{\text{inner}} = 0.$$

The magnitude of the field at the outer surface of the sphere is

$$E(0.50 \text{ m})_{\text{outer}} = \frac{9 \times 10^{-6}}{4\pi\epsilon_0 (0.50)^2} = \frac{9 \times 10^{-6}}{\pi\epsilon_0}$$
$$= 3.2 \times 10^5 \text{ N} \cdot \text{C}^{-1}.$$

At any point outside of the sphere the magnitude of the electric field is the same as that due to a point charge of the same magnitude located at the center of the sphere. Therefore,

$$E(1.00 \text{ m}) = \frac{9 \times 10^{-6}}{4\pi\epsilon_0 (1.00)^2} = \frac{9 \times 10^{-6}}{4\pi\epsilon_0} = 8.1 \times 10^4 \text{ N} \cdot \text{C}^{-1}.$$

4.4 GAUSS' LAW FOR GRAVITATIONAL INTERACTIONS

The gravitational force is another example of an inverse square force.[1] Therefore, Gauss' law is also applicable to gravitational interactions and may be written as

$$\Phi_G = \int_S g_n \, dS = -4\pi G m$$

where g_n is the component of the **gravitational field** normal to the element of surface area dS, G is the universal gravitational constant, and m is the mass enclosed by the Gaussian surface. The gravitational field **g** is defined as the gravitational force per unit mass. The minus sign indicates that the gravitational force is always attractive. The form of the right-hand side of the equation is a consequence of Newton's law of gravitation (see Problem 13).

If we apply Gauss' law to a homogeneous spherical mass distribution of radius R, we will obtain a graph identical to that of Fig. 4.4 for the variation of the magnitude of the gravitational field as a function of the distance r from the center of the distribution.

The variation of g in the earth's interior

In elementary physics texts[2] it is sometimes assumed that the earth may be approximated by a homogeneous spherical mass distribution. Figure 4.4 then gives a linear variation of the acceleration g due to gravity with the distance from the earth's center for points beneath the earth's surface. This result leads to the intriguing possibility that rapid intercontinental travel could be achieved by crisscrossing the earth with frictionless subterranean passages.[3]

[1] *MWTP*, Section 7.6.
[2] *MWTP*, Section 11.5.
[3] P. W. Cooper, "Through the Earth in Forty Minutes," *American Journal of Physics*, **34** (January 1966), 68.

The model of a homogeneous earth is, however, in serious contradiction with the experimental facts obtained by **geophysicists**[4] through the observation of **seismic waves** triggered by earthquakes. Three main types of seismic waves are readily identified on **seismograms**. There are primary (P) waves that cause compression and expansion of the medium in the direction in which the wave is traveling and that travel through both solid and liquid parts of the earth. There are secondary (S) waves that cause particles in the medium to oscillate perpendicular to the direction of travel of the wave and that travel only through the solid portion of the earth. Finally, there are surface waves confined to the upper 0.5% of the earth's surface. The speed of P waves differs from the speed of S waves and they both vary with depth in the earth. As a result, the path of a wave usually curves upward. At any boundary layer the waves may be reflected or refracted. Either a P or an S wave may give rise to both a P and an S wave at any boundary. Therefore, a seismogram obtained by a particular recording station from a particular earthquake is complex. The first significant result of analyses of this sort of data was the identification of a large liquid-like **core** through which S waves will not propagate. In fact, the core contains two distinct regions. All of the earth outside the core is called the **mantle**. Both P and S waves propagate through the mantle indicating that everywhere it is essentially solid. The upper 0.5% of the mantle is known as the **crust**. Seven distinct regions in all have been identified in the earth.

The speeds of P and S waves are determined by the density, compressibility, and rigidity of the material through which they pass. Speed measurements, however, do not provide sufficient information to solve exact equations for those parameters. Therefore, in order to obtain the physical characteristics of the earth's interior, we must use indirect clues provided by the earth's mass and moment of inertia, laboratory experiments on rocks, and mathematical theories of elasticity and gravitation. By these means the earth's density has been deduced (see Fig. 4.9). From the density variation and Gauss' law the variation of the acceleration g due to gravity within the earth can be calculated. It is particularly interesting

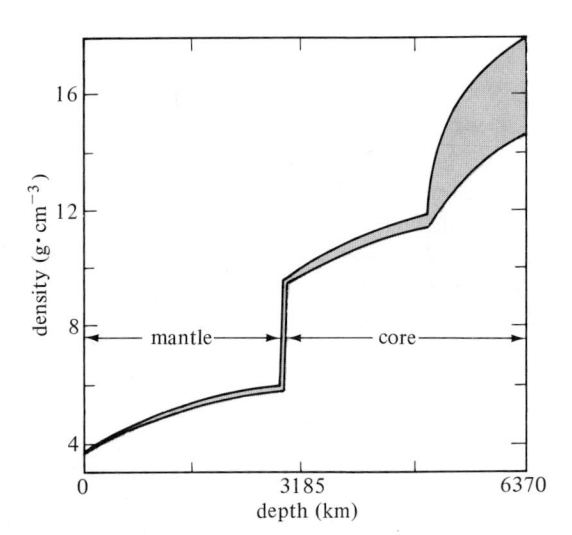

Fig. 4.9. The variation of the density in the earth's interior. The probable range of uncertainty is indicated by the shaded area.

---

[4] K. E. Bullen "The Interior of the Earth," *Scientific American*, September, 1955. Available as *Scientific American Offprint 804* (San Francisco: W. H. Freeman and Co., Publishers).

QUESTIONS AND PROBLEMS

1. Discuss one of the contributions to the sciences made by Karl Friedrich Gauss other than that discussed in this chapter.

2. A nonconducting cube of side L is in a uniform electric field of magnitude E. Find the electric flux through the surface of the cube.

3. A surface encloses an electric quadrupole. Comment on Φ_E for this surface. Calculate Φ_E for a large spherical surface of radius R centered on the quadrupole.

4. Show that Gauss' law holds only if the exponent in Coulomb's law is exactly two.

5. Consider the equation

$$\frac{d}{dt}\int_v \rho\, dv = -\int_S \mathbf{n}\cdot(\rho\mathbf{V})\, dS$$

where $\mathbf{V}$ is a velocity field and ρ is a scalar density field. The surface S encloses the volume v; $\mathbf{n}$ is a unit vector normal to the surface at every point. What interpretation can you attach to this equation? What interpretation would you give if the flux on the right were zero?

6. A point charge q is introduced into the cavity of a hollow conducting object. Use Gauss' law to show that a charge $-q$ is induced on the inner surface of the conductor.

7. Show that the surface charge density at a point on a conductor is related to the electric field at that point by the relation $\sigma = \epsilon_0 E$.

8. A hollow closed conductor carries a charge $+5q$. An object carrying a charge $-3q$ is introduced into the interior of the conductor. A second object carrying a charge $-6q$ is brought near the outside of the conductor. Deduce the net charge on the inner and outer surfaces of the conductor.

9. A closed conducting surface produces perfect electrical shielding of its interior. Discuss this statement.

10. Compare the frequency of an electron traveling in a circular path of radius r about a fixed point charge q and the frequency of an electron oscillating along the diameter of a uniform spherical charge distribution of magnitude q and of radius r. These two cases correspond to the atomic models proposed by Ernest Rutherford and Joseph J. Thompson.

11. Use Gauss' law to show that the potential difference between two concentric

cylindrical conductors of radii r_1 and r_2 ($r_1 < r_2$) each of which has a net charge per unit length of λ C·m^{-1} is given by

$$V_1 - V_2 = \frac{\lambda}{2\pi\epsilon_0} \ln\left(\frac{r_1}{r_2}\right).$$

Assume that the outer surface is negatively charged and the inner one positively charged.

12. Calculate the potential difference between two concentric conducting spheres of radii r_1 and r_2 ($r_1 < r_2$). Assume that the inner sphere has a net charge q on it and the outer sphere is grounded.

13. Gauss' law for the gravitational interaction may be written as

$$\Phi_G = \int_S g_n \, dS = -4\pi Gm.$$

Use this equation to derive Newton's law of gravitation.

14. Derive an expression for the gravitational field within a uniform spherical cloud of dust with density ρ.

*15. Assume that the density of a hypothetical star is given by

$$\rho = \frac{Ma^2}{2\pi r(r^2 + a^2)^2},$$

where M is the total mass, r is the distance from the center, and a is a constant that determines the size of the star. Find the gravitational field as a function of r.

16. Find the gravitational field at a distance d from a large plane sheet of density σ per unit area. Where would such sheets of anomolous mass density occur in the earth's crust?

17. What is the physical basis for the inability of the liquid core of the earth to propagate S waves?

18. What are Rayleigh and Love waves?

*19. Use Fig. 4.9 to deduce the variation of the acceleration due to gravity in the interior of the earth.

20. How are the earth's mass and moment of inertia determined?

*21. Assuming that the gravitational force is constant near the surface of the earth, calculate the mean density of rock at the surface. The radius of the earth is 6.37×10^6 m, the gravitational acceleration at the surface is 9.81 m·sec^{-2}, and G is 6.67×10^{-11} N·m^2·kg^{-2}. Compare your answer with the value given in Fig. 4.9.

22. The electric field at the earth's surface is approximately 200 V·m^{-1} directed toward the earth's center. At a height of 1400 m it is only 20 V·m^{-1}, again directed toward the earth's center. What is the average charge density in the atmosphere below 1400 m? (Hint: Apply Gauss' law at both locations.)

MICHAEL FARADAY

By permission of J. F. Lehmanns Verlag.

Capacitance 5

5.1 INTRODUCTION

Capacitance is a measure of the ability of a conductor or group of conductors to store charge. It can be shown quite generally that the potential V of a given conductor is directly dependent on the total charge q that it possesses. Therefore, we can write

$$V \propto q$$

or

$$V = \frac{q}{C}$$

or

$$q = VC$$

where the constant C is the capacitance of the conductor. The capacitance in coulomb·volt^{-1} (C·V^{-1}) or **farads** (F)[1] is just the charge in coulombs required to raise the potential of the conductor by one volt. A conductor having a large capacitance will contain a larger charge than a conductor of smaller capacitance having the same potential. Since external work is required to assemble a collection of charges on a conductor, the capacitance will depend on the form of the electric field created by the charges placed upon the conductor. In the absence of any dissipative forces, an amount of energy equal to the external work is added to the system in the form of energy stored in the electric field.

5.2 THE PARALLEL-PLATE CAPACITOR

When two or more conductors are brought close enough together so that the potential of each conductor is significantly affected by the presence of the other(s), then the collection of conductors form a **capacitor**. Most useful capacitors consist of a pair of conductors, with the parallel-plate capacitor (in several variations) being the most common form. The simplest form of parallel-plate capacitor is that shown in Fig. 5.1. A positive charge q is placed on a flat conductor or plate of area A. An identical conductor is positioned paralled to the first and a distance d away. The second plate is connected to the earth and is said to be **grounded**. The earth is a relatively good conductor and, being large, can be considered as an (almost) infinite source of charge. This then fixes the bottom plate at a constant reference potential which we call "zero" for convenience.

When the charge $+q$ is placed on the first conductor, an equal but opposite charge $-q$ collects on the grounded plate thereby minimizing the

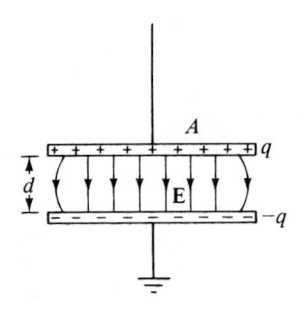

Fig. 5.1. A parallel-plate capacitor with plate area A, plate separation d, and charge q.

[1] Capacitances that are met in practice are expressed in microfarads (μF) or picofarads (pF) where 1 μF = 10^{-6} F and 1 pF = 10^{-12} F.

energy of the system. Thus, an electric field is formed between the plates, the field being uniform except near the edges of the plates as shown in Fig. 5.1. The field E can be calculated in the following manner using Gauss' law. Let us consider a section of the capacitor such as shown in Fig. 5.2. We choose a cylindrical Gaussian surface S whose axis is parallel to **E** and whose ends of area A' are within the upper plate and between the plates, respectively. If $\sigma = q/A$ is the density of charge on the surface of the upper plate, then from Gauss' law

$$\int_S E_n \, dS = E \int_S dS = EA' = \frac{\sigma A'}{\epsilon_0}$$

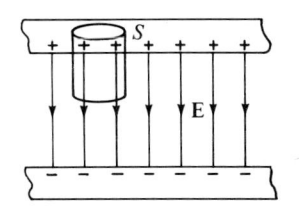

Fig. 5.2. Calculation of the electric field **E** between the plates of a parallel-plate capacitor.

or

$$E = \frac{\sigma}{\epsilon_0} = \frac{q}{\epsilon_0 A}.$$

In evaluating this integral we have used the fact that the electric field within a conductor is zero (Section 4.3). Also, we have assumed that the nonuniformity of electric field near the edges of the capacitor is small enough so that the assumption that the charge q is uniformly distributed over the area A is a good one. This is a reasonable approximation if d is sufficiently small compared to the linear extent of the plate.

The potential V of the upper plate can be calculated quite easily since the field between the plates is uniform so that

$$V = -\int_0^d \mathbf{E} \cdot d\mathbf{s} = E \int_0^d ds = \frac{qd}{\epsilon_0 A} \text{ volts.}$$

Therefore,

$$q = \frac{\epsilon_0 A V}{d} \text{ coulombs}$$

and

$$C = \frac{\epsilon_0 A}{d} \text{ farads}$$

is the capacitance of the parallel-plate capacitor. The capacitance is proportional to the area and inversely proportional to the separation of the plates (so long as the area does not become too small or the separation too large).

5.3 THE OSCILLOSCOPE

One of the most versatile electronic measuring instruments is the **oscilloscope**. This instrument provides a means of obtaining a graphical display of the time variation of an electrical signal.

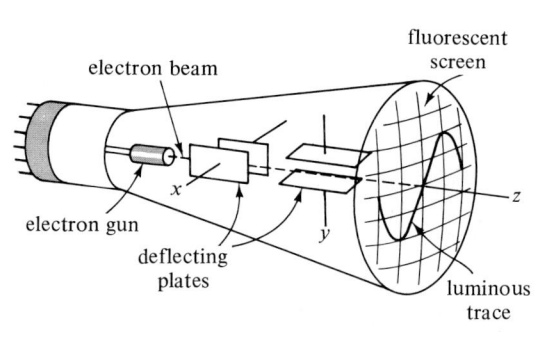

Fig. 5.3. A cathode-ray tube.

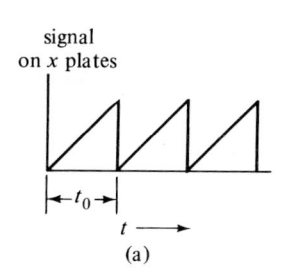

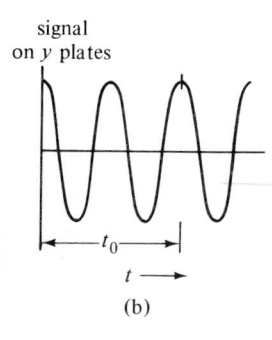

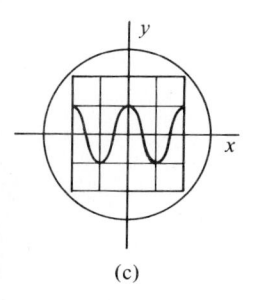

Fig. 5.4. (a) The potential across the x plates increases linearly from zero to a maximum value in a time t_0. (b) The potential across the y plates varies sinusoidally with time with a period $t_0/2$. (c) The steady display on the oscilloscope screen shows two complete cycles of the sine wave.

The central element of the oscilloscope is the cathode-ray tube illustrated in Fig. 5.3. An electron beam is generated by the electron gun and accelerated along the z axis, its position in the x–y plane is controlled by the potentials on the two pairs of deflecting parallel plates, and it is focused on the fluorescent screen. Each set of parallel plates constitutes a parallel-plate capacitor in which a uniform electric field is established. The intensity of the luminous trace is dependent on the length of time that an element on the screen is illuminated and by the electron beam power. As a result, any cathode-ray tube has a finite **writing rate** measured in cm·μsec^{-1}. If the beam is caused to move across the screen at a faster rate, no visible trace will appear. The **deflection sensitivity** of the oscilloscope in the x–y plane is inversely proportional to the potential used to accelerate the electrons along the z axis. A common laboratory oscilloscope may employ an accelerating potential of $\simeq$ 5 kV and have a deflection sensitivity of $\simeq$ 10 V·cm^{-1}. The x plates allow for the horizontal deflection of the beam. The potential across the x plates may be made to increase linearly from zero to some maximum value, then returned to zero and the cycle repeated indefinitely. This procedure provides a **linear time base** as shown in Fig. 5.4(a). That is, it causes the beam to move horizontally across the screen at a constant rate. The y plates allow for the vertical deflection of the beam. If an input signal is used to vary the potential across the y plates periodically with time [see Fig. 5.4(b)], we can obtain a steady display on the screen by synchronizing the frequency of the time base to the signal frequency. This is illustrated in Fig. 5.4(c).

The beam intensity may also be varied. This is known as **z-axis modulation** and is most familiar to us in the television tube.

In a **dual-beam oscilloscope** two electron beams are produced by two separate electron guns. The beams are deflected by different y plates so that two related signals may be displayed simultaneously by means of a common time base.

When a signal is repetitive, it may be studied a large number of times

SEC. 5.3 THE OSCILLOSCOPE

to determine its shape. This procedure is used in the **sampling oscilloscope**. The time base in this case is a set of equally spaced steps. Each time the beam switches from one step to the next an electronic gate is opened for a short time and a minute portion of the signal is allowed through. The charge in this sample is stored on a capacitor, amplified, and used to provide a y coordinate corresponding to the x position. The trace consists of a large number of dots. Sampling oscilloscopes provide a practical means to observe very high frequency signals.

Example. A schematic diagram of the electron gun and vertical deflecting system of an oscilloscope is shown in Fig. 5.5. Calculate the vertical deflection D experienced by an electron.

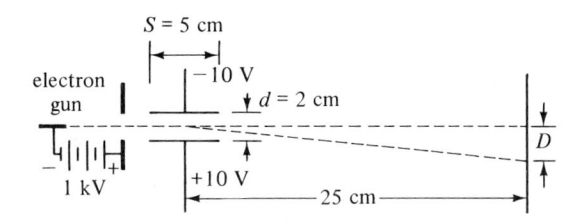

Fig. 5.5. Schematic diagram of the electron gun and vertical deflecting system of an oscilloscope.

Solution. The vertical velocity v of the electron as it emerges from the parallel plate capacitor has magnitude

$$v_y = \left(\frac{eE}{m_e}\right) t$$

where eE/m_e is the force per unit mass (acceleration) experienced by the electron of mass m_e and charge e while it is in the constant field $E = V/d$ of the capacitor. The time t that the electron spends in the capacitor is

$$t = \frac{s}{u}$$

where s is the horizontal dimension of the capacitor and u the horizontal velocity of the electron as given by

$$eV' = \frac{1}{2} m_e u^2$$

with V' the accelerating potential in the electron gun. Therefore,

$$u = \left(\frac{2\,eV'}{m_e}\right)^{1/2}$$

$$= \left(\frac{2 \times 1.60 \times 10^{-19} \times 10^3}{9.11 \times 10^{-31}}\right)^{1/2}$$

$$= 1.88 \times 10^7 \text{ m·sec}^{-1}$$

and

$$v = \frac{e}{m_e}\frac{V}{d}\frac{s}{u}$$

$$= \frac{1.60 \times 10^{-19}}{9.11 \times 10^{-31}} \times \frac{20}{2.0} \times \frac{5}{1.88 \times 10^7}$$

$$= 4.68 \times 10^5 \text{m} \cdot \text{sec}^{-1}.$$

Therefore, the deflection D is approximately,

$$D \simeq \frac{4.68 \times 10^5}{1.88 \times 10^7} \times 25.0 = 0.62 \text{ cm} = 6.2 \text{ mm}.$$

5.4 THE CYLINDRICAL CAPACITOR

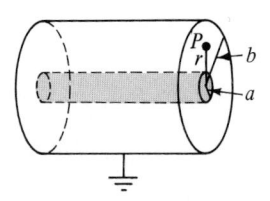

The cylindrical capacitor pictured in Fig. 5.6 consists of two coaxial cylinders of radii a and b, the outer cylinder being hollow and the inner cylinder usually solid. The outer cylinder is grounded. The field between the cylinders is just that due to a linear uniform charge distribution which, in the approximation of a very long capacitor with b–a small compared to the length, is

$$E = \frac{\mu}{2\pi\epsilon_0 r}$$

(see Section 4.3) at all points P of distance r from the axis of the capacitor where μ is the charge per unit length on the inner conductor. This field is directed radially outward so that the potential difference V between the conductors is

$$V = -\int_b^a \mathbf{E} \cdot d\mathbf{r}$$

$$= -\frac{\mu}{2\pi\epsilon_0}\int_b^a \frac{dr}{r}$$

$$= \frac{\mu}{2\pi\epsilon_0}\int_a^b \frac{dr}{r} = \frac{\mu}{2\pi\epsilon_0}\ln\left(\frac{b}{a}\right).$$

The capacitance per unit length is

$$C = \frac{\mu}{V} = \frac{2\pi\epsilon_0}{\ln\left(\frac{b}{a}\right)}.$$

Cylindrical capacitors are of great interest since, for example, high-frequency electromagnetic signals are often transmitted by means of coaxial cables. The cable capacitance affects the quality of transmission and must be taken into account in the design of high-frequency equipment.

Example. A 50 cm section of coaxial line to be used in a cryostat designed for low-temperature physics experiments consists of two concentric cylinders, the outer one 2.0 cm in diameter and the inner one 2.0 mm in diameter. Calculate the capacitance of this section of coaxial line.

Solution. The capacitance C per unit length of coaxial line is

$$C = \frac{2\pi\epsilon_0}{\ln\left(\frac{b}{a}\right)}$$

$$= \frac{2\pi \times 8.85 \times 10^{-12}}{\ln\left(\frac{20}{2}\right)}$$

$$= 24 \text{ pF} \cdot \text{m}^{-1}.$$

Therefore, the capacitance of a 50 cm section will be 12 pF.

5.5 COMBINATIONS OF CAPACITORS

Capacitors are often combined in **series** or in **parallel** in electric circuits, or in some combination of series and parallel connections. In this way capacitances that are different from those of any individual capacitor may be obtained.

Parallel connection

The parallel connection of N capacitors is shown in Fig. 5.7. If a charge q is placed on the upper plate of any one of the capacitors, charge will flow between the capacitors, since they are connected by conductors, until the potential difference between the plates of each capacitor is the same. If the charges on each capacitor are $q_1, q_2, \ldots, q_N$, then

$$V = \frac{q_1}{C_1} = \frac{q_2}{C_2} = \ldots = \frac{q_N}{C_n}.$$

If we regard the collection of capacitors as one large capacitor of capacitance C and total charge $q = q_1 + q_2 + \ldots + q_N$, then

$$V = \frac{q}{C} = \frac{q_1 + q_2 + \ldots + q_N}{C}$$

$$= \frac{V(C_1 + C_2 + \ldots + C_N)}{C}$$

and

$$C = C_1 + C_2 + \ldots + C_N.$$

Therefore, when capacitors are connected in parallel, the resulting capacitance is equal to the sum of the individual capacitances.

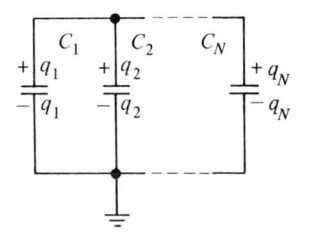

Fig. 5.7. Capacitors connected in parallel.

Series connection

The series connection of N capacitors is pictured in Fig. 5.8. We assume that the capacitors are initially uncharged. If a charge of $+q$ is placed on the upper plate of C_1, a charge of $-q$ will be attracted to the lower plate of C_1. This will leave a charge $+q$ on the upper plate of C_2 which will induce a charge of $-q$ on the lower plate of C_2, etc. In the equilibrium configuration, each capacitor has a charge of magnitude q on each plate. The potential differences between the plates of the separate capacitors are

$$V_1 = \frac{q}{C_1}, \quad V_2 = \frac{q}{C_2}, \quad \ldots, \quad V_N = \frac{q}{C_N}.$$

If we regard the collection of capacitors as a single capacitor of capacitance C with potential difference V between its plates, we see that

$$V = V_1 + V_2 + \ldots + V_N = \frac{q}{C}$$

or

$$\frac{q}{C_1} + \frac{q}{C_2} + \ldots + \frac{q}{C_N} = \frac{q}{C}$$

and

$$\frac{1}{C} = \frac{1}{C_1} + \frac{1}{C_2} + \ldots + \frac{1}{C_N}.$$

When capacitors are connected in series, the reciprocal of the resulting capacitance is equal to the sum of the reciprocals of the separate capacitances.

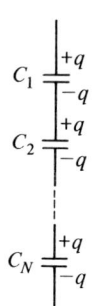

Fig. 5.8. Series connection of capacitors.

Example. For the circuit shown in Fig. 5.9, (a) deduce the equivalent capacitance, (b) find the magnitude of the charge on each of the plates if the potential difference between points A and B is 100 V.

Solution.

(a) The capacitors C_2 and C_3 are equivalent to a single capacitor C_5 where

$$C_5 = 4 + 6 = 10 \; \mu\text{F}.$$

With the substitution of C_5 the circuit reduces to the form shown in Fig. 5.10. The capacitors C_1, C_5, and C_4 may be replaced by a single capacitor C where

$$\frac{1}{C} = \frac{1}{C_1} + \frac{1}{C_5} + \frac{1}{C_4}$$

$$= \frac{1}{5} + \frac{1}{10} + \frac{1}{5} = \frac{1}{2}.$$

That is, the equivalent capacitance of the circuit shown in Fig. 5.9 is 2 μF.

(b) The magnitude q of the charge on each of the capacitor plates indicated in

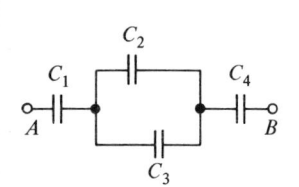

$C_1 = C_4 = 5 \; \mu\text{F}$, $C_2 = 4 \; \mu\text{F}$, $C_3 = 6 \; \mu\text{F}$

Fig. 5.9. A circuit composed of four capacitors. The potential between points A and B is 100 V.

Fig. 5.10. The capacitor C_5 is equivalent to the capacitors C_2 and C_3 in Fig. 5.9.

Fig. 5.10 is
$$q = CV$$
$$= 2 \times 10^{-6} \times 100$$
$$= 2 \times 10^{-4} \text{ C}.$$

The potential differences between the plates of the separate capacitors are given by
$$V_i = \frac{q}{C_i}.$$

Therefore,
$$V_1 = \frac{2 \times 10^{-4}}{5 \times 10^{-6}}$$
$$= 40 \text{ V} = V_4,$$
$$V_5 = \frac{2 \times 10^{-4}}{10 \times 10^{-6}}$$
$$= 20 \text{ V}.$$

Returning to Fig. 5.9, we see that the magnitudes of the charges on the plates of the individual capacitors are given by
$$q_i = C_i V_i.$$

Therefore,
$$q_1 = 5 \times 10^{-6} \times 40$$
$$= 2 \times 10^{-4} \text{ C} = q_4 = q,$$
$$q_2 = 4 \times 10^{-6} \times 20$$
$$= 0.8 \times 10^{-4} \text{ C},$$
$$q_3 = 6 \times 10^{-6} \times 20$$
$$= 1.2 \times 10^{-4} \text{ C},$$

and
$$q_2 + q_3 = 2.0 \times 10^{-4} \text{ C} = q.$$

5.6 ENERGY STORED IN A CAPACITOR

Work must be done to accumulate like charges on a conductor because of the electrostatic repulsion between them. Moreover, as the number of charges increases, the amount of work required to bring an additional charge to the conductor increases. The potential V of the conductor when it has charge q is
$$V = \frac{q}{C}.$$

The change in potential dV associated with a change in charge dq is thus

$$dV = \frac{dq}{C}$$

or

$$dq = C\,dV.$$

The work done in bringing the additional charge dq to the conductor which is initially at potential V is

$$dW = V\,dq$$
$$= VC\,dV.$$

The total energy stored in a capacitor after it has acquired a charge q is

$$U = W = \int_0^q dW = C\int_0^V V\,dV = \tfrac{1}{2}CV^2 \text{ joules.}$$

Alternate forms for the energy stored are

$$U = \tfrac{1}{2}qV = \tfrac{1}{2}\frac{q^2}{C} \text{ joules.}$$

Let us consider the parallel-plate capacitor. When charge is added to the capacitor, an electric field appears between the plates where no field existed originally. A charge in this field would experience a force. The field can do work in accelerating charges and, for this reason, we speak of the energy stored in the capacitor as being stored in the field. Neglecting edge effects, we note that the parallel-plate capacitor has a uniform field

$$E = \frac{\sigma}{\epsilon_0} = \frac{q}{\epsilon_0 A}$$

and capacitance

$$C = \frac{\epsilon_0 A}{d}$$

(see Section 5.2), where

$$\sigma = \frac{q}{A}$$

is the charge density on the plates. Therefore,

$$q = \sigma A = \epsilon_0 EA$$

and the energy stored is

$$U = \frac{1}{2}\frac{q^2}{C} = \frac{1}{2}(\epsilon_0^2 E^2 A^2)\frac{d}{\epsilon_0 A}$$
$$= \frac{1}{2}(Ad)\epsilon_0 E^2.$$

The field occupies a volume Ad between the plates of the capacitor so that the energy density (energy per unit volume) in the field is

$$u = \frac{U}{Ad} = \frac{1}{2}\epsilon_0 E^2.$$

The energy density in the field is proportional to the square of the magnitude of the field. This relation is true also for nonuniform fields and the general relation for the total energy stored in an electric field is

$$U = \int_v u \, dv = \frac{1}{2}\epsilon_0 \int_v E^2 \, dv$$

where the integration is over the volume in which the field exists.

Example. Calculate the total electric potential energy of a thin spherical conducting shell of radius R and carrying charge q.

Solution. The electric energy density u is

$$u = \frac{1}{2}\epsilon_0 E^2.$$

The magnitude E of the electric field for the charged spherical shell is

$$E = \frac{1}{4\pi\epsilon_0}\frac{q}{r^2} \quad r \geq R$$
$$= 0 \quad r < R.$$

The electric potential energy dU contained within a spherical shell of radius r, thickness dr and concentric with the spherical conducting shell is

$$dU = u 4\pi r^2 \, dr = 2\pi\epsilon_0 E^2 r^2 \, dr.$$

Therefore, the total energy in the electric field surrounding the conducting sphere is

$$U = \int dU = \int_R^\infty 2\pi\epsilon_0 \left(\frac{q}{4\pi\epsilon_0 r^2}\right)^2 r^2 \, dr$$
$$= \frac{q^2}{8\pi\epsilon_0} \int_R^\infty \frac{dr}{r^2}$$
$$= \frac{q^2}{8\pi\epsilon_0 R} \text{ joules.}$$

QUESTIONS AND PROBLEMS

1. The unit of capacitance, the farad, is named after the English physicist Michael Faraday. Write a note concerning Faraday's contributions to physics.
2. Explain the use of a **guard ring** to eliminate edge effects in a capacitor.
3. The capacitor indicated in Fig. 5.11 finds practical application as a frequency-tuning element in radio receivers. Alternate plates are connected together,

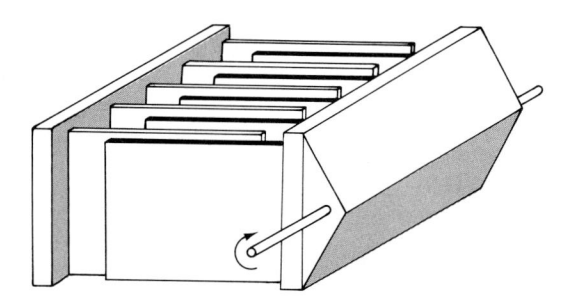

Fig. 5.11. A variable capacitor.

with one group fixed in position and the other group free to rotate to vary the capacitance. Assuming n plates of area A, each separated by a distance d from its neighbors, calculate the maximum capacitance of the variable capacitor.

4. What is an **electrolytic** capacitor? Describe its method of construction as well as its advantages and disadvantages when used in electronic circuits.

5. The circuit shown in Fig. 5.12 is comprised of an input voltage source V_1, an integrated circuit amplifier which produces an output voltage 10^4 times larger than the input voltage and a capacitor C connected between the input and output terminals. Show that the current I is the same as the current that would flow if the amplifier and the capacitor were replaced by a capacitor of capacitance 10,001 C. This arrangement is frequently used in electronic applications when a very large capacitance is needed.

6. The **rise time** of an oscilloscope is the time required to impress a large enough voltage across the deflection plates to cause a full scale deflection. The amplifier used to drive the vertical deflection plates of the cathode-ray tube (CRT) shown in Fig. 5.5 can produce a maximum output current of 100 μA before overloading, the area of each plate is 10 cm^2, and the full scale deflection occurs when there is a 20 V potential difference across the plates. Calculate the rise time and estimate the maximum frequency that can be observed on the oscilloscope.

7. Suppose you put sinusoidal signals of equal amplitude on both the horizontal and vertical deflection plates of an oscilloscope. If the frequencies are in a 2:1 ratio, deduce the pattern that will appear on the screen.

8. Suppose you put sinusoidal signals on both the horizontal and vertical deflection plates as in Problem 7, but this time they are both of the same frequency and amplitude, except the vertical signal leads the horizontal by a phase angle ϕ. Sketch the patterns observed on the screen for various values of ϕ between 0 and π.

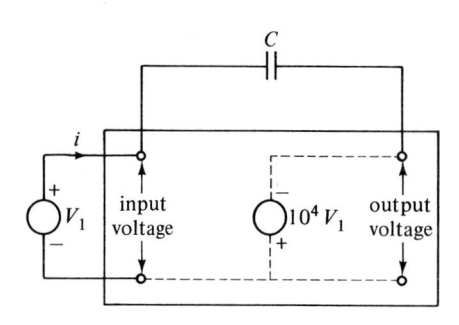

Fig. 5.12. A circuit utilizing an integrated circuit amplifier to produce a large effective capacitance.

9. Calculate the magnitude of the electric field required to produce a deflection of 2.5 cm in the CRT shown in Fig. 5.5. What potential difference on the plates would produce such a field? What is the surface charge density on each plate?

10. Calculate the capacitance of the earth considered as a spherical conductor.

11. A spherical capacitor consists of two concentric spherical shells of radii a and b with $b > a$. Calculate the capacitance of this capacitor assuming that the outer sphere is grounded.

*12. Calculate the capacitance of two parallel and oppositely charged wires of radius a and separation d where $d \gg a$. Assume that the charge on each wire is uniformly distributed and that the wires are of infinite length.

13. For the circuit shown in Fig. 5.13, (a) deduce the equivalent capacitance, (b) find the magnitude of the charge on each of the plates if the potential between A and B is 200 V, (c) find the energy stored in each of the capacitors.

*14. Deduce the equivalent capacitance for the circuit shown in Fig. 5.14.

*15. Twelve 1.0 μF capacitors are arranged along the edges of a cube. Calculate the effective capacitance of the circuit from one corner to the opposite. (Hint: It is easier to tackle this problem from first principles than to try to apply the rules for parallel and series connections derived in Section 5.5.)

16. Suppose you are given a group of 5 μF capacitors, each of which can withstand 200 V. How could you construct a composite capacitor of 10 μF that could withstand 1000 V?

17. Calculate the energy stored per unit length in a coaxial cable with inner radius $r_a = 0.060$ cm and outer radius $r_b = 0.50$ cm when a potential difference $V_0 = 500$ V exists between the conductors.

18. Calculate the electric energy density at the surface of a spherical conducting shell of diameter 1.25 m maintained at a potential of 8.0×10^6 V.

19. A potential difference V_0 is maintained across the plates of a parallel plate capacitor of width a, length b, and separation d as shown in Fig. 5.15. A slab of **dielectric** material of just the right size is inserted between the plates; the effect of a dielectric material is to increase the capacitance by a factor of K; that is, $C' = KC_0$ where C_0 is the capacitance in the absence of the dielectric.
    (a) Calculate the capacitance if the dielectric is displaced a distance x from one of the sides of width a. (Hint: treat the system as two capacitors.)
    (b) Calculate the force pulling the dielectric back into the center of the capacitor if it is displaced a distance x as shown.

20. In Section 5.2 we spoke of an equal and opposite charge accumulating at the grounded plate of a parallel-plate capacitor. It was stated that this minimizes the energy of the system. Explain why this would be so.

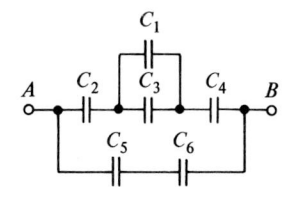

Fig. 5.13. A collection of capacitors connected together.

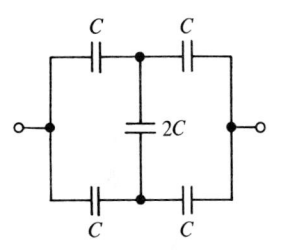

$C_1 = C_2 = C_4 = 2\ \mu\text{F}$,
$C_3 = C_5 = C_6 = 4\ \mu\text{F}$

Fig. 5.14. A circuit made up of capacitors.

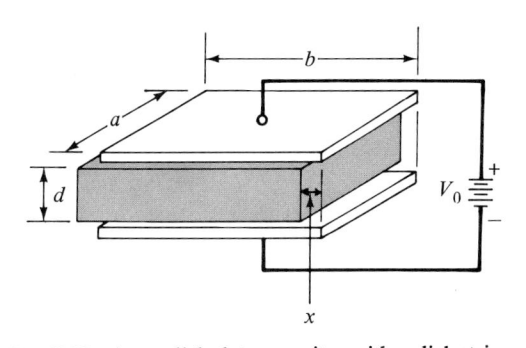

Fig. 5.15. A parallel-plate capacitor with a dielectric slab between its plates.

21. Although the subject of **electrostatics**, the study of charged particles at rest, has been well understood for many years, new applications are still being suggested.[2] Discuss some examples in support of this statement.

---

[2]A. D. Moore, "Electrostatics," *Scientific American*, (March 1972).

WILHELM WEBER

The magnetic field 6

6.1 INTRODUCTION

The space around a magnet or a current can be considered as the site of a **magnetic field** just as the space around a charge is considered as the site of an electric field. The basic magnetic field vector **B** has been known by a variety of names—the **magnetic induction**, the **magnetic field strength**, the **magnetic flux density**. The origin of these names is historical and need not concern us here. The common practice today is to refer to **B** as the **magnetic field** just as we refer to **E** as the electric field. Indeed, **B** plays the same role in magnetism as **E** plays in electrostatics. For example, we define a **magnetic flux** Φ_B by

$$\Phi_B = \int_S \mathbf{B} \cdot d\mathbf{S}$$

where the integral is taken over the surface S (which may be closed or open); the electric flux Φ_E is defined in an identical manner (Section 4.2).

The magnetic field is represented by lines of force whose density is proportional to the magnitude of **B**. The tangent to a magnetic line of force at any point gives the direction of **B** at that point. There are two major differences between magnetic and electric lines of force. First, the electric force on a charged particle is in the direction of $+\mathbf{E}$ or $-\mathbf{E}$ so that the force is tangent to the electric lines. The magnetic force on a charged particle moving with velocity **v** is in the direction of $+\mathbf{v} \times \mathbf{B}$ or $-\mathbf{v} \times \mathbf{B}$; that is, the magnetic force is perpendicular to the plane containing **v** and **B** and is not tangent to the magnetic lines. Second, magnetic lines are continuous whereas electric lines begin on positive charges and end on negative charges. The continuity of magnetic lines could be taken as an argument against the existence of **magnetic charges**. Such magnetic charges or **monopoles** are not forbidden theoretically;[1] however, they have never been observed, although many experiments have been conducted for this purpose.[1,2]

6.2 MAGNETIC FORCE ON A MOVING CHARGE

A charge moving in a magnetic field experiences a force in addition to any force that it may experience due to gravity or to the presence of an electric field. Gravitational forces are almost always several orders of magnitude smaller than magnetic forces in problems of interest and are hereafter assumed to be negligible. If no force acts on a charge at rest, then no electric field is present (this we can always arrange). We can now examine the properties of the magnetic field by observing the behavior of

[1] K. W. Ford, "Magnetic Monopoles," *Scientific American*, December, 1963.
[2] H. H. Kolm, "Search for Magnetic Monopole in Deep-Sea Sediment" *Physics Today*, October, 1967, 69.

SEC. 6.2 MAGNETIC FORCE ON A MOVING CHARGE

moving charged particles. If a charge experiences a force in a direction perpendicular to the direction of motion of the charge at some point P, a magnetic field is present at the point P. The following experimental behavior is then observed.

1. As the direction of the velocity **v** through the point P changes, the direction of the force on the charge remains perpendicular to **v** but its magnitude varies; for two particular orientations of **v** (antiparallel to each other) the force becomes zero. We will choose one of these orientations as a reference direction and label it $\mathbf{v}_r$.
2. The force is observed to be a maximum when **v** is in a direction perpendicular to the direction of $\mathbf{v}_r$ which gives zero force.
3. As the direction of **v** varies from the direction of $\mathbf{v}_r$, the magnitude of the force is found to be proportional to $\sin\theta$ where θ is the angle between **v** and $\mathbf{v}_r$.
4. For a given value of θ, the force is found to be proportional to the product of the magnitude of the charge and the magnitude of the velocity for all directions.
5. The forces experienced by charges of opposite sign are in opposite directions.

These observations can be summarized by writing the equation

$$\mathbf{F} = q\mathbf{v} \times \mathbf{B}$$

for the force experienced by a charge q moving with velocity **v** in a magnetic field **B**. The magnitude of the force is

$$F = qvB\sin\theta$$

where θ is the angle between **v** and **B**. The force is zero when **v** is parallel to **B** ($\theta = 0$) and when **v** is antiparallel to **B** ($\theta = 180°$) and is a maximum when **v** is perpendicular to **B** ($\theta = 90°$). Conditions (1) and (2) are, therefore, satisfied.

The direction of **B** is defined to be the direction of $\mathbf{v}_r$. The sense of **B** (that is, its orientation along the line of zero force) is defined by the right-hand rule.[3] For q positive, **F** is in the direction $\mathbf{v} \times \mathbf{B}$ while for q negative, **F** is in the direction $-\mathbf{v} \times \mathbf{B}$, satisfying condition (5). The spatial relation between the vectors **v**, **B**, and **F** for both positive and negative charges is shown in Fig. 6.1.

The magnitude of **B** for the point P has the value

$$B = \frac{F_\perp}{qv}$$

where $F_\perp$ is the force experienced by the charge when **v** is perpendicular to **B**. For a given value of θ, F is proportional to qv satisfying condition

[3] *MWTP*, Section 6.5.

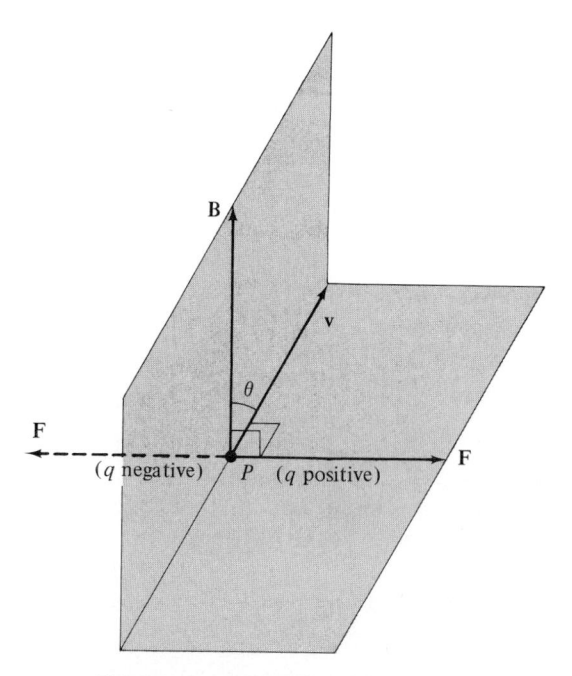

Fig. 6.1. Spatial relation of **v**, **B**, and **F**.

(4). As the direction of **v** is varied for a given q, F is proportional to $\sin \theta$, satisfying condition (3).

From the defining equation we see that **B** has the unit $N \cdot C^{-1} \cdot (m \cdot sec^{-1})^{-1}$ or $N \cdot A^{-1} \cdot m^{-1}$. Historically, **B** was referred to as the magnetic flux density and the special unit **weber per square meter** ($Wb \cdot m^{-2}$) was assigned to **B** where the weber is the unit of magnetic flux. In fundamental units we have

$$1 \text{ Wb} \cdot m^{-2} = 1 \text{ kg} \cdot sec^{-1} \cdot C^{-1}.$$

The CGS emu unit for magnetic field is the **gauss** (G) and is often used. The conversion factor is

$$1 \text{ Wb} \cdot m^{-2} \equiv 10^4 \text{ G}.$$

Since **F** is perpendicular to **v**, the charge experiences no acceleration in the direction of its motion. Therefore, the magnitude of **v** remains constant while only the direction of **v** is altered. Since there is no force in the direction of motion, there is no work done and the kinetic energy remains unchanged.

The lines of force of a magnetic field are defined as those paths through space along which a charged particle can move without experiencing a magnetic force. A small permanent magnet placed in a magnetic field aligns itself along these lines.

If both an electric field and a magnetic field are present in a region of space, the resultant force on a moving charge is the vector sum of the electric and magnetic forces and

$$\mathbf{F} = q\mathbf{E} + q\mathbf{v} \times \mathbf{B}.$$

This is known as the **Lorentz force**.

6.3 COSMIC RAYS

The existence of **cosmic radiation** (penetrating radiation coming from outside the earth's atmosphere)[4,5,6,7,8] was first demonstrated conclusively by the balloon flights of the young Austrian physicist, Victor Hess, in 1912. The particles arriving at the top of the atmosphere are called the **primary cosmic rays**. These high-energy particles interact with the molecules making up the atmosphere to produce **secondary cosmic rays**. The production of secondaries can be seen in cosmic-ray showers and is understood theoretically. The flux of cosmic rays arriving at the earth's surface provided the main source of high-energy particles for the early elementary particle physicists. For example, cosmic-ray research led to the discovery of the **positron** in 1932, the **muon** in 1937, and the **pion** in 1947. Cosmic rays are of considerable interest in biology, since by producing mutations in genes they are thought to play an important role in the evolution of life on our planet. During the past fifteen years these energetic particles have come to be regarded as one of the main components of the universe along with the stars, gas and dust, and electromagnetic radiation.

Primary cosmic rays include the same particles as are general throughout the universe but in somewhat different proportions. The major consitutent is high-energy protons.

The energy spectrum of cosmic-ray particles shows a general reduction in the numbers of particles at higher energies. The most energetic particles detected have energies of the order of 10^{11} GeV. By way of comparison, the world's largest particle accelerators attain $\simeq 10^2$ GeV. Very high-energy cosmic particles are extremely rare, the intensity of particles with

[4] B. Rossi, "Where Do Cosmic Rays Come From?" *Scientific American*, September, 1953. Available as *Scientific American Offprint 239* (San Francisco: W. H. Freeman and Co., Publishers).

[5] D. Ter Haar "On the Origin of Cosmic Rays," *Contemporary Physics*, **6** (June 1965), 338.

[6] G. Burbridge, "The Origin of Cosmic Rays," *Scientific American*, August, 1966.

[7] V. L. Ginsburg, "The Astrophysics of Cosmic Rays," *Scientific American*, February, 1969.

[8] R. Cowsik and P. B. Price, "Origins of Cosmic Rays, *Physics Today*, (October 1971).

energy in excess of 10^6 GeV being $\simeq 3$ particles per 10 cm² per day, compared with a total of $\simeq 3 \times 10^5$ particles per 10 cm² per day of average energy 6 GeV.

The identification of the sources of the primary radiation is rendered difficult by the substantial scattering of the particles that arise from their interaction with the magnetic fields present in our galaxy. As a result the primary cosmic radiation arrives isotropically in the vicinity of the earth and we have no means of correlating various cosmic rays with their sources. The problem of the origin of cosmic rays has therefore been one of considerable speculation.

The total energy density of primary cosmic-ray particles in the neighborhood of the earth is $\simeq 1$ eV·cm⁻³ or $\simeq 10^{-19}$ J·cm⁻³ corresponding to about 10^{-10} particles·cm⁻³ with average energy 10 GeV. It follows that the total cosmic-ray energy in our galaxy is $\simeq 10^{49}$ J. The energy density of the cosmic rays is of the same order as the average energy density of the magnetic field of the galaxy and of the thermal motion of the interstellar gas. Cosmic-ray particles constitute a gas of particles traveling at relativistic speeds that exerts a pressure proportional to its energy density. It therefore follows that cosmic rays must play an important role in the dynamics of the interstellar gas.

Cosmic-ray particles escape from our galaxy through a diffusion process and an average lifetime of about 10^{16} sec is associated with them. Therefore, the total power needed to maintain a constant cosmic-ray energy of $\simeq 10^{49}$ J in our galaxy is $\simeq 10^{33}$ J·sec⁻¹. One possible source of such vast amounts of energy is those spectacular cosmological outbursts known as **supernovae**. Supernovae are known to be strong sources of electromagnetic radiation in the $10^6 - 10^{10}$ Hz band. It is thought that the radiation is emitted by high-energy charged particles spiraling in a magnetic field. On the basis of this assumption it is possible to infer the total cosmic-ray density and the magnitude of the magnetic field of the sources. For the radio source in the Crab Nebula, the remnant of the supernova observed by the Chinese in 1054 A.D., the cosmic-ray energy is $\simeq 10^{42}$ J and the magnetic field is 10^{-3} G. Taking an average frequency of supernovae outbursts of one per century, we thereby obtain a power of 10^{33} J·sec⁻¹. Therefore, supernovae constitute a possible source for the energy required to account for the cosmic-ray production in our galaxy.

As charged particles approach the earth their motion is influenced by the magnetic field of the earth. The earth's magnetic field lines are shown in Fig. 6.2. Although the earth's field is only $\simeq 0.5$ G, it extends far out into space and therefore can influence the motion of even high-energy particles. Charged particles approaching the earth in the polar regions travel along magnetic field lines; therefore, they experience no force and are not deflected. On the other hand, charged particles approaching the earth in the equatorial regions cut across magnetic field lines, experience

SEC. 6.4   MOTION OF A CHARGE IN A MAGNETIC FIELD   79

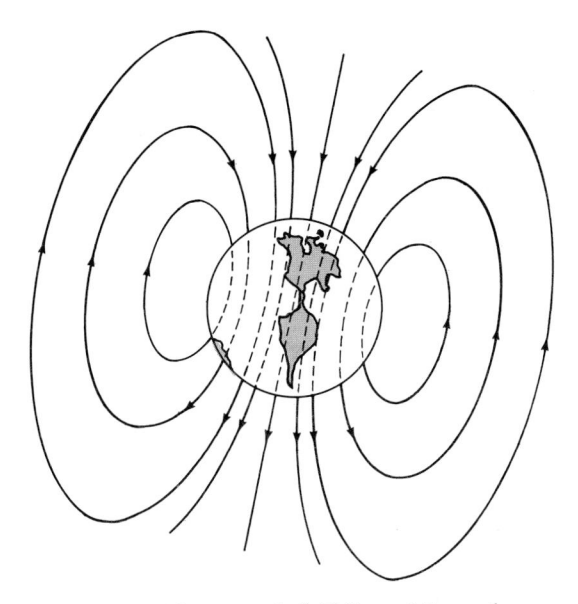

Fig. 6.2. The magnetic field lines of the earth.

a magnetic force, and are deflected. Indeed, many particles are so severely deflected that they never strike the earth's atmosphere. Experimental measurements of the variation of intensity of cosmic radiation with latitude reveal that the intensity is greatest in the polar regions and least in the equatorial regions. This is known as the **latitude effect** and provides convincing evidence that the incoming particles carry electrical charges.

The direction of the magnetic force acting on particles approaching the equator is such that it deflects positively charged particles toward the east and negatively charged particles toward the west. It is observed experimentally that the particles arriving at the equator come preferentially from the west. This is known as the **east–west effect** and provides evidence that the primary particles carry positive charges.

The actual orbits of the cosmic-ray particles can be adequately calculated by numerical solutions of the equation

$$\mathbf{F} = q\mathbf{v} \times \mathbf{B}$$

with the earth's field substituted for **B**.

6.4 MOTION OF A CHARGE IN A MAGNETIC FIELD

We consider a positive charge moving in a uniform magnetic field. The force experienced by the charge is

$$\mathbf{F} = q\mathbf{v} \times \mathbf{B}.$$

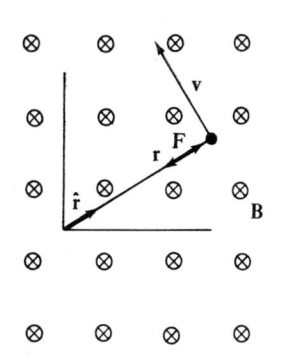

Fig. 6.3. A charge undergoes circular motion in a uniform magnetic field.

If **B** is constant in magnitude and $\mathbf{v} \perp \mathbf{B}$, the motion is in a plane perpendicular to **B** and is circular since **F** is constant in magnitude. The force **F** is in the direction $-\hat{\mathbf{r}}$ (see Fig. 6.3) and

$$\mathbf{F} = -qvB\hat{\mathbf{r}}$$

or

$$F = qvB.$$

The centripetal force on the charge is

$$F = \frac{mv^2}{r}$$

where m is the mass of the charge. Equating the two expressions for F gives

$$qvB = \frac{mv^2}{r}$$

and

$$r = \frac{mv}{qB} = \frac{p}{qB}$$

is the radius of the circular motion.

Example. The wake of small bubbles produced by a charged particle traveling in a bubble chamber[9] can be photographed. From measurements on a particle's track the momentum of the particle can be computed. A photograph of a bubble chamber shows a proton moving in a circular arc 15 cm in radius and perpendicular to a magnetic field of 3000 G. Deduce the momentum and energy of the proton.

Solution. The charge e carried by a proton is 1.6×10^{-19} C. The momentum p of the proton is

$$\begin{aligned} p &= erB \\ &= 1.6 \times 10^{-19} \times 0.15 \times 3000 \times 10^{-4} \\ &= 7.2 \times 10^{-21} \text{ kg} \cdot \text{m} \cdot \text{sec}^{-1}. \end{aligned}$$

The kinetic energy (nonrelativistic) is given by

$$K = \frac{p^2}{2m_p}$$

where $m_p =$ mass of the proton $= 1.67 \times 10^{-27}$ kg. Therefore,

$$K = \frac{(7.2 \times 10^{-21})^2}{2(1.67 \times 10^{-27})} = 1.6 \times 10^{-14} \text{ J} = 100 \text{ keV}.$$

Since the rest mass of a proton is equivalent to an energy of $m_p c^2 = 938$ MeV, a relativistic calculation[10] would give the same result.

---
[9] D. A. Glaser, "The Bubble Chamber," *Scientific American*, February, 1955. Available as *Scientific American Offprint 214* (San Francisco: W. H. Freeman and Co. Publishers).

[10] *MWTP*, Section 12.4.

If the velocity **v** is not perpendicular to **B**, the motion is readily determined by writing

$$\mathbf{v} = \mathbf{v}_\| + \mathbf{v}_\perp$$

where $\mathbf{v}_\|$ and $\mathbf{v}_\perp$ are the components of the velocity parallel and perpendicular to **B**, respectively. Now,

$$\begin{aligned}\mathbf{F} &= q\mathbf{v} \times \mathbf{B} = q(\mathbf{v}_\| + \mathbf{v}_\perp) \times \mathbf{B} \\ &= q\mathbf{v}_\perp \times \mathbf{B}\end{aligned}$$

since $\mathbf{v}_\| \times \mathbf{B} = 0$. Then

$$F = qv_\perp B$$

and

$$r = \frac{mv_\perp}{qB} = \frac{p_\perp}{qB}$$

gives the radius for circular motion. Now, however, there is also motion in the direction of **B** (or −**B**) due to the component $\mathbf{v}_\|$. The magnitude of $\mathbf{v}_\|$ is constant. Therefore, in a uniform field, the charge undergoes circular motion of radius $p_\perp/qB$ superimposed upon uniform linear motion at right angles to the circular motion. The resultant path of the charge is helical as shown in Fig. 6.4. The axis of the helix is defined by the magnetic line of force which is coincident in space with the (instantaneous) center of the circular motion. The helical motion takes place about a magnetic line of force. If the field is not uniform, the radius of the helix changes and the helical motion still follows the magnetic line of force (which is no longer a straight line).

The cyclotron frequency

We have seen that a charged particle injected into a uniform magnetic field travels at constant speed in a helical path about a magnetic field line. We can write the velocity component $v_\perp$ as

$$v_\perp = \omega r$$

with ω the angular frequency of the particle and r the radius of the helical path. Solving for ω yields

$$\omega = \frac{v_\perp}{r} = \frac{q}{m} B.$$

The characteristic frequency f associated with the particle's motion is then

$$f = \frac{1}{2\pi} \frac{q}{m} B$$

independent of the particle's speed and of the radius of the helix. This frequency is known as the **cyclotron frequency**.

Fig. 6.4. A charge undergoes helical motion in a uniform magnetic field when **v** is not perpendicular to **B**.

Example. Calculate the cyclotron frequency for electrons in a magnetic field of 0.300 Wb·m⁻².

Solution. For electrons

$$\frac{e}{m_e} = \frac{1.60 \times 10^{-19}}{9.11 \times 10^{-31}} = 1.76 \times 10^{11} \text{ C·kg}^{-1}.$$

Therefore, the cyclotron frequency for the electrons will be

$$f = \frac{1.76 \times 10^{11} \times 0.300}{2 \times 3.14}$$

$$= 8.41 \times 10^9 \text{ Hz}$$

$$= 8.41 \text{ GHz}.$$

If a sample containing free electrons and situated in a magnetic field of 0.300 Wb·m⁻² is irradiated with microwave radiation of frequency 8.41 GHz, the electrons will absorb energy from the microwave field. This effect is known as **cyclotron resonance.**

6.5 MASS SPECTROMETERS[11]

In essence, the mass spectrometer is a device for sorting and identifying atoms and molecules. The sample to be analyzed is first ionized, that is, the constituent particles are given electrical charges. The ions are then accelerated and injected into a magnetic field where they travel in circular paths, the radii of which depend upon the masses of the particles and their charges. Since it is possible to measure accurately the speed of the ions, the strength of the magnetic field, and the radii of the paths traversed, it is an easy matter to calculate precisely the relative masses of the ions present.

A simple form of mass spectrometer is shown in Fig. 6.5. The atoms or molecules to be analyzed are ionized in the source and accelerated to a speed v by falling through the potential difference V. As a result the particles have kinetic energy

$$K = \tfrac{1}{2}mv^2 = qV$$

and momentum

$$p = mv = (2mqV)^{1/2}$$

when they enter the region of magnetic field **B**. Since the motion of the particles is in the plane perpendicular to **B**, a particle with momentum p will travel in a circular path of radius r where

$$p = qBr.$$

[11] A. O. C. Nier, "The Mass Spectrometer," *Scientific American*, March, 1953. Available as *Scientific American Offprint 256* (San Francisco, W. H. Freeman and Co., Publishers).

SEC. 6.5   MASS SPECTROMETERS

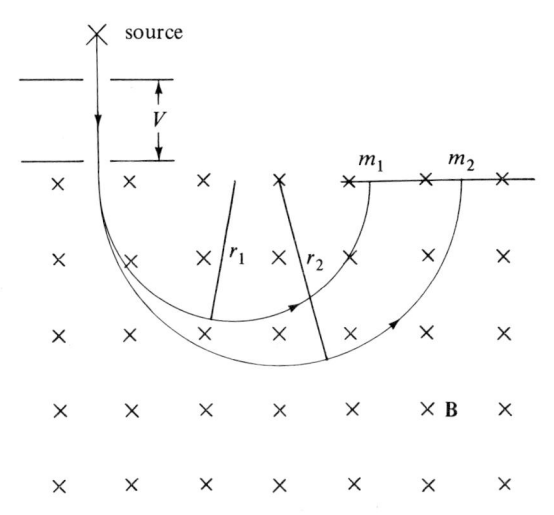

Fig. 6.5. A simple form of mass spectrometer.

Equating the two expressions for the particle's momentum gives

$$qBr = (2mqV)^{1/2}$$

or

$$\frac{m}{q} = \frac{B^2 r^2}{2V}.$$

Each ion usually carries one unit of charge. This equation predicts that the larger the mass of the ion the greater the radius of curvature of the ion's path in the magnetic field. That is, the device achieves **mass dispersion**.

Age determination of rocks

One of the most interesting applications of the mass spectrometer is to age determinations of rocks through measurements of the isotopic abundances of the isotopes of lead. Very slowly in the course of time and at accurately known rates, radioactive uranium-238 decays to lead-206, uranium-235 to lead-207, and thorium-232 to lead-208. By carrying out a chemical analysis of the amounts of uranium, thorium, and lead in a mineral sample and a mass spectrometric analysis for the amounts of the three isotopes of lead, three independent estimates of the age of the mineral can be had, assuming that the decay rates of these three radio-isotopes are constants independent of time. From such measurements, estimates of the ages of the earth, the moon, and even the universe, have been made. The analysis is complicated by the possibility that the sample may contain natural lead as well as that formed by the radioactive decay of uranium and thorium. An interesting property of natural lead samples is that although they all are found to have precisely the same atomic weight, the relative abundances of the various isotopes vary considerably from sample to sample. This is shown in Fig. 6.6.

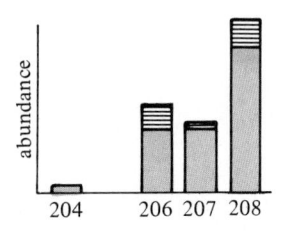

Fig. 6.6. Mass spectra of natural lead show differences in isotopic content. The isotopic content may vary from one deposit to another by as much as the shaded area in each bar. However, the variation is always such as to result in the same atomic weight for any natural lead sample.

Mass spectrometer in biology

The mass spectrometer finds use in biological tracer experiments employing stable isotopes. Stable isotopes are in many instances preferable to radioactive isotopes for studies of metabolism. In some instances, no suitable radioactive isotopes are available but stable ones are. For example, nitrogen-15 and oxygen-18 are useful stable isotopes of the biologically important nitrogen and oxygen atoms.

The mass spectrograph has been used to provide continuous records of the composition of respiratory gases under conditions such as sudden changes in pressure and in the treatment of patients suffering certain lung disorders.

6.6 VAN ALLEN RADIATION ZONES[12,13]

Conclusive experimental evidence for the existence of significant numbers of charged particles trapped by the earth's magnetic field was obtained early in 1958 by James A. Van Allen and collaborators. The early measurements suggested that the trapped particles were divided into two regions which came to be known as the **inner** and **outer radiation zones**, respectively. The region between the two zones was called the **slot region** (see Fig. 6.7). Later measurements have shown that it is best to consider the zones as one trapping region extending out to the limit of the earth's magnetic field. Within this region the spatial distributions of particles of different energies are not the same.

Trapped electrons and protons spiral along magnetic field lines as indicated in Fig. 6.8. The radii of the helices are negligible compared to the earth's radius of 6370 km (see Table 6.1). A spiraling particle approaches a region where the magnitude of the magnetic field is increasing, that is, a region in which the magnetic field lines are converging. The particle experiences a retarding force, is brought to rest and then spirals back along the same magnetic field line, crosses the geomagnetic equato-

---

[12] J. A. Van Allen, "Radiation Belts Around the Earth," *Scientific American*, March, 1959. Available as *Scientific American Offprint 248* (San Francisco: W. H. Freeman and Co., Publishers).

[13] W. G. V. Rosser, "The Van Allen Radiation Zones," *Contemporary Physics*, **5** (February 1964), 198.

SEC. 6.6 VAN ALLEN RADIATION ZONES

rial plane, again enters a region of increasing magnetic field, and is again reflected. Converging lines of **B** act as **magnetic mirrors** for the reflection of charged particles. To a good approximation the particle is reflected at a point where the earth's magnetic field is of magnitude

$$B = \frac{B_0}{\sin^2 \alpha_0}$$

where B_0 is the earth's magnetic field in the geomagnetic equatorial plane and α_0 the angle that the velocity **v** of the charged particle makes with B_0

Table 6.1 Radii of curvature for particles trapped by the earth's magnetic field

Energy (keV)	Radius Electron (m)	Proton (km)
10	87	3.7
100	287	11.7
1000	1220	37.1

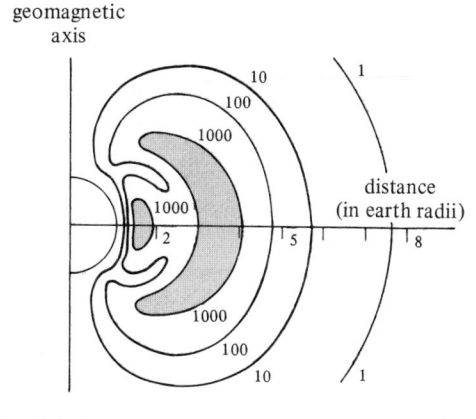

Fig. 6.7. Intensity structure of the trapped particles around the earth. Contours of constant intensity are labeled with numbers 1, 10, 100, and 1000 indicating relative numbers of particles.

as the particle crosses the equatorial plane. The points of reflection are called **mirror points**. Particles that are injected into orbits where they have small values of α_0 have mirror points at low altitudes, where the atmosphere causes substantial scattering, or even within the earth. Such particles are lost from the radiation zones.

Because of the small variation of the earth's field over the helical orbits of the charged particles, it follows that, in addition to spiraling along magnetic field lines and oscillating between mirror points, trapped particles will undergo longitudinal drifts. The drift will be toward the west for protons and toward the east for electrons.

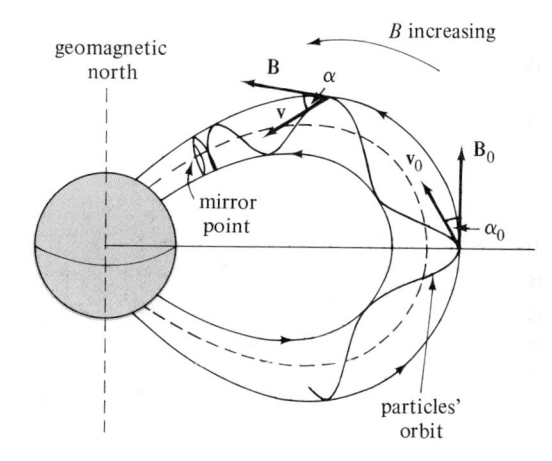

Fig. 6.8. The motion of a charged particle trapped by the earth's magnetic field. When it approaches the earth, the particle enters a region of increasing **B** and is reflected.

It has been stated that charged particles trapped by the earth's magnetic field exhibit three types of oscillatory motion. The cyclotron periods, due to the particles spiraling around magnetic field lines, are generally in the microsecond range for trapped electrons and in the millisecond range for trapped protons. At a geocentric distance of two earth radii, the period for oscillation between mirror points is 0.1 to 1.0 seconds for trapped electrons and 1 to 50 seconds for trapped protons. The period for longitudinal drift around the earth at the same radial distance is $\simeq 3$ hours for trapped 100 keV electrons and protons.

The theory of the trapping of charged particles in the earth's magnetic field has been checked by a series of high-altitude nuclear explosions. In this manner a known quantity of electrons was produced at a known point in space. Some of these electrons were trapped by the earth's magnetic field. The spatial distribution, longitudinal drift, and temporal changes in intensity of these trapped electrons were then studied systematically.

6.7 MAGNETIC FORCE ON A CURRENT

A magnetic field should exert a transverse force on a current since a current consists of a flow of charges. We shall consider first a current I flowing at right angles to a uniform field **B**. Let us suppose that the current consists of a flow of positive charges q, all moving at speed v as shown in Fig. 6.9. The force on each charge q is

$$F = qvB.$$

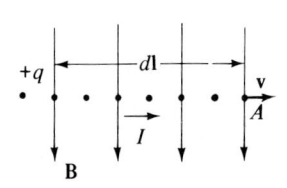

Fig. 6.9. A current I in a magnetic field **B**.

The total force on all the charges in a length dl is

$$dF = nqvB$$

where n is the number of charges in length dl. The n charges in length dl will all pass by the point A in Fig. 6.9 in the time interval dt where

$$dt = \frac{dl}{v}.$$

But the current I is just the total number of charges passing point A per unit time so that

$$I = \frac{nq}{dt} = \frac{nqv}{dl}$$

or

$$nqv = I\,dl.$$

Therefore, the force on the element of length dl of the current is

$$dF = I\,dl\,B.$$

This equation is valid only for **B** at right angles to the current. A more general expression is

$$d\mathbf{F} = I\,d\mathbf{l} \times \mathbf{B}$$

where $d\mathbf{l}$ is a vector element of length in the direction of the (positive) current. For a flow of negative charges a minus sign must be inserted into the equation. Integration of the equation $d\mathbf{F}$ over the length will give the total force on a nonlinear current (for example, a current in a conductor).

Example. A semicircular loop of wire carrying current I is situated in a magnetic field $\mathbf{B}$ perpendicular to the plane defined by the loop. Deduce the force exerted on the loop.

Solution. (See Fig. 6.10.)
A segment of wire of length $d\mathbf{l}$ has a force $d\mathbf{F}$ on it of magnitude

$$dF = IB\,dl = IBR\,d\theta$$

directed toward the center of the semicircle. The component $dF\cos\theta$ is canceled by the corresponding contribution from the arc segment located at $\pi - \theta$. Only the component $dF\sin\theta$ when summed over the complete semicircle gives a nonzero result. The resultant force is therefore

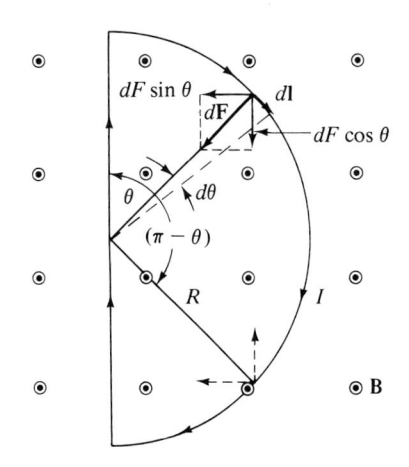

$$\begin{aligned}F &= \int_0^\pi dF\sin\theta \\ &= IBR\int_0^\pi \sin\theta\,d\theta \\ &= 2IBR.\end{aligned}$$

Fig. 6.10. A semicircular loop of wire carrying a current I in a magnetic field $\mathbf{B}$.

Note that the force experienced by the semicircle is the same as would be experienced by a straight wire of length $2R$.

QUESTIONS AND PROBLEMS

1. Describe one experiment designed to observe magnetic monopoles.
2. In light of what you know about magnetic monopoles, discuss the equation

$$\Phi_B = \int_S \mathbf{B} \cdot d\mathbf{S}$$

where integration is over a closed surface S.

*3. In classical mechanics it is customary to describe forces in terms of potentials. What potential, if any, can be ascribed to the magnetic force?

4. Scientists find many interesting interactions between magnetic fields and biological systems.[14] Write an account of **biomagnetism**.

*5. Find an expression for the most general path for a charged particle moving in a uniform magnetic field. Show that the field can do no work on the charge.

[14] A. Kolin, "Magnetic Fields in Biology," *Physics Today*, November, 1968.

6. Write a short biography of Victor Hess.

7. A 100 GeV cosmic-ray proton moves vertically downward through a region of horizontal uniform magnetic field near the earth's surface. What force does the proton experience if the field is directed from South to North and is of magnitude 5.0×10^{-5} Wb·m^{-2}?

8. Secondary cosmic rays, produced by collisions of high-energy protons and molecules of the air, contain large numbers of high-energy μ-mesons. The **half-life** of a μ-meson is 1.54×10^{-6} seconds before it decays into an electron, a neutrino, and an antineutrino. If we consider 1000 μ-mesons approaching the ground at a speed 0.995 c starting from an altitude of 5.0 km, what fraction will reach the surface? What fraction would be predicted to reach the surface if relativistic effects could be neglected?

9. The earth's magnetic field at the equator is horizontal, points North, and is of magnitude 7.0×10^{-5} Wb·m^{-2}. What velocity must a proton be given in order that it encircle the earth traveling at a constant speed? Can the gravitational force be safely neglected?

10. Show that the kinetic energy of a charged particle moving at right angles to a constant magnetic field is proportional to the square of the radius of the path.

11. An electron is traveling at a speed such that its mass is 2.00 times its rest mass. Calculate the radius of curvature of the electron's path in a perpendicular magnetic field of magnitude $B = 0.200$ Wb·m^{-2}.

12. An electron is accelerated through 15 kV and then injected into a uniform magnetic field for which $B = 0.200$ Wb·m^{-2}. Calculate the cyclotron frequency of the electron.

13. A Wien **velocity filter** is a device that combines an electric field and a magnetic field in such a way that ions of one particular velocity are not deflected in passage through them. Explain how the two fields must be arranged.

*14. Consider the motion of an electron in crossed electric and magnetic fields. Such a situation arises in the **parallel-plate magnetron** as shown in Fig. 6.11. If an electron is emitted with negligible velocity from the lower plate what is the minimum potential, $V_{\min}$, at which the electron would just reach the upper plate?

15. Charged particles traveling along magnetic lines of force are undeflected. Show that if the particles make a small angle with the lines of force the magnetic field will act in the sense to cause the particles to move along the lines of force.

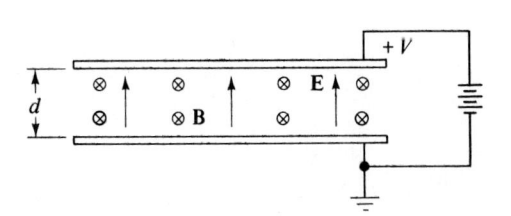

Fig. 6.11. A parallel-plate magnetron.

*16. An electron moves in a circular orbit with angular frequency ω_0 due to the presence of a centripetal force $m\omega_0^2 r$. If a small magnetic field is applied perpendicular to the plane of the orbit, deduce the new frequency (assume that the radius of the electron motion remains the same).

17. Discuss the relation of the **Zeeman effect** to the calculation performed in the previous problem.
18. The **cyclotron** is a particle accelerator based on the fact that the cyclotron frequency is independent of the radius of curvature. Write a short account of the operation of a cyclotron. In your account show that the frequency of the radio-frequency voltage should be given by
$$f = \frac{qB}{2\pi m} \text{ Hz}$$
where q/m is the charge to mass ratio for the particle being accelerated.
19. Suggest two ways of measuring the mass of a charged particle other than by means of a mass spectrometer.
20. A beam of singly ionized copper atoms enters the uniform magnetic field of a mass spectrometer traveling at a speed of 1.0×10^5 m·sec^{-1}. After being bent through 180°, the ions impinge on a photographic plate. If the field strength is 0.50 Wb·m^{-2}, calculate the separation between two isotopes having masses 63 u and 65 u (1 u = 1.6604×10^{-27} kg).
21. The mass spectrometer is used in potassium-argon dating of geological samples. Write an essay describing how the mass spectrometer is applied to this important field of geophysics.
22. Some mass spectrometers use a magnetic field which is variable. The mass of an ion can be determined then by noting at which value of B the ions fall on a collecting plate (see Fig. 6.12). When an ion falls on the collecting plate, it releases its charge which goes through a large resistor and then into the earth.
    (a) At what value of B will a singly ionized atom of mass m having been accelerated through a potential V land on the collector a distance D from the source?
    (b) What current will flow through the resistor if there are N such particles entering the magnetic field per second?
    (c) If a sample of air is being analyzed, plot a graph of voltage across the resistor ($V = IR$) against magnetic-field strength.

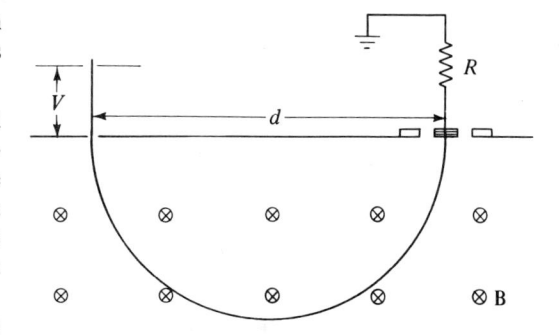

Fig. 6.12. A variable-field mass spectrometer.

23. Write a short essay on **sunspots**.
24. In recent years it has become clear that most solar activity is an expression of changes in the magnetic field of the sun.[15] Discuss this statement.
*25. Show that protons trapped in the Van Allen radiation zones will undergo a longitudinal drift to the West.
26. A 0.50 MeV proton is traveling back and forth between the poles along a magnetic field line. As it crosses the magnetic equator it makes an angle

[15] W. C. Livingston, "Magnetic Fields on the Quiet Sun," *Scientific American*, November, 1966.

of 30° with the geomagnetic axis. At this point the magnitude of the earth's field is 2.0×10^{-5} Wb·m^{-2}. What is the radius of circular motion (**Larmor radius**)? At what magnitude of the magnetic field will it be reflected as it approaches one pole?

*27. The **magnetic moment** of a gyrating charged particle is defined by:

$$\mu = AI$$

where I is the equivalent current and A is the area enclosed by the gyrating particle. Show that the magnetic moment of a particle gyrating along one of the earth's lines of magnetic force is given by:

$$\mu = \frac{K_\perp}{B}$$

where $K_\perp$ is the kinetic energy associated with the component of its velocity $v_\perp$. A general approach to the problem of charged particles moving in a plasma shows that the magnetic moment is a constant of the motion. Using this fact and knowing that the total energy of the particle is a constant, derive the expression (see Section 6.6) for the field at a reflection point. (Hint: As B increases along the line so must $K_\perp$.)

28. Calculate the force on a wire 50 m in length carrying a current of 10 A in a northerly direction that is due to the earth's magnetic field. Take the field to be 8.0×10^{-5} Wb·m^{-2} making an angle 60° to the horizontal.

29. Show that the net force on a closed wire carrying a current I in a uniform magnetic field is zero.

30. A conductor having a mass of 20 g and a length 5.0 m is oriented east-west. If a current of 10 A passes through the conductor toward the east, what magnetic field would exactly balance the gravitational force?

31. If a magnetic field is directed at right angles to a current-carrying conductor, an electric potential is found to exist across the conductor perpendicular to both **B** and **I**. This is called the **Hall Effect**. How could you determine the sign of the charge on free-charge carriers by measuring the Hall potential?

32. A physics student dreamed that he had been convicted of copying problem sets and sentenced to be executed. He was placed in a uniform magnetic field (1 Wb·m^{-2}) directed upward and a wire ($R = 1$ ohm) was wrapped once around his neck in a manner such that from end A to end B it was clockwise when viewed from above. Being given a last wish, he chose to be the one to connect the battery (500 V) himself. To which terminal, $+$ or $-$, should he connect end A? If he guessed wrong, how much pressure must his neck be able to withstand in order to survive? The wire was 1 mm in diameter. Assume his neck has a diameter of 10 cm.

ANDRÉ M. AMPÈRE

Ampère's law 7

7.1 INTRODUCTION

In Chapter 2 we discussed the experimental evidence concerning the force between two very long, parallel currents. We noted that the force per unit length is proportional to the product of the currents and is inversely proportional to the distance separating them. The constant of proportionality is $\mu_0/2\pi$, where $\mu_0 = 4\pi \times 10^{-7}$ Wb·A^{-1}·m^{-1} is called **the permeability of free space**. We are now in a position to examine the force between current elements in more detail. In so doing we shall discover the proper expression for the magnetic field due to a current element and shall then be able to calculate the magnetic field for several simple current arrangements. A consideration of the magnetic field around a long, straight current will lead us to Ampère's law which is one of the basic equations of electromagnetism.

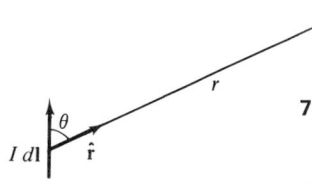

Fig. 7.1. Two parallel current elements.

7.2 THE FORCE BETWEEN CURRENT ELEMENTS

Let us consider two parallel current elements of magnitude $I\,dl$ and $I'\,dl'$ separated by a distance r as shown in Fig. 7.1. From experiment, the force between these current elements is given by

$$dF = \frac{\mu_0}{4\pi} \frac{I\,dl\,I'\,dl'\,\sin\theta}{r^2}$$

where θ is the angle between the direction of the current I and the direction of the vector separating the two current elements. In practice, this force law is deduced by a study of the force between current-carrying conductors which must of necessity be part of closed circuits. Consideration of the forces between entire circuits leads to this expression for the force between current elements so that in the following discussion we are justified in speaking of current elements as separate entities.

In the last chapter we discovered that a current experiences a force when placed in a magnetic field. Since currents produce forces on each other, we conclude immediately that we can associate a magnetic field with each current element mentioned above, with each field producing a force on the other current element. Let the field due to current element $I\,dl$ at the position of current element $I'\,dl'$ be dB in magnitude. The force on $I'\,dl'$ is then (see Section 6.7)

$$dF = I'\,dl'\,dB\,\sin\phi$$

where ϕ is the angle between the direction of the current element $I'\,dl'$ and $d\mathbf{B}$. However, it is noted experimentally that the forces between parallel currents act in the plane containing the currents which means that the field produced by one current element must be perpendicular to the second current element at the position of the second current element. Thus the angle ϕ must be 90° and $\sin\phi = 1$. Therefore,

$$dF = I'\,dl'\,dB.$$

Comparison with the expression for the force between the two current elements shows that

$$dB = \frac{\mu_0}{4\pi} \frac{I\,dl \sin\theta}{r^2}.$$

This equation can be written as a vector equation in the following manner. Let $\hat{\mathbf{r}}$ be a unit vector pointing from $I\,dl$ to $I'\,dl'$. Let $d\mathbf{l}$ be a vector element of length dl in the direction of the current I; $I\,d\mathbf{l}$ then becomes a vector current element (see Fig. 7.1). The vector equation for $d\mathbf{B}$ is

$$d\mathbf{B} = \frac{\mu_0}{4\pi} \frac{I\,d\mathbf{l} \times \hat{\mathbf{r}}}{r^2}.$$

This expression is known as the **Biot–Savart law** after Jean Biot and Felix Savart. The spatial relation of the vectors $I\,d\mathbf{l}$, $\hat{\mathbf{r}}$, and $d\mathbf{B}$ is as shown in Fig. 7.2 where I is assumed to be a current of **positive** charges.

From a consideration of Fig. 7.2 we see that the magnetic field lines around a long, straight current must form closed circles as shown in Fig. 7.3. If a compass needle is placed near the current, the needle orients itself parallel to the magnetic lines and perpendicular to the current. This effect was originally observed by Hans Christian Oersted.

The force on a current element $I'\,d\mathbf{l}'$ placed at the point P in Fig. 7.2 would be

$$d\mathbf{F} = I'\,d\mathbf{l}' \times d\mathbf{B}.$$

If $I'\,d\mathbf{l}'$ is parallel to $I\,d\mathbf{l}$, $d\mathbf{B} \perp I'\,d\mathbf{l}'$ and

$$dF = |d\mathbf{F}| = I'\,dl\,dB$$

and $d\mathbf{F}$ is a vector in the plane containing $I'\,d\mathbf{l}$ and $I\,d\mathbf{l}$. Referring again to Fig. 7.1 we see that the force between the two current elements $I\,d\mathbf{l}$ and $I'\,d\mathbf{l}'$ is given by the general expression

$$d\mathbf{F} = \frac{\mu_0}{4\pi} \frac{I'\,d\mathbf{l}' \times (I\,d\mathbf{l} \times \hat{\mathbf{r}})}{r^2},$$

where $\hat{\mathbf{r}}$ points from $I\,d\mathbf{l}$ toward $I'\,d\mathbf{l}'$.

The force between two long, parallel currents is found by integrating the general expression over the lengths of the currents. To do this we first calculate the magnetic field near a long, straight current and then consider the second current to be located in the field of the first.

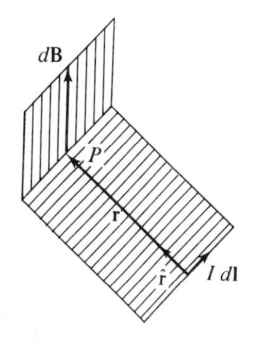

Fig. 7.2. Spatial relation of $I\,d\mathbf{l}$, $\hat{\mathbf{r}}$, and $d\mathbf{B}$.

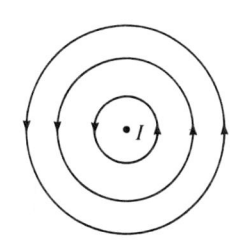

Fig. 7.3. Magnetic lines around a long, straight (positive) current moving upward out of the page.

Magnetic field near a long, straight current

The contribution to the field at a point P a distance R from the current due to the current element $I\,dl$ (see Fig. 7.4) is

$$dB = \frac{\mu_0}{4\pi} \frac{I\,dl \sin\theta}{r^2}.$$

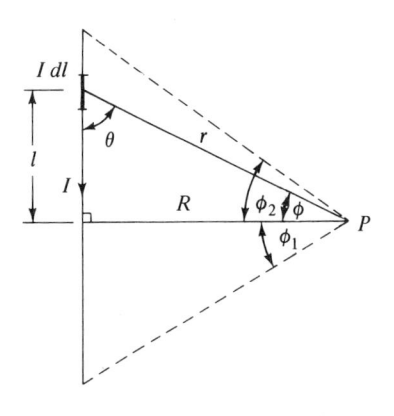

Fig. 7.4. Calculation of the field near a long, straight current.

From Fig. 7.4 we have the following relations:

$$\sin \theta = \cos \phi; \quad r = R \sec \phi; \quad l = R \tan \phi.$$

Differentiating the last equation gives

$$dl = R \sec^2 \phi \, d\phi.$$

Substitution into the expression for dB gives

$$dB = \frac{\mu_0}{4\pi} \frac{IR \sec^2 \phi \, d\phi \cos \phi}{R^2 \sec^2 \phi}$$

$$= \frac{\mu_0}{4\pi} \frac{I}{R} \cos \phi \, d\phi.$$

When the current is of finite length, the angle ϕ will vary between limits ϕ_1 and ϕ_2, say. Then, the total field at P is

$$B = \frac{\mu_0 I}{4\pi R} \int_{-\phi_1}^{\phi_2} \cos \phi \, d\phi$$

$$= \frac{\mu_0 I}{4\pi R} (\sin \phi_2 + \sin \phi_1) \text{Wb} \cdot \text{m}^{-2}.$$

When the current is very long, both ϕ_1 and ϕ_2 approach 90° and the field at a distance R from a very long, straight current is

$$B = \frac{\mu_0 I}{2\pi R} \text{Wb} \cdot \text{m}^{-2}.$$

Force between two long, parallel currents

The situation for two long, parallel currents is as shown in Fig. 7.5. The general expression for the force on current element $I' \, d\mathbf{l}'$ due to current element $I \, d\mathbf{l}$ is

$$d\mathbf{F}' = \frac{\mu_0}{4\pi} \frac{II' \, d\mathbf{l}' \times (d\mathbf{l} \times \hat{\mathbf{r}})}{r^2}$$

and is in the direction shown. Examination of Fig. 7.5 and the expression for $d\mathbf{F}'$ shows that the force $d\mathbf{F}$ has the direction shown and is equal in magnitude to $d\mathbf{F}'$.

The magnetic field at $I' \, d\mathbf{l}'$ due to the current I is

$$B = \frac{\mu_0 I}{2\pi R}.$$

The force experienced by $I' \, d\mathbf{l}'$ in the field B is given by

$$dF' = I' \, dl' \, B$$

$$= I' \, dl' \left(\frac{\mu_0 I}{2\pi R} \right).$$

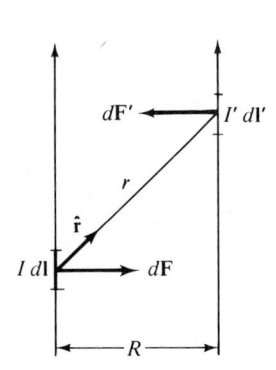

Fig. 7.5. Two long parallel currents exert attractive forces on each other; if the currents are antiparallel, the forces are repulsive.

The magnetic force per unit length exerted on the second current is

$$F = \frac{dF'}{dl'} = \frac{\mu_0 I I'}{2\pi R} \text{ N} \cdot \text{m}^{-1},$$

which is the expression for the force between two long, parallel wires that we quoted in Section 2.4.

Example. A conducting loop in the form of a square of side $2L$ carries a current I. Deduce the magnetic field at the center O of the square, assuming the current to flow in the counterclockwise sense.

Solution. (See Fig. 7.6.) Each of the four sides of the square will produce the same contribution to the magnetic field at the center; the magnetic field at O points out of the page. We can calculate the field at O arising from the straight conductors using the formula

$$B = \frac{\mu_0 I}{4\pi L} \int_{-\phi_1}^{\phi_2} \cos\phi \, d\phi$$

with $\phi_1 = \phi_2 = \pi/4$. Therefore,

$$B_0 = 4B = 4\left(\frac{\mu_0 I}{4\pi L}\right) \int_{-\pi/4}^{\pi/4} \cos\phi \, d\phi = \frac{0.866 \mu_0 I}{\pi L}.$$

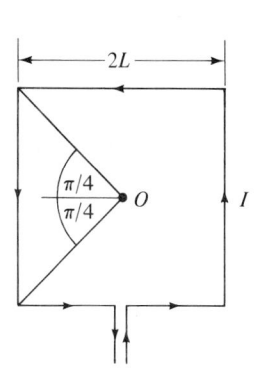

Fig. 7.6. A current I flows in a square conducting loop.

7.3 AMPÈRE'S LAW

We next inquire into the value of the quantity $\mathbf{B} \cdot d\mathbf{l}$ integrated along a closed magnetic line around a long, straight current (see Fig. 7.7). Since $\mathbf{B}$ and $d\mathbf{l}$ are everywhere parallel and $dl = R\, d\theta$, the integral becomes

$$\oint \mathbf{B} \cdot d\mathbf{l} = \frac{\mu_0 I}{2\pi R} \oint R\, d\theta = \mu_0 I.$$

This result is known as **Ampère's law**. It has been derived for the special case of a long, straight current but it is a general law and holds for **all current arrangements and for all closed paths** (see Problem 16). The general law can be stated as follows: the line integral of the magnetic field around any closed path is equal to the permeability μ_0 multiplied by the current passing through (threading) the closed path.

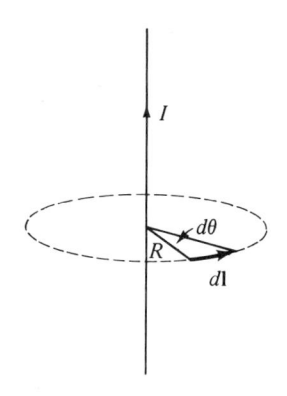

Fig. 7.7. Path for integration of $\mathbf{B}$ around a circular field line.

Ampère's law is one of the basic equations of electromagnetism and is incorporated into the theory of Maxwell as one of his equations. In doing this, however, Maxwell was forced to modify the equation to take account of the fact that a time-varying electric field produces a magnetic field. We shall discuss this important modification in Section 9.4.

Example. Calculate the magnetic field inside an ideal infinite **solenoid**. A solenoid consists of a long wire wound into a close-packed helix. When the wire carries a current, the ideal solenoid is equivalent to an infinitely long, cylindrical current sheet. The magnetic field outside such a solenoid is zero.

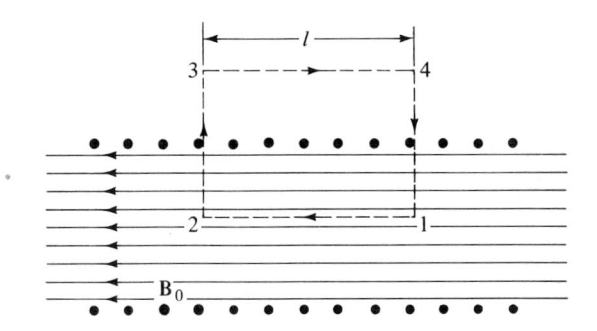

Fig. 7.8. A cross-sectional view of an ideal solenoid.

Solution. We can find the field inside the solenoid by applying Ampère's law to the rectangular path indicated in Fig. 7.8. Therefore,

$$\oint \mathbf{B} \cdot d\mathbf{l} = \int_1^2 \mathbf{B} \cdot d\mathbf{l} + \int_2^3 \mathbf{B} \cdot d\mathbf{l} + \int_3^4 \mathbf{B} \cdot d\mathbf{l} + \int_4^1 \mathbf{B} \cdot d\mathbf{l}.$$

The first integral on the right-hand side is $B_0 l$ where B_0 is the magnitude of the magnetic field inside the solenoid and l is the length of the path between points 1 and 2. The other three integrals on the right-hand side are all zero. The second and fourth integrals are zero because **B** is perpendicular to $d\mathbf{l}$ for these paths. The third integral is zero because we have taken **B** to be zero for all points external to an ideal solenoid. Therefore,

$$\oint \mathbf{B} \cdot d\mathbf{l} = B_0 l.$$

The net current that passes through the area defined by the path of integration is

$$nlI$$

where n is the number of turns per unit length of the solenoid and I is the current flowing in the wire. Ampère's law then becomes

$$B_0 l = \mu_0 n l I$$

or

$$B_0 = \mu_0 n I.$$

Note that B_0 is constant over the cross-sectional area of the solenoid and independent of the diameter of the solenoid. A long solenoid provides a practical means to establish a calculable uniform magnetic field for experimentation, just as a parallel-plate capacitor provides a practical way to set up a uniform electric field.

Example. Calculate the magnetic field at points within a **toroid**. A toroid may be thought of as a solenoid of finite length bent into the shape of a doughnut (see Fig. 7.9).

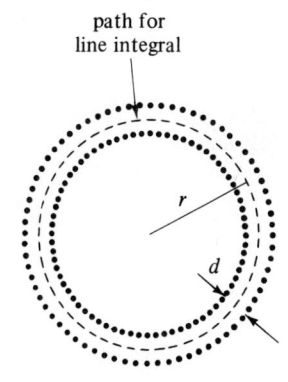

Fig. 7.9. A cross-sectional view of a toroid.

Solution. From the symmetry of the configuration it follows that the magnetic lines of force form **concentric** circles inside the toroid. We can readily find the magnetic field inside the toroid by applying Ampère's law to the circular path indicated in Fig. 7.9 to obtain

$$\oint \mathbf{B} \cdot d\mathbf{l} = B 2\pi r.$$

The net current that passes through the area defined by the path of integration is

$$NI$$

where N is the total number of turns. Ampère's law then becomes

$$B 2\pi r = \mu_0 NI$$

or

$$B = \frac{\mu_0}{2\pi} \frac{NI}{r}.$$

Note that in contrast to the solenoid B is not constant over the cross-sectional area of the toroid.

It is often convenient to define an average field $\bar{B}$ for a toroid as

$$\bar{B} = \mu_0 n I$$

where

$$n = \frac{N}{2\pi r}.$$

The variation of B within the toroid is not large if $r \gg d$.

7.4 THE RELATION BETWEEN ELECTRIC AND MAGNETIC FIELDS

We begin our exploration of the relation between the vector fields **E** and **B** by considering the simple problem of two charges q_1 and q_2, with separation vector **r**, each moving with velocity **v** (see Fig. 7.10).

An observer in a frame of reference moving with velocity **v** would measure an electric force $\mathbf{F}_e$ on charge q_2 where

$$F_e = \frac{1}{4\pi\epsilon_0} \frac{q_1 q_2}{r^2};$$

he would observe no magnetic force.

An observer at rest in the laboratory measures a magnetic field $d\mathbf{B}$ at the position of charge q_2 due to the motion of charge q_1 where

$$d\mathbf{B} = \frac{\mu_0}{4\pi} \frac{I\, d\mathbf{l} \times \hat{\mathbf{r}}}{r^2} = \frac{\mu_0}{4\pi} \frac{q_1 \mathbf{v} \times \hat{\mathbf{r}}}{r^2}$$

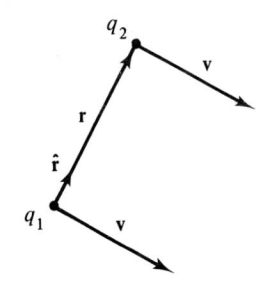

Fig. 7.10. Two charges q_1 and q_2 each moving with velocity **v**.

since $I\, d\mathbf{l} = q_1 \mathbf{v}$.

The magnetic force on charge q_2 due to the field $d\mathbf{B}$ is

$$\mathbf{F}_m = q_2 \mathbf{v} \times d\mathbf{B}$$

and has magnitude (note that **r**, **v**, and $d\mathbf{B}$ are mutually perpendicular)

$$F_m = q_2 v \left(\frac{\mu_0}{4\pi} \frac{q_1 v}{r^2} \right)$$

$$= \left(\frac{\mu_0}{4\pi} \right) \frac{q_1 q_2 v^2}{r^2}.$$

The ratio of the electric force to the magnetic force is

$$\frac{F_e}{F_m} = \frac{1}{4\pi\epsilon_0} \frac{q_1 q_2}{r^2} \cdot \frac{4\pi}{\mu_0} \frac{r^2}{q_1 q_2 v^2} = \frac{1}{\epsilon_0 \mu_0 v^2}.$$

Since this ratio must be dimensionless, we see that the constant $(\epsilon_0 \mu_0)^{-1}$ has the dimension (speed)^2.

We write

$$\frac{1}{\epsilon_0 \mu_0} = c^2$$

where c is a constant of value

$$c = (\epsilon_0 \mu_0)^{-1/2}.$$

Measured values of ϵ_0 and μ_0 predict c to have a value very close to the speed of light (see also Section 9.7). Recognition of this fact by scientists in the last century led to a great increase in research in optics, electricity, and magnetism which finally led to the identification of light as an electromagnetic wave.

The transformation of the electric and the magnetic fields

Our simple example has shown us that the motion of the observer must be taken into account in the description of the fields **E** and **B** surrounding charges. In general, a field that is purely magnetic or purely electric in one frame of reference will have both magnetic and electric components in another. Electric and magnetic fields really have no separate meaning but are contained within the concept of a single electromagnetic field.

The force on a charge q moving with velocity **v** in an electric field **E** and magnetic field **B** is (Section 6.2)

$$\mathbf{F} = q(\mathbf{E} + \mathbf{v} \times \mathbf{B}).$$

We consider observers in two frames of reference R and R' (see Fig. 7.11) where frame R' is moving with velocity **u** with respect to frame R along the common x, x' axes. We shall consider the special case of the charge q being instantaneously at rest in frame R'. The components of the force as observed in the two frames for this particular case are related by the equations[1]

[1] *MWTP*, Section 12.5.

SEC. 7.4 THE RELATION BETWEEN ELECTRIC AND MAGNETIC FIELDS

$$F_x = F'_x$$
$$F_y = \frac{F'_y}{\gamma} = F'_y\left(\frac{1-u^2}{c^2}\right)^{1/2}$$
$$F_z = \frac{F'_z}{\gamma} = F'_z\left(\frac{1-u^2}{c^2}\right)^{1/2}.$$

Since the charge q is taken to be instantaneously at rest in R', the force on q in R' will be

$$\mathbf{F}' = q\mathbf{E}'$$

where $\mathbf{E}'$ is the electric field in R' while in frame R, the force on q is

$$\mathbf{F} = q(\mathbf{E} + \mathbf{u} \times \mathbf{B})$$

since q has velocity $\mathbf{u}$ as measured in R. We consider x components of the force first. In R,

$$F_x = q[E_x + (\mathbf{u} \times \mathbf{B})_x] = q[E_x + u_y B_z - u_z B_y]$$
$$= qE_x$$

since

$$u_y = u_z = 0.$$

In R',

$$F'_x = qE'_x.$$

Since

$$F'_x = F_x,$$

we have

$$E'_x = E_x$$

for the transformation equation for the x component of the electric field.

The y component of the force is given by

$$F_y = \frac{F'_y}{\gamma} = q[E_y + (\mathbf{u} \times \mathbf{B})_y]$$
$$= q[E_y + (u_z B_x - u_x B_z)]$$
$$= q(E_y - uB_z).$$

Therefore,

$$E'_y = \frac{F'_y}{q} = \gamma(E_y - uB_z).$$

The z component of the force when transformed yields a similar equation for E'_z; that is,

$$E'_z = \gamma(E_z + uB_y).$$

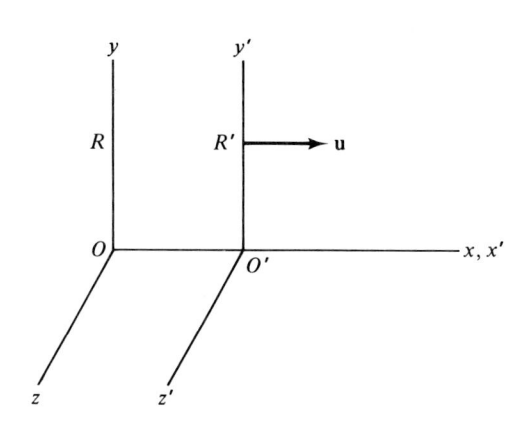

Fig. 7.11. Coordinate systems attached to reference frames R and R' in uniform relative motion along the common x, x' axes.

In summary, we see that the components of the electric field transform according to the equations

$$E'_x = E_x$$
$$E'_y = \gamma(E_y - uB_z)$$
$$E'_z = \gamma(E_z + uB_y)$$

or, conversely,

$$E_x = E'_x$$
$$E_y = \gamma(E'_y + uB'_z)$$
$$E_z = \gamma(E'_z - uB'_y).$$

The result is the same for any arbitrary direction of the velocity **u**. In such a case, we write

$$\mathbf{E} = \mathbf{E}_\| + \mathbf{E}_\perp$$

and

$$\mathbf{E}' = \mathbf{E}'_\| + \mathbf{E}'_\perp$$

where

$$\mathbf{E}'_\| = \mathbf{E}_\|$$

and

$$\mathbf{E}'_\perp = \gamma[E_\perp + (\mathbf{u} \times \mathbf{B})_\perp].$$

The component vector of **E** parallel to the direction of relative motion is unaffected, while the component vectors perpendicular to the direction of relative motion transform to mixed electric and magnetic fields.

The transformation of the components of the magnetic field is most easily seen by considering the frames of reference R and R' of Fig. 7.11 and assuming a charge q to move in the R' frame with velocity $\mathbf{v}' = v'_z$. The force on the charge in the R' frame is

$$F' = q(\mathbf{E}' + \mathbf{v}' \times \mathbf{B}')$$

which has components

$$F'_x = q[E'_x + (\mathbf{v}' \times \mathbf{B}')_x]$$
$$= q[E'_x + (v'_y B'_z - v'_z B'_y)] = q(E'_x - v'B'_y),$$
$$F'_y = q(E'_y + v'B'_x)$$

and

$$F'_z = qE'_z.$$

The velocity of charge q in the R frame is obtained through the equations[2] for the transformation of the components of the velocity $\mathbf{v}'$ which give

$$v_x = \frac{v'_x + u}{1 + uv'_x/c^2} = u$$

[2] *MWTP*, Section 5.2.

SEC. 7.4 THE RELATION BETWEEN ELECTRIC AND MAGNETIC FIELDS

and
$$v_y = \frac{v'_y}{\gamma(1 + uv'_x/c^2)} = 0$$

$$v_z = \frac{v'_z}{\gamma(1 + uv'_x/c^2)} = \frac{v'_z}{\gamma} = \frac{v'}{\gamma}.$$

Therefore, the force
$$\mathbf{F} = q(\mathbf{E} + \mathbf{v} \times \mathbf{B})$$

in frame R has components
$$F_x = q[E_x + (\mathbf{v} \times \mathbf{B})_x]$$
$$= q[E_x + (v_y B_z - v_z B_y)] = q\left(E_x - \frac{v'}{\gamma} B_y\right),$$
$$F_y = q\left(E_y + \frac{v'}{\gamma} B_x - uB_z\right)$$

and
$$F_z = q(E_z + uB_y).$$

In order to find the relations between the components of the magnetic field as measured in R and R' we must make use of the general force-transformation equations[3] (see Problem 17)

$$F_x = \frac{F'_x + (u/c^2)\mathbf{v} \cdot \mathbf{F}'}{1 + uv'_x/c^2} \qquad F'_x = \frac{F_x - (u/c^2)\mathbf{v} \cdot \mathbf{F}}{1 - uv_x/c^2}$$

$$F_y = \frac{F'_y}{\gamma(1 + uv'_x/c^2)} \qquad F'_y = \frac{F_y}{\gamma(1 - uv_x/c^2)}$$

$$F_z = \frac{F'_z}{\gamma(1 + uv'_x/c^2)} \qquad F'_z = \frac{F_z}{\gamma(1 - uv_x/c^2)}$$

as well as the relations between the components of the electric field derived earlier in this section. The result is (see Problem 18)

$$B_x = B'_x \qquad\qquad B'_x = B_x$$
$$B_y = \gamma\left(B'_y - \frac{u}{c^2} E'_z\right) \qquad B'_y = \gamma\left(B_y + \frac{u}{c^2} E_z\right)$$
$$B_z = \gamma\left(B'_z + \frac{u}{c^2} E'_y\right) \qquad B'_z = \gamma\left(B_z - \frac{u}{c^2} E_y\right).$$

Again, if the relative velocity $\mathbf{u}$ were in some arbitrary direction, we could write $\mathbf{B} = \mathbf{B}_\| + \mathbf{B}_\perp$ and $\mathbf{B}' = \mathbf{B}'_\| + \mathbf{B}'_\perp$ where

$$\mathbf{B}'_\| = \mathbf{B}_\|$$

and
$$\mathbf{B}'_\perp = \gamma\left[\mathbf{B}_\perp - \frac{1}{c^2}(\mathbf{u} \times \mathbf{E})_\perp\right]$$

---

[3] W. G. V. Rosser, "Electric and Magnetic Fields of a Charge Moving with Uniform Velocity", *Contemporary Physics*, 1 (August, 1960), 453.

7.5 THE FIELD OF A POINT CHARGE IN UNIFORM MOTION

We consider a charge q moving with uniform velocity $\mathbf{u}$ in frame R. In order to calculate the fields $\mathbf{E}$ and $\mathbf{B}$ in R produced by this moving charge, we shall first calculate the fields $\mathbf{E}'$ and $\mathbf{B}'$ in a frame R' in which the charge q is at rest at the origin (see Fig. 7.12) and then use the transformation equations developed in Section 7.4. We assume that O and O' coincide at time $t = 0$. Therefore, charge q has coordinates $(ut, 0, 0)$ in R.

Since q is at rest in R', the field in R' is purely electric so that

$$\mathbf{E}' = \frac{q\hat{\mathbf{r}}'}{4\pi\epsilon_0 r'^2}, \qquad \mathbf{B}' = 0$$

where

$$r' = (x'^2 + y'^2 + z'^2)^{1/2}$$

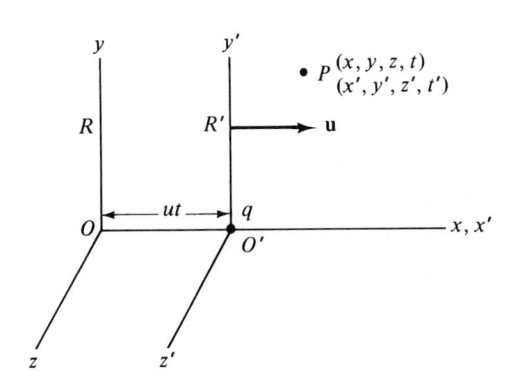

Fig. 7.12. A charge q is at rest at the origin of system R' which moves with uniform velocity in the x, x' direction.

is the distance from the origin O' to the field point P. Since we must express the fields $\mathbf{E}$ and $\mathbf{B}$ in R in terms of the spatial coordinates x, y, and z, we must use the Lorentz transformation equations[4]

$$x' = \gamma(x - ut)$$
$$y' = y$$
$$z' = z$$
$$t' = \gamma\left(t - \frac{ux}{c^2}\right)$$

and write

$$r' = [\gamma^2(x - ut)^2 + y^2 + z^2]^{1/2}.$$

The components of $\mathbf{E}'$ in R' are

$$E'_x = \frac{qx'}{4\pi\epsilon_0 r'^3}, \qquad E'_y = \frac{qy'}{4\pi\epsilon_0 r'^3}, \qquad E'_z = \frac{qz'}{4\pi\epsilon_0 r'^3}.$$

[4]*MWTP*, Section 4.7.

SEC. 7.5 THE FIELD OF A POINT CHARGE IN UNIFORM MOTION

Using the transformation equations

$$E_x = E'_x$$
$$E_y = \gamma(E'_y + uB'_z) = \gamma E'_y$$
$$E_z = \gamma(E'_z - uB'_y) = \gamma E'_z$$

we obtain

$$E_x = \frac{q\gamma(x - ut)}{4\pi\epsilon_0[\gamma^2(x - ut)^2 + y^2 + z^2]^{3/2}}$$

$$E_y = \frac{q\gamma y}{4\pi\epsilon_0[\gamma^2(x - ut)^2 + y^2 + z^2]^{3/2}}$$

$$E_z = \frac{q\gamma z}{4\pi\epsilon_0[\gamma^2(x - ut)^2 + y^2 + z^2]^{3/2}}.$$

The transformation equations for the magnetic field are

$$B'_x = B_x = 0$$
$$B'_y = \gamma\left[B_y + \left(\frac{u}{c^2}\right)E_z\right] = 0$$
$$B'_z = \gamma\left[B_z - \left(\frac{u}{c^2}\right)E_y\right] = 0$$

giving

$$B_x = 0$$
$$B_y = -\frac{u}{c^2} \cdot \frac{q\gamma z}{4\pi\epsilon_0[\gamma^2(x - ut)^2 + y^2 + z^2]^{3/2}}$$
$$B_z = \frac{u}{c^2} \cdot \frac{q\gamma y}{4\pi\epsilon_0[\gamma^2(x - vt)^2 + y^2 + z^2]^{3/2}}.$$

Note that the fields are time-dependent, as well as position-dependent.

At time $t = 0$ in R we have

$$\mathbf{E} = \mathbf{E}_x + \mathbf{E}_y + \mathbf{E}_z = \frac{q\gamma\mathbf{r}}{4\pi\epsilon_0[\gamma^2 x^2 + y^2 + z^2]^{3/2}}.$$

The field is radial in R and the magnitude of $\mathbf{E}$ falls off as r^{-2}; however, the magnitude at a given r depends on the direction with respect to the x axis and the field is no longer radially symmetric (see Fig. 7.13 and Problem 19).

Finally, we note that charge should be invariant under a Lorentz transformation. Since, from Gauss' law (Section 4.2), we can write

$$q = \epsilon_0 \int_S \mathbf{E} \cdot d\mathbf{S},$$

the electric flux over a surface enclosing the charge must be invariant. We leave it as an exercise for the reader to show that this is so (see Problem 20).

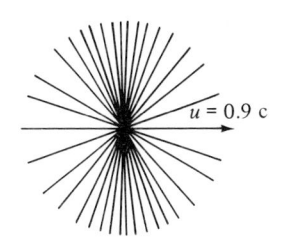

Fig. 7.13. Electric field around a charge moving with speed $u = 0.9\,c$ with respect to an observer.

7.6 FUSION POWER REACTORS[5,6,7,8]

The achievement of controlled nuclear fusion would provide an essentially infinite source of energy for the people of the world.[9] The energy output of a fusion process arises from the reduction of the total mass of a nuclear system that accompanies the merger of two or more light nuclei.[10] Deuterium is the most likely fuel for a fusion power reactor. This abundant isotope of hydrogen is easily separated from seawater.

The uncontrolled release of a massive amount of fusion energy was achieved in 1952 with the first thermonuclear test explosion. This experiment proved that fusion energy could be released on a large scale by raising the temperature of a gas of charged particles (a **plasma**) to about $5 \times 10^7 °C$. Almost immediately the search for a controlled means of releasing fusion energy began in the United States, Britain, and the Soviet Union. The basic problem is to find a practical way of maintaining a comparatively low-density plasma at a sufficiently high temperature so that the output of fusion energy derived from the plasma is greater than the input of some other kind of energy supplied to the plasma. Since no solid materials can exist at the necessary temperature, the principle emphasis has been on the use of magnetic fields to confine the plasma.

In principle, an ideal containing magnetic field can be produced by passing a strong electric current through an infinite solenoid. In a solenoid the path of a charged particle is a helix with a magnetic line of force as axis. Since very few particles would ever touch the walls, the infinite solenoid would be an ideal bottle for a fusion reactor. Unfortunately, a real solenoid has ends at which the lines of force must emerge and at which the gas must therefore come in contact with solid matter. One method of eliminating the end effects is to bend the solenoid into a circle and join the two ends to form a closed toroidal coil. Although the lines of force are now circles, the toroidal coil has a fatal defect. Because of the curvature, the strength of the magnetic field is greater near the inside wall of the tube than near the outside wall (see Fig. 7.14). This inhomogeneity alters the paths of the charged particles. Near the inside wall the relatively stronger field curves the path of the particle more sharply than near the outside wall. As a result the charged particles drift across the field with the positively

---

[5]T. K. Fowler and R. F. Post, "Progress Toward Fusion Power," *Scientific American*, December, 1966.

[6]J. B. Adams, "Progress Towards Nuclear Fusion Reactor," *Contemporary Physics*, **10** (January, 1969), 1.

[7]W. C. Gough and B. J. Eastlund, "The Prospects of Fusion Power," *Scientific American*, February, 1971.

[8]M. J. Lubin and A. P. Fraas, , "Fusion by Laser," *Scientific American*, June, 1971.

[9]M. K. Hubbert, "The Energy Resources of the Earth," *Scientific American*, September, 1971.

[10]*MWTP*, Section 12.6.

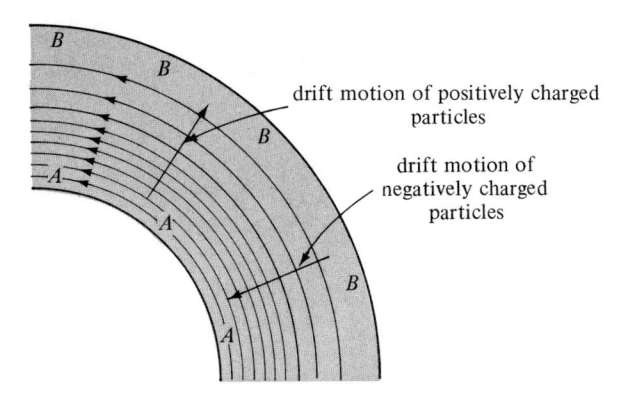

Fig. 7.14. The magnetic field in a toroid is greater at region A than at region B. This causes charged particles to drift across the field.

charged nuclei drifting from A toward B and electrons drifting from B toward A. The resultant separation of electric charges produces a large electric field that disrupts the particle paths completely, throwing the entire plasma into the walls of the tube.

The remedy to this problem is to twist the magnetic field about its circular axis. This counteracts the tendency of the charges to separate since a single line of force followed for a distance leads from region A to region B. In the early models of the **stellarator**,[11] developed at the Princeton Plasma Physics Laboratory, the lines of force were twisted by simply bending the torus into a figure eight. In later models the toroidal field is twisted by the interaction with an additional transverse magnetic field generated by a set of helical windings. In the **tokamak**, originally developed at the Institute of Atomic Energy near Moscow, the toroidal field is twisted by a magnetic field generated by an electric current flowing along the lines of magnetic force of the toroidal field.

Other attempts to develop an appropriate magnetic bottle have also been tried; one example is the **pinch effect**.[12] In this scheme a heavy flow of current through the plasma generates a strong magnetic field which both contains the gas and brings it to a high temperature by compressing or pinching it.

There are two fundamental problems in the development of a fusion power reactor. The problem of achieving the ignition temperature was first overcome in 1953. The problem of confining the plasma for a long

---

[11] L. Spitzer, Jr., "The Stellarator," *Scientific American*, October, 1958. Available as *Scientific American Offprint 246* (San Francisco: W. H. Freeman and Co., Publishers).

[12] R. F. Post, "Fusion Power," *Scientific American*, December, 1957. Available as *Scientific American Offprint 236* (San Francisco: W. H. Freeman and Co., Publishers).

enough time to release a significant net amount of energy has proved to be more difficult[13] but has been overcome recently. Unfortunately, the two problems have so far not been solved in one machine but in quite different machines, each specifically designed to maximize the conditions for reaching one goal or the other. At present there are several fusion devices which show promise of simultaneously overcoming the problems of ignition temperature and confinement times. It is not yet clear, however, whether or not any of these devices could be developed into an economical fusion reactor in the future (say, the next 30 years).

7.7 CURRENT LOOPS; THE MAGNETIC DIPOLE MOMENT

The magnetic field at a point P on the axis of a circular current loop can be calculated readily from

$$d\mathbf{B} = \frac{\mu_0}{4\pi} \frac{I \, d\mathbf{l} \times \hat{\mathbf{r}}}{r^2}.$$

As we see from Fig. 7.15, $\hat{\mathbf{r}} \perp I \, d\mathbf{l}$ for all current elements so that $|d\mathbf{l} \times \hat{\mathbf{r}}| = dl$. From the symmetry, all components of $d\mathbf{B}$ perpendicular to the axis will cancel so that the field at P is axial and is given by

$$B = \frac{\mu_0 I}{4\pi} \frac{\sin \alpha}{r^2} \oint dl$$

since α and r are constant for all current elements around the loop. The integral is simply $2\pi a$. Also $\sin \alpha = a/r$ and $r^2 = a^2 + b^2$ so that the field at P is

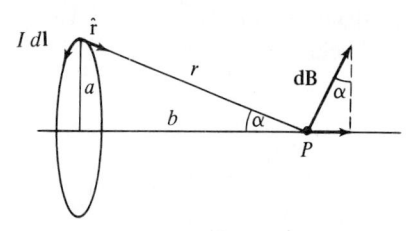

Fig. 7.15. Calculation of the magnetic field along the axis of a circular current loop.

$$B = \frac{\mu_0 I}{4\pi} \frac{a}{(a^2 + b^2)^{3/2}} 2\pi a$$

$$= \frac{\mu_0}{2\pi} \frac{IA}{(a^2 + b^2)^{3/2}} \text{ Wb} \cdot \text{m}^{-2}$$

where $A = \pi a^2$ is the area of the current loop.

For points P far removed from the current loop, $b \gg a$ and $b \simeq r$. The axial field then becomes

$$B = \frac{\mu_0}{2\pi} \frac{IA}{r^3} \text{ Wb} \cdot \text{m}^{-2}.$$

This expression is identical in form with the expression for the electric

[13] F. F. Chen, "The Leakage Problem in Fusion Reactors," *Scientific American*, July, 1967.

SEC. 7.7　CURRENT LOOPS; THE MAGNETIC DIPOLE MOMENT

field on the axis of an electric dipole

$$E = \frac{1}{4\pi\epsilon_0} \frac{p}{r^3} \text{ N} \cdot \text{C}^{-1}$$

(see Section 3.4) which in vector form is

$$\mathbf{E} = \frac{1}{4\pi\epsilon_0} \frac{\mathbf{p}}{r^3} \text{ N} \cdot \text{C}^{-1}$$

where $\mathbf{p}$ is the electric dipole moment. This suggests that we should write for the magnetic field on the axis of a current loop (and far removed from it)

$$\mathbf{B} = \frac{\mu_0}{2\pi} \frac{\boldsymbol{\mu}}{r^3} \text{ Wb} \cdot \text{m}^{-2}$$

where $\boldsymbol{\mu}$ is the **magnetic dipole moment** of magnitude IA and has direction along the axis of the current loop. The magnetic field due to a current loop will be identical in form with the electric field for an electric dipole for points far removed from the current loop.

A current loop will experience a torque in a magnetic field. We will carry out the calculation for the rectangular loop shown in Fig. 7.16 where $\mathbf{B}$ is perpendicular to the sides of the loop of length L_1, since it is somewhat easier than that for a circular loop. Let us consider the four current elements marked ① → ④ in Fig. 7.16. The forces on elements ① and ③ are $d\mathbf{F}'$ and $-d\mathbf{F}'$ as shown; they produce no torque since they act in opposite directions along the same line. This is true for all elements along the sides of length L_2 and there is no torque produced by these forces.

The forces on current elements ② and ④ are $d\mathbf{F}$ and $-d\mathbf{F}$ where

$$dF = I\,dl\,B.$$

In this case, however, $d\mathbf{F}$ and $-d\mathbf{F}$ are not collinear and produce a total torque of magnitude

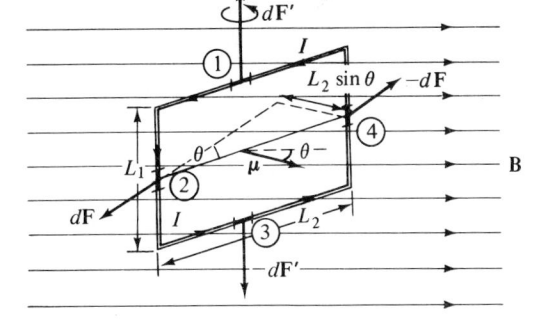

Fig. 7.16. A current loop experiences a torque when placed in a magnetic field.

$$d\tau = dF\,L_2 \sin\theta = IBL_2 \sin\theta\,dl.$$

Corresponding pairs of current elements along the sides L_1 will produce a similar torque and the total torque is easily seen to be

$$\tau = IBL_1L_2 \sin\theta.$$

But $L_1L_2 = A$, the area of the current loop. The magnetic dipole moment $\boldsymbol{\mu}$ has magnitude IA and θ is the angle between $\boldsymbol{\mu}$ and $\mathbf{B}$. Therefore, the equation for the torque can be written in vector form as

$$\boldsymbol{\tau} = \boldsymbol{\mu} \times \mathbf{B}.$$

This is identical with the expression $\boldsymbol{\tau} = \mathbf{p} \times \mathbf{E}$ obtained in Section 3.4

for the torque on an electric dipole placed in an electric field. A calculation similar to that carried out in Section 3.4 for the potential energy of an electric dipole placed in an external electric field indicates that the potential energy of a magnetic dipole placed in an external magnetic field is

$$U = -\boldsymbol{\mu} \cdot \mathbf{B}.$$

The proof of this relation is left as an exercise for the reader (see Problem 25).

7.8 ELECTRON, NUCLEAR, AND ATOMIC MAGNETIC DIPOLE MOMENTS

Magnetic moment of an orbiting electron

The single electron traversing a circular orbit within a hydrogen atom according to the Bohr model[14] is equivalent to a current of magnitude

$$I = \frac{e}{T}$$

where e is the elementary charge and T the period of the motion. The Bohr condition for a permitted orbit is

$$m_e v r = \frac{nh}{2\pi} = n\hbar$$

where $m_e v r$ is the angular momentum of the electron of mass m_e traveling with speed v in an orbit of radius r, h is Planck's constant, $\hbar = h/2\pi$, and n is an integer. Since

$$v = \frac{2\pi r}{T},$$

it follows from the Bohr condition that

$$T = \frac{2\pi m_e r^2}{n\hbar}.$$

Therefore,

$$I = \frac{ne\hbar}{2\pi m_e r^2}$$

or

$$IA = \mu = \frac{ne\hbar}{2m_e}$$

with μ the magnetic moment associated with the orbiting electron. For

[14] *MWTP*, Section 18.4.

SEC. 7.8   ELECTRON, NUCLEAR, AND ATOMIC MAGNETIC DIPOLE MOMENTS

the ground state of the hydrogen atom $n = 1$ and

$$\mu = \frac{eh}{2m_e} = 9.2733 \times 10^{-24} \text{ A} \cdot \text{m}^2 = \beta.$$

The parameter β, called the **Bohr magneton**, is the common unit for tabulating atomic magnetic moments. It is interesting to note that we can write

$$\mu = \frac{e}{2m_e}(m_e v r)$$

indicating that the magnetic moment of the orbiting electron is proportional to its orbital angular momentum.

According to quantum mechanics[15] the orbital angular momentum associated with an electron in any atom is $\mathbf{l}\hbar$ which has magnitude $[l(l+1)]^{1/2}\hbar$ and the quantum number l takes values $0, 1, 2, \ldots$. In vector notation

$$\boldsymbol{\mu} = -\frac{e}{2m_e}\mathbf{l}\hbar = -\beta\mathbf{l}$$

The minus sign indicates that the current in an electron orbit is in the opposite direction to the electron's motion.

Magnetic moment of a spinning electron

Electrons possess spin angular momentum.[16] There is a **spin magnetic moment** $\boldsymbol{\mu}_s$ associated with the spin angular momentum $\mathbf{s}\hbar$ of magnitude $[s(s+1)]^{1/2}\hbar$ with $s = 1/2$. The relation between these two quantities is

$$\boldsymbol{\mu}_s = -g_s\beta\mathbf{s}$$

where g_s is called the **electron spin g-factor**. It has been shown[17] both theoretically and experimentally that $g_s = 2.0023193$. This represents one of the most precise measurements in all of physics!

Magnetic moment of a nucleus

All nuclei of odd mass number possess spin angular momentum $\mathbf{I}\hbar$ of magnitude $[I(I+1)]^{1/2}\hbar$ with $I = 1/2, 3/2, \ldots$. Nuclei of even mass number and odd charge posses spin angular momentum $\mathbf{I}\hbar$ of magnitude $[I(I+1)]^{1/2}\hbar$ with $I = 1, 2, \ldots$. Nuclei of even mass number and even charge have no spin angular momentum. The possesion of both spin angular momentum and charge results in a magnetic moment $\boldsymbol{\mu}_N$ for a

---
[15] *MWTP*, Section 18.6.
[16] *MWTP*, Section 18.7.
[17] H. R. Crane, "The g-Factor of the Electron," *Scientific American*, January, 1968.

nucleus where

$$\boldsymbol{\mu}_N = g_N \beta_N \mathbf{I}.$$

The **nuclear magneton** β_N is defined as

$$\beta_N = \frac{e\hbar}{2m_p} = 5.0506 \times 10^{-27} \text{ A} \cdot \text{m}^2$$

where m_p is the mass of the proton. Some typical **nuclear g-factors** are listed in Table 7.1. Note that g_N can be either positive or negative.

Table 7.1 Nuclear g-factors

Nucleus	I	g_N
^{1}H	$\frac{1}{2}$	5.585
^{12}C	0	—
^{13}C	$\frac{1}{2}$	1.405
^{14}N	1	0.403
^{15}N	$\frac{1}{2}$	−0.567
^{16}O	0	—
^{17}O	$\frac{5}{2}$	−0.757
^{35}Cl	$\frac{3}{2}$	0.548

Magnetic moment of a free atom

For an atom which is composed of N electrons and a nucleus we can write for the **atomic magnetic moment** $\boldsymbol{\mu}_A$ the vector sum

$$\boldsymbol{\mu}_A = -\beta \sum_{i=1}^{N}(\mathbf{l}_i + g_s \mathbf{s}_i) + g_N \beta_N \mathbf{I}.$$

In practice, for many atoms it is a good approximation to sum the individual orbital angular momenta $\mathbf{l}_i \hbar$ to form an atomic orbital angular momentum $\mathbf{L}\hbar$ and to sum the individual spin angular momenta $\mathbf{s}_i \hbar$ to form an atomic spin angular momentum $\mathbf{S}\hbar$. That is,

$$\mathbf{L}\hbar = \sum_{i=1}^{N} \mathbf{l}_i \hbar, \qquad \mathbf{S}\hbar = \sum_{i=1}^{N} \mathbf{s}_i \hbar.$$

The magnitudes of $\mathbf{L}\hbar$ and $\mathbf{S}\hbar$ are $[L(L+1)]^{1/2}\hbar$ and $[S(S+1)]^{1/2}\hbar$, respectively, where L and S are the quantum numbers associated with the orbital and spin angular momentum of the atom. Therefore, we can write

$$\boldsymbol{\mu}_A = -\beta(\mathbf{L} + g_s \mathbf{S}) + g_N \beta_N \mathbf{I}.$$

In addition, it is frequently a good approximation to sum $\mathbf{L}\hbar$ and $\mathbf{S}\hbar$ to form a total electronic angular momentum $\mathbf{J}\hbar$ for the atom. That is

$$\mathbf{J}\hbar = \mathbf{L}\hbar + \mathbf{S}\hbar$$

where $\mathbf{J}\hbar$ has magnitude $[J(J+1)]^{1/2}\hbar$ with J the electronic angular momentum quantum number. The quantum state of an atom is specified in terms of the quantum numbers L, S, and J. In performing the vector addition of $L\hbar$ and $S\hbar$ we must take account of the fact that the constant of proportionality between the magnetic moment and the angular momentum is different for $L\hbar$ and $S\hbar$. As a result the vector $\mathbf{\mu}_J$ representing the total electronic magnetic moment is not parallel to $J\hbar$. By analogy with the definitions of the magnetic moments associated with $L\hbar$ and $S\hbar$ we define the magnitude of $\mathbf{\mu}_J$ to be

$$|\mathbf{\mu}_J| = g_J \beta [J(J+1)]^{1/2}.$$

The factor g_J, called the **Landé g-factor**, may be shown to be

$$g_J = \frac{3J(J+1) - L(L+1) + S(S+1)}{2J(J+1)}$$

(see Problem 34).

We should note that to deduce the magnetic moment of an atom it is only necessary to consider partly filled shells of electrons since for completely filled shells the magnitudes of $L\hbar$, $S\hbar$, and $J\hbar$ are all zero.

7.9 MAGNETIC DIPOLE MOMENT IN AN INHOMOGENEOUS MAGNETIC FIELD

In Section 7.7 we introduced the concept of a magnetic dipole moment. For some purposes it is convenient to represent a magnetic dipole as two fictitious magnetic poles m and $-m$ separated by a distance d in analogy with the definition of the electric dipole. In Fig. 7.17(a) we see a fictitious magnetic dipole in a homogeneous magnetic field. Magnetic pole m will experience a force $\mathbf{F}$ and magnetic pole $-m$ an equal and oppo-

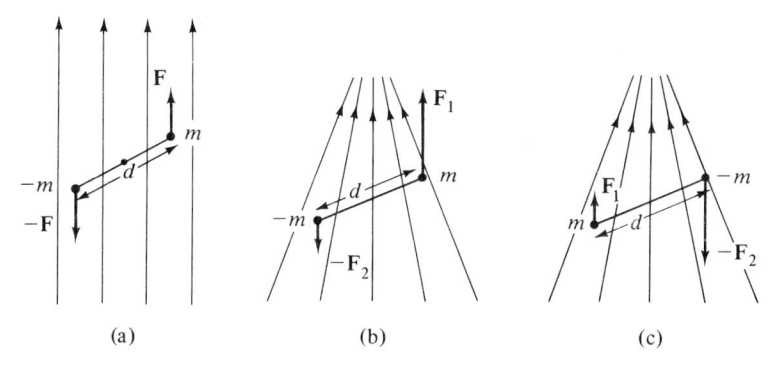

Fig. 7.17. (a) A fictitious magnetic dipole in a homogeneous magnetic field. (b) (c) A fictitious dipole in a converging field experiences a translational force as well as a torque.

site force $-\mathbf{F}$. Therefore, although the magnetic dipole will experience a torque that will tend to align it in the field, its center of mass will remain stationary since the opposing forces will cancel. In Fig. 7.17(b) we see the dipole in a converging magnetic field. The magnetic lines of force are closer together in the vicinity of magnetic pole m and farther apart in the vicinity of magnetic pole $-m$. The forces $\mathbf{F}_1$ and $-\mathbf{F}_2$ acting on m and $-m$, respectively, are not of equal magnitude. In particular, $F_1 > F_2$ and the dipole, as well as tending to align in the direction of the field, will move in the direction of convergence of the field lines. In Fig. 7.17(c) we see the dipole reversed in direction relative to Fig. 7.17(b). In this case $F_1 < F_2$ and the dipole, as well as tending to align in the direction of the field, will move in the direction of divergence of the field lines. That is, if initially we place two antiparallel dipoles side by side in an inhomogeneous magnetic field, they will experience magnetic forces that will tend to move them apart.

The Stern–Gerlach experiment

In 1925 Otto Stern and Walter Gerlach carried out an experiment to demonstrate **spatial quantization** of the orientation of the magnetic moments of silver atoms in a magnetic field. In a silver atom all of the electrons but one are contained in filled shells. These give no contribution to the magnetic moment of the atom. The remaining electron is a $5s$ electron with $l = 0$ and, consequently, it has zero orbital angular momentum. Therefore, a silver atom has spin angular momentum only, of magnitude $(3)^{1/2}\hbar/2$, and the magnetic dipole moment of a silver atom should have only two possible orientations with respect to an external magnetic field;[18] the components of the dipole moment in the direction of the external field and in the direction opposite to the field should each have magnitude β (to a very good approximation—see Section 7.8).

In the Stern–Gerlach experiment, illustrated schematically in Fig. 7.18, a beam of silver atoms from a small oven was defined by a number

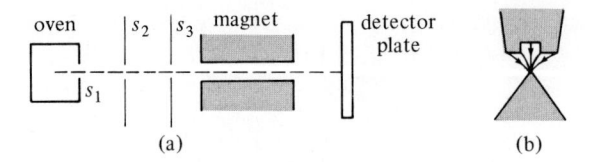

Fig. 7.18. (a) Experimental arrangement for the Stern–Gerlach experiment. (b) An end view of the magnetic pole pieces producing the inhomogeneous field.

[18]*MWTP*, Section 18.6.

of slits and then allowed to pass through an inhomogeneous magnetic field. As the atoms left the field their spatial position was recorded on a photographic plate. With the field off, the beam of atoms defined a single spot on the plate; with the field turned on two spots appeared on the plate corresponding to the two allowed orientations of the magnetic moment of the silver atoms in the magnetic field. The magnetic moments of different orientations experienced different net forces, thereby resulting in the splitting of the beam into two components.

QUESTIONS AND PROBLEMS

1. Compare and contrast the Biot–Savart law with Coulomb's law.
2. Discuss the magnetic interaction in the two situations shown in Fig. 7.19. Do these systems violate the laws of conservation of linear and angular momentum?
3. Calculate the magnetic field at points lying between two wires separated by a distance d and carrying equal currents in opposite directions.
4. Calculate the magnetic field at the center of a rectangle of length L, width W, and carrying a current I.
*5. A loop of wire defines a square of side L and carries a current I. How does the magnetic field vary along a diagonal of the square?
6. By integrating the contribution due to thin circular loops, deduce the relation for the magnetic field at the center of a long solenoid.
7. A nonconducting disc of radius R carries a charge q uniformly spread over its surface The disc is rotated at frequency Ω about an axis through the center of and perpendicular to the disc. Calculate the magnetic field at the center of the disc and its magnetic moment.
*8. Calculate the magnetic field at a point P which is a distance R (measured perpendicularly) from the center of a long, flat metallic strip of width w and negligible thickness. Assume the strip to carry a current I. Show that at points far from the strip the field becomes indistinguishable from that produced by a cylindrical conductor.
9. Find the magnetic field near a thin conducting sheet carrying a uniform current in one direction.
*10. Show that the magnetic field at the point $(0, a)$ due to a charge q moving with speed v along the x axis is given by

$$B = \frac{\mu_0 qva}{4\pi}(d^2 + v^2 t^2)^{-3/2}$$

if the particle is at the origin at $t = 0$.

11. A square loop of side L is placed a distance d from a wire carrying a current I as shown in Fig. 7.20. Calculate the total flux through the loop.

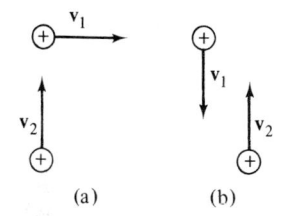

Fig. 7.19. Two magnetically interacting charges: (a) traveling in mutually perpendicular directions, (b) traveling in antiparallel directions.

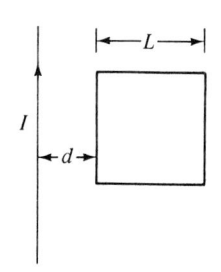

Fig. 7.20. A square loop a distance d from a wire carrying a current I.

12. Two thin coaxial coils, each having the same radius and the same number of turns and carrying the same current can be used to obtain a nearly constant magnetic field over a considerable distance along the axis. This is known as the **Helmholtz arrangement**. Deduce the optimal coil separation.

13. A circuit is composed of N straight wires in the form of a regular polygon. The distance from the center of the polygon to the center of any wire is a. Calculate the magnetic field at the center when a current I flows through the wires. Show that B approaches that of a circular loop as $n \rightarrow \infty$.

14. Use Ampère's law to show that the field outside an ideal toroid is zero.

15. A solenoid is constructed to be 1.0 m in length, 5.0 cm in diameter, and to have five layers of windings of 1000 turns each. What current must flow in the windings of the solenoid in order to produce a field of 5.0×10^{-2} Wb·m^{-2} at the center of the coil? If the solenoid is bent into a toroid and the same current passed through it, what is the field at the inner and outer edges of the coil?

*16. Show that Ampère's law holds for all current arrangements and for all closed paths.

17. Show that the components of the forces $\mathbf{F}$ and $\mathbf{F'}$ measured in two frames of reference R and R', respectively, where R' moves with velocity $\mathbf{u}$ in the x, x' direction, are related by the transformation equations

$$F_x = \frac{F'_x + (u/c^2)\mathbf{v'} \cdot \mathbf{F'}}{1 + uv'_x/c^2} \qquad F'_x = \frac{F_x - (u/c^2)\mathbf{v} \cdot \mathbf{F}}{1 - uv_x/c^2}$$

$$F_y = \frac{F'_y}{\gamma(1 + uv'_x/c^2)} \qquad F'_y = \frac{F_y}{\gamma(1 - uv_x/c^2)}$$

$$F_z = \frac{F'_z}{\gamma(1 + uv'_x/c^2)} \qquad F'_z = \frac{F_z}{\gamma(1 - uv_x/c^2)}$$

where $\mathbf{v}$ and $\mathbf{v'}$ are the velocities of the particle experiencing the force as measured in R and R', respectively.

18. Show that the components of the magnetic fields $\mathbf{B}$ and $\mathbf{B'}$ measured in the frames of reference R and R' (see Problem 17), respectively, are related by the transformation equations

$$B_x = B'_x \qquad B'_x = B_x$$

$$B_y = \gamma\left(B'_y - \frac{u}{c^2}E'_z\right) \qquad B'_y = \gamma\left(B_y + \frac{u}{c^2}E_z\right)$$

$$B_z = \gamma\left(B'_z + \frac{u}{c^2}E'_y\right) \qquad B'_z = \gamma\left(B_z - \frac{u}{c^2}E_y\right).$$

*19. Show that the electric field about a moving charge is not radially symmetric.

20. Show that the electric flux over a surface enclosing a charge is invariant with respect to the motion of the charge.

21. A series of particles of equal charge are distributed uniformly along a straight line and are each moving with the same speed along that straight line. A lengthy series of such particles constitutes a long, straight current. Using

the transformation equations for the electric and magnetic fields of a moving particle, show that the magnetic field near this series of moving particles is

$$B = \frac{\mu_0 I}{2\pi R}$$

where R is the perpendicular distance from the field point to the line of motion of the particles.

22. Write an essay on the application of the pinch effect to the development of fusion power reactors.

23. The oceans of the earth contain about 10^{21} kg of water in which there is one atom of deuterium ^2_1H for every 6000 atoms of hydrogen ^1_1H. What amount of energy could be harvested if all this deuterium could be extracted and reacted in a magnetic bottle? One possible reaction would be

$$^2_1H + ^2_1H \longrightarrow ^3_2He + {}_0n$$

where

$${}_0n = 1.008665\ u$$
$$^3_2He = 3.016030\ u$$
$$^2_1H = 2.014102\ u$$
$$1\ u = 1.6604 \times 10^{-27}\ \text{kg}.$$

24. Discuss the relative merits in producing electric power by means of fossil fuel, hydro, fission, and fusion processes.

25. Show that the potential energy of a magnetic dipole placed in an external magnetic field is

$$U = -\boldsymbol{\mu} \cdot \mathbf{B}.$$

26. A coil is 5.0 cm in diameter, carries a current of 50 mA, and has 1000 turns. What torque will this coil experience if it makes an angle of 30° to a magnetic field of 0.10 Wb·m^{-2}?

27. The maximum torque on a dipole in a magnetic field is 1.0×10^{-4} N·m. How much energy is required to rotate the dipole from alignment with the field to an angle of 150° with the field?

*28. A magnetic dipole consisting of a single loop of wire of mass m and radius r carries a current I. If the dipole is rotated by a small angle from its equilibrium position in a magnetic field and then released, it oscillates in simple harmonic motion. Obtain an expression for the period of oscillation.

29. An electron with magnetic moment $\boldsymbol{\mu}_0$ in a magnetic field $\mathbf{B}$ can be in one of two states: parallel or antiparallel to the field. The probability of a system being in a state i is given by statistical arguments[19] as being proportional to $\exp(-E_i/kT)$ where E_i is the energy of state i, k is the Boltzmann constant, and T is the absolute temperature. What is the expected value of the dipole moment vector?

30. According to the Bohr model of the hydrogen atom the electron travels

[19] *MWTP*, Section 19.4.

about the nucleus at a distance of 5.3×10^{-11} m at a frequency of 6.8×10^{15} rev·sec^{-1}. Calculate the magnetic field at the center of the orbit. What is the equivalent dipole moment?

*31. A quantum-mechanical picture of the hydrogen atom is that of an electron cloud rotating about a fixed proton. The charge distribution is given by

$$\rho(r) = \frac{-e}{10\pi a_0^3} \exp\left(-\frac{r}{2a_0}\right)$$

where r is the distance from the nucleus, a_0 is the radius of the first Bohr orbit, and $-e$ is the charge on an electron. Assuming that this cloud of charge rotates at the frequency of an electron in the first Bohr orbit, calculate the field due to the electron at the center and the equivalent dipole moment.

32. A beam of electrons passes through a magnetic field of 0.50 Wb·m^{-2}. Calculate the difference in energy between the two possible alignments of the electrons.

33. The quantum numbers describing the ground state of the Ce^{3+} ion are $L = 3$, $S = 1/2$, $J = 5/2$. Calculate the magnetic moment of this ion in units of the Bohr magneton. The experimental value is 2.4β. Repeat the calculation for the Gd^{3+} ion for which $L = 0$, $S = 7/2$, $J = 7/2$. The experimental value is 8.0β.

*34. Show that the Landé g-factor is given by

$$g_J = \frac{3J(J+1) - L(L+1) + S(S+1)}{2J(J+1)}.$$

35. Consider a square circuit of side l in which a current I flows and which is placed in a magnetic field. (See Fig. 7.21.) If the magnetic field is constant, the net force on the loop is zero (see Chapter 6, Problem 29). Suppose however that the field is increasing in the x direction such that $\mathbf{B}(x, y, z) = \mathbf{B}(0, 0, 0) + Bx\mathbf{i}$. Show that the net force on the circuit is given by

$$\mathbf{F} = \boldsymbol{\mu} \, \Delta B$$

where $\boldsymbol{\mu}$ is the magnetic dipole moment of the circuit.

36. Write an essay on the scientific achievements of Otto Stern.

*37. In a Stern–Gerlach experiment silver atoms were heated to 1500°K and were then passed for a distance of 15 cm through a magnetic field whose gradient was 10 Wb·m^{-3}. What will be the separation between the spots on the photographic plate? Use the fact that the average kinetic energy of the silver atoms is $(3/2)kT$ ($k = 1.38 \times 10^{-23}$ J·°K^{-1}) and assume that each atom is given the average amount of energy. What modifications might you make to the experiment so that it would not be necessary to make this approximation?

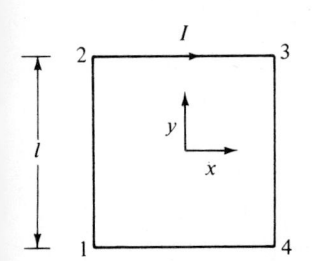

Fig. 7.21. A square circuit carrying current I in a magnetic field.

JOSEPH HENRY

Electromagnetic induction 8

8.1 INTRODUCTION

We have seen in previous chapters that a magnetic field is associated with a current. This suggests the possibility that magnetism might produce electrical effects. A method for demonstrating such effects was discovered by Michael Faraday and, independently, by Joseph Henry. Faraday, however, published his results first and the law of **electromagnetic induction** is named after him.

Faraday's law was developed after observation of the current in an electrical circuit in the presence of a varying magnetic field. Before we can proceed to a discussion of the results of such observations, however, we must discuss briefly the nature of current flow in an electrical circuit.

When current flows through a conducting medium, such as a wire, some energy is dissipated (converted to thermal energy) due to the interaction of free electrons with the crystal lattice of the conductor. This **resistance** to current flow will be discussed in detail in Chapter 11. Any real electrical circuit contains resistive elements and, therefore, must also contain a source of energy if a current is to be maintained. There are several suitable sources of energy that will produce current in electrical circuits, for example, a battery. For the present, however, we need only introduce a quantity that characterizes a source of energy, the **electromotive force**.

8.2 ELECTROMOTIVE FORCE (EMF)

The electromotive force $\mathcal{E}$ of a source of energy is defined by the equation

$$\frac{dW}{dt} = \mathcal{E}I$$

where dW/dt is the rate at which energy from the source is converted into electrical energy. Since the current I is just the rate of flow of charge dq/dt, we can also write the equation defining $\mathcal{E}$ as

$$\mathcal{E} = \frac{dW}{dq} \; \text{J} \cdot \text{C}^{-1} \quad \text{or} \quad \text{V}.$$

Therefore, the emf is the work per unit charge done by the source of energy. It is important not to confuse the emf with electric potential difference for the units are the same. The electric field **E** associated with a distribution of electric charges is conservative and hence derivable from an electric potential difference (see Section 3.6). It follows that the line integral of **E** around a closed loop is zero; that is

$$\oint \mathbf{E} \cdot d\mathbf{l} = 0.$$

SEC. 8.3 INDUCED EMF

On the other hand, the electric field $\mathbf{E}_s$ associated with an energy source is nonconservative; its line integral around a closed loop is just the emf of the circuit (see Section 11.6). That is,

$$\oint \mathbf{E}_s \cdot d\mathbf{l} = \mathcal{E}.$$

8.3 INDUCED EMF

The basic electrical effects produced by changing magnetic fields are the following:

1. In Fig. 8.1(a) a helical coil H_1 is connected to a measuring device. As the permanent magnet is thrust into the coil a current is observed to flow in the coil; as the permanent magnet is withdrawn from the coil a current is again observed to flow in the coil. The directions of the current flow are opposite in the two cases.
2. In Fig. 8.1(b) the permanent magnet is replaced by a second helical coil H_2 connected to a source of emf so that a current flows through coil H_2. If coil H_2 is moved toward or away from coil H_1, a current is observed to flow in coil H_1. The directions of the current flow are opposite in the two cases.
3. In Fig. 8.1(c) a switch is added to the circuit containing coil H_2. A current is observed to flow in coil H_1 while the switch in the circuit containing coil H_1 is in the process of closing or opening, that is, while the current is changing in the circuit containing the coil H_1. The directions of the current flow are opposite in the two cases.

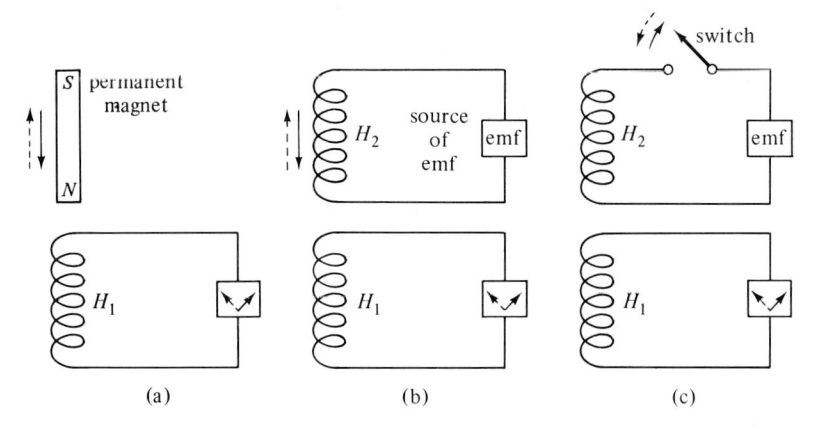

Fig. 8.1. Three different types of apparatus for demonstrating induced currents.

No current flows through coil H_1 while the switch remains closed or remains open.

In statements 1–3 we have spoken of currents produced by changing magnetic fields. Since emf is required to produce a current, we see that changing magnetic fields induce emf's in electrical circuits influenced by them. The emf, in turn, produces a current if the circuit is a closed one.

Faraday's law of electromagnetic induction

We consider now the circuit shown in Fig. 8.2. $ACPQ$ forms a rectangular circuit in which one side PQ is able to slide in the direction shown while remaining parallel to CA and preserving good electrical contact. For simplicity we assume a uniform magnetic field **B** perpendicular to the plane of the circuit, as indicated. Let us assume that PQ moves to the right at some speed $v = dx/dt$, reaching the position $P'Q'$ after time dt. Each element of charge dq in PQ will experience a force $F = vB\,dq$ in magnitude during the motion. Although it is the negative charges (electrons) in conductors which actually move, we shall speak of the positive charges as if they move since the directions of current flow and magnetic field are related by a right-hand rule (see Section 7.2, especially Fig. 7.2). The direction of the force **F** on $+dq$ is given by

$$\mathbf{F} = dq\,\mathbf{v} \times \mathbf{B}$$

and acts from P toward Q in Fig. 8.2.

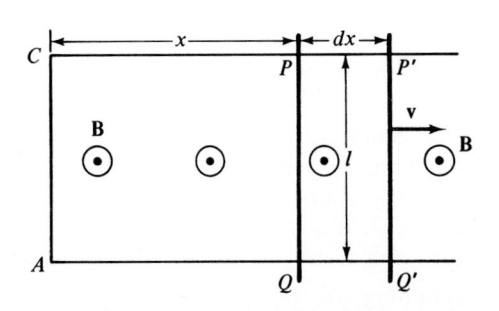

Fig. 8.2. A circuit for calculation of induced emf.

Work is done by **F** on dq as long as the motion persists. The total work dW required to move a charge dq from P to Q is

$$dW = Fl = vBl\,dq$$

or

$$\frac{dW}{dq} = vBl = Bl\frac{dx}{dt}.$$

But $dW/dq = \mathcal{E}$, an emf, by definition. Therefore, the motion produces an emf of magnitude

$$\mathcal{E} = lB\frac{dx}{dt}$$

in the circuit.

Magnetic flux is defined by the equation (Section 6.1)

$$\Phi_B = \int_S \mathbf{B} \cdot d\mathbf{S}.$$

SEC. 8.3 INDUCED EMF

Here **B** is everywhere perpendicular to the plane surface $ACPQ$ so that the magnetic flux through $ACPQ$ is simply

$$\Phi_B = Blx.$$

The change in flux through the circuit during the time dt during which the movable side slides from PQ to $P'Q'$ is

$$d\Phi_B = Bl\,dx.$$

In Fig. 8.2 the flux through the circuit is increasing with time. The (positive) current generated by the motion flows from P to Q, that is, in a clockwise direction around the loop. This clockwise induced current gives rise to a (induced) magnetic field that is directed downward into the page. The rule is that the direction of the induced current is such that the induced flux opposes the time rate of chage of the external flux. This is known as **Lenz's principle** (after Heinrich Lenz). Therefore, if the external flux is increasing, the induced emf is negative, while if the external flux is decreasing, the induced emf is positive. Lenz's principle can also be deduced by applying conservation of energy. This is left as an exercise for the reader (see Problem 3).

Therefore, we may write that the induced emf is given by

$$\mathcal{E} = -\frac{d\Phi_B}{dt}$$

where $\mathcal{E}$ is in volts, Φ_B is in **webers** (Wb), and the minus sign has been inserted in order to specify the direction of the induced current. The unit of flux in the MKS system, the weber, was chosen in order to ensure a constant of proportionality equal to unity in the equation for the induced emf.

If the magnetic flux passes through N identical loops or circuits (for example, a helix) rather than a single loop, then the emf is given by

$$\mathcal{E} = -N\frac{d\Phi_B}{dt}.$$

This latter equation is the general form for **Faraday's law of electromagnetic induction**.

In the above discussion we considered the external field to be constant while the change in flux through the circuit was due to change in dimensions of the circuit. In experiments 1–3 above, the dimensions of the helices were constant and the change in flux was due to change in magnetic field due to motion of the source. According to Faraday's law, the induced emf is dependent only on the time rate of change of flux linking a circuit and it matters not whether it is the external field or the dimensions of the circuit, or both, that is varying with time.

Example. A simple electric generator consists of an N-turn coil of cross-sectional area A in a uniform magnetic field **B** (see Fig. 8.3). The coil is rotated

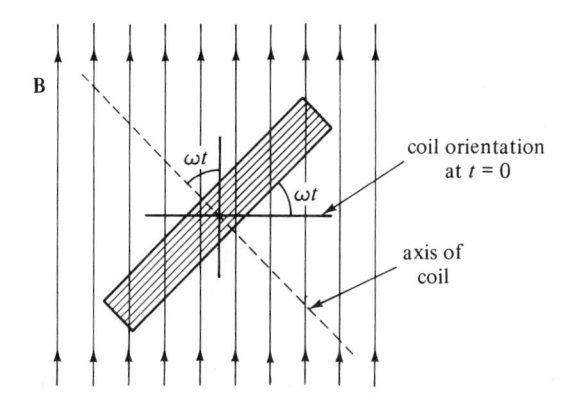

Fig. 8.3. An N-turn coil is rotated in a uniform magnetic field at constant angular frequency ω.

about a diameter at constant angular frequency ω. Calculate the emf induced in the coil. Assume that at $t = 0$ the plane of the coil is perpendicular to **B**.

Solution. At time t the axis of the coil makes an angle ωt with the direction of **B**. The magnetic flux through one turn of the coil is

$$\Phi_B = AB \cos \omega t.$$

The induced emf is

$$\mathcal{E} = -N\frac{d\Phi_B}{dt} = \omega NAB \sin \omega t = \mathcal{E}_0 \sin \omega t.$$

That is, the induced emf in the coil varies sinusoidally in time at the frequency of rotation of the coil. Note that the amplitude of the induced emf is proportional to the frequency of rotation of the coil.

Faraday's law and the electric field

In Section 8.2 we noted that

$$\mathcal{E} = \oint \mathbf{E} \cdot d\mathbf{l}$$

where **E** is the electric field associated with the energy source. In this section we have seen that

$$\mathcal{E} = -\frac{d\Phi_B}{dt}.$$

A comparison of these results yields

$$\oint \mathbf{E}_{\text{ind}} \cdot d\mathbf{l} = -\frac{d\Phi_B}{dt}.$$

This equation contains the fundamental electromagnetic induction effect —a changing magnetic flux generates a nonconservative electric field.

We use the designation $\mathbf{E}_{\text{ind}}$ to emphasize that the electric field results from electromagnetic induction. The work done on a unit charge taken around a closed path is not zero but is equal to the emf in the loop. The electric field lines from a changing magnetic flux form closed loops.

As an illustrative example we consider a uniform magnetic field increasing in magnitude with time and confined to a cylindrical region of space of radius R (see Fig. 8.4). The symmetry of the situation tells us immediately that the electric field lines will define concentric circles. From Lenz's principle the direction of the induced electric field is clockwise.

For a circular path of radius $r > R$

$$E_{\text{ind}}(2\pi r) = -\left(\frac{d\Phi_B}{dt}\right)_r = -\left(\frac{d\Phi_B}{dt}\right)_R$$

or

$$E_{\text{ind}} = -\frac{1}{2\pi r}\left(\frac{d\Phi_B}{dt}\right)_R.$$

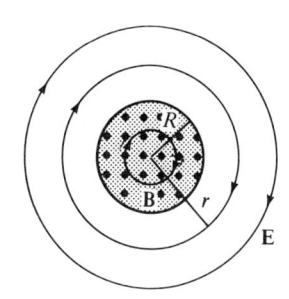

Fig. 8.4. A uniform magnetic field pointing out of the page and increasing in magnitude with time produces concentric induced electric field lines.

Note that $(d\Phi_B/dt)_R$ represents the total rate of magnetic flux change within the region of radius R. For a circular path of radius $r < R$

$$E_{\text{ind}}(2\pi r) = -\left(\frac{d\Phi_B}{dt}\right)_r = -\left(\frac{r}{R}\right)^2\left(\frac{d\Phi_B}{dt}\right)_R$$

or

$$E_{\text{ind}} = -\frac{1}{2\pi r}\left(\frac{r}{R}\right)^2\left(\frac{d\Phi_B}{dt}\right)_R = -\frac{r}{2\pi R^2}\left(\frac{d\Phi_B}{dt}\right)_R.$$

These results are shown graphically in Fig. 8.5.

8.4 THE BETATRON

The acceleration of charged particles by means of the electric field induced by a changing magnetic field was first realized in the **betatron**, a machine developed in 1941 by Donald W. Kerst. Such a machine had been suggested in principle as early as 1928 but difficulties in magnet design were not overcome until Kerst built his 2 MeV machine. The betatron received its name from the fact that it was designed to accelerate electrons (at that time usually called beta particles). In a betatron the electrons travel in circular orbits in an evacuated, glass, doughnut-shaped tube located between the poles of an electromagnet as indicated in Fig. 8.6. The electron's orbit is a circle perpendicular to the page. The magnetic field also produces the radial force required to maintain the electrons in circular orbits of constant radius. The electrons are injected

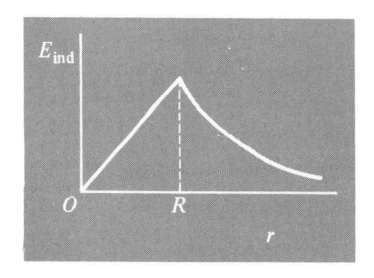

Fig. 8.5. Induced electric field as a function of the distance from the center of the cylindrical field for the physical situation depicted in Fig. 8.4.

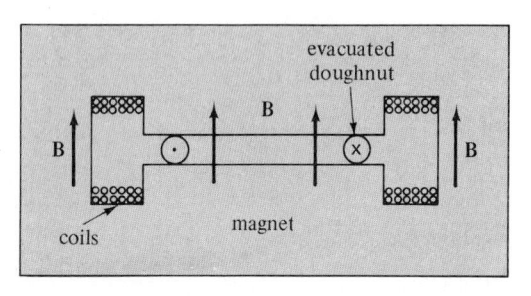

Fig. 8.6. Cross-sectional view of a betatron.

Fig. 8.7. The magnetic flux through the orbit of the betatron during one cycle of the current in the coils.

into the tube at the right ($\times$) and extracted at the left ($\cdot$). The current in the coils is altered sinusoidally, typically at 60 Hz, to produce a time-varying flux Φ_B through the orbit. If the electrons are to circulate and be accelerated in the counter-clockwise sense as viewed from above, then Φ_B must be increasing and in the correct direction. Let us consider Fig. 8.7 in which Φ_B is taken as positive when **B** has the direction indicated in Fig. 8.6. Let us assume that acceleration takes place during the first quarter cycle labeled AB, during which Φ_B is increasing. According to Lenz's principle the induced electric field is directed out of the page at the right ($\times$) and into the page at the left ($\cdot$) of Fig. 8.6. Since an electron is negatively charged, it experiences a force in the direction opposite to the induced electric field. Therefore, the electrons will experience a tangential force acting in their direction of motion and they will be accelerated during the one-quarter cycle AB.

In order for an electron to travel in a circular path of fixed radius r its momentum p at any instant must satisfy the condition

$$p = Ber$$

where B is the magnetic field at radius r at the same instant. Since the electron experiences a tangential force F, its subsequent acceleration will be governed by Newton's law, that is,

$$F = \frac{dp}{dt} = er\frac{dB}{dt}.$$

The force F is related to the time-varying magnetic flux $d\Phi_B/dt$ through Faraday's law and

$$F = eE = \frac{e\mathcal{E}}{2\pi r} = \left(\frac{e}{2\pi r}\right)\frac{d\Phi_B}{dt}.$$

(The negative sign is omitted in the last step since we are concerned with the magnitude of the force only; we have previously specified the direction.) The total magnetic flux Φ_B through the circle of radius r is

$$\Phi_B = \pi r^2 \bar{B}$$

where $\bar{B}$ is the average magnetic field over the circle. Therefore,

$$F = \left(\frac{e}{2\pi r}\right)\pi r^2 \frac{d\bar{B}}{dt}.$$

Equating the two expressions for F in terms of the time variation of the magnetic fields B and $\bar{B}$ gives

$$\frac{dB}{dt} = \frac{1}{2}\frac{d\bar{B}}{dt}$$

or, since the magnetic field changes at the same rate at all points,

$$B = \frac{1}{2}\bar{B}.$$

That is, in order for the electrons to travel at constant radius r independent of their speed, the magnetic field at the location of the electron beam must be exactly one-half the average magnetic field over the region within the orbit. It is important to note that this derivation is applicable even for speeds close to the speed of light. The pole pieces of the magnet are constructed to provide a relatively strong central field and a weaker one at the position of the orbit to satisfy this condition.

When the magnetic flux reaches its maximum value (point B in Fig. 8.7), the electron orbit is disturbed by a current sent through an auxiliary winding on the magnet. In this manner an external beam of high-energy electrons may be obtained. The limitation on the maximum energy that may be achieved is set by the loss of energy through radiation. Whenever a charged particle is accelerated, it radiates energy (see Section 9.8). When the energy loss per revolution through radiation equals the gain through the induction acceleration mechanism, the electrons cannot be further accelerated. The problem can be alleviated somewhat by increasing the radius of the orbit. However, the size of the magnet soon becomes a new limitation on the maximum available energy.

In a 100 MeV betatron a typical value of $d\Phi_B/dt$ is 430 V so that an electron will increase its energy by 430 eV per revolution. Therefore, an electron must make about 2.3×10^5 revolutions to gain 100 MeV. At this energy the electron speed is 0.999987 c.

8.5 THE MAGNETIC FIELD OF THE EARTH

To a first approximation the shape of the earth's magnetic field is illustrated in Fig. 8.8 with the axis for the field aligned approximately along the axis of rotation of the earth. Such a field can be generated by the presence of circular electric currents closed upon themselves and flowing in the interior of the earth. Such currents can be ac-

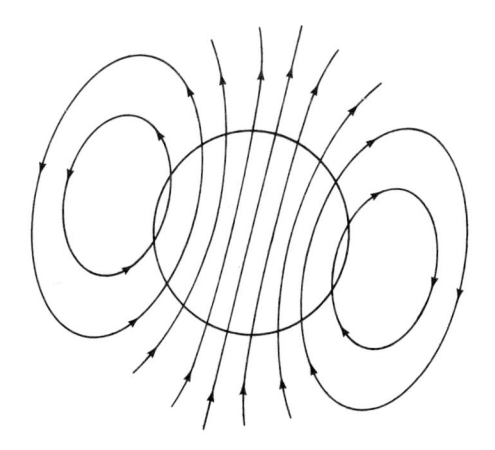

Fig. 8.8. Approximate nature of the earth's magnetic field.

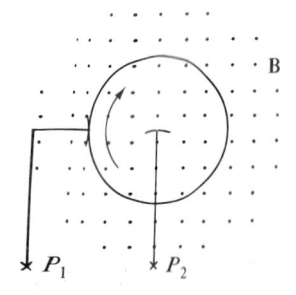

Fig. 8.9. A Faraday disc dynamo.

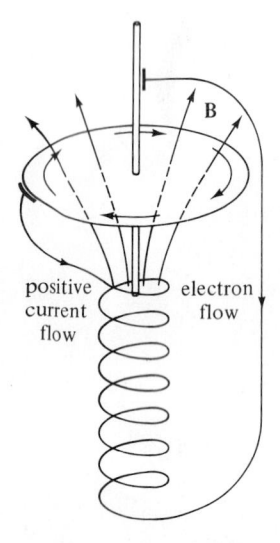

Fig. 8.10. The current generated by the disc dynamo can be used to provide a self-contained system for generating current merely by spinning the disc.

counted for by using a dynamo theory with the earth's rotation providing the driving force.[1]

To begin we consider the simple disc dynamo invented by Michael Faraday to convert the energy of mechanical motion into electric current. A schematic view of Faraday's machine is shown in Fig. 8.9. A copper disc is mounted with its axis parallel to a uniform magnetic field **B**. As the disc rotates an emf appears between points P_1 and P_2 which are connected to brushes, as shown, one making contact with the axis of the disc and the other with its circumference.

If we could use the current induced in the disc to produce the magnetic field **B**, we would have a self-contained system for generating current simply by spinning the disc (see Fig. 8.10). We would require a current in the coil initially so that there is a field present when the disc commences to spin. Once the emf is established in the spinning disc, the induced current alone could maintain the magnetic field. Unfortunately, this scheme is rendered impractical by the finite resistance of the disc to the flow of induced current. The very weak induced current in a disc of small diameter in unable to sustain itself even in a copper disc because of energy loss due to the resistance of the conductor. However, such a scheme could be made self-sustaining by scaling up the diameter of the disc to the order of the diameter of the earth's core—the bigger such a dynamo, the slower the speed of rotation required to make it self-sustaining.

Seismological studies tell us that the interior of the earth contains a liquid core (see Section 4.4), the main constituent of which is iron. That is, the earth's core is just the kind of medium required for the Faraday type of dynamo—it allows both mechanical motion and current flow.

Mechanical motions in the core could arise from small differences in the chemical constitution of different parts of the core, or from convection currents arising due to the flow of heat outward from the center of the earth, or because the earth's interior has not yet reached an equilibrium state. If motions do occur, variations in the earth's magnetic field are to be expected. Indeed, the earth's field is found to vary with time in a complicated manner. From the measured changes in the field it has been deduced that the material in the core moves at a rate of approximately 3 cm·sec^{-1}. It is thought that the more or less random motions within the core produce localized regions of current flow and that these localized regions have their orientations ordered by the earth's rotation (as depicted schematically in Fig. 8.11), thereby giving rise to a single current flowing in a large circle around the entire core. This current then generates a reasonably stable magnetic field of the form indicated in Fig. 8.8. The

---

[1]W. M. Elsasser, "The Earth as a Dynamo," *Scientific American*, May, 1958. Available as *Scientific American Offprint 825* (San Francisco: W. H. Freeman and Co., Publishers).

more or less random growth and decay of the localized regions of current cause the earth's field to vary in an unpredictable fashion.

It should be noted that additional transient variations of the earth's magnetic field result from currents induced in the earth's upper atmosphere by such things as solar atmospheric tides.[2]

Finally, it might be noted that the direction of the earth's magnetic field has reversed nine times in the past 3.6×10^6 years.[3] It has been proposed that the earth's field has two stable states: it can point either toward the North Pole or toward the South Pole. The mechanism of reversal is not understood.

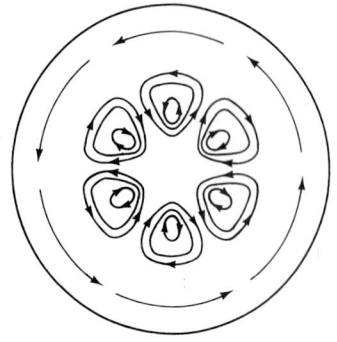

Fig. 8.11. A section through the earth's equator showing schematically the circular current in the core which is driven by localized regions of current flow and is responsible for the earth's field.

8.6 INDUCTANCE

Let us consider the two helical coils shown in Fig. 8.12. So long as the source of emf produces a constant current I flowing through coil 1 nothing unusual happens. However, if the current in coil 1 varies with time, emf's are induced **in both coils**. We shall discuss that induced in coil 1 first.

Self-inductance

The emf induced in coil 1 by a changing current through the coil is given by

$$\mathcal{E}_1 = -N_1 \frac{d\Phi_B(1)}{dt}$$

where $\Phi_B(1)$ is the flux linking coil 1. Since $\Phi_B(1)$ is produced by current I flowing through coil 1, we can write

$$N_1 \Phi_B(1) = LI$$

where L is a constant. The equation for the emf is then written as

$$\mathcal{E}_1 = -L \frac{dI}{dt};$$

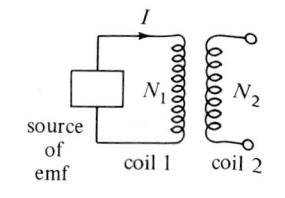

Fig. 8.12. An arrangement of coils for the discussion of inductance.

---
[2]S. H. Hall, "The Magnetic Field of the Ionospheric Dynamo," *Contemporary Physics*, **7** (November, 1966), 430.
[3]A. Cox, G. B. Dalrymple, and R. R. Doell, "Reversals of the Earth's Magnetic Field," *Scientific American*, February, 1967.

the constant L is known as the **self-inductance** (often just called **inductance**) of coil 1. The numerical value of L depends only on the geometry of the coil and the material contained within it. Since the induced emf produces a current that opposes the changing current in the coil (Lenz's principle), it is sometimes called the **back emf**.

Example. Calculate the self-inductance per unit length of an infinitely long solenoid having cross-sectional area A and n turns per unit length.

Solution. The magnetic field B within the solenoid is

$$B = \mu_0 n I$$

where I is the current flowing in the coil (see Section 7.3). The magnetic flux Φ_B through each turn is

$$\Phi_B = BA = \mu_0 n I A.$$

The self-inductance L per unit length of the coil is therefore

$$L = \frac{n\Phi_B}{I} = \mu_0 n^2 A.$$

Note that the self-inductance is proportional to the square of the number of turns per unit length; this behavior is characteristic of coils, regardless of their configuration in space.

Mutual-inductance

We now consider coil 2 of Fig. 8.12. The emf induced in coil 2 by a variable current I flowing through coil 1 is

$$\mathcal{E}_2 = -N_2 \frac{d\Phi_B(2)}{dt}$$

where $\Phi_B(2)$ is the flux linking coil 2. Since this flux is produced by the current I flowing in coil 1, we can write

$$N_2 \Phi_B(2) = MI$$

where M is a constant. The equation for the emf then becomes

$$\mathcal{E}_2 = -M \frac{dI}{dt};$$

the constant M is known as the **mutual-inductance**. It is characteristic of the pair of coils and it is immaterial in which of the coils the current is flowing.

Example. Deduce the mutual-inductance of two parallel coils of radii r_1 and r_2 having N_1 and N_2 turns, respectively, and separated by distance d which is large compared to both r_1 and r_2.

Solution. (See Fig. 8.13.) If a current I flows in the coil of radius r_1, then a magnetic field B is established at the center of the coil of radius r_2 where

$$B = \frac{\mu_0 I r_1^2 N_1}{2d^3}$$

(see Section 7.7). Since $d \gg r_1, r_2$, it follows that to a reasonable approximation the magnetic field through the coil of radius r_2 will be uniform and the magnetic flux $\Phi_B(2)$ linking coil 2 will be given by

$$\Phi_B(2) = \pi r_2^2 B$$
$$= \frac{\mu_0 \pi N_1 r_1^2 r_2^2 I}{2d^3}.$$

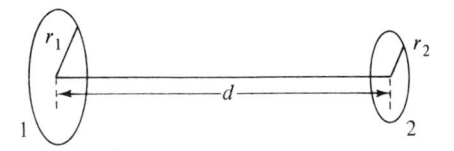

Fig. 8.13. Two parallel coils separated by distance d.

The mutual-inductance M is therefore

$$M = \frac{N_2 \Phi_B(2)}{I} = \frac{\mu_0 \pi N_1 N_2 r_1^2 r_2^2}{2d^3}.$$

Unit of inductance

The unit of inductance is the **henry** (H). A single coil has a self-inductance (or simply, inductance) of 1 henry if a current through it changing at the rate of 1 ampere per second gives rise to an induced emf of 1 volt. The mutual-inductance of a pair of coils (or circuits) is 1 H if a current changing in one of the coils at a rate of 1 A·sec^{-1} gives rise to an emf of 1 V in the second coil.

Example. Calculate the approximate inductance of a 5000-turn solenoid of length $l = 50$ cm and radius 3 cm.

Solution. Using the formula derived earlier in this section for an infinite solenoid, we have

$$L = \mu_0 n^2 Al = \mu_0 \left(\frac{N}{l}\right)^2 (\pi r^2) l$$
$$= 4\pi \times 10^{-7} \times (10^4)^2 \times \pi \times (3 \times 10^{-2})^2 \times 0.5$$
$$= 0.18 \text{ H}.$$

8.7 ENERGY STORED IN A MAGNETIC FIELD

If we increase the current flowing through a coil from zero, a back emf

$$\varepsilon = -L\frac{dI}{dt}$$

is produced which opposes the rise in current. In time dt, the emf which is establishing the current must do work

$$dW = -\varepsilon \, dq = -\varepsilon I \, dt = LI \, dI$$

against the opposing back emf. The total work required to raise the current to I amperes is

$$W = \int dW = L \int_0^I I\, dI = \frac{1}{2} L I^2.$$

Since this work is done in building up a magnetic field in the coil, it can be considered as **magnetic energy**, that is, energy stored in a magnetic field. This is similar to the electrical case (Section 5.6).

We shall now develop an expression for the energy density in a magnetic field. We shall do this with reference to a toroidal coil since the calculation is quite easy. The result, however, is a general one, independent of the form of the magnetic field under consideration.

The inductance of the coil is, by definition,

$$L = \frac{N\Phi_B}{I} = \frac{N}{I} \int_S B\, dS.$$

The magnetic field within a toroid of large radius and small cross section is (see Section 7.3)

$$B = \mu_0 n I$$

where $n = N/2\pi r =$ number of turns per unit length of the toroid. Therefore, to a good approximation,

$$\int_S B\, dS = \mu_0 n I A$$

where A is the cross-sectional area of the toroid, and the inductance of the toroid is

$$L = \frac{N}{I} \mu_0 n I A = \mu_0 n^2 A l$$

where $l = 2\pi r$ is the length of the toroid.

The energy stored in the coil is

$$U = \frac{1}{2} L I^2 = \frac{1}{2} \mu_0 n^2 A l I^2$$

so that the energy density u_B is

$$u_B = \frac{U}{Al} = \frac{1}{2} \mu_0 n^2 I^2 = \frac{1}{2} \mu_0 \frac{B^2}{\mu_0^2} = \frac{1}{2} \frac{B^2}{\mu_0} \ \text{J} \cdot \text{m}^{-3}.$$

The energy density in a magnetic field is proportional to the square of the field. This can be compared to the energy density in an electric field which is (see Section 5.6)

$$u_E = \frac{1}{2} \epsilon_0 E^2.$$

The total energy density in a region in which both electric and magnetic fields are present is

$$u = u_E + u_B = \frac{1}{2}\left(\epsilon_0 E^2 + \frac{B^2}{\mu_0}\right).$$

Example. Much larger amounts of energy can be stored in magnetic fields than in electric fields of magnitudes that are readily obtainable in the laboratory. To illustrate this fact compare the energy per unit volume in a uniform magnetic field of 1.0 Wb·m^{-2} with that in a uniform electric field of 1.0×10^5 V·m^{-1}.

Solution. For the magnetic case

$$u_B = \frac{B^2}{2\mu_0} = \frac{(1.0)^2}{2 \times 4\pi \times 10^{-7}}$$
$$= 4.0 \times 10^5 \text{ J}.$$

For the electric case

$$u_E = \frac{1}{2}\epsilon_0 E^2 = 0.5 \times 8.9 \times 10^{-12} \times (1.0 \times 10^5)^2 = 4.5 \times 10^{-2} \text{ J}.$$

In this example the energy stored in the magnetic field exceeds that stored in the electric field by a factor of 10^7.

8.8 OSCILLATIONS IN *LC* CIRCUITS

A combination of capacitors and inductors in an electrical circuit possesses a characteristic frequency of oscillation. In this section we shall discuss the behavior of the very simple circuit of Fig. 8.14 consisting of a single capacitor C and a single inductor L. The effect of resistance to current flow in such a circuit is discussed in Section 11.7. We suppose that the switch is initially open and that the capacitor carries a charge q_0. The energy stored in the capacitor is initially

$$U_i = \frac{1}{2}\frac{q_0^2}{C}$$

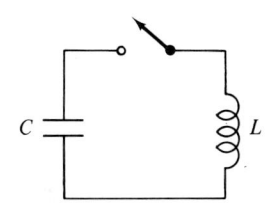

Fig. 8.14. A simple *LC* circuit.

while the energy initially stored in the inductor is zero since no current flows.

When the switch is closed, charge begins to flow from one plate of the capacitor through the inductor to the second plate. The energy stored in the inductor when the current is I is

$$U_B = \frac{1}{2}LI^2;$$

the energy stored in the capacitor at this instant is

$$U_E = \frac{1}{2}\frac{q^2}{C}$$

where $q < q_0$. The total energy at any instant is

$$U = U_B + U_E = \frac{1}{2}LI^2 + \frac{1}{2}\frac{q^2}{C}.$$

Since stored energy is conserved (no resistance),

$$\frac{dU}{dt} = \frac{1}{2}L\frac{d(I^2)}{dt} + \frac{1}{2C}\frac{d(q^2)}{dt}$$

$$= LI\frac{dI}{dt} + \frac{q}{C}\frac{dq}{dt} = 0.$$

Since

$$I = \frac{dq}{dt},$$

$$\frac{dI}{dt} = \frac{d^2q}{dt^2}.$$

Therefore,

$$L\frac{d^2q}{dt^2} + \frac{q}{C} = 0$$

or

$$\frac{d^2q}{dt^2} + \frac{1}{LC}q = 0.$$

This equation is mathematically identical to the equation for the simple harmonic motion of a mass oscillating on a spring[4]

$$\frac{d^2x}{dt^2} + \frac{k}{m}x = 0.$$

By analogy, we can immediately write down the frequency of oscillation of the LC circuit to be

$$\omega = \frac{1}{(LC)^{1/2}}.$$

The general solution of the equation for q is

$$q = q_0 \cos(\omega t + \delta)$$

where δ is the initial phase. Since we had $q = q_0$ at time $t = 0$, we must have $\delta = 0$. Therefore,

$$q = q_0 \cos \omega t.$$

The electric energy stored in the capacitor at any time t is

$$U_E = \frac{1}{2}\frac{q^2}{C} = \frac{1}{2}\frac{q_0^2}{C}\cos^2 \omega t.$$

---

[4] *MWTP*, Section 11.3.

The magnetic energy at time t is

$$U_B = \frac{1}{2}LI^2 = \frac{1}{2}L\left(\frac{dq}{dt}\right)^2$$

$$= \frac{1}{2}L\omega^2 q_0^2 \sin^2 \omega t = \frac{1}{2}\frac{q_0^2}{C}\sin^2 \omega t.$$

U_E and U_B both have maximum values of q_0^2/C which is the energy initially stored in the capacitor. During the oscillations energy is periodically transferred from the inductor to the capacitor and back again. This behavior is illustrated in Fig. 8.15. The analogy with the mass-spring system is complete.

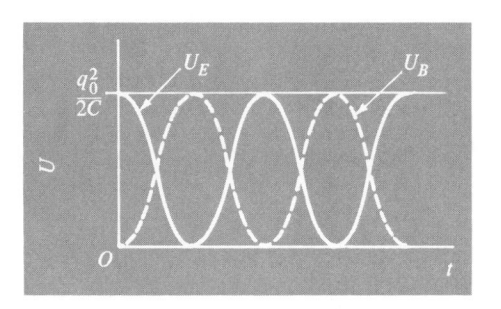

Fig. 8.15. Stored energy in an LC circuit oscillates between the inductor and the capacitor.

QUESTIONS AND PROBLEMS

1. It has been said that "it was largely because of the electrical discoveries of Michael Faraday that industry in Britain progressed so rapidly during the latter half of the nineteenth century." Discuss this statement.

2. Write a short biography of Heinrich Lenz.

3. Use the principle of conservation of energy to deduce Lenz's principle.

4. What are **eddy currents**?

5. Suppose that a conducting sheet lies in a plane perpendicular to a magnetic field **B** (see Fig. 8.16). If the magnetic field varies periodically with time at a high frequency, what will occur in the vicinity of point P?

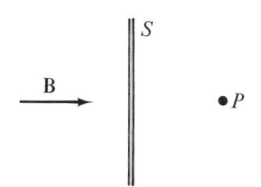

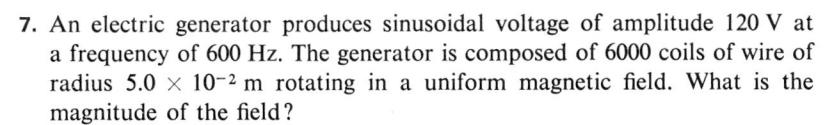

Fig. 8.16. The conducting sheet S is in the plane perpendicular to **B**.

6. A force of 5.0 N is required to move a conducting loop through a non-uniform magnetic field at a speed of 50.0 m·sec^{-1}. What power is being dissipated in the loop? How is it dissipated?

7. An electric generator produces sinusoidal voltage of amplitude 120 V at a frequency of 600 Hz. The generator is composed of 6000 coils of wire of radius 5.0×10^{-2} m rotating in a uniform magnetic field. What is the magnitude of the field?

8. A long wire lies in a plane defined by a square loop of side l. The wire is parallel to one side of the loop and is a distance s from it. Calculate the emf induced in the loop if the current in the wire is a known function of time $f(t)$. Compute the emf induced in the loop if the current is I and the loop is moving toward the wire with speed v.

9. Calculate the emf induced in a rectangular loop located in the plane defined by two parallel wires each carrying a current I that changes with time. The distances of the wires from the loop are shown in Fig. 8.17.

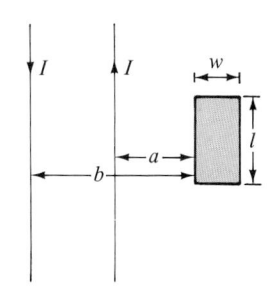

Fig. 8.17. A rectangular loop lies in the plane defined by two parallel wires, each carrying current I.

*10. A conducting rod of length L and mass m slides down a ramp formed of two conducting bars inclined at an angle θ to the horizontal as shown in Fig. 8.18. A constant magnetic field B acts downward. The bars are connected at the bottom by a resistor $R(\mathcal{E} = IR)$. Calculate the terminal velocity of the rod assuming friction.

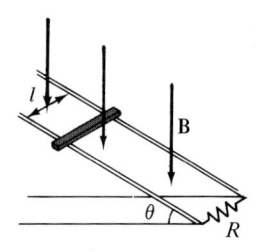

Fig. 8.18. A conducting rod slides down a ramp consisting of two conducting bars.

*11. A circular conducting loop of area A lies along the axis of a bar magnet of dipole moment μ. The loop is rotated at frequency ν about an axis through its center and perpendicular to the axis of the magnet. Calculate the emf induced in the loop. Take the distance between the center of the loop and the midpoint of the magnet to be R, where R is much larger than the diameter of the loop and the length of the bar magnet.

*12. The magnetic field in some region of space is described by $B_x = B_0 \sin(ky) \sin(\omega t)$ where $\omega = kv$. A square loop of side $\pi/2k$ with sides parallel to the y and z axes moves along the y axis with constant speed v. Calculate the emf induced in the loop if the trailing edge of the loop is at $y = 0$ at $t = 0$.

13. Discuss the use of the betatron for cancer therapy.

14. What change in the central flux in a betatron will make the electrons spiral outward?

15. A **beta-ray spectrograph** is a device, similar to a mass spectrometer, used in determining electron energies. Calculate the magnetic field strength that would be required to produce a separation of 0.5 cm between spots on a photographic plate due to beams of 50 MeV electrons and 55 MeV electrons.

16. An electron is moving in a circular orbit 1.0 m in radius in a betatron. If it gains 700 eV per orbit, calculate the rate of change of magnetic flux within the orbit.

17. The magnetic field of a betatron is uniform in a cylindrical region of space 0.50 m in diameter. The field changes at a rate of 1.0×10^{-2} Wb·m^{-2}·sec^{-1}. Calculate the electric field at points located 0.10 m, 0.50 m, and 1.0 m from the center.

18. The current through an electromagnet oscillates at a frequency of 1000 Hz producing a uniform magnetic field over an area of 2.0×10^{-2} m^2 which has a maximum magnitude of 10.0 Wb·m^{-2}. Derive an expression for the electric field as a function of time for a point 0.25 m from the axis of the magnet.

*19. A magnetic field is confined to a cylindrical region of radius R and is parallel to the axis of the cylinder at every point. How must B vary with distance r from the center in order that the electric field induced by uniform changes in B be independent of r. Assume that $B(r, t) = B'(r)B''(t)$.

20. The earth's magnetic field is known to have been changing slowly since the first observations were made of it. Paleomagnetic studies of rock samples also indicate that periodically it has reversed direction. Discuss these long-term changes with particular reference to their origin.

21. Discuss the transient variations of the earth's magnetic field that result from currents induced in the earth's upper atmosphere.

CHAP. 8   QUESTIONS AND PROBLEMS   135

22. If the disc of a Faraday disc dynamo (Fig. 8.9) rotates at a rate ω rad·sec^{-1}, find the emf appearing between points P_1 and P_2.

*23. The earth's magnetic field has a predominant dipole component that could be attributed to a dipole of moment 8.0×10^{24} C·m at the center of the earth. What circulating equatorial current at the boundary of the core would produce such a field?

24. Deduce an expression for the mutual-inductance M of two coaxial toroidal coils wound on the same form and characterized by self-inductances L_1 and L_2.

25. Consider a long solenoid of n_1 turns per unit length and a second short coil of n_2 turns per unit length wound on the same coil form but insulated from the first coil. Deduce an expression for the mutual-inductance of the two coils.

26. Toroids of rectangular cross section are sometimes constructed. What is the inductance of the toroid shown in Fig. 8.19 if it has 1500 turns of wire?

*27. Consider a long coaxial cable constructed of two concentric cylinders of radii r_1 and r_2, respectively. The central conductor carries a steady current I. Calculate the energy stored in the magnetic field for unit length of the cable. Deduce the inductance per unit length of the cable.

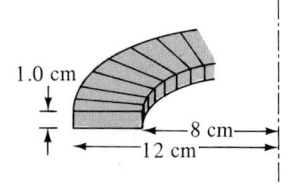

Fig. 8.19. A toroid of rectangular cross section.

28. What is the magnetic energy density a distance r from a long, straight conductor carrying current I?

29. Calculate the magnetic energy per unit length of a long solenoid carrying a current 5.0 A of radius 2.0 cm with 2000 turns per meter of length.

30. Derive an expression for the effective inductance of a series combination of two inductors.

31. Derive an expression for the effective inductance of a parallel combination of two inductors.

32. Calculate the effective inductance of the circuits shown in Fig. 8.20.

33. An inductor as used in electronic circuits is often called a **high-frequency choke**. Why?

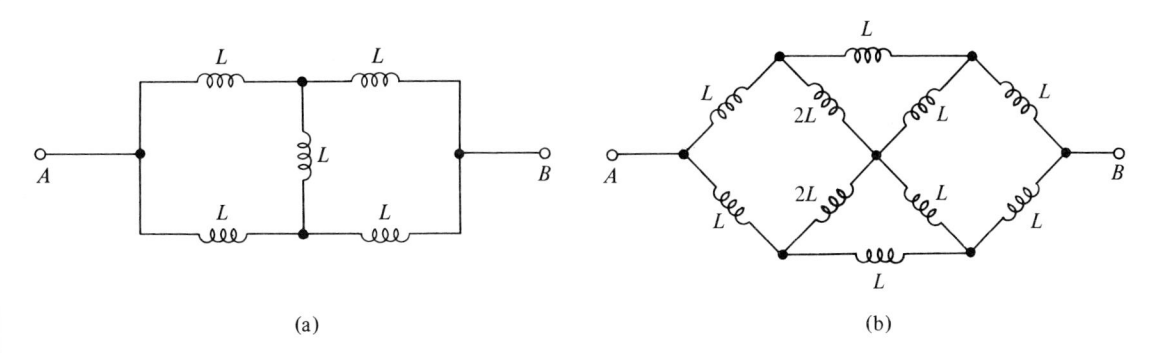

(a)   (b)

Fig. 8.20. Two circuits composed of inductors.

JAMES C. MAXWELL

9 Maxwell's equations: electromagnetic waves

9.1 INTRODUCTION

In previous chapters we have discussed three basic laws of electricity and magnetism, those of Gauss, Ampère, and Faraday. James Clerk Maxwell was the first to realize that these three laws, when written in a suitably more general mathematical form and with a necessary modification to Ampère's law to take account of the possibility of varying electric fields, yield a set of four equations from which all electric and magnetic phenomena can be derived. Maxwell's equations, which were published in 1865, predicted the existence of electromagnetic waves that travel in a vacuum with a speed which is equal to that of light in a vacuum, to within experimental limits. From this Maxwell inferred that light must consist of electromagnetic radiation. It was not until the period 1886–88 that Heinrich Hertz first produced electromagnetic waves in the laboratory and showed that they possessed the properties predicted by Maxwell.

9.2 SOME USEFUL VECTOR MATHEMATICS

In Section 3.6 we introduced the vector differential operator $\mathbf{\nabla}$ (del) defined as

$$\mathbf{\nabla} = \mathbf{i}\frac{\partial}{\partial x} + \mathbf{j}\frac{\partial}{\partial y} + \mathbf{k}\frac{\partial}{\partial z}$$

in Cartesian coordinates. When it operates directly on a scalar quantity V, it produces a vector quantity $\mathbf{\nabla}V$ ($\equiv$ grad V) called the **gradient** of V where

$$\mathbf{\nabla}V = \text{grad } V = \mathbf{i}\frac{\partial V}{\partial x} + \mathbf{j}\frac{\partial V}{\partial y} + \mathbf{k}\frac{\partial V}{\partial z}.$$

In addition, $\mathbf{\nabla}$ may operate on a vector through either a scalar (dot) or vector (cross) product. For example, writing the electric field $\mathbf{E}$ as

$$\mathbf{E} = \mathbf{i}E_x + \mathbf{j}E_y + \mathbf{k}E_z$$

we can define the quantity $\mathbf{\nabla} \cdot \mathbf{E}$ ($\equiv$ div $\mathbf{E}$), the **divergence of E**, as

$$\mathbf{\nabla} \cdot \mathbf{E} = \left(\mathbf{i}\frac{\partial}{\partial x} + \mathbf{j}\frac{\partial}{\partial y} + \mathbf{k}\frac{\partial}{\partial z}\right) \cdot (\mathbf{i}E_x + \mathbf{j}E_y + \mathbf{k}E_z)$$
$$= \frac{\partial E_x}{\partial x} + \frac{\partial E_y}{\partial y} + \frac{\partial E_z}{\partial z}.$$

We also define the quantity $\mathbf{\nabla} \times \mathbf{E}$ ($\equiv$ curl $\mathbf{E}$), the **curl** of $\mathbf{E}$ as

$$\mathbf{\nabla} \times \mathbf{E} = \left(\mathbf{i}\frac{\partial}{\partial x} + \mathbf{j}\frac{\partial}{\partial y} + \mathbf{k}\frac{\partial}{\partial z}\right) \times (\mathbf{i}E_x + \mathbf{j}E_y + \mathbf{k}E_z)$$
$$= \mathbf{i}\left(\frac{\partial E_z}{\partial y} - \frac{\partial E_y}{\partial z}\right) + \mathbf{j}\left(\frac{\partial E_x}{\partial z} - \frac{\partial E_z}{\partial x}\right)$$
$$+ \mathbf{k}\left(\frac{\partial E_y}{\partial x} - \frac{\partial E_x}{\partial y}\right).$$

There are two theorems we require in order to develop Maxwell's equations. The theorems will be stated here without proof, but they are developed in detail in Appendix A.

The divergence theorem

The divergence theorem transforms a surface integral into a volume integral. The theorem is stated in symbols as follows:

$$\int_S \mathbf{A} \cdot d\mathbf{S} = \int_v \mathbf{\nabla} \cdot \mathbf{A} \, dv.$$

The left-hand side of the equation represents the integration of the vector quantity **A** over a **closed** surface, while the right-hand side is an integral of the divergence of **A** throughout **the volume enclosed by the surface considered in the surface integral.**

Stokes' theorem

Stokes' theorem deals with the transformation of a line integral into a surface integral. In symbols, the theorem is stated as follows:

$$\int_S (\mathbf{\nabla} \times \mathbf{A}) \cdot d\mathbf{S} = \oint \mathbf{A} \cdot d\mathbf{l}.$$

The left-hand side of the equation represents the integration of the curl of a vector quantity **A** over some surface, while the right-hand side is an integral of **A around the boundary of the surface.**

9.3 GAUSS' LAW: MAXWELL'S FIRST AND FOURTH EQUATIONS

Gauss' law was developed in Chapter 4 in connection with the electric field and charge. It is possible also to apply the law to the magnetic field so that from Gauss' law come two of Maxwell's equations. We shall consider the two in turn.

Gauss' law and the electric field

Gauss' law for the general case of a continuous charge distribution is written as

$$\int_S \mathbf{E} \cdot d\mathbf{S} = \frac{1}{\epsilon_0} \int_v \rho \, dv.$$

According to the divergence theorem stated in Section 9.2 it is always possible to transform the surface integral in Gauss' law into a volume integral as follows:

$$\int_S \mathbf{E} \cdot d\mathbf{S} = \int_v \nabla \cdot \mathbf{E}\, dv$$

where the integration is throughout that volume contained by the surface. The volume integral in Gauss' law is also taken throughout the volume enclosed by the surface so that we can write

$$\int_v \nabla \cdot \mathbf{E}\, dv = \frac{1}{\epsilon_0} \int_v \rho\, dv.$$

The equality of these two integrals for an arbitrary volume implies that the integrands are identical; that is,

$$\nabla \cdot \mathbf{E} = \frac{\rho}{\epsilon_0}.$$

This is **Maxwell's first equation**.

Gauss' law and the magnetic field

We interpret the surface integral in Gauss' law as the electric flux or total number of electric lines going through the surface S (see Section 4.2). Since electric lines originate or terminate on electric charges, the number of lines, and hence the surface integral of $\mathbf{E}$, is connected with the presence of charge within the surface in question. Magnetic lines, on the contrary, are closed lines, neither beginning nor ending on magnetic "charges." It is thus not possible to construct a closed surface through which more magnetic lines emerge than enter, or vice-versa. This means that the surface integral of the magnetic field over any closed surface is zero, and Gauss' law for magnetism is

$$\int_S \mathbf{B} \cdot d\mathbf{S} = 0.$$

Application of the divergence theorem to this equation gives

$$\int_v \nabla \cdot \mathbf{B}\, dv = 0$$

which, since the volume is arbitrary, implies that

$$\nabla \cdot \mathbf{B} = 0.$$

This is **Maxwell's fourth equation**.

9.4 AMPÈRE'S LAW: MAXWELL'S SECOND EQUATION

In Section 7.3 we stated Ampère's law as

$$\oint \mathbf{B} \cdot d\mathbf{l} = \mu_0 I$$

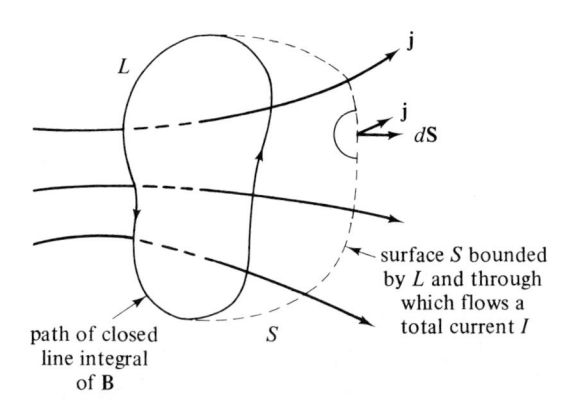

Fig. 9.1. Ampère's law relates the line integral of **B** to the total current threading the path of integration.

where I is the total current threading the closed path (see Fig. 9.1). Ampère's law may be written in an alternative form in terms of current density **j** where **j** has dimensions $A \cdot m^{-2}$ (that is, it is current per unit area). The current dI through any element of surface area $d\mathbf{S}$ of Fig. 9.1 is

$$dI = \mathbf{j} \cdot d\mathbf{S}$$

and the total current through the surface is

$$I = \int_S \mathbf{j} \cdot d\mathbf{S}.$$

The alternative form for Ampère's law becomes

$$\oint \mathbf{B} \cdot d\mathbf{l} = \mu_0 \int_S \mathbf{j} \cdot d\mathbf{S}$$

where the integral on the right is taken over any surface through which the charge flow corresponding to the current I takes place.

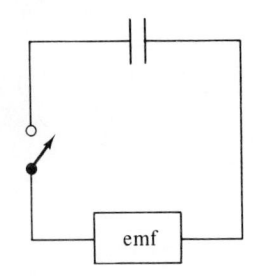

Fig. 9.2. A circuit consisting of a capacitor, a source of emf, and a switch.

Displacement current

A difficulty arises in connection with Ampère's law when an electric field that varies with time is present. The problem may be illustrated by considering the electrical circuit shown in Fig. 9.2 consisting of a capacitor, a source of emf, and a switch. When the switch is closed, current begins to flow in the circuit. The plates of the capacitor accumulate charge (equal amounts of positive and negative charge, respectively) until the accumulation becomes sufficiently large that the source of emf is no longer able to move in more charge against the repulsive force due to the charge already collected.

SEC. 9.4 AMPÈRE'S LAW: MAXWELL'S SECOND EQUATION

When the charge density on the capacitor plates is σ, the electric field is (Section 5.2)

$$E = \frac{\sigma}{\epsilon_0}$$

or

$$\sigma = \epsilon_0 E.$$

While the capacitor is accumulating charge, the charge density on the plates is changing at the rate

$$\frac{d\sigma}{dt} = \epsilon_0 \frac{dE}{dt}.$$

Maxwell showed that it was necessary to consider the quantity $d\sigma/dt$ to be a current flowing in the space between the plates of the capacitor. It is called the **displacement current**. (The reason for choosing this name is explained in Section 10.3.)

The necessity for considering a current to flow in the gap between the plates of the capacitor can be seen readily by referring to Fig. 9.3. According to Ampère's law

$$\oint \mathbf{B} \cdot d\mathbf{l} = \mu_0 \int_S \mathbf{j} \cdot d\mathbf{S} = \mu_0 \int_{S'} \mathbf{j} \cdot d\mathbf{S'} = \mu_0 I.$$

This is true only if there is a total current through S' equal to the total current I through S. Now, the displacement current is

$$\frac{d\sigma}{dt} = \epsilon_0 \frac{dE}{dt}.$$

The total current between the plates (assumed to have area A) is

$$A \frac{d\sigma}{dt} = \frac{dq}{dt} = I$$

which is the current flowing in the circuit of which the capacitor is a part.

The interpretation of the displacement current is difficult since there is no physical movement of charge associated with it. However, there are magnetic affects associated with the displacement current, as we shall see in Section 9.7, which are the same as would arise from a real current equal to the displacement current. That is, the time-varying electric field between the plates of the capacitor produces a magnetic field. This is true in general; whenever an electric field varies with time, a magnetic field is produced.

Maxwell proposed extending Ampère's law to include displacement current. The law becomes

$$\oint \mathbf{B} \cdot d\mathbf{l} = \mu_0 \int_S \left(\mathbf{j} + \epsilon_0 \frac{\partial \mathbf{E}}{\partial t} \right) \cdot d\mathbf{S}$$

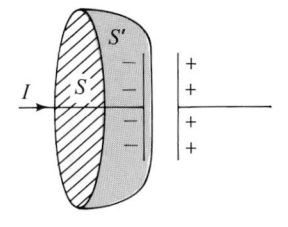

Fig. 9.3. Ampère's law is true for surfaces S and S' only if a current flows in the space between the plates of the capacitor.

which has general validity. The partial derivative is used since **E** may be a function of position as well as time.

The line integral can be transformed into a surface integral using Stokes' theorem stated in Section 9.2. According to this theorem

$$\oint \mathbf{B} \cdot d\mathbf{l} = \int_S (\mathbf{\nabla} \times \mathbf{B}) \cdot d\mathbf{S}.$$

Therefore, Ampère's law may be written as

$$\int_S (\mathbf{\nabla} \times \mathbf{B}) \cdot d\mathbf{S} = \mu_0 \int_S \left(\mathbf{j} + \epsilon_0 \frac{\partial \mathbf{E}}{\partial t}\right) \cdot d\mathbf{S}$$

which implies

$$\mathbf{\nabla} \times \mathbf{B} = \mu_0 \left(\mathbf{j} + \epsilon_0 \frac{\partial \mathbf{E}}{\partial t}\right).$$

This is Maxwell's **second equation**.

Example. A parallel-plate capacitor consists of two circular plates of radius $R = 10.0$ cm. Suppose that the capacitor is being charged at a uniform rate so that the electric field between the plates changes at the constant rate $dE/dt = 10^{13}$ V·m^{-1}·sec^{-1}. Find the displacement current for the capacitor. Derive an expression for the magnitude B of the induced magnetic field at a distance r from the center of the capacitor in a direction parallel to the plates. Evaluate B at $r = R$.

Solution. The total displacement current I_D is

$$I_D = \epsilon_0 \frac{\partial E}{\partial t} \pi R^2$$
$$= 8.9 \times 10^{-12} \times 10^{13} \times 3.14 \times (0.1)^2$$
$$= 2.8 \text{ A}.$$

From Ampère's law we can write

$$\oint \mathbf{B} \cdot d\mathbf{l} = \mu_0 \epsilon_0 \int \frac{\partial \mathbf{E}}{\partial t} \cdot d\mathbf{S}.$$

For $r \leq R$

$$B(2\pi r) = \mu_0 \epsilon_0 \frac{dE}{dt} (\pi r^2)$$

or

$$B = \frac{\mu_0 \epsilon_0}{2} r \frac{dE}{dt}.$$

For $r \geq R$

$$B(2\pi r) = \mu_0 \epsilon_0 \frac{dE}{dt} (\pi R^2)$$

or

$$B = \frac{\mu_0 \epsilon_0}{2} \frac{R^2}{r} \frac{dE}{dt}.$$

The variation of the induced magnetic field with r is shown in Fig. 9.4. The magnitude of the field at $r = R$ is

$$B = \frac{\mu_0 \epsilon_0}{2} R \frac{dE}{dt}$$
$$= \frac{1}{2} \times 4\pi \times 10^{-7} \times 8.9 \times 10^{-12} \times 0.1 \times 10^{13}$$
$$= 5.6 \times 10^{-6} \text{ Wb} \cdot \text{m}^{-2}.$$

Note that even though the displacement current is reasonably large, it produces only a small magnetic field.

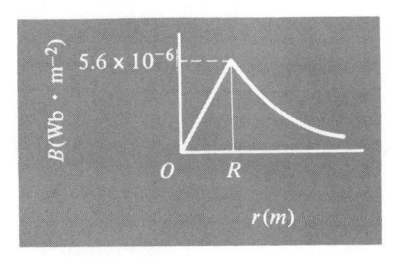

Fig. 9.4. Variation of the induced magnetic field between the plates of the capacitor due to the changing electric field.

9.5 FARADAY'S LAW: MAXWELL'S THIRD EQUATION

Faraday's law for the emf induced in a closed circuit is

$$\mathcal{E} = -\frac{d\Phi_B}{dt}.$$

But

$$\Phi_B = \int_S \mathbf{B} \cdot d\mathbf{S}$$

by definition of magnetic flux, so that

$$\mathcal{E} = -\int_S \frac{\partial \mathbf{B}}{\partial t} \cdot d\mathbf{S}$$

where the partial derivative is used since $\mathbf{B}$ may be a function of position also.

The emf is also defined as the work done per unit charge in transporting a unit charge around the closed circuit. That is,

$$\mathcal{E} = \oint \mathbf{E} \cdot d\mathbf{l}$$

where $\mathbf{E}$ is the electric field associated with the induced emf. From these two equations for $\mathcal{E}$ we obtain

$$\oint \mathbf{E} \cdot d\mathbf{l} = -\int_S \frac{\partial \mathbf{B}}{\partial t} \cdot d\mathbf{S}.$$

Applying Stokes theorem to the line integral yields

$$\oint \mathbf{E} \cdot d\mathbf{l} = \int_S (\nabla \times \mathbf{E}) \cdot d\mathbf{S}$$

so that

$$\int_S (\nabla \times \mathbf{E}) \cdot d\mathbf{S} = -\int_S \frac{\partial \mathbf{B}}{\partial t} \cdot d\mathbf{S}$$

which implies

$$\nabla \times \mathbf{E} = -\frac{\partial \mathbf{B}}{\partial t}.$$

This is Maxwell's **third equation**.

9.6 SUMMARY OF MAXWELL'S EQUATIONS

The four differential equations of Maxwell express the relationships that exist between the electric and magnetic field vectors **E** and **B**. These equations are

$$\nabla \cdot \mathbf{E} = \frac{\rho}{\epsilon_0}$$

$$\nabla \times \mathbf{B} = \mu_0 \left(\mathbf{j} + \epsilon_0 \frac{\partial \mathbf{E}}{\partial t} \right)$$

$$\nabla \times \mathbf{E} = -\frac{\partial \mathbf{B}}{\partial t}$$

$$\nabla \cdot \mathbf{B} = 0.$$

They have been developed without consideration of the electrical or magnetic properties of any material in which charge may reside. To take account of the electrical and magnetic properties of matter, Maxwell's equations must be modified, not in form, but by the introduction of two new field vectors **D** and **H** which are related to **E** and **B**, respectively. This will be carried out in Section 14.2.

In principle, any problem in electricity and magnetism including circuit theory, electrostatics, and magnetism could be treated as an application of Maxwell's equations. This is, however, an unnecessary complication at an introductory level. Maxwell's equations take on their simplest form in free space, where the charge density ρ and the current density **j** are both zero (by definition). The equations for free space are then

$$\nabla \cdot \mathbf{E} = 0$$

$$\nabla \times \mathbf{B} = \mu_0 \epsilon_0 \frac{\partial \mathbf{E}}{\partial t}$$

$$\nabla \times \mathbf{E} = -\frac{\partial \mathbf{B}}{\partial t}$$

$$\nabla \cdot \mathbf{B} = 0.$$

The second and third equations establish the connection between electric and magnetic fields. **E** and **B** are each generated by time variations of the other field vector.

9.7 ELECTROMAGNETIC WAVES

Maxwell's equations for free space point directly to the existence of electromagnetic waves. In this section we shall show this and deduce some of the properties of the waves. First, however, we shall recall the expression for a transverse sinusoidal traveling wave and show that the expression is the solution of a differential equation for traveling waves.

The wave equation

The equation of a plane transverse sinusoidal wave traveling along the x axis may be written in the form[1]

$$\Delta y(x, t) = A \sin 2\pi \left(vt - \frac{x}{\lambda} \right)$$

where $\Delta y(x, t)$ is the displacement at position x and time t in the transverse direction, v is the frequency of oscillation, and λ is the wavelength. The speed v of the wave is given by

$$v = v\lambda.$$

Let us differentiate the equation for Δy twice with respect to time t. We obtain

$$\frac{\partial^2}{\partial t^2}(\Delta y) = -(2\pi v)^2 A \sin 2\pi \left(vt - \frac{x}{\lambda} \right).$$

Now let us differentiate the equation for Δy twice with respect to position x. This yields

$$\frac{\partial^2}{\partial x^2}(\Delta y) = -\left(\frac{2\pi}{\lambda} \right)^2 A \sin 2\pi \left(vt - \frac{x}{\lambda} \right).$$

Inspection of these two derivatives shows that

$$\frac{\partial^2}{\partial t^2}(\Delta y) = v^2 \lambda^2 \frac{\partial^2}{\partial x^2}(\Delta y)$$

or

$$\frac{\partial^2}{\partial t^2}(\Delta y) = v^2 \frac{\partial^2}{\partial x^2}(\Delta y)$$

or

$$\frac{\partial^2}{\partial x^2}(\Delta y) = \frac{1}{v^2} \frac{\partial^2}{\partial t^2}(\Delta y).$$

It is left as an exercise (see Problem 19) for the reader to show that the wave equation for a transverse sinusoidal wave, when the direction of

---

[1] *MWTP*, Section 15.3.

motion is not parallel to one of the axes of a Cartesian coordinate system, is

$$\frac{\partial^2 s}{\partial x^2} + \frac{\partial^2 s}{\partial y^2} + \frac{\partial^2 s}{\partial z^2} = \frac{1}{v^2}\frac{\partial^2 s}{\partial t^2}$$

where $s = s(x, y, z, t)$ is the displacement at position (x, y, z) at time t.

We recall now the vector differential operator

$$\nabla = \mathbf{i}\frac{\partial}{\partial x} + \mathbf{j}\frac{\partial}{\partial y} + \mathbf{k}\frac{\partial}{\partial z}.$$

The scalar product of ∇ with itself is

$$\nabla \cdot \nabla = \nabla^2 = \left(\mathbf{i}\frac{\partial}{\partial x} + \mathbf{j}\frac{\partial}{\partial y} + \mathbf{k}\frac{\partial}{\partial z}\right) \cdot \left(\mathbf{i}\frac{\partial}{\partial x} + \mathbf{j}\frac{\partial}{\partial y} + \mathbf{k}\frac{\partial}{\partial z}\right)$$
$$= \frac{\partial^2}{\partial x^2} + \frac{\partial^2}{\partial y^2} + \frac{\partial^2}{\partial z^2}.$$

We see immediately that the wave equation can be written as

$$\nabla^2 s = \frac{1}{v^2}\frac{\partial^2 s}{\partial t^2}.$$

Electromagnetic waves

Maxwell's third equation is

$$\nabla \times \mathbf{E} = -\frac{\partial \mathbf{B}}{\partial t}.$$

Let us take the curl of both sides of this equation to obtain

$$\nabla \times (\nabla \times \mathbf{E}) = -\frac{\partial}{\partial t}(\nabla \times \mathbf{B})$$
$$= -\mu_0 \epsilon_0 \frac{\partial^2 \mathbf{E}}{\partial t^2}$$

where we have used Maxwell's second equation. We can reduce the left-hand side of this equation by using the vector identity (see Problem 2)

$$\nabla \times (\nabla \times \mathbf{E}) = \nabla(\nabla \cdot \mathbf{E}) - \nabla^2 \mathbf{E}$$

and noting that the first term on the right-hand side of the identity is zero from Maxwell's first equation. Therefore,

$$\nabla^2 \mathbf{E} = \mu_0 \epsilon_0 \frac{\partial^2 \mathbf{E}}{\partial t^2}.$$

A similar calculation leads to the equation

$$\nabla^2 \mathbf{B} = \mu_0 \epsilon_0 \frac{\partial^2 \mathbf{B}}{\partial t^2}$$

for the magnetic field.

These equations for $\mathbf{E}$ and $\mathbf{B}$ both have the form of the wave equation developed above and we can conclude immediately that there exists plane, transverse, electromagnetic waves of speed

$$v = (\mu_0 \epsilon_0)^{-1/2}.$$

Since the frequency or wavelength of the wave does not appear in the wave equation, we conclude that all electromagnetic waves regardless of frequency have this same speed.

Since ϵ_0 and μ_0 are constants that can be determined experimentally, we can calculate the speed of the waves. Using the values

$$\epsilon_0 = 8.8538 \times 10^{-12} \text{ C}^2 \cdot \text{N}^{-1} \cdot \text{m}^{-2}$$
$$\mu_0 = 4\pi \times 10^{-7} \text{ Wb} \cdot \text{A}^{-1} \cdot \text{m}^{-1}$$

we obtain

$$v = (4\pi \times 10^{-7} \times 8.8538 \times 10^{-12})^{-1/2}$$
$$= 2.9980 \times 10^8 \text{ m} \cdot \text{sec}^{-1}.$$

The speed of light in vacuum is 2.9979×10^8 m·sec^{-1} which is the same as the predicted speed of electromagnetic waves within the accuracy of this calculation. A similar calculation prompted Maxwell to declare that light is a special case of propagation of electromagnetic waves. The speed of electromagnetic waves in vacuum is denoted by c.

It was more than twenty years after Maxwell predicted the existence of electromagnetic waves that Heinrich Hertz produced them in the laboratory and showed that they had just the properties predicted of them.[2] In a sense the subsequent invention of wireless telegraphy, radio, and television were anticlimactic—the practical results of a basic discovery!

The relation between **E** and **B** in the wave

Obvious solutions to the wave equations for **E** and **B** are the following:

$$\mathbf{E} = \mathbf{E}_0 \sin 2\pi \left(vt - \frac{x}{\lambda} \right)$$

$$\mathbf{B} = \mathbf{B}_0 \sin 2\pi \left(vt - \frac{x}{\lambda} \right)$$

where $c = v\lambda$ and $\mathbf{B}_0$ and $\mathbf{E}_0$ are constant vectors. We can take the curl of the second equation and obtain

$$\nabla \times \mathbf{B} = \nabla \times \left[\mathbf{B}_0 \sin 2\pi \left(vt - \frac{x}{\lambda} \right) \right].$$

The operator ∇ operates on both $\mathbf{B}_0$ and the sine term in the brackets.

---

[2] M. M. Shamos [ed.], *Great Experiments in Physics* (New York: Holt, Rinehart & Winston, Inc., 1960), Chapter 13.

We use the vector identity (see Problem 3)
$$\nabla \times (\mathbf{B}_0 \phi) = \phi (\nabla \times \mathbf{B}_0) - \mathbf{B} \times \nabla \phi$$
to write
$$\nabla \times \mathbf{B} = \sin 2\pi \left(vt - \frac{x}{\lambda} \right) (\nabla \times \mathbf{B}_0) - \mathbf{B}_0 \times \nabla \left[\sin 2\pi \left(vt - \frac{x}{\lambda} \right) \right].$$

Since $\mathbf{B}_0$ is a constant vector, $\nabla \times \mathbf{B}_0 = 0$ and
$$\nabla \times \mathbf{B} = -\mathbf{B}_0 \times \nabla \left[\sin 2\pi \left(vt - \frac{x}{\lambda} \right) \right].$$

Now
$$\nabla \left[\sin 2\pi \left(vt - \frac{x}{\lambda} \right) \right] = \left(\mathbf{i} \frac{\partial}{\partial x} + \mathbf{j} \frac{\partial}{\partial y} + \mathbf{k} \frac{\partial}{\partial z} \right) \left[\sin 2\pi \left(vt - \frac{x}{\lambda} \right) \right]$$
$$= -\mathbf{i} \frac{2\pi}{\lambda} \cos 2\pi \left(vt - \frac{x}{\lambda} \right)$$

so that
$$\nabla \times \mathbf{B} = \mathbf{B}_0 \times \mathbf{i} \frac{2\pi}{\lambda} \cos 2\pi \left(vt - \frac{x}{\lambda} \right) = \mu_0 \epsilon_0 \frac{\partial \mathbf{E}}{\partial t}.$$

The vector $\mathbf{E}$ is perpendicular to both $\mathbf{B}_0$ (or $\mathbf{B}$) and $\mathbf{i}$, which is a unit vector in the direction of propagation. Therefore, $\mathbf{E}$ is perpendicular to the direction of propagation. A similar development shows that $\mathbf{B}$ is perpendicular to both $\mathbf{E}$ and the direction of propagation. Both $\mathbf{E}$ and $\mathbf{B}$ are normal to the direction of propagation so that the wave is transverse. This wave is pictured in Fig. 9.5.

Similar results may be obtained for propagation in the y and z directions. In any event, for a plane electromagnetic wave, it is always possible to choose the coordinate axes so that one of the axes lies along the direction of propagation. The wave then has no component along this axis. If it is possible to choose one of the remaining two axes in such a way that the vector $\mathbf{E}$ is parallel to it, the vector $\mathbf{B}$ is then parallel to the remaining axis. This type of wave in which the $\mathbf{E}$ and $\mathbf{B}$ vectors are always parallel to a fixed direction is called a **plane-polarized wave**. Unpolarized plane waves involve a superposition of polarized waves in which $\mathbf{E}$ and $\mathbf{B}$ are randomly distributed in the plane perpendicular to the direction of propagation.

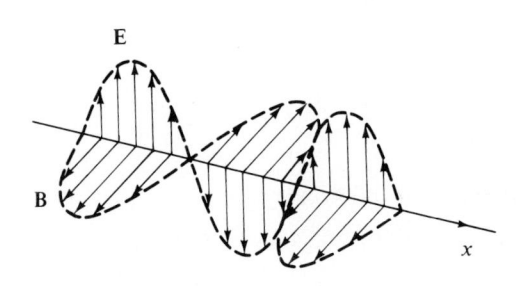

Fig. 9.5. A schematic representation of a plane electromagnetic wave traveling in the x direction.

The relation between the amplitudes $\mathbf{E}_0$ and $\mathbf{B}_0$ of the plane wave can be obtained by differentiating the expression for $\mathbf{E}$ with respect to time to obtain
$$\frac{\partial \mathbf{E}}{\partial t} = 2\pi v \mathbf{E}_0 \cos 2\pi \left(vt - \frac{x}{\lambda} \right)$$

SEC. 9.7 ELECTROMAGNETIC WAVES

$$= \frac{1}{\mu_0 \epsilon_0} \mathbf{B}_0 \times \mathbf{i} \frac{2\pi}{\lambda} \cos 2\pi \left(vt - \frac{x}{\lambda} \right).$$

The relation between the magnitudes of $\mathbf{E}_0$ and $\mathbf{B}_0$ is then

$$vE_0 = \frac{B_0}{\mu_0 \epsilon_0 \lambda}$$

or

$$\frac{E_0}{B_0} = \frac{E}{B} = \frac{1}{(\mu_0 \epsilon_0)^{1/2}} = c.$$

Energy flow

The direction of propagation of an electromagnetic wave is the direction $\mathbf{E} \times \mathbf{B}$. The energy density in free space when electric and magnetic fields are present is (see Section 8.7)

$$u = \frac{1}{2} \left(\epsilon_0 E^2 + \frac{B^2}{\mu_0} \right).$$

Energy must be propagated through space by an electromagnetic wave since energy is stored in electric and magnetic fields. We can obtain an expression for the time variation of the energy density by taking the divergence of the quantity $\mathbf{E} \times \mathbf{B}$ and using the following vector identity (see Problem 4)

$$\nabla \cdot (\mathbf{E} \times \mathbf{B}) = \mathbf{B} \cdot (\nabla \times \mathbf{E}) - \mathbf{E} \cdot (\nabla \times \mathbf{B}).$$

Substituting from Maxwell's second and third equations gives

$$\nabla \cdot (\mathbf{E} \times \mathbf{B}) = -\mathbf{B} \cdot \frac{\partial \mathbf{B}}{\partial t} - \mu_0 \epsilon_0 \mathbf{E} \cdot \frac{\partial \mathbf{E}}{\partial t}$$

$$= -\mu_0 \frac{\partial}{\partial t} \left(\frac{\epsilon_0 E^2}{2} + \frac{B^2}{2\mu_0} \right)$$

$$= -\mu_0 \frac{\partial u}{\partial t}$$

or

$$\nabla \cdot \left(\mathbf{E} \times \frac{\mathbf{B}}{\mu_0} \right) = -\frac{\partial u}{\partial t}.$$

This equation gives the rate at which the energy density is decreasing.

If we consider a specific volume v, then the stored energy in this volume is decreasing at a rate given by

$$\int_v \nabla \cdot \left(\mathbf{E} \times \frac{\mathbf{B}}{\mu_0} \right) dv = -\frac{\partial}{\partial t} \int_v u \, dv$$

Applying the divergence theorem to the integral on the left-hand side gives

$$\int_S \left(\mathbf{E} \times \frac{\mathbf{B}}{\mu_0} \right) \cdot d\mathbf{S} = -\frac{\partial}{\partial t} \int_v u \, dv$$

The vector $\mathbf{E} \times \mathbf{B}/\mu_0$ is a vector representing energy flow, for integrating it over a closed surface gives us the rate at which energy is being lost from the volume within the surface. Since the vector $\mathbf{E} \times \mathbf{B}/\mu_0$ points in the direction in which the electromagnetic wave is traveling, an electromagnetic wave transports energy in the direction in which it is traveling.

Example. The intensity of solar radiation at a point in space that is at the mean distance of the earth from the sun is 1.35×10^3 J·m^{-2}·sec^{-1}. If radiation from the sun were all at one wavelength, what would be the amplitude of electromagnetic waves from the sun at the position of the earth?

Solution. The vector $\mathbf{E} \times \mathbf{B}/\mu_0$ gives the energy flow in terms of the instantaneous values of $\mathbf{E}$ and $\mathbf{B}$. Since both $\mathbf{E}$ and $\mathbf{B}$ vary sinusoidally with time, their average values are $(2)^{-1/2}\mathbf{E}_0$ and $(2)^{-1/2}\mathbf{B}_0$, respectively, where $\mathbf{E}_0$ and $\mathbf{B}_0$ are the amplitudes of the electric and magnetic components of the wave. Therefore, the average value of the energy flow is

$$\frac{1}{2}\mathbf{E}_0 \times \frac{\mathbf{B}_0}{\mu_0} \text{ J·m}^{-2}\text{·sec}^{-1}.$$

Since $E_0 = cB_0$, the magnitude of the average energy flow is

$$\frac{1}{2}E_0\frac{B_0}{\mu_0} = \frac{1}{2}\frac{E_0^2}{\mu_0 c} = 1.35 \times 10^3 \text{ J·m}^{-2}\text{·sec}^{-1}.$$

Therefore,
$$E_0^2 = 2(4\pi \times 10^{-7})(3.00 \times 10^8)(1.35 \times 10^3)$$
$$= 1.02 \times 10^6$$

and
$$E_0 = 1.01 \times 10^3 \text{ V·m}^{-1}$$

is the amplitude of the electric component of the wave. The amplitude of the magnetic component is

$$B_0 = \frac{E_0}{c} = \frac{1.01 \times 10^3}{3.00 \times 10^8} = 3.37 \times 10^{-6} \text{ Wb·m}^{-2}.$$

The spectrum of electromagnetic waves

Electromagnetic waves were initially classified according to the most common source of production or method of detection of waves within a given range. The classification is not rigid since waves of a particular frequency may be produced in more than one way. The common classes of electromagnetic waves, and their methods of production, are given in Table 9.1.

9.8 THE INVARIANCE OF MAXWELL'S EQUATIONS

Since the speed of light is the same in all frames of reference, we require that Maxwell's equations be invariant under the Lorentz transfor-

SEC. 9.8 THE INVARIANCE OF MAXWELL'S EQUATIONS

Table 9.1 Classification of electromagnetic waves

Class	λ range (m)	ν range (Hz)	Method of production
radiofrequency (rf)	$>10^4$–0.3	$<3 \times 10^4$–10^8	electronic devices
microwave	0.3–10^{-3}	10^8–3×10^{11}	electronic devices
infrared	10^{-3}–7.80×10^{-7} (780 nm)	3×10^{11}–3.8×10^{14}	hot objects, molecular transitions
visible light	7.8×10^{-7}–3.80×10^{-7} (780 nm) (380 nm)	3.8×10^{11}–7.9×10^{14}	electron transitions in atoms and molecules
ultraviolet (uv)	3.8×10^{-7}–6×10^{-10}	7.9×10^{14}–5×10^{17}	atomic and molecular transitions in electric discharges
x-ray	10^{-9}–6×10^{-12}	3×10^{17}–5×10^{19}	transitions of inner electrons in atoms, deceleration of charged particles
gamma ray (γ-ray)	10^{-10}–$<10^{-14}$	3×10^{18}–$>3 \times 10^{22}$	nuclear transitions

mation which was derived under this assumption. Maxwell's equations for free space are

$$\nabla \cdot \mathbf{E} = 0$$

$$\nabla \times \mathbf{B} = \mu_0 \epsilon_0 \frac{\partial \mathbf{E}}{\partial t}$$

$$\nabla \times \mathbf{E} = -\frac{\partial \mathbf{B}}{\partial t}$$

$$\nabla \cdot \mathbf{B} = 0.$$

We shall deal only with the first two equations to show the general procedure for demonstrating the invariance of Maxwell's equations.

First, we rewrite the equations in component form as

$$\frac{\partial E_x}{\partial x} + \frac{\partial E_y}{\partial y} + \frac{\partial E_z}{\partial z} = 0$$

and

$$\frac{\partial B_z}{\partial y} - \frac{\partial B_y}{\partial z} = \mu_0 \epsilon_0 \frac{\partial E_x}{\partial t}$$

$$\frac{\partial B_x}{\partial z} - \frac{\partial B_z}{\partial x} = \mu_0 \epsilon_0 \frac{\partial E_y}{\partial t}$$

$$\frac{\partial B_y}{\partial x} - \frac{\partial B_x}{\partial y} = \mu_0 \epsilon_0 \frac{\partial E_z}{\partial t}.$$

These equations involve the components of the fields as observed at a point $P(x, y, z, t)$ in a frame of reference R. In a frame of reference R' moving at velocity $\mathbf{u}$ with respect to R in the common x, x' direction, we require that the first two equations of Maxwell be written in component form as

$$\frac{\partial E'_x}{\partial x'} + \frac{\partial E'_y}{\partial y'} + \frac{\partial E'_z}{\partial z'} = 0$$

and

$$\frac{\partial B'_z}{\partial y'} - \frac{\partial B'_y}{\partial z'} = \mu_0 \epsilon_0 \frac{\partial E'_x}{\partial t'}$$

$$\frac{\partial B'_x}{\partial z'} - \frac{\partial B'_z}{\partial x'} = \mu_0 \epsilon_0 \frac{\partial E'_y}{\partial t'}$$

$$\frac{\partial B'_y}{\partial x'} - \frac{\partial B'_x}{\partial y'} = \mu_0 \epsilon_0 \frac{\partial E'_z}{\partial t'}$$

where the coordinates (x', y', z', t') of P in R' are related to the coordinates (x, y, z, t) of P in R by the Lorentz transformation[3]

$$\begin{aligned} x' &= \gamma(x - ut) & x &= \gamma(x' + ut') \\ y' &= y & y &= y' \\ z' &= z & z &= z' \\ t' &= \gamma\left(t - \frac{ux}{c^2}\right) & t &= \gamma\left(t' + \frac{ux'}{c^2}\right). \end{aligned}$$

To carry out the transformation we need to know the relation between partial derivatives in the frames R and R'. Since

$$x = \gamma(x' + ut'),$$

x is a function of both x' and t'. Therefore, we can write

$$\frac{\partial}{\partial x} = \frac{\partial x'}{\partial x'} \cdot \frac{\partial}{\partial x} + \frac{\partial t'}{\partial t'} \cdot \frac{\partial}{\partial x} = \frac{\partial x'}{\partial x} \cdot \frac{\partial}{\partial x'} + \frac{\partial t'}{\partial x} \cdot \frac{\partial}{\partial t'}.$$

Now,

$$\frac{\partial x'}{\partial x} = \frac{\partial}{\partial x}[\gamma(x - ut)] = \gamma$$

[3] *MWTP*, Section 4.4.

SEC. 9.8 THE INVARIANCE OF MAXWELL'S EQUATIONS

and
$$\frac{\partial t'}{\partial x} = \frac{\partial}{\partial x}\left[\gamma\left(t - \frac{ux}{c^2}\right)\right] = -\frac{\gamma u}{c^2}.$$

Therefore,
$$\frac{\partial}{\partial x} = \gamma\frac{\partial}{\partial x'} - \frac{\gamma u}{c^2}\frac{\partial}{\partial t'}.$$

Similarly,
$$\frac{\partial}{\partial y} = \frac{\partial}{\partial y'}$$
$$\frac{\partial}{\partial z} = \frac{\partial}{\partial z'}$$

and
$$\frac{\partial}{\partial t} = \gamma\frac{\partial}{\partial t'} - \gamma u\frac{\partial}{\partial x'}.$$

Applying these results to the equation
$$\frac{\partial B_x}{\partial z} - \frac{\partial B_z}{\partial x} = \mu_0\epsilon_0\frac{\partial E_y}{\partial t}$$

we obtain
$$\frac{\partial B_x}{\partial z'} - \gamma\frac{\partial B_z}{\partial x'} + \frac{\gamma u}{c^2}\frac{\partial B_z}{\partial t'} = \gamma\mu_0\epsilon_0\frac{\partial E_y}{\partial t'} - \gamma u\mu_0\epsilon_0\frac{\partial E_y}{\partial x'}$$

or
$$\frac{\partial B_x}{\partial z'} - \frac{\partial}{\partial x'}[\gamma(B_z - u\mu_0\epsilon_0 E_y)] = \frac{\partial}{\partial t'}\left[\gamma\left(\mu_0\epsilon_0 E_y - \frac{\gamma u}{c^2}B_z\right)\right].$$

For the equation to be invariant in form we require
$$\frac{\partial B'_x}{\partial z'} - \frac{\partial B'_z}{\partial x'} = \mu_0\epsilon_0\frac{\partial E'_y}{\partial t'}.$$

Noting that $\mu_0\epsilon_0 = c^{-2}$ (see Section 9.7) we see that invariance requires that
$$B'_x = B_x$$
$$B'_z = \gamma\left(B_z - \frac{uE_y}{c^2}\right)$$

and
$$E'_y = \gamma(E_y - uB_z).$$

These relations are identical to those found previously (see Section 7.4) for the components of the electromagnetic field assuming invariance of charge and the Lorentz force
$$\mathbf{F} = q(\mathbf{E} + \mathbf{u} \times \mathbf{B})$$

where we used the general relations for the transformation of force.

By substitution into the remaining component equations for **E** and **B** we can deduce the remaining relations between components

$$B'_y = \gamma\left(B_y + \frac{uE_z}{c^2}\right)$$
$$E'_x = E_x$$

and

$$E'_z = \gamma(E_z + uB_y).$$

We conclude that Maxwell's first two equations are invariant under a Lorentz transformation provided that the fields transform according to the relations just derived. These relations have been derived previously using the force transformation equations of relativistic mechanics plus the invariance of charge. In this respect, the special theory of relativity has proved to be internally consistent; the special theory was derived primarily to account for difficulties in the interpretation of electromagnetic phenomena, including the noninvariance of Maxwell's equations under a Galilean transformation.[4]

9.9 ELECTROMAGNETIC RADIATION

An electromagnetic wave consists of oscillating electric and magnetic fields. It follows that we should be able to generate electromagnetic waves by producing simultaneous oscillations in electric and magnetic fields where it is arranged that the **E** and **B** fields are so oriented that the product **E** × **B** is nonzero. Thus energy will flow in the direction of **E** × **B** and an electromagnetic wave will result. This can be accomplished most simply through an **oscillating dipole**. Indeed, this is the most common method for the production of electromagnetic waves. Radiation produced by other oscillating multipoles (both electric and magnetic) is not unknown, however, and is often observed, for example, in the production of γ-rays within atomic nuclei.

Electric dipole radiation (qualitative)

In theory, an oscillating electric dipole consists of two equal and opposite charges of dipole moment **p** where **p** oscillates sinusoidally with time as shown in Fig. 9.6. In practice, an oscillating current in a conducting wire is equivalent to an oscillating dipole as indicated in Fig. 9.7. This figure shows the first three-eighths of a cycle of an oscillating current. In Fig. 9.7(a) there is no separation of charge (**p** = 0) and the current flow

[4] *MWTP*, Section 4.6.

SEC. 9.9   ELECTROMAGNETIC RADIATION

is a maximum. A magnetic field surrounds the wire as indicated. One eighth of a cycle later, I has fallen to 0.707 of its maximum value, $|\mathbf{p}|$ has risen to 0.707 of its maximum value giving rise to the dipole electric field shown in Fig. 9.7(b), and there is a net flow of energy outward in the direction $\mathbf{E} \times \mathbf{B}$. After one-quarter of a cycle, I has become zero, $|\mathbf{p}|$ has its maximum value, and there is no magnetic field as indicated in Fig. 9.7(c). After three-eighths of a cycle, I has again increased to 0.707 of its maximum value but is flowing in the opposite direction as shown in Fig. 9.7(d). The vector $\mathbf{E} \times \mathbf{B}$ now points inward since $\mathbf{B}$ has reversed direction while the direction of $\mathbf{E}$ is unchanged since $\mathbf{p}$ has not yet reversed direction.

Since electromagnetic effects are not transmitted instantly from point to point in space, but rather at the finite speed c, there is a time lag between changes in charge and current distributions on the dipole and corresponding changes in the electric and magnetic fields at some distance from the dipole. This time lag allows some of the energy to continue flowing outward even though conditions at the dipole may have changed to indicate an inward flow of energy. In a sense, it is as if some of the electric and magnetic field has become detached from the dipole or "shaken off" by the oscillation. It is obvious from this very qualitative description of the generation of an electromagnetic wave that the intensity of the emitted wave should be very dependent on the frequency of oscillation and the time lag should become more significant as the frequency increases. In fact, for a given dipole oscillator, the power radiated varies as the fourth power of the frequency.

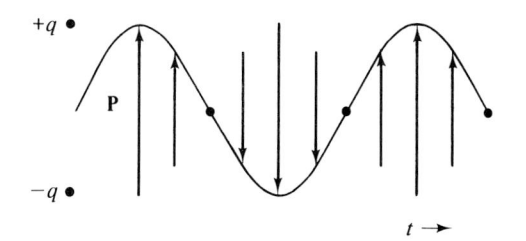

Fig. 9.6. An oscillating electric dipole.

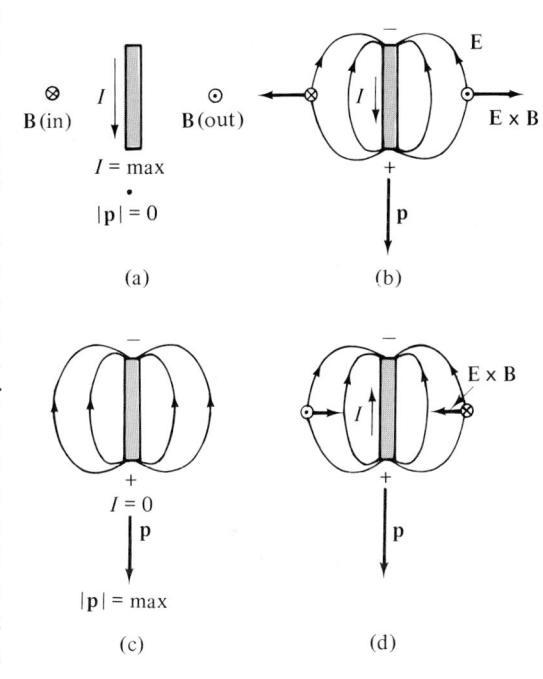

Fig. 9.7. Electric and magnetic fields around an oscillating dipole.

Electric dipole radiation (quantitative)

The current in an oscillating electric dipole is given by the equation
$$I = I_0 \cos \omega t.$$
Since
$$I = \frac{dq}{dt},$$

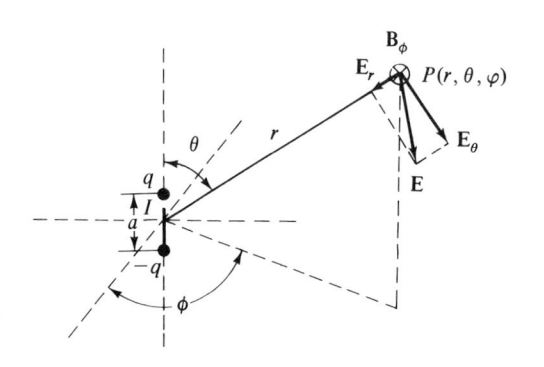

Fig. 9.8. The field of an oscillating dipole at a point $P(r, \theta, \phi)$ has components E_r, E_θ, and B_ϕ only.

the charge at one end of the dipole as a function of time t is given by

$$q = \int_0^t I\, dt = \int_0^t I_0 \cos \omega t\, dt = \frac{I_0}{\omega} \sin \omega t$$

and the dipole moment has magnitude

$$p = qa = \frac{I_0 a}{\omega} \sin \omega t.$$

We assume that the dipole is so short that $a \ll r$, the distance from the dipole to a point P where the field is to be evaluated (see Fig. 9.8).

Since electromagnetic effects travel with the speed c, the electric field at P at any instant of time is determined by the state of the dipole a time r/c earlier. Therefore, we write

$$\tau = t - \frac{r}{c}$$

where t is the time at which the field is evaluated at P and τ is the **retarded time**. The retarded time should be used in the expressions for the current and dipole moment. That is,

$$I = I_0 \cos \omega \tau$$

and

$$p = \frac{I_0 a}{\omega} \sin \omega \tau$$

describe the state of the electric dipole that gives rise to the field at P at time t.

We discovered in Section 3.7 that the **static** electric field of an electric dipole for $a \ll r$ is given by

$$E_r = \frac{2p \cos \theta}{4\pi \epsilon_0 r^3}$$

$$E_\theta = \frac{p \sin \theta}{4\pi \epsilon_0 r^3}$$

$$E_\phi = 0.$$

Substituting the time-dependent expression for p stated above gives

$$E_r = \frac{2 I_0 a \cos \theta}{4\pi \epsilon_0 \omega r^3} \sin \omega \tau$$

$$E_\theta = \frac{I_0 a \sin \theta}{4\pi \epsilon_0 \omega r^3} \sin \omega \tau$$

$$E_\phi = 0.$$

SEC. 9.9 ELECTROMAGNETIC RADIATION

The magnetic field at P generated by the current I is found from applying the Biot–Savart law (see Section 7.2)

$$d\mathbf{B} = \frac{\mu_0}{4\pi} \frac{I \, d\mathbf{l} \times \hat{\mathbf{r}}}{r^2}.$$

The vector cross product $d\mathbf{l} \times \hat{\mathbf{r}}$ of magnitude $dl \sin\theta$ has no components in the θ and r directions as indicated in Fig. 9.8 so that only the component B_ϕ is nonzero. Since $a \ll r$, we take

$$dl \simeq a$$

so that

$$B_\phi \simeq dB$$

and we obtain

$$B_\phi = \frac{\mu_0 I a \sin\theta}{4\pi r^2} = \frac{\mu_0 I_0 a \sin\theta}{4\pi r^2} \cos\omega\tau$$

$$B_\theta = 0$$

$$B_r = 0.$$

Note that B_ϕ is $\pi/2$ rad out of phase with both E_r and E_θ (that is, $\mathbf{B}$ is $\pi/2$ rad out of phase with $\mathbf{E}$).

The field components are **only an approximation** to the correct field, however, since we have neglected one important fact. As indicated by Maxwell's second and third equations, a changing electric field produces a changing magnetic field and a changing magnetic field produces a changing electric field. Both of these extra fields will in turn generate time-varying fields, and so on. When these extra contributions are added to the time-varying dipole field, the components become[5]

$$E_r = \frac{2I_0 a \cos\theta}{4\pi\epsilon_0 \omega r^3} \sin\omega\tau + \frac{2I_0 a}{4\pi\epsilon_0 c} \frac{\cos\theta}{r^2} \cos\omega\tau$$

$$E_\theta = \frac{I_0 a \sin\theta}{4\pi\epsilon_0 \omega r^3} \sin\omega\tau + \frac{I_0 a \sin\theta}{4\pi\epsilon_0 c r^2} \cos\omega\tau - \frac{\omega I_0 a \sin\theta}{4\pi\epsilon_0 c^2 r} \sin\omega\tau$$

$$E_\phi = 0$$

and

$$B_r = 0$$

$$B_\theta = 0$$

$$B_\phi = \frac{\mu_0 I_0 a \sin\theta}{4\pi r^2} \cos\omega\tau - \frac{\mu_0 \omega I_0 a \sin\theta}{4\pi c r} \sin\omega\tau.$$

Because of the contributions to the electric field due to the changing magnetic field, the electric field about the dipole no longer is the same as

---

[5] Cf. V. Rojansky, *Electromagnetic Fields and Waves* (Englewood Cliffs, N.J.: Prentice-Hall, Inc., 1971), Chapter 24.

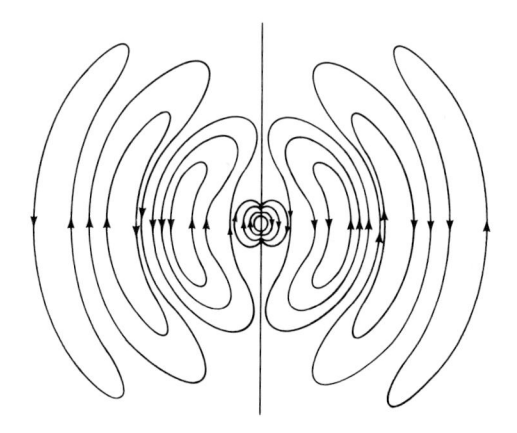

Fig. 9.9. An instantaneous representation of the electric field near an oscillating dipole.

the static field. In fact, the time-dependent field even includes closed lines that are detached from the dipole. An instantaneous representation of the electric field near an oscillating dipole is given in Fig. 9.9.

The exact field reduces to the local dipole field for r small enough that, for example, the second term in the component E_r is negligible compared to the first. That is,

$$\omega r^3 \ll cr^2$$

or

$$r \ll \frac{c}{\omega} = \frac{c}{2\pi\nu} = \frac{\lambda}{2\pi}$$

where λ is the wavelength of the radiation. Values of λ are of the order of 300 m in the standard broadcast (AM) band and of the order of 6×10^{-7} m for visible light.

When r is very large compared to the wavelength λ, the terms in r^{-1} in the exact field dominate. That is, the radial component E_r of the electric field becomes negligible compared to the component E_θ, and we can write for the distant field

$$\mathbf{E} = \frac{\omega I_0 a \sin\theta}{4\pi\epsilon_0 c^2 r} \sin\left(\omega\tau - \frac{2\pi r}{\lambda}\right) \mathbf{i}_\theta$$

$$\mathbf{B} = \frac{\mu_0 \omega I_0 a \sin\theta}{4\pi c r} \sin\left(\omega\tau - \frac{2\pi r}{\lambda}\right) \mathbf{i}_\phi$$

where $\mathbf{i}_\theta$ and $\mathbf{i}_\phi$ are unit vectors in the θ and ϕ directions, respectively. The $\mathbf{E}$ and $\mathbf{B}$ vectors are now in phase.

The vector $\mathbf{E} \times \mathbf{B}/\mu_0$ has the value

$$\frac{\mathbf{E} \times \mathbf{B}}{\mu_0} = \frac{\omega^2 I_0^2 a^2 \sin^2\theta}{(4\pi)^2 \epsilon_0 c^3 r^2} \sin^2\left(\omega\tau - \frac{2\pi r}{\lambda}\right) \mathbf{i}_r$$

where $\mathbf{i}_r$ is a vector in the radial direction. The average value of $\mathbf{E} \times \mathbf{B}/\mu_0$ is

$$\frac{\omega^2 I_0^2 a^2 \sin^2\theta}{2(4\pi)^2 \epsilon_0 c^3 r^2} \mathbf{i}_r$$

since the average value of $(\sin)^2$ is $\frac{1}{2}$. The coefficient of $\mathbf{i}_r$ in this expression is always positive so that energy always flows radially outward from the oscillating dipole. The energy flow varies as

$$\frac{\sin^2\theta}{r^2};$$

that is, it is angle dependent and has an inverse square dependence upon r. Because of the $\sin^2\theta$ dependence, the energy flow is zero along the axis of

SEC. 9.9   ELECTROMAGNETIC RADIATION   159

the dipole and maximum at right angles to it. Values of $\sin^2 \theta$ are given in Table 9.2 and are plotted in Fig. 9.10. Since the energy flow varies as $\sin^2 \theta$ for a given r, Fig. 9.10 gives the energy flow as a function of θ at a given distance r from the oscillating dipole. The solid line of Fig. 9.10 is called the **radiation pattern** of the oscillating dipole.

Table 9.2 Values of $\sin^2 \theta$

θ (deg)	$\sin^2 \theta$	θ (deg)	$\sin^2 \theta$	θ (deg)	$\sin^2 \theta$
0	0	70	0.879	140	0.410
10	0.029	80	0.963	150	0.250
20	0.126	90	1.00	160	0.126
30	0.250	100	0.963	170	0.029
40	0.410	110	0.879	180	0
50	0.583	120	0.748		
60	0.748	130	0.583		

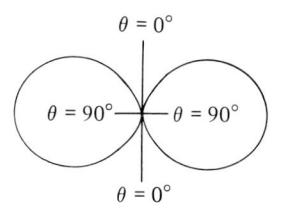

Fig. 9.10. The radiation pattern of an oscillating dipole. The solid line gives the magnitude of the energy flow as a function of θ at a given distance r from the dipole.

Magnetic dipole radiation (qualitative)

A static magnetic dipole produces only a constant magnetic field. However, if the magnetic field changes periodically with time, an electric field is produced as we mentioned earlier in this chapter when discussing Maxwell's equations. The field around an oscillating magnetic dipole consisting of a small current loop is shown in Fig. 9.11. Following reasoning similar to that for the electric dipole we can see that an oscillating magnetic dipole should emit radiation also. Note, however, that the roles of the electric and magnetic fields are opposite for the two dipoles. For the magnetic dipole, the electric lines are perpendicular to the magnetic dipole moment vector while for the electric dipole, the magnetic lines are perpendicular to the electric dipole moment vector. In both cases the intensity of the emitted radiation is maximum in the equatorial plane passing through the center of the dipole and is zero along the axis of the dipole.

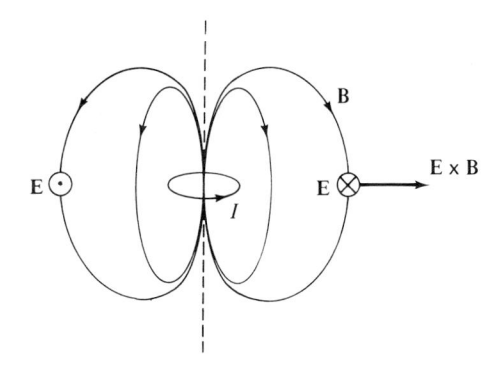

Fig. 9.11. An oscillating magnetic dipole emits radiation in a manner similar to that of an oscillating electric dipole.

A detailed mathematical analysis shows that the ratio of the energy U_m radiated per unit time by a magnetic dipole to the energy U_e radiated per unit time by an electric dipole is

$$\frac{U_m}{U_e} = \left(\frac{A\omega}{s_0 c}\right)^2$$

where A is the area of the current loop of the magnetic dipole, ω is the angular frequency of the radiation, s_0 is the length of the electric dipole, and c is the speed of light. Since

$$\frac{\omega}{c} = \frac{2\pi\nu}{\lambda\nu} = \frac{2\pi}{\lambda}$$

and

$$A \simeq s_0^2,$$

therefore,

$$\frac{U_m}{U_e} \simeq \left(\frac{2\pi s_0}{\lambda}\right)^2.$$

At broadcast frequencies, λ is normally much larger than s_0 (for example, for $\nu = 1$ MHz, $\lambda = 300$ m) so that magnetic dipole radiation is much weaker than electric dipole radiation. As the frequency increases, λ decreases and the production of electromagnetic waves via magnetic dipole oscillations becomes more favorable.

Example. Compare the intensities of electric and magnetic dipole radiation emitted by atoms and nuclei.

Solution. In an atom $s_0 \simeq 10^{-10}$ m and the radiation emitted is in the visible region of the spectrum and of wavelength $\simeq 6 \times 10^{-7}$ m. Therefore, the ratio of the intensity of magnetic to electric dipole radiation is

$$\frac{U_m}{U_e} \simeq \left(\frac{2\pi \times 10^{-10}}{6 \times 10^{-7}}\right)^2 \simeq 10^{-6}.$$

That is, magnetic dipole radiation from atoms is six orders of magnitude weaker than electric dipole radiation. Cases do exist, however, in which symmetry precludes the emission of electric dipole radiation during a transition. In such cases magnetic dipole radiation can be significant.

In a nucleus $s_0 \simeq 10^{-14}$ m and the radiation emitted is in the γ-ray region of the spectrum and of wavelength $\simeq 6 \times 10^{-13}$ m. Therefore, the ratio of the intensity of magnetic to electric dipole radiation is

$$\frac{U_m}{U_e} \simeq \left(\frac{2\pi \times 10^{-14}}{6 \times 10^{-13}}\right)^2 \simeq 10^{-2}.$$

That is, magnetic dipole radiation from nuclei is typically about two orders of magnitude weaker than electric dipole radiation. However, there are many cases where electric dipole (or electric quadrupole) radiation is forbidden due to quantum mechanical selection rules, so that magnetic dipole radiation is often produced in nuclear transitions.

Radiation from an accelerated charge

A charge moving with uniform velocity is surrounded by both an electric and a magnetic field. The electric field lines are directed radially outward from the charge while the magnetic field lines form concentric, transverse circles about the line of motion of the charges as indicated in Fig. 9.12. The vector **E** × **B** is in the direction of motion and has no net component in any other direction. (Fig. 9.12 is actually a simplified picture of the fields since the charge is moving and the magnetic field lines are not confined to a single plane containing the charge.)

In order to determine whether any electromagnetic energy is radiated from the charge, we could enclose the charge by a suitable surface S and consider the value of the integral

$$\int_S \mathbf{E} \times \mathbf{B} \cdot d\mathbf{S}$$

which, if different from zero, would indicate radiation of energy. A spherical surface is a suitable surface for a point charge. For such a surface about a charge in uniform motion, the vector **E** × **B** is everywhere tangent to the surface. This is easy to see since **E** is everywhere perpendicular to the surface and **E** × **B** is perpendicular to **E** and, therefore, tangent to the surface. The surface integral is then zero and there is no electromagnetic radiation from a charge in uniform motion.

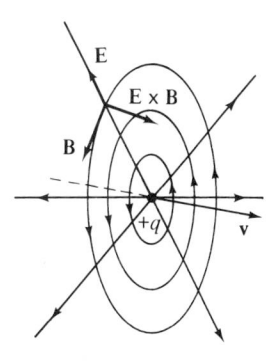

Fig. 9.12. Electric and magnetic lines about a charge in uniform motion.

When a charge is accelerated, the electric field lines are no longer directed radially outward from the charge but appear somewhat as indicated in Fig. 9.13. The field ahead of the charge is increased over the field to the rear of the charge. Moreover, since the charge is constantly increasing its speed, the increase in the field ahead of the charge is greater than the decrease in the field behind (which corresponds to an earlier, slower speed). Therefore, some energy must be transported into the space around the charge to continually increase the field ahead of the charge. An accelerated charge radiates electromagnetic energy.

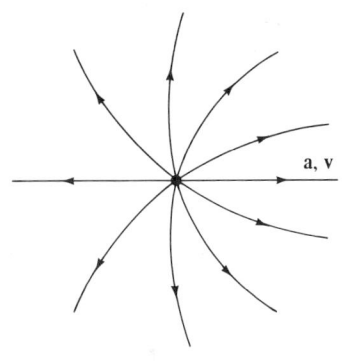

Fig. 9.13. Electric lines about a positive charge accelerated in the direction of motion.

A detailed analysis shows the total energy U radiated per unit time is given by

$$\frac{dU}{dt} \propto \frac{a^2}{c^3}$$

while the intensity of radiation in the direction making the angle θ with the velocity is

$$I(\theta) \propto \left(\frac{a^2}{c^3}\right) \frac{\sin^2 \theta}{r^2}$$

where a is the acceleration and c the speed of light. This last expression is valid only for $v \ll c$. As the speed of the particle approaches c, the direction of maximum intensity for the radiation moves away from 90° toward the direction of motion of the particle.

We note also that the acceleration of a charge in an oscillating electric dipole is given by

$$a = -\omega^2 s$$

where s is the displacement. Therefore,

$$\frac{dU}{dt} \propto \frac{\omega^4 s^2}{c^3}$$

which verifies the statement, given in the discussion of electric dipole radiation, that the radiated power varies as the fourth power of the frequency.

Acceleration also occurs when the direction of the velocity vector changes with time. A charged particle moving in a circular path experiences a centripetal acceleration and should be expected to emit electromagnetic radiation. The loss of energy in circular motion due to radiation becomes significant only when the particle speeds are very high so that the centripetal acceleration becomes very large. This mode of energy loss first became important in circular particle accelerators called **synchrotrons**. For this reason, the loss of energy by radiation from charges experiencing centripetal acceleration is called **synchrotron radiation.**

9.10 ELECTROMAGNETIC WAVES FROM SPACE

Electromagnetic radiation at 21.1 cm in our galaxy[6]

In 1951 three independent research groups in the United States, Holland, and Australia reported, within a few months of one another, the discovery of radio emission at a frequency of 1420.406 MHz (a wavelength of 21.1061 cm) from neutral hydrogen atoms in our galaxy. These announcements heralded the arrival of a fruitful new branch of radio astronomy. For some years this remained the only line available to the radio astronomers. In recent years, however, scientists have identified radio waves from an ever increasing number of molecular species, such as water, ammonia, the hydroxyl radical, formaldehyde, carbon monoxide, and cyanogen. Throughout 1970 new species were reported at the rate of one per month. Although the concentrations of these molecules in space are far less than the concentration of neutral hydrogen atoms, a large gain in the intensity per emitter results from the fact that the radiation from the molecules is electric dipole while that from neutral hydrogen atoms is magnetic dipole.

[6]R. D. Davies, "Radio Emission from Interstellar Neutral Hydrogen," *Contemporary Physics*, **2** (August, 1961), 428.

The 21.1 cm radio emission from neutral hydrogen comes from atoms in their lowest energy state,[7] which in fact consists of two closely spaced **hyperfine levels** resulting from the fact that the atom has slightly different energies depending on whether the electron magnetic moment is parallel or antiparallel to the nuclear magnetic moment. The symmetry properties of the energy states (or levels) forbid the emission of electric dipole radiation during a transition between these states. The transition is often referred to as a **spin-flip** transition, since the directions of the electron spin angular momenta are opposite in the two states. The thermal equilibrium populations are maintained by collisions between hydrogen atoms which occur at the rate of about one every 50 years. An atom in the upper state will spontaneously fall to the lower level with the emission of a photon of wavelength 21.1 cm. The average time spent by an atom in the upper state before the atom will radiate is 11×10^6 years. Since the density of interstellar neutral hydrogen is only $\simeq 1$ atom$\cdot$cm^{-3}, there will be $\simeq 1$ photon per 11×10^6 years$\cdot$cm^{-3} in our galaxy. The fact that such a slow rate of energy release is detectable at all attests to the immense size of our galaxy. For example, neutral hydrogen atoms extend for $\simeq 6 \times 10^{22}$ cm in the direction of the galactic center.

As early as 1928 astronomers began to realize that our galaxy was a rotating disc-shaped assemblage of stars, some 50,000 light years across, and in many ways similar to the great spiral galaxies. In such a galaxy the inner stars move as a solid disc while the outer ones move in orbits as if they were independently moving about a central point mass. The 21.1 cm radiation provided the ideal probe to trace the spiral structure of the galaxy. Concentrations of neutral hydrogen are detected at many different frequencies which are Doppler-shifted[8] relative to the frequency of stationary neutral hydrogen due to movement in the line of sight. The frequency shifts measured in kHz may be converted to velocities in km$\cdot$sec^{-1} by multiplying by -0.211. A positive velocity denotes recession. The velocities in turn can be converted into distances by use of the **law of red shifts**, or **Hubble's law**.[9] When data for all directions are correlated, a picture of the distribution of neutral hydrogen throughout our galaxy emerges.

Magnetic dipole radiation from pulsars[10,11]

An exciting new type of radio-emitting object was reported by a group of radio astronomers in Britain in 1968. The strange feature characterizing

---

[7] *MWTP*, Section 18.4.
[8] *MWTP*, Section 15.8.
[9] A. R. Sandage, "The Red Shift," *Scientific American*, September, 1956. Also available as *Scientific American Offprint 240* (San Francisco: W. H. Freeman and Co., Publishers).
[10] A. Hewish, "Pulsars," *Scientific American*, October, 1968.
[11] J. P. Ostriker, "The Nature of Pulsars," *Scientific American*, January, 1971.

these objects is the emission of short bursts of radio noise at regular intervals. The repetition periods for these first **pulsars** ranged from 0.25 to 1.3 sec, which is very much shorter than any previously known periodic astronomical phenomenon. The distances of the first pulsars from the earth were estimated to be in the range of tens to hundreds of light years. The amount of radio energy that they emitted was found to average between 10^{-6} and 10^{-4} of the total energy emitted by the sun. That is, pulsars are intrinsically very weak sources of radiation. However, whereas the sun emits almost all of its energy at visible frequencies, the pulsars emit almost all of their energy at radio frequencies.

Astrophysicists now mostly agree that pulsars are **neutron stars.** Early in the 1930's it had been predicted theoretically that if a star of about twice the sun's mass began to collapse under its own gravitational forces, the electrons would eventually be absorbed into the nuclei and a **neutron gas** would appear. At sufficiently high neutron density the repulsive nuclear forces among the neutrons would balance the gravitational forces and an equilibrium state would again occur. The resulting **neutron star** would resemble a giant atomic nucleus. Its radius would be about 10 km and its mean density in the range 10^{14} to 10^{18} kg·m^{-3}! The gravitational forces associated with such objects are so large that they must be discussed in the framework of the general theory of relativity.

Even the initial observations contained important clues suggesting that pulsars were neutron stars. The narrow pulse widths implied a small source for the electromagnetic waves. Their intrinsic faintness indicated that they had to be either small or cold or both. The periodic nature of the emission could be attributed to the existence of a small radiating region fixed on the surface of a rapidly rotating neutron star. Once each rotation period, the beam of electromagnetic radiation would sweep by the observer who would see a pulse.

The parent stars from which neutron stars could be formed have magnetic fields that are dipole in character and of magnitude 10^{-2} to 1 Wb·m^{-2} at the star's surface. As the parent star collapses the magnetic lines of force would be effectively frozen into the stellar material so that the field strength, which is proportional to the number of lines of force intersecting unit area, would increase as the inverse square of the decreasing stellar radius. Since the radius of the resulting neutron star is $\simeq 10^{-5}$ of the radius of its parent, it might be expected to have a magnetic field at its surface of 10^8 to 10^{10} Wb·m^{-2}.

The most likely parent stars have rotational periods $\simeq 10^5$ sec. Therefore, if angular momentum were conserved during the collapse, the rotation period of the neutron star would be $\simeq 10^{-5}$ sec. In fact, a neutron star rotating at this rate would be mechanically unstable. It is thought that some loss of angular momentum, in the form of **gravitational radiation** occurs during the collapse, leaving the new-born neutron star with a period $\simeq 10^{-2}$ sec.

A rotating magnetic dipole will emit magnetic dipole radiation. The magnetic field surrounding a neutron star is pictured in Fig. 9.14. Near the star the magnetic field lines have the conventional dipole pattern and would rotate with the star at speeds proportional to their distance from the rotation axis. The distance at which the speed of rotation would equal the speed of light defines the **speed of light cylinder**. As this surface is approached the magnetic field lines depart from those of a simple magnetic dipole. Far from the rotating neutron star the magnetic field would be accompanied by an electric field perpendicular to it. That is, far from the star the electromagnetic field would assume the character of a low-frequency electromagnetic wave. Such radiation would carry off both energy and angular momentum, thereby resulting in a lengthening of the rotation period of the star. Indeed, careful observations of the pulsar discovered in the Crab nebula have verified

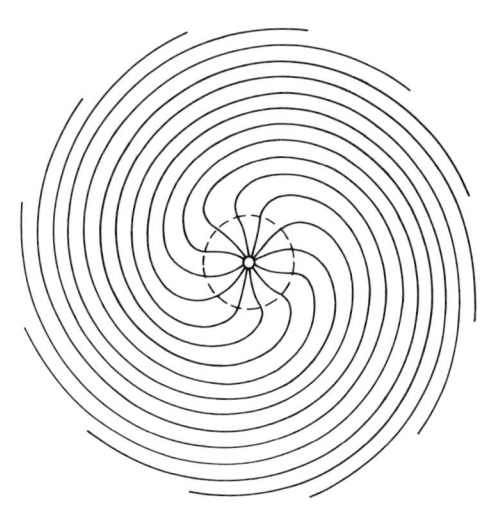

Fig. 9.14. The magnetic field surrounding a neutron star. The dotted circle represents the speed of light cylinder.

this prediction. The Crab nebula is the gaseous remnant of a supernova explosion observed and reported by oriental astronomers in 1054 A.D. The young pulsar associated with the Crab nebula has a period of about 0.03 sec and the period has been observed to be increasing at the rate of 3.8×10^{-8} sec (38 nsec) per day. In fact, it is now thought that it is just the energy loss from the spinning neutron star in the Crab nebula that is keeping it glowing.

Electromagnetic radiation from the primeval fireball[12]

Edwin P. Hubble's discovery that other galaxies are moving away from ours, and are doing so at speeds that increase with the distance to the galaxy, is the single observation upon which most contemporary cosmological theories rest. For example, Hubble's law is the basis for both the **big-bang cosmology**, which contends that the universe originated in a superdense state some 7×10^9 years ago, and the **steady-state cosmology**, which claims that the universe looks now as it always has in the past and always will in the future.

In 1965 an observation of a second basic cosmological fact was made. At that time cosmic radio-frequency electromagnetic radiation which apparently fills the universe and arrives at the earth equally from all directions was detected. Such radiation is consistent with the big-bang theories. These theories suggest that about 7×10^9 years ago all of the matter in the

[12] P. J. E. Peebles and D. T. Wilkinson, "The Primeval Fireball," *Scientific American*, June, 1967.

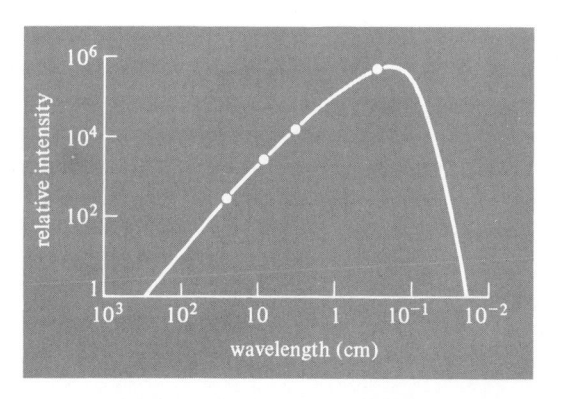

Fig. 9.15. Intensity of electromagnetic radiation presumed to have originated with the primeval fireball. The solid curve is a blackbody radiation curve for 3°K.

universe was packed together in an inferno of particles and electromagnetic radiation. As time elapsed the universe expanded, the matter cooling and condensing to form stars and galaxies. At the outset the electromagnetic radiation consisted of extremely energetic gamma rays. With the expansion of the universe the wavelength of the radiation increased until at the present it appears mainly in the radio-frequency and microwave ranges. If the observed radiation indeed has its origin in the primeval fireball, it must survive two severe tests. First, it must be blackbody radiation[13], since it was emitted by a source in thermal equilibrium. Second, it must be isotropic, since it is presumed to fill the universe. At present all of the evidence is positive. Four measurements of the relative intensity of the radiation at different frequencies have been made. All lie on a blackbody curve consistent with a temperature of 3°K (see Fig. 9.15). The measurement at 2.6 mm was obtained by observing absorption of radiation by cyanogen molecules. There is also support for the isotropy of the radiation in space. An upper limit of 0.5% anisotropy has been established.

Quasars and synchrotron radiation[14,15,16,17]

In 1962 **quasi-stellar radio sources** or **quasars** were first identified. They are small diameter, very intense sources of electromagnetic radiation. They are associated with stars which emit visible radiation that is, at first glance, quite similar to that from blue stars. However, closer examination of individual emission lines shows them to have very large Doppler shifts[18] toward lower frequencies. If the Doppler shifts are taken as a measure of the distance to the quasars, then quasars must be located at the very edge of the universe and are therefore of considerable cosmological importance.

[13] *MWTP*, Section 18.2.
[14] J. L. Greenstein, "Quasi-Stellar Radio Sources," *Scientific American*, December, 1963.
[15] G. Burbidge and F. Hoyle, "The Problem of the Quasi-Stellar Objects," *Scientific American*, December, 1966. Also available as *Scientific American Offprint 305* (San Francisco: W. H. Freeman and Co., Publishers).
[16] M. S. Longair, "Quasi-Stellar Radio Sources," *Contemporary Physics*, **8** (July, 1967), 357.
[17] M. Schmidt and F. Bello, "The Evolution of Quasars," *Scientific American*, May, 1971.
[18] *MWTP*, Section 15.8.

Furthermore, if the quasars are at such great distances, it follows that they are emitting of the order of ten times as much energy in the photographic part of the spectrum (that is, the ultraviolet, visible, and infrared regions) as is the brightest galaxy containing $\simeq 10^{11}$ stars! It is little wonder that the discovery of quasars caused such excitement among astronomers and astrophysicists.

It is generally believed that the radio-frequency radiation from quasars is produced by the synchrotron mechanism—the radiation produced by relativistic electrons spiraling in a magnetic field. The calculated distribution of radio-frequency energy as a function of frequency is in qualitative agreement with the observations (see Fig. 9.16). One reason for the decrease in energy at high frequencies is that electrons radiate away their energy in proportion to the square of the energy they possess. The decrease at low frequencies is due to the self-absorption of the radiation by the electrons.

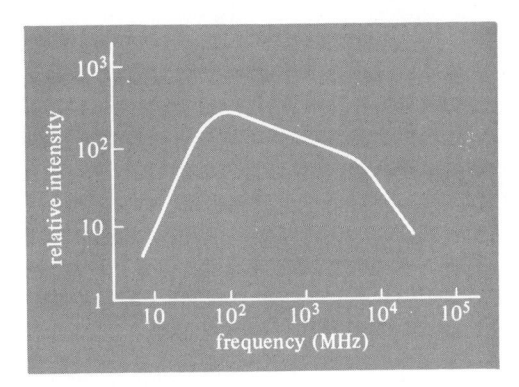

Fig. 9.16. The distribution of radio-frequency energy as a function of frequency due to the synchrotron mechanism.

The source of the energy necessary to produce the relativistic particles in the quasar is of fundamental importance. On the basis of the synchrotron hypothesis it is possible to deduce that an amount of energy $\simeq 10^{54}$ J is required to account for the radio emission from a quasar. This is an enormous amount of energy when one realizes that the rest energy of one solar mass of material is $\simeq 10^{47}$ J. That is, the disappearance of about 10^7 solar masses of matter must occur during the time of the explosion that gives birth to the quasar.

The problem of understanding quasars remains one of the most important and fascinating tasks in all of physics.

QUESTIONS AND PROBLEMS

1. Write an account of the career of Heinrich Hertz.

2. Prove the vector identity
$$\nabla \times (\nabla \times \mathbf{E}) = \nabla(\nabla \cdot \mathbf{E}) - \nabla^2 \mathbf{E}.$$

3. Prove the vector identity
$$\nabla \times (\mathbf{B}_0 \phi) = \phi(\nabla \times \mathbf{B}_0) - \mathbf{B}_0 \times \nabla\phi.$$

4. Prove the vector identity
$$\nabla \cdot (\mathbf{E} \times \mathbf{B}) = \mathbf{B} \cdot (\nabla \times \mathbf{E}) - \mathbf{E} \cdot (\nabla \times \mathbf{B}).$$

*5. A circular disc rotates with angular velocity $\omega\mathbf{k}$ about its axis. Show that
$$\nabla \times \mathbf{v} = 2\omega\mathbf{k}$$
where $\mathbf{v}$ is the velocity of any point, $\mathbf{k}$ is a unit vector along the axis of rotation, and the sense of the rotation is related by the right-hand rule to $\mathbf{k}$. If $\nabla \times \mathbf{v} = 0$, show that the disc must be nonrigid and that ω is inversely proportional to r^2 where r is the distance from the center of the disc.

*6. If $\mathbf{A}$ is a vector such that $\mathbf{A} = f(r)\mathbf{r}$, show that $\nabla \cdot \mathbf{A} = 0$ implies
$$f(r) = \frac{C}{r^3}$$
with C a constant and that $\nabla \times \mathbf{A} = 0$.

*7. Evaluate the following integrals:
   (a) $\int_S \mathbf{A} \cdot d\mathbf{S}$ where $\mathbf{A} = x\mathbf{i} + y\mathbf{j} + z\mathbf{k}$, and S is the surface of the tetrahedron bounded by $x = 0$, $y = 0$, $z = 0$, and $x + y + z = a$.
   (b) $\int_S \mathbf{A} \cdot d\mathbf{S}$ where $\mathbf{A} = 2x^3y\mathbf{i} + 3xz\mathbf{j} + xy^2z\mathbf{k}$, and S is a cube of side 2 centered at the origin with sides parallel to the axes.
   (c) $\int_S \nabla \times \mathbf{A} \cdot d\mathbf{S}$ where $\mathbf{A} = x\mathbf{i} + y\mathbf{j} + z\mathbf{k}$, and S is a surface bounded by
$$\frac{x^2}{4} + \frac{y^2}{9} = 1$$
   in the xy plane.
   (d) $\int_S \nabla \times \mathbf{A} \cdot d\mathbf{S}$ where $\mathbf{A} = y^2\mathbf{i} + xy\mathbf{j} - yz\mathbf{k}$, and S is the hemisphere $z = +(4 - x^2 - y^2)^{1/2}$.

8. Show that the electrostatic potential V satisfies **Poisson's equation**
$$\nabla^2 V = -\frac{\rho}{\epsilon_0}.$$

9. In a parallel-plate capacitor the displacement current I_D can be written as
$$I_D = C\frac{dV}{dt}.$$
   Prove this result.

10. Write the symmetrical counterpart of Faraday's law of electromagnetic induction
$$\oint \mathbf{E} \cdot d\mathbf{l} = -\frac{d\Phi_B}{dt}.$$
   Why is Faraday's law of induction the more familiar of the two equations?

11. It has been stated that Maxwell did for electromagnetic phenomena what Newton had accomplished for mechanics. Discuss this statement.

12. Starting with Maxwell's equations derive an expression for the electric field due to a point charge q.

13. A hypothetical universe contains magnetic monopoles as well as electric charges. Write a set of equations describing electromagnetic phenomena in this universe. Compare these equations with Maxwell's equations.[19]

[19]D. Garrick and R. Kunselman, "Magnetic Monopoles," *The Physics Teacher*, **9** (October 1971), 366.

14. For very high-frequency fields currents are limited essentially to the surface of a conductor, while **E** and **B** are zero within the conductor. If **E** and **B** outside the conductor are tangential, **E** being directed along the x axis and **B** along the y axis, show that

$$\frac{\partial E}{\partial z} = -\frac{\partial B}{\partial t}.$$

15. Write Maxwell's equations in integral form.

16. Construct a table of four columns to contain the following information about Maxwell's equations:
    1) First column—basic law of electromagnetism
    2) Second column—corresponding Maxwell equation in differential form
    3) Third column—corresponding Maxwell equation in integral form
    4) Fourth column—experimental evidence for the law.

*17. Show that Maxwell's equations are invariant under the following substitutions

$$\mathbf{E} = -\boldsymbol{\nabla} V - \frac{\partial \mathbf{A}}{\partial t}$$

$$\mathbf{B} = \boldsymbol{\nabla} \times \mathbf{A}.$$

V and **A** are known as the **scalar** and **vector potential**, respectively.

18. Show that for a general direction of propagation the wave equation for a plane, transverse sinusoidal wave can be written

$$\nabla^2 s = \frac{1}{v^2} \frac{\partial^2 s}{\partial t^2}.$$

19. Explain the necessity of the term $\epsilon_0(\partial \mathbf{E}/\partial t)$ in Ampère's equation for an understanding of the propagation of electromagnetic waves.

20. Derive the equation

$$\nabla^2 \mathbf{B} = \mu_0 \epsilon_0 \frac{\partial^2 \mathbf{B}}{\partial t^2}.$$

21. Show that $g(u, v)$ is a solution of the wave equation in one dimension where $u = x - ct$ and $v = x + ct$ and g is any function such that $\partial^2 g/\partial u\, \partial v = 0$.

22. A light wave is traveling along the z axis. At some instant the electric field has a magnitude of 50 V·m^{-1} along the x axis. What is the magnitude and direction of the magnetic field? Calculate the energy flux.

23. Show that the time-averaged energies of the electric and magnetic fields of a light wave are equal.

24. Show by dimensional analysis[20] that the vector $\mathbf{E} \times \mathbf{B}/\mu_0$ has units J·m^{-2}·sec^{-1}.

25. The average energy flux of a light wave is 2×10^5 J·m^{-2}·sec^{-1}. Calculate the magnitudes of the electric and magnetic fields. If the wave is composed

[20] *MWTP*, Section 15.4.

of monochromatic light of wavelength 5×10^{-7} m, calculate the photon flux.[21]

26. Describe the **headlight effect**[22] of special relativity.

27. Show that substitution of the relations

$$B'_x = B_x$$
$$B'_z = \gamma \left(B_z - \frac{uE_y}{c^2} \right)$$

and

$$E'_y = \gamma(E_y - uB_z)$$

into the appropriate component equations derived from Maxwell's first two equations yields the relations

$$B'_y = \gamma \left(B_y + \frac{uE_z}{c^2} \right)$$
$$E'_x = E_x$$

and

$$E'_z = \gamma(E_z + uB_y).$$

28. Demonstrate that Maxwell's third and fourth equations are invariant under a Lorentz transformation.

29. A transmitter is radiating electric dipole radiation at a frequency of 7.5×10^5 Hz. Calculate the difference in phase between the oscillating dipole moment and the components of the electric field at a distance of 10 km from the transmitter at any instant of time.

*30. Using Maxwell's equations, suggest a first-order correction to the electric and magnetic fields around an oscillating electric dipole which will take account of the magnetic and electric fields generated by the varying electric and magnetic components of the primary dipole field.

31. Discuss the significance of the radio signals received from hydroxyl radicals in space.[23]

32. Verify that the correct numerical factor for converting the frequency shifts associated with the 21.1 cm radiation in our galaxy to velocities is -0.211.

33. By making use of the articles referred to in the section on quasars discuss the possibility that they are relatively close to us rather than at the edge of the universe.

34. The same principle that we have applied to account for the large magnetic

[21]*MWTP*, Section 18.3.
[22]E. F. Taylor and J. A. Wheeler, *Spacetime Physics* (San Francisco: W. H. Freeman and Co., Publishers, 1966).
[23]A. H. Barrett, "Radio Signals from Hydroxyl Radicals," *Scientific American*, December, 1968.

fields associated with pulsars has been used to generate large magnetic fields in the laboratory.[24] Discuss the production of large laboratory fields.

35. Write a note on the techniques of measurement of electromagnetic radiation from the primeval fireball.

[24]F. Bitter, "Ultrastrong Magnetic Fields," *Scientific American*, July, 1965.

PETER J. W. DEBYE

10 Dielectric materials

10.1 INTRODUCTION

In our discussion of electricity and magnetism to this point we have neglected the presence of matter almost completely. This was done deliberately to emphasize that electromagnetism is associated with charged particles and not with matter in bulk and to show particularly that electromagnetic waves can exist in space independent of the presence of matter. With this chapter we begin our investigation of the interaction of electric and magnetic fields with matter. In particular, we discuss the interaction of static electric fields with those materials that are **insulators** or **nonconductors** of electric current.

When the space between the plates of a capacitor is completely filled with some insulating material, such as glass, mica, or even air, the capacitance is found to be increased by a factor K over that for a vacuum between the plates, where K is characteristic of the material between the plates. K is known as the **dielectric constant** of the insulator. Values of the dielectric constant vary widely (for example, $K = 1.006$ for air, 81 for water, and $\simeq 10^4$ for some ferroelectric crystals) but, for a given material, are independent of the size and shape. This chapter is devoted to a discussion of the properties of **dielectrics**, which are all those materials possessing a dielectric constant greater than 1.

10.2 POLARIZATION AND SUSCEPTIBILITY

Matter consists of complex arrays of atoms and molecules. Although an atom is normally electrically neutral, it consists of charged particles (electrons and protons) as well as uncharged particles (neutrons). The positively charged protons are fixed in the nucleus of the atom while the negatively charged electrons are distributed throughout most of the volume of the atom. When atoms combine to form molecules, one or more of the electrons from a given atom are shared by neighboring atoms. In some solid materials one or more of the electrons from each atom are able to move freely throughout the volume of the solid; these solids are called **conductors**. However, in most solids (and in most liquids and gases), the electrons are not able to move far from the parent atom, even under the influence of external electric fields; these are the dielectric materials. However, in such materials external fields can cause small relative displacements of electric charge over atomic dimensions. The magnitude of the charge separation depends on the forces that act between the charged particles in the atoms or molecules of the material. When such a charge separation occurs in a material it is said to be **polarized**.

We shall now view a dielectric material from the macroscopic point of view before proceeding on the atomic level. We can consider the dielectric

174   DIELECTRIC MATERIALS   CHAP. 10

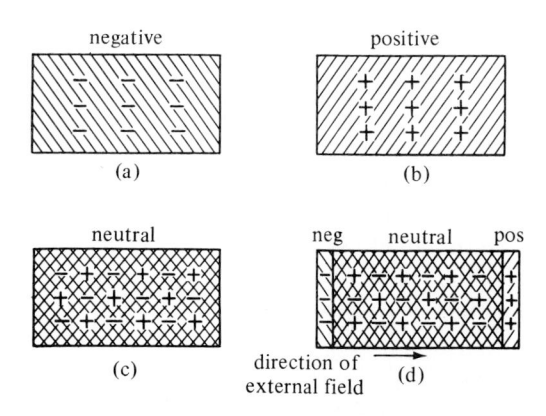

Fig. 10.1. A dielectric can be considered to consist of two interpenetrating charge distributions (a) and (b). Normally the dielectric has no net charge (c). In the presence of an external field, polarization occurs (d).

to consist of two interpenetrating distributions of charge, one negative (associated with the electrons) and one positive (associated with the protons in the nuclei) [Fig. 10.1(a) and (b)]. Normally, there are equal numbers of charges in the two distributions and these charges occupy the same volume so that the net charge is zero [Fig. 10.1(c)]. When an external static electric field is applied to the dielectric, a charge separation takes place to the extent allowed by the atomic or molecular forces. Over most of the volume of the dielectric there are still equal numbers of positive and negative charges. However, uncompensated positive and negative charges appear on the surfaces of the material as shown in Fig. 10.1(d). This charge is called **polarization charge** or **bound charge** and is essentially a surface charge since it is only within the surface layer of atoms that a net charge can arise.

We note that the dielectric material in an external field has the characteristics of an electric dipole [see Fig. 10.1(d)]. There are two uncompensated surface charges, one positive and one negative, separated by the main body of the dielectric. The constituent atoms or molecules of the dielectric become electric dipoles under the action of the external field. It is convenient to define a vector quantity **P**, called the **polarization**, as the dipole moment per unit volume. From the foregoing discussion we should expect the polarization to be related to the polarization charge on the surface of the dielectric.

We can develop the relation between **P** and the polarization charge by reference to Fig. 10.2, which is a generalization of Fig. 10.1(d) to three dimensions. The volume of the dielectric box shown in Fig. 10.2 is AL so that the dipole moment of the dielectric is PAL. The dipole moment can also be calculated by considering the dielectric as one dipole with surface charges $\sigma_p A$ and $-\sigma_p A$ and separation L. This gives a dipole moment of $\sigma_p AL$. The magnitude of **P** is then given by

$$P = |\sigma_p|.$$

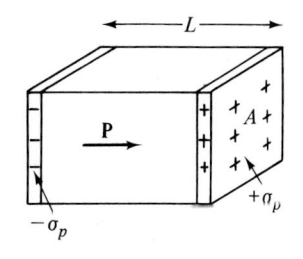

Fig. 10.2. A polarized dielectric, with polarization charge density σ_p.

Note, however, that **P** is a vector that is directed from the negative polarization charge toward the positive polarization charge. In fact, **P** is a vector field quantity that has properties very similar to the electric field **E**. However, the electric field is directed from positive charge toward negative charge. Therefore, in order to apply Gauss' law, for example, to a dielectric, we must take account of the fact that the lines of **P** emerge from a volume containing negative polarization charge. We write the relation between

SEC. 10.3 ELECTRIC DISPLACEMENT

P and σ_p as

$$P = -\sigma_p$$

to allow for this property of **P**.

The relation between P and σ_p was developed with respect to a dielectric in which **P** is normal to the surface. The relation holds for all configurations if the **component of P normal to the surface** is used. The proof of this is left as a problem (see Problem 1). We then write in general that

$$P_n = -\sigma_p$$

for all dielectrics where P_n is the normal component of **P**.

If both free and polarization charge are present in a region of space, we write Gauss' law for the flux of electric lines out of the volume concerned as

$$\int_S \mathbf{E} \cdot d\mathbf{S} = \frac{q}{\epsilon_0} + \frac{q_p}{\epsilon_0} = \frac{q_t}{\epsilon_0}$$

where q and q_p are charges of opposite sign and $q_t = q + q_p$ is the net charge within the Gaussian surface and is less than the free charge q. Thus, the electric flux is reduced by the presence of the dielectric. An alternative form of Gauss' law for electric flux in a dielectric material is given in the next section.

For dielectrics, it is usually a good assumption that the polarization is proportional to the external field. In addition, if the dielectric is isotropic (uniform electrical properties in all directions), **P** has the same direction as the external field. These conditions allow us to write

$$\mathbf{P} = \epsilon_0 \chi_e \mathbf{E}$$

where χ_e is a dimensionless quantity called the **electric susceptibility**. We should note that the presence of the polarization charge on the surface of a dielectric will modify the external applied field. It is this modified field that appears in the equation relating **P** and **E**.

10.3 ELECTRIC DISPLACEMENT

Let us consider a parallel-plate capacitor with plate area A and separation d, with the space between the plates completely filled with a dielectric material. In the absence of the dielectric, the field between the plates would have magnitude

$$E = \frac{\sigma}{\epsilon_0}$$

where σ is the "free" charge density on the plates. When the dielectric is present, it becomes polarized in the presence of the external field as shown in Fig. 10.3. The effective field between the plates of the capacitor is altered by the presence of the polarization charge. If we apply Gauss' law as we did in Section 5.2, we note that the effective charge density within the appropriate Gaussian surface is not σ but $\sigma + \sigma_p$ and the electric field becomes

$$E = \frac{1}{\epsilon_0}(\sigma + \sigma_p).$$

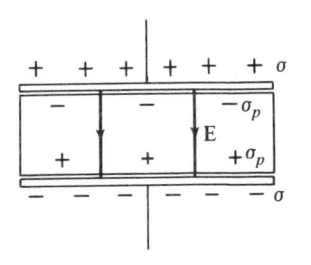

Fig. 10.3. Polarization of a dielectric within a capacitor.

The polarization charge resides on the surface of the dielectric which is adjacent to the plate of the capacitor. The close proximity of the free and polarization charges leads to an effective charge $\sigma + \sigma_p$ on the capacitor. Since σ_p is of the opposite sign to σ, the field E is reduced in magnitude by the presence of the dielectric.

We can rewrite the equation for E to read

$$\sigma = \epsilon_0 E - \sigma_p = \epsilon_0 E + P.$$

The quantity on the left-hand side of this equation is the free charge density on the plates. The quantity on the right-hand side therefore depends on the free charge only and is defined to be the **electric displacement** D where

$$D = \epsilon_0 E + P$$

or, in vector form,

$$\mathbf{D} = \epsilon_0 \mathbf{E} + \mathbf{P}.$$

Note that **D** is not altered by the presence of the dielectric; the polarization **P** of the dielectric is compensated by a reduction in **E** due to the presence of the dielectric such that **D** remains constant.

Capacitance

The capacitance of the capacitor is defined to be

$$C = \frac{q}{V}$$

where q is the charge on the plates and V is the potential between the plates. The charge on the plates is the free charge $q = \sigma A$ while the potential between the plates becomes $V = Ed$ where E is the field which is reduced due to the presence of the dielectric. The capacitance with the dielectric present is

$$C = \frac{\sigma A}{Ed} = \frac{(\epsilon_0 E + P)A}{Ed} = \frac{(\epsilon_0 E + \epsilon_0 \chi_e E)A}{Ed}$$
$$= \frac{\epsilon_0(1 + \chi_e)A}{d}.$$

The capacitance with no dielectric present is simply $\epsilon_0 A/d$ so that the dielectric constant K, which was defined in Section 10.1, is

$$K = 1 + \chi_e.$$

The capacitance with a dielectric present can also be written in the form

$$C = \frac{\epsilon A}{d}$$

where ϵ is known at the **permittivity** of the dielectric. The quantities ϵ, K, and χ_e are related through the equations

$$\epsilon = \epsilon_0(1 + \chi_e)$$

and

$$K = \frac{\epsilon}{\epsilon_0}.$$

This latter relation shows that the dielectric constant is the ratio of the permittivity of the dielectric material to that of free space.

Let us now look a little more closely at the vectors **P**, **D**, and **E**. The relation between them is given by the equations

$$\mathbf{D} = \epsilon \mathbf{E} = \epsilon_0 \mathbf{E} + \mathbf{P}.$$

In isotropic dielectrics **P**, **D**, and **E** are all in the same direction (see Fig. 10.4). The vector **D** is independent of whether or not there is a dielectric present. When there is no dielectric,

$$\mathbf{D} = \epsilon_0 \mathbf{E}$$

since $\mathbf{P} = 0$; when a dielectric is present **E** is reduced by the factor ϵ/ϵ_0 and

$$\mathbf{D} = \epsilon \mathbf{E} = \epsilon_0 \mathbf{E} + \mathbf{P}.$$

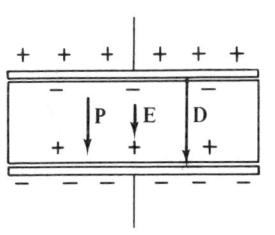

Fig. 10.4. The vectors **D**, **P**, and **E** all act in the same direction in an isotropic dielectric.

Example. A parallel-plate capacitor has plates of surface area $A = 100$ cm² which are separated by $d = 0.50$ cm. Calculate the capacitance. A slab of pyrex glass, dielectric constant 5.6, which just fills the space between the plates is slid between them. Deduce the capacitance with the slab in place. If the capacitor is charged and then disconnected before the dielectric slab is slid into place, what happens to the electric field E between the plates as a result of the insertion of the dielectric?

Solution. The capacitance without the dielectric slab in place is

$$C_0 = \frac{\epsilon_0 A}{d} = \frac{8.9 \times 10^{-12} \times 10^{-2}}{0.50 \times 10^{-2}}$$

$$= 18 \text{ pF}.$$

The capacitance with the dielectric slab in place is

$$C = \frac{\epsilon A}{d} = \frac{\epsilon_0 K A}{d}$$

$$= \frac{8.9 \times 10^{-12} \times 5.6 \times 10^{-2}}{0.50 \times 10^{-2}}$$

$$= 100 \text{ pF}.$$

If the source of charge is disconnected before the dielectric slab is inserted, the charge q on the plates must remain constant. Therefore, since

$$q = CV,$$

if C increases by a factor of 5.6, the potential difference V must decrease by the same factor. But

$$E = \frac{V}{d}$$

so that E must also be reduced by a factor of 5.6.

Example. Given that the dielectric constant of teflon is $K = 2.0$ determine values for the electric susceptibility and the permittivity. Deduce the magnitude of the dipole moment and the surface charge density of a slab of teflon that just fills the space between the plates of the capacitor described in the previous example for a potential difference of 1000 V between the plates of the capacitor.

Solution. The electric susceptibility of teflon is

$$\chi_e = K - 1$$
$$= 1.0$$

and the permittivity is

$$\epsilon = K\epsilon_0$$
$$= 2.0 \times 8.9 \times 10^{-12}$$
$$= 1.8 \times 10^{-11} \text{ C}^2 \cdot \text{N}^{-1} \cdot \text{m}^{-2}.$$

The polarization P of the teflon slab in the capacitor is

$$P = \epsilon_0 \chi_e E = \epsilon_0 \chi_e \frac{V}{d}$$
$$= \frac{8.9 \times 10^{-12} \times 1.0 \times 10^3}{0.50 \times 10^{-2}}$$
$$= 1.8 \times 10^{-6} \text{ C} \cdot \text{m}^{-2}.$$

The surface charge density of the teflon slab is therefore

$$\sigma_p = 1.8 \times 10^{-6} \text{ C} \cdot \text{m}^{-2}.$$

The magnitude of the dipole moment of the slab is

$$PAL = 1.8 \times 10^{-6} \times 10^{-2} \times 0.50 \times 10^{-2}$$
$$= 9.0 \times 10^{-11} \text{ C} \cdot \text{m}.$$

Stored energy

We are now in a position to write down a more general expression for the energy stored in an electric field. We have already shown in Section 5.6 that the energy density in an electric field when no dielectric is present is

$$u = \tfrac{1}{2}\epsilon_0 E^2 \text{ J} \cdot \text{m}^{-3}.$$

Since $D = \epsilon_0 E$ in this case, we can also write

$$u = \tfrac{1}{2} DE \text{ J} \cdot \text{m}^{-3}.$$

SEC. 10.3 ELECTRIC DISPLACEMENT

The insertion of a dielectric material between the plates of the capacitor changes the relation between D and E to

$$D = \epsilon E.$$

We could proceed to calculate the stored energy as in Section 5.6, replacing ϵ_0 by ϵ to take account of the presence of the dielectric. The result would again be (see Problem 13)

$$u = \tfrac{1}{2} DE \text{ J} \cdot \text{m.}^{-3}.$$

Example. Calculate the change in the stored energy of the capacitor described in the example given on p. 177 as a result of inserting the dielectric slab.

Solution. The initial energy U_0 is

$$U_0 = \frac{1}{2} D_0 E_0 (Ad)$$
$$= \frac{1}{2} \epsilon_0 \left(\frac{V_0}{d}\right)^2 (Ad)$$

The stored energy U after the dielectric slab is inserted is

$$U = \frac{1}{2} DE(Ad)$$
$$= \frac{1}{2} \epsilon \left(\frac{V}{d}\right)^2 (Ad)$$
$$= \frac{1}{2} K\epsilon_0 \left(\frac{V_0}{dK}\right)^2 (Ad)$$
$$= \frac{U_0}{K}.$$

The energy is reduced by a factor K when the dielectric slab is inserted. The change in energy ΔU is

$$\Delta U = U_0 - U = U_0 \left(1 - \frac{1}{K}\right).$$

For $K = 5.6$ (pyrex glass),

$$\frac{\Delta U}{U_0} = 1 - \frac{1}{5.6} = 0.82.$$

You could easily understand this "disappearance" of energy if you were to try to insert the slab into the capacitor without acceleration. You would find that it would be necessary to exert a restraint upon the slab.

Displacement current

In Section 9.4 we showed that the displacement current flowing in the space between the plates of a capacitor is $d\sigma/dt$. Now

$$\sigma = D = \epsilon_0 E + P$$

so that
$$\frac{d\sigma}{dt} = \frac{dD}{dt} = \epsilon_0 \frac{dE}{dt} + \frac{dP}{dt}.$$
When there is no dielectric between the plates
$$\frac{d\sigma}{dt} = \epsilon_0 \frac{dE}{dt}$$
in agreement with our result of Section 9.4. The additional term dP/dt is just the rate of change of polarization in a dielectric. It is associated with the actual motion of charge in the dielectric during the formation, rotation, or destruction of atomic or molecular dipoles. The reason for the term displacement current now becomes apparent; this current is just the rate of change of the electric displacement and is associated with the rearrangement or "displacement" of charge in a dielectric.

Force between charges in a dielectric

The force between charges in a dielectric is found to be reduced by the factor K, the dielectric constant. The proof of this is left as a problem (see Problem 14). The force between two charges q_1 and q_2 becomes
$$\mathbf{F} = \frac{1}{4\pi\epsilon_0} \frac{q_1 q_2}{K r^2} \hat{\mathbf{r}}.$$
This equation is valid only for r large enough that the polarization charge density about each charge is spherically symmetric. The dielectric constant can be written as
$$K = \frac{\epsilon}{\epsilon_0}$$
so that the force law also has the form
$$\mathbf{F} = \frac{1}{4\pi\epsilon} \frac{q_1 q_2}{r^2} \hat{\mathbf{r}}$$
which is identical to that in free space with ϵ replacing ϵ_0.

Gauss' law for D

Since
$$\mathbf{D} = \epsilon_0 \mathbf{E} + \mathbf{P},$$
$$\int_S \mathbf{D} \cdot d\mathbf{S} = \epsilon_0 \int_S \mathbf{E} \cdot d\mathbf{S} + \int_S \mathbf{P} \cdot d\mathbf{S}$$
$$= q_t - q_p = q$$
where q is the free charge, q_p the polarization charge and q_t the total charge.

ELECTRIC DISPLACEMENT

The proof of the result of the integral for **P** is left as a problem (see Problem 15). Gauss' law for **D** is then

$$\int_S \mathbf{D} \cdot d\mathbf{S} = q.$$

This confirms that **D** is a vector quantity that depends only on free charge and is independent of the presence of dielectrics.

Boundary conditions at dielectric surfaces

We consider now the interface $ABCD$ between two dielectrics of dielectric constants K_1 and K_2 as pictured in Fig. 10.5. First we apply Gauss' law to the "pillbox." We assume that there is no free charge on the interface so that there is no free charge inside the pillbox. Gauss' law applied to this surface gives

$$\int_S \mathbf{D} \cdot d\mathbf{S} = 0.$$

If we assume the pillbox to have negligible dimensions perpendicular to the interface, the integral reduces to

$$-D_1(n)S + D_2(n)S = 0$$

with S the area of one of the ends of the pillbox, and $D_1(n)$ and $D_2(n)$ the components of D_1 and D_2, respectively, normal to the interface. That is,

$$D_1(n) = D_2(n).$$

This equation states that the **normal component of the electric displacement is continuous across the boundary between two dielectrics**.

The work done in carrying a charge around a closed path in a static electric field is zero (that is, a static electric field is a conservative field). This is also expressed by the equation

$$\oint \mathbf{E} \cdot d\mathbf{l} = 0.$$

Evaluating this integral around the closed path $abcda$, which has sides ad and bc of negligible length and sides $ab = cd = l$ gives

$$-E_1(t)l + E_2(t)l = 0$$

or

$$E_1(t) = E_2(t)$$

where $E_1(t)$ and $E_2(t)$ are the tangential components of $\mathbf{E}_1$ and $\mathbf{E}_2$, respectively, at the interface. Therefore, the **tangential component of E is continuous across the boundary between two dielectrics**.

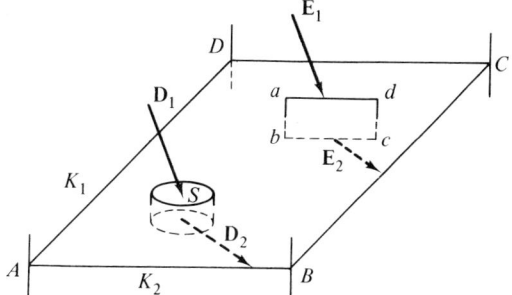

Fig. 10.5. The boundary conditions on **D** and **E** at the interface between two dielectrics are determined by application of Gauss' law to the "pillbox" and by evaluation of a line integral along the path $abcda$.

Refraction of electric field lines

As an example of the application of the boundary conditions on **D** and **E** at dielectric surfaces we consider the change in direction of the electric field lines at the interface between two isotropic dielectric regions characterized by dielectric constants K_1 and K_2. The problem is illustrated in Fig. 10.6. Note that **D** and **E** are parallel to each other in an isotropic dielectric. The boundary conditions on **D** and **E** at such an interface can be expressed by the equations

$$D_1 \cos \theta_1 = D_2 \cos \theta_2$$

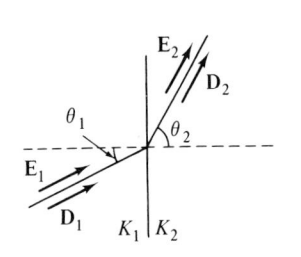

Fig. 10.6. E and D change direction at the boundary between two dielectrics.

and

$$E_1 \cos(90 - \theta_1) = E_2 \cos(90 - \theta_2)$$

or

$$E_1 \sin \theta_1 = E_2 \sin \theta_2.$$

Now $D_1 = K_1 \epsilon_0 E_1$ and $D_2 = K_2 \epsilon_0 E_2$. Therefore,

$$\frac{1}{K_1 \epsilon_0} \tan \theta_1 = \frac{1}{K_2 \epsilon_0} \tan \theta_2$$

or

$$\frac{\tan \theta_1}{\tan \theta_2} = \frac{K_1}{K_2}$$

which describes the refraction of the electric lines at the boundary between two dielectrics.

Example. Deduce the change in direction of the electric field lines incident upon a vacuum-teflon interface if the field lines in vacuum make an angle of 30° with the normal to the surface.

Solution. The dielectric constant of vacuum is $K_1 = 1.0$ and that of teflon is $K_2 = 2.0$. Therefore, since $\theta_1 = 30°$, substitution in the relation

$$\frac{\tan \theta_1}{\tan \theta_2} = \frac{K_1}{K_2}$$

yields

$$\tan \theta_2 = 2.0 \times \tan 30°$$
$$= 2.0 \times 0.577 = 1.154$$
$$\theta_2 = 49°.$$

The electric field lines have their direction changed by 19° at the interface.

10.4 STATIC DIELECTRIC CONSTANTS OF POLAR AND NONPOLAR MEDIA

The polarization of a dielectric, as we have already stated, actually takes place at the atomic level. We wish now to discuss the polarization of individual atoms or molecules. We shall talk only about molecules in this section, but we should remember that the theory holds also for atoms.

The macroscopic field in a dielectric is only an average field over the entire volume of the dielectric and is not a good representation of the field at the location of individual molecules. We define a **molecular polarizability** α by

$$\mathbf{p} = \alpha \mathbf{E}_l$$

where E_l is the magnitude of the local field at the position of the molecule and $\mathbf{p}$ is the induced molecular dipole moment due to the local field. If there are n molecules per unit volume, the polarization $\mathbf{P}$ is given by

$$\mathbf{P} = n\mathbf{p} = n\alpha \mathbf{E}_l.$$

The induced molecular dipole moment is in the direction of the applied field. The molecular configurations involved in its production are complicated and will not be discussed here. Some molecules possess permanent dipole moments; they are called **polar molecules**. These permanent dipole moments tend to orient themselves in the direction of an applied field and thereby contribute to the polarization of the dielectric.

Nonpolar dielectric media

Let us evaluate the local field $\mathbf{E}_l$ at some point A in an isotropic dielectric material as shown in Fig. 10.7. The local field is in general the sum of four components

$$\mathbf{E}_l = \mathbf{E}_0 + \mathbf{E}_1 + \mathbf{E}_2 + \mathbf{E}_3.$$

$\mathbf{E}_0$ is the field that would exist at A in the absence of any dielectric material. $\mathbf{E}_1$ is the field due to the polarization charge σ_p on the surface of the dielectric. For the flat-surfaced dielectric shown in Fig. 10.7, E_1 has magnitude $\sigma_p/\epsilon_0 = P/\epsilon_0$ (neglecting edge effects) and is in the direction opposite to $\mathbf{E}_0$. Therefore

$$\mathbf{E}_1 = -\frac{\mathbf{P}}{\epsilon_0}.$$

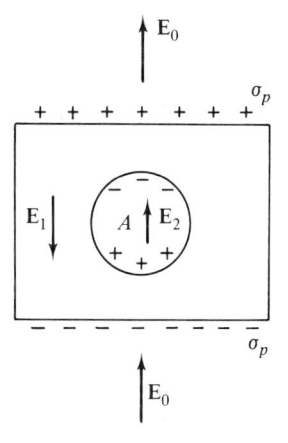

Fig. 10.7. Contributions to the local field at a point A in a dielectric.

The field $\mathbf{E}_1$ represents contributions to $\mathbf{E}_l$ from all polarized molecules which are sufficiently far from A that they appear as part of a continuous medium. However, molecules sufficiently close to A must be treated individually; the spherical boundary around A in Fig. 10.7 divides these molecules from the rest of the dielectric. The field $\mathbf{E}_2$ is the field that would exist at A due to polarization charge that would be formed on the surface of the sphere if the dielectric material within the spherical surface were removed. This induced field is given by

$$\mathbf{E}_2 = \frac{\mathbf{P}}{3\epsilon_0};$$

the proof of this is left as a problem (see Problem 29). Finally, $\mathbf{E}_3$ is the field due to polarized molecules within the sphere. For a cubic crystal lattice, $\mathbf{E}_3$ is identically zero, while for gases and weak solutions in which the molecules are moving at random $\mathbf{E}_3$ is so small that it can be neglected. For these dielectrics we can write

$$\mathbf{E}_l = \mathbf{E}_0 - \frac{\mathbf{P}}{\epsilon_0} + \frac{\mathbf{P}}{3\epsilon_0}$$

$$= \frac{\mathbf{D}}{\epsilon_0} - \frac{\mathbf{P}}{\epsilon_0} + \frac{\mathbf{P}}{3\epsilon_0} = \mathbf{E} + \frac{\mathbf{P}}{3\epsilon_0}.$$

The magnitude of the polarization of the dielectric can now be written as

$$P = np = n\alpha E_l = n\alpha\left(E + \frac{P}{3\epsilon_0}\right) = E(\epsilon - \epsilon_0).$$

Therefore,

$$E(\epsilon - \epsilon_0) = n\alpha E\left(1 + \frac{\epsilon - \epsilon_0}{3\epsilon_0}\right)$$

or

$$\frac{n\alpha}{3\epsilon_0} = \frac{(\epsilon - \epsilon_0)}{(\epsilon + 2\epsilon_0)} = \left(\frac{K-1}{K+2}\right)$$

which relates the molecular polarizability to the dielectric constant of the material.

The values of the dielectric constants of some common nonpolar gases are listed in Table 10.1. It is seen that K differs from unity by amounts of the order of 1 part in 10^3 and that K increases with the size of the molecule.

Table 10.1 Measured dielectric constants of some nonpolar gases at *NTP*.

Gas	K
Helium (He)	1.000071
Hydrogen (H$_2$)	1.000270
Oxygen (O$_2$)	1.000531
Nitrogen (N$_2$)	1.000588
Methane (CH$_4$)	1.000948
Ethylene (C$_2$H$_4$)	1.00138

Polar dielectric media

The discussion presented above should hold for polar as well as nonpolar molecules since the application of an electric field will always cause a distortion of the molecule and thereby produce an induced dipole moment. For a polar molecule, an additional effect arises from the pres-

SEC. 10.4 STATIC DIELECTRIC CONSTANTS

ence of the permanent dipole moments. In zero applied electric field they have random orientations so that no net polarization of the dielectric occurs. When a nonzero electric field is present, it is energetically more favorable for the molecules to point in the direction of the field than against it so that a net polarization results.

The contribution to the polarization due to permanent dipole moments can be evaluated assuming a Maxwell–Boltzmann distribution[1] for the energy states of the polar molecules in an external field. The potential energy of an electric dipole of moment $\mathbf{p}'$ in an electric field $\mathbf{E}_l$ is $-\mathbf{p}' \cdot \mathbf{E}_l$ (see Section 3.4). The probability $W(\theta)$ that the molecular dipole will take up an orientation θ with respect to the direction of the external field is then

$$W(\theta) \propto \exp\left(\frac{\mathbf{p}' \cdot \mathbf{E}_l}{kT}\right)$$

or

$$W(\theta) \propto \exp\left(\frac{p' E_l \cos \theta}{kT}\right)$$

where T is the temperature of the dielectric.

In order to determine the average electric moment in the direction of the external field, we need to know the number of moments n_θ per unit volume with orientation θ. Obviously, we can write

$$\frac{n_\theta}{n_{\pi/2}} = \frac{\exp(p' E_l \cos \theta / kT)}{\exp\left(p' E_l \cos \dfrac{\pi}{2} \Big/ kT\right)}$$

or

$$n_\theta = n_{\pi/2} \exp\left(\frac{p' E_l \cos \theta}{kT}\right).$$

The number of moments per unit volume with orientations between θ and $\theta + d\theta$ is

$$dn_\theta = \frac{d}{d\theta}(n_\theta) = -n_{\pi/2} \exp\left(\frac{p' E_l \cos \theta}{kT}\right) \frac{p' E_l}{kT} \sin \theta \, d\theta.$$

The average value of the electric moment in the direction of the external field is

$$\overline{p' \cos \theta} = \frac{\int_0^\pi (p' \cos \theta) \, dn_\theta}{\int_0^\pi dn_\theta} = \frac{p'^2 E_l}{3kT}$$

for $E_l p' \ll kT$. The evaluation of $\overline{p' \cos \theta}$ in this approximation is left as an exercise for the reader (see Problem 28).

The contribution of permanent dipole moments to the polarization is

---

[1] *MWTP*, Section 19.4.

seen to be temperature dependent. This contribution is in addition to the **induced** polarization. For dielectrics with permanent dipole moments, we should write for the electric dipole moment of a molecule in a dielectric in an external field E_l

$$p = p_i + \overline{p' \cos \theta} = \left(\alpha_i + \frac{p'^2}{3kT}\right) E_l = \alpha E_l$$

or

$$\alpha = \alpha_i + \frac{p'^2}{3kT},$$

where p_i and α_i are the induced dipole moment and polarizability, respectively. The relation between molecular polarizability and dielectric constant now becomes

$$\frac{n}{3\epsilon_0}\left(\alpha_i + \frac{p'^2}{3kT}\right) = \left(\frac{K-1}{K+2}\right)$$

or

$$\frac{K-1}{K+2} = \frac{np'^2}{9\epsilon_0 k}\left(\frac{1}{T}\right) + \frac{n\alpha_i}{3\epsilon_0}.$$

A plot of $(K-1)/(K+2)$ vs $1/T$ is a straight line of slope $np'^2/9\epsilon_0 k$ and intercept $n\alpha_i/3\epsilon_0$. The permanent dipole moment is determined from the slope of the line. Nonpolar materials have no permanent dipole moments and the resultant line is parallel to the $1/T$ axis (see Fig. 10.8). Such graphical procedures allow us to determine the permanent dipole moments and induced molecular polarizabilities of polar gases where the contribution to the local field from nearby molecules is negligible.

The values of the dielectric constants and permanent dipole moments of several polar gases are listed in Table 10.2. A comparison of these dielectric constants with those for nonpolar gases given in Table 10.1 shows that dielectric constants for polar gases are markedly higher than for nonpolar gases.

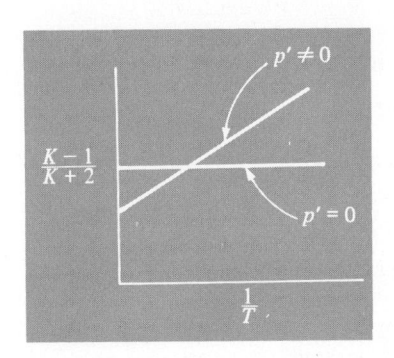

Fig. 10.8. Plots of $(K-1)/(K+2)$ vs $(1/T)$ for a gas.

Table 10.2 Measured dielectric constants of some polar gases at *NTP*

Gas	Dipole moment [2] (debyes)	K
Carbon monoxide (CO)	0.10	1.000692
Nitrous oxide (N$_2$O)	0.17	1.00108
Ammonia (NH$_3$)	1.45	1.00834
Sulphur dioxide (SO$_2$)	1.59	1.00993

[2] 1 debye (D) = 3.336×10^{-30} C·m

The application of the above theory to polar liquids and polar solids leads to nonsensical results because the local field is now very large and a description for it as given above is completely inadequate.

10.5 FERROELECTRIC CRYSTALS

A crystal that exhibits a spontaneous macroscopic electric dipole moment is by definition a **ferroelectric crystal**. In these crystals the center of positive charge does not coincide with the center of negative charge.

The simplest crystal structure that allows ferroelectricity is the perovskite structure. Barium titanate ($BaTiO_3$) is an example of a ferroelectric crystal having this structure. At a temperature known as the **Curie temperature** T_c, the crystal undergoes a structural transition from the low temperature, polarized state to the high temperature, unpolarized state. In the high temperature phase $BaTiO_3$ has the cubic structure shown in Fig. 10.9. The Ba^{2+} ions occur at the cube corners, the O^{2-} ions at the face centers, and the Ti^{4+} ion at the body center. Below the Curie temperature the structure is slightly deformed so that the crystal has tetragonal symmetry.[3] In general, when a crystal is cooled through the Curie temperature, it is found that the entire crystal does not have the same tetragonal axis. That is, different regions of the crystal have different directions of spontaneous polarization. A region in which the spontaneous polarization is all in the same direction is called a **ferroelectric domain**.

The occurrence of ferroelectricity in substances having the perovskite structure is thought to result from a **polarization catastrophe**. That is, the local electric fields that result from the polarization itself increase faster than the electric restoring forces on the ions in the crystal. This leads to an asymmetrical shift in the ionic positions. The anharmonic contribution[4] to the restoring forces limits the shift to a finite displacement.

The spontaneous polarization P_s of $BaTiO_3$ at room temperature has a maximum value of $0.25\ C \cdot m^{-2}$. Since the volume of a unit cube is $\simeq 5 \times 10^{-29}\ m^3$, the dipole moment per unit cube is $\simeq 10^{-29}\ C \cdot m$. This polarization could be accounted for by a displacement of the Ti^{4+} ion parallel to one edge of the cube by an amount $\simeq 0.25 \times 10^{-10}\ m$.

In Fig. 10.10 we show the variation of the dielectric constant of $BaTiO_3$ with temperature. We see that the dielectric constant increases rapidly at the Curie temperature T_c. Above T_c the decrease in the dielectric constant with increasing temperature is almost hyperbolic and is reasonably well represented by an equation of the form

$$K(T) = K_0 + \frac{C}{T - \theta}$$

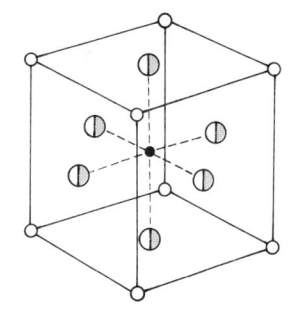

Fig. 10.9. Structure of $BaTiO_3$ above its Curie temperature.

[3] *MWTP*, Section 21.2.
[4] *MWTP*, Section 22.7.

where K_0, C, and $\theta \simeq T_c$ are constants. BaTiO$_3$ exhibits a **hysteresis effect** which is evident in Fig. 10.10. That is, the curve of K vs T measured as the temperature is lowered through the transition region does not coincide exactly with the curve of K vs T measured as the temperature is raised through the transition region.

We should also note that the dielectric constant is usually very much larger when measured perpendicular to the tetragonal axis (called the c axis) than when measured parallel to it.

In a ferroelectric crystal the electric displacemet **D** is not proportional to the electric field **E** except for very small values of **E**. In this case we define the permittivity ϵ_c parallel to the c axis, say, as

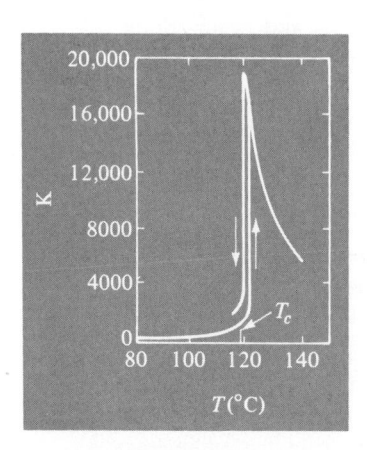

$$\epsilon_c = \frac{D_c}{E_c}.$$

Fig. 10.10. Dielectric constant as a function of temperature for BaTiO$_3$.

That is, the permittivity (or the dielectric constant) is a function of direction, of temperature, and of the applied electric field.

QUESTIONS AND PROBLEMS

1. Show that in general

$$P_n = -\sigma_P$$

where P_n is the component of the polarization **P** normal to a surface and σ_P is the polarization surface charge density.

2. In Section 3.7 it was shown that the electric potential due to a dipole **p** is given by

$$V = \frac{1}{4\pi\epsilon_0} \frac{\mathbf{p} \cdot \hat{\mathbf{r}}}{r^2}.$$

(a) Show that the electric potential at a point external to a dielectric of polarization **P** is given by

$$V = \int_v \frac{1}{4\pi\epsilon_0} \mathbf{P} \cdot \nabla\left(\frac{1}{r}\right) dv.$$

(b) By using first the identity $\nabla \cdot \phi\mathbf{A} = \phi\nabla \cdot \mathbf{A} - \mathbf{A} \cdot \nabla\phi$ and then the divergence theorem, show that

$$V = \frac{1}{4\pi\epsilon_0} \int_S \frac{\mathbf{P} \cdot d\mathbf{S}}{r} - \frac{1}{4\pi\epsilon_0} \int_v \frac{\nabla \cdot \mathbf{P}}{r} dv.$$

(c) This potential can be ascribed to a surface density and a volume density of charge. What are these densities?

(d) Explain physically why a volume charge density is associated with a nonuniform polarization.

3. Using Maxwell's first equation and the results of the preceeding problem, show that

$$\nabla \cdot \mathbf{D} = \rho.$$

4. Show that for a homogeneous, linear medium of dielectric constant K

$$\rho' = -\frac{(K-1)}{K}\rho$$

where ρ' is the volume polarization charge density and ρ is the density of unbound charge.

5. Describe the construction of several types of commercial capacitors.

6. A parallel-plate capacitor is filled with two dielectrics as indicated in Fig. 10.11. Derive an expression for the capacitance. What does the expression become for the special case of two dielectrics of the same dimensions?

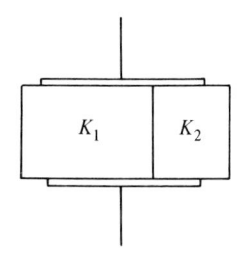

Fig. 10.11. A capacitor with two dielectrics between its plates.

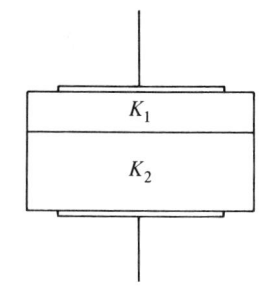

7. A parallel-plate capacitor is filled with two dielectrics as indicated in Fig. 10.12. Derive an expression for the capacitance. What does the expression become for the special case of two dielectrics of the same dimensions?

8. A coaxial cable has inner and outer conductors of 0.10 cm and 0.30 cm radii, respectively. Calculate the capacitance per unit length if the region between the inner and outer conductors is filled with a dielectric having a dielectric constant 7.0.

9. The dielectric constant varies linearly from K_1 to K_2 between the plates of a parallel-plate capacitor. Assuming the separation between the plates is d and the area A, calculate the capacitance.

10. A parallel-plate capacitor has plate area 150 cm² and separation 1.5 cm. A potential difference of 200 V is applied to the capacitor to charge it. The source of the charge is then disconnected from the capacitor and a dielectric slab of dielectric constant 6.0 and thickness 0.75 cm is inserted midway between the plates as indicated in Fig. 10.13.
    (a) Calculate the free charge.
    (b) Calculate **E**, **D**, and **P** in the air gap.
    (c) Calculate **E**, **D**, and **P** in the dielectric.
    (d) Calculate the potential difference between the plates.
    (e) Calculate the capacitance with the slab in place.

Fig. 10.12. A capacitor with two dielectrics between its plates.

11. For the capacitor described in Problem 10, calculate the energy stored in the air gaps and in the dielectric slab.

12. What happens to the energy stored in a parallel-plate capacitor when a dielectric of constant K is inserted in the space between the plates
    (a) if a constant potential is maintained between the plates.
    (b) if a constant charge is maintained on the plates.

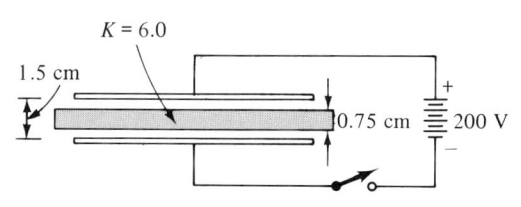

Fig. 10.13. A parallel-plate capacitor partially filled with a dielectric slab.

13. Using the example of a parallel-plate capacitor show that the energy density in an electric field in the presence of a dielectric is
$$u = \tfrac{1}{2} DE \text{ J} \cdot \text{m}^{-3}.$$

14. Show that the force between two point charges in a dielectric is reduced from that in free space by the factor K, the dielectric constant.

15. Show that (as stated on page 180)
$$\int_S \mathbf{P} \cdot d\mathbf{S} = -q_p$$
where q_p is the polarization charge.

*16. A hydrogen atom in its ground state[5] is situated in an electric field of 2.0×10^6 V·m^{-1}. Use the Bohr theory to estimate the resulting electric dipole moment of the atom. This phenomenon is known as the **Stark effect**.

17. When an atom of boron is diffused into a silicon crystal, four of its electrons bind covalently with adjacent silicon atoms. The fifth, however, does not and can be treated as a nearly free electron in a Bohr orbit about a B^{1+} ion. Calculate the binding energy and the radius of this electron orbit. The dielectric constant of silicon is 11.7. Impurity atoms, such as described above, have wide use in semiconductors as we shall see in Chapter 12.

18. Show that the polarization charge σ_P at the interface of a dielectric and conductor is related to the surface charge σ of the conductor by:
$$-\sigma_P = \frac{K-1}{K} \sigma.$$

*19. A dielectric sphere of constant K is placed in a uniform electric field E_0. Show that the field outside the sphere is equal to the field E_0 plus a term due to a dipole at the center of the sphere. Deduce the magnitude of the dipole moment and the magnitude E_1 of the field within the sphere. Hint: requiring continuity of the potential at the boundary of the sphere is equivalent to requiring continuity of the tangential component of $\mathbf{E}$.

*20. Calculate the potential at the center of a dielectric sphere of radius R which contains a uniform charge density ρ.

21. Show that the speed v of an electromagnetic wave in a dielectric material of constant K is given by
$$v = \frac{c}{K^{1/2}}$$
where c is the speed of an electromagnetic wave in vacuum.

22. Calculate the average value of the energy flux of a plane electromagnetic wave in a nonmagnetic medium of dielectric constant $K = 4.0$ if the maximum magnetic field is 6.5×10^{-7} Wb·m^{-2}.

23. Consider a dielectric sphere of radius R and dielectric constant K with a point charge q at the center. Calculate $\mathbf{E}$, $\mathbf{P}$, and $\mathbf{D}$ as functions of r, the radial

[5] *MWTP*, Section 18.4.

distance. What is the total bound charge on the outer surface of the sphere? What is the bound charge on the surface adjacent to the point charge q? What is the net charge at the center of the sphere?

24. A certain material is composed of dielectric sheets, thickness a, stacked on top of each other as shown in Fig. 10.14. The dielectric constant of each sheet is given by the relation $K_n = nK$ where n is the number of the sheet counting from the top. An electric line is incident on the material at the origin at an angle θ_0. Calculate the coordinates of the points where the line intersects the boundaries of the sheets.

25. Deduce the polarizabilities of the helium and hydrogen molecules from their dielectric constants as given in Table 10.1.

26. The static dielectric constants of ammonia (NH_3) as measured at 0°C and 100°C, and a pressure of 1 atmosphere, are 1.00834 and 1.00487, respectively. Deduce the permanent electric dipole moment and polarizability of an NH_3 molecule. Estimate the radius a of an NH_3 molecule assuming the polarizability to be given by $\alpha = 4\pi\epsilon_0 a^3$.

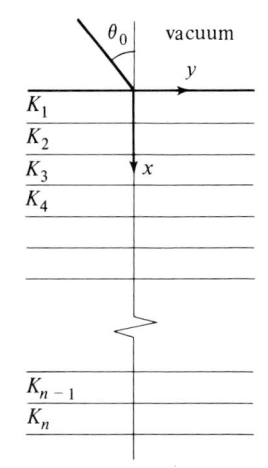

Fig. 10.14. An electric line incident on a stack of dielectric sheets.

27. The water molecule has a dipole moment of 6.2×10^{-30} C·m. Calculate the dipole moment per unit volume of water at a temperature of 250°C and at a pressure[6] of 760 mm of Hg and in an electric field $E_l = 1.0 \times 10^4$ V·m^{-1}.

28. Show that, for a distribution of permanent electric dipole moments $\mathbf{p}'$ in thermal equilibrium in an electric field $\mathbf{E}_l$, the average value of the electric moment in the direction of the external field is $p'^2 E_l/3kT$ provided that $E_l p' \ll kT$.

29. Show that the contribution $\mathbf{E}_2$ to the local field in a dielectric as defined in Section 10.4 is given by

$$\mathbf{E}_2 = \frac{\mathbf{P}}{3\epsilon_0}.$$

30. Discuss the relation between **ferroelectric, piezoelectric,** and **pyroelectric** crystals.

31. **Piezoelectricity** is a property manifested by various dielectric materials whereby mechanical pressure exerted upon the crystal results in the appearance of electric charge on its surfaces. Write an essay on piezoelectricity and include a discussion of its practical applications.

*32. Assume that the Curie point in a ferroelectric crystal is determined approximately by the interaction energy of an electric dipole moment with the local internal electric field caused by the polarization itself. The interaction for self-energy problems is $-\frac{1}{2}(\mathbf{p} \cdot \mathbf{E})$. Roughly estimate T_c for $BaTiO_3$. The observed value of T_c is 380°K.

[6] *MWTP*, Section 23.4.

GEORG S. OHM

11 Current, resistance, and the free electron theory of metals

11.1 INTRODUCTION

The atoms of a metal are joined together in such a way that the valence electrons of the individual atoms may be common to the entire atomic lattice of the metal. This basic concept underlies the theory of metals. We can think of these electrons as forming a "gas" in the metal. This gas interacts strongly with the positive ions of the metallic structure, providing a strong cohesive force. When an external electric field is applied to the metal, the "free" electrons experience a force and a current or flow of electrons results. If the crystal lattice were rigid and free from irregularities, the electrons would pass through it without encountering resistance. The resistance of a metal to the passage of electrical current is due to deviations from a perfect lattice caused by lattice defects, impurity atoms, and thermal vibrations of the atoms comprising the lattice. In this chapter we shall concern ourselves with the motion of the free electrons in a metal including the resistance offered by the lattice to electron movement. Finally, we shall discuss the flow of current in circuits containing resistance, capacitance, and inductance. In the next chapter we shall investigate more closely the reasons for the presence of free electrons in metals.

11.2 OHM'S LAW

We assume that the electrons in a metallic conductor are free to move in random directions and make periodic collisions with other electrons and with the lattice ions. In the absence of an external electric field the motion of a typical electron might be like that shown in Fig. 11.1(a).

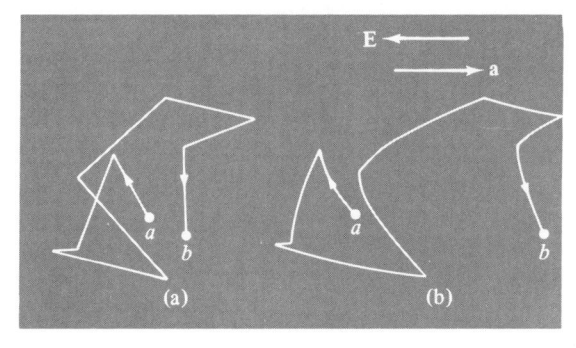

Fig. 11.1. A typical electron path in a metal: (a) with no electric field; (b) with an electric field. The drift component of velocity to the right in (b) has been exaggerated for clarity.

When an electric field E is present, each electron experiences a force $F = eE$ with a resultant acceleration

$$a = \frac{F}{m_e} = \frac{eE}{m_e}.$$

The electron paths between collisions become curved as shown qualitatively in Fig. 11.1(b). If τ is the average time between collisions, the average increase in velocity Δv to the right in Fig. 11.1(b) is

$$\Delta v = a\tau$$
$$= \frac{eE\tau}{m_e}.$$

We assume that each time an electron makes a collision it starts off again in some random direction. Therefore, on the average, after a collision, an electron has zero velocity while at the end of the time interval τ it has velocity Δv in the direction opposite to the field. We see that the **average drift speed** v_d through the conductor is

$$v_d = \frac{\Delta v}{2} = \frac{eE\tau}{2m_e}.$$

In vector notation we write

$$\mathbf{v}_d = -\frac{e\tau}{2m_e}\mathbf{E}.$$

Let us suppose that we have a wire of length L, cross-sectional area A, with n electrons per unit volume. The current flowing in this wire is given by

Current = free electron density × cross-sectional area × charge per electron × electron drift speed.

Therefore,

$$I = nAev_d$$
$$= \frac{nAe^2E\tau}{2m_e}.$$

If the potential difference between the ends of the wire is V, then

$$E = \frac{V}{L}$$

and

$$I = \frac{nAe^2\tau}{2m_e L}V.$$

Experiments carried out by Georg Simon Ohm, long before the nature of metals was understood, had shown that the current through a conductor is proportional to the potential difference between its ends, provided that the temperature remains constant. This relation is known as **Ohm's law** and is usually written as

$$V = IR$$

where the constant of proportionality R is called the **resistance** of the conductor. According to our free electron model, the resistance is given by the expression

$$R = \frac{2m_e}{ne^2\tau} \cdot \frac{L}{A}$$

$$= \rho \frac{L}{A}$$

where

$$\rho = \frac{2m_e}{ne^2\tau}$$

is called the **resistivity** of the conductor. The resistivity is a function only of the properties of the conductor and not of its geometric shape. The unit of resistance is the **ohm** (Ω); a conductor has a resistance of one ohm if a potential difference of 1 volt applied across it causes a current of 1 ampere to flow.

Conductors are sometimes identified according to their **conductivity** σ which is just the reciprocal of the resistivity ρ. Therefore,

$$\sigma = \frac{1}{\rho} = \frac{ne^2\tau}{2m_e}.$$

Both ρ and σ are often expressed in terms of the mean random electron speed $\bar{v}$ between collisions and the **mean free path** λ traveled by electrons between collisions. Obviously

$$\tau = \frac{\lambda}{\bar{v}}$$

so that

$$\rho = \frac{2m_e \bar{v}}{ne^2 \lambda} = \frac{1}{\sigma}.$$

(It might be noted that a more sophisticated treatment of the random electron motion leads to a similar result but without the factor 2.) Some typical values of the resistivity of pure metals at 20° C are listed in Table 11.1.

In general, the resistance R of an element in a circuit is defined as

$$R = \frac{V}{I}$$

where V is the potential difference across the element and I is the current passing through it. If R is a constant, the element is **ohmic**. Metallic ele-

Table 11.1 Resistivities of Some Pure Metals at 20°C

Metal	Resistivity (Ohm·m)
Silver	1.6×10^{-8}
Copper	1.7×10^{-8}
Gold	2.4×10^{-8}
Aluminum	2.7×10^{-8}
Zinc	5.9×10^{-8}
Platinum	10.6×10^{-8}

ments maintained at constant temperature are ohmic. If R is not constant, the element is **nonohmic**. Elements with this property include diodes and transistors which are incorporated into essentially every electronic circuit.

11.3 THE SEPARATION OF LIVING CELLS ACCORDING TO THEIR VOLUMES

Biophysicists have come to realize the potential importance of being able to separate living cells according to such physical characteristics as their size. One method that can be used to separate cells according to their volume is based on the simple physical principle that for a given source of emf the flow of current in a conductor is inversely proportional to the resistance of the conducting medium. To illustrate how the method works let us consider Fig. 11.2. A fluid of resistivity ρ flows through a cylindrical orifice of radius b. The section of fluid forms part of an electrical circuit and therefore its resistance determines in part the current flowing in the circuit. We suppose that there are spherical cells of infinite resistivity suspended in the liquid. Each time that a cell appears in the orifice the resistance of the circuit will increase and the current flow through the circuit will temporarily decrease. That is, by monitoring the current as a function of time it is possible to identify each cell as it passes through the orifice through the observation of a momentary decrease in the current (a current pulse).

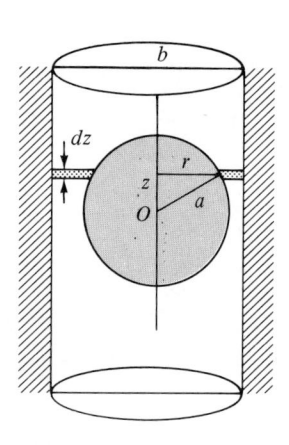

Fig. 11.2. A cell suspended in a fluid flowing through a cylindrical orifice of radius b. An annular ring of fluid of thickness dz is indicated. The size of the cell has been exaggerated for clarity.

The question that we now wish to answer is how, if at all, the current pulses reflect the sizes of the cells. We calculate the change ΔR in the resistance of the section of fluid that results from the presence of a spherical cell of radius a. We choose the origin of the coordinate system to be at the center of the cell. The resistance dR of the annular ring of fluid of thickness dz indicated in Fig. 11.2 is

$$dR = \frac{\rho \, dz}{\pi(b^2 - r^2)}.$$

Therefore,

$$\Delta R = \int_{-a}^{a} \frac{\rho dz}{\pi(b^2 - r^2)} - \frac{\rho}{\pi b^2}\int_{-a}^{a} dz.$$

Note that we are only concerned with values of z satisfying $-a \leq z \leq a$, since the contribution to the resistance for other values of z is unchanged by the presence of the cell. The integrals may be evaluated to yield for $b \gg a$,

$$\Delta R \simeq \frac{2\rho a^3}{\pi b^4} = \frac{3\rho V}{2\pi^2 b^4}$$

where V is the volume of the cell. That is, the change in resistance and therefore the height of the current pulse due to the presence of the cell in the orifice are proportional to the volume of the cell.

A scheme has been devised whereby the cells acquire an electrical charge proportional to the height of the current pulse that they produce in passing through the orifice. The charged cells are then passed through an electric field that deflects each cell by an amount proportional to the charge that it carries. In this manner the cells may be separated according to their volume.

Blood cells in sickness and in health

It has been found that the profile of the distribution of blood cells according to their volume in a healthy human remains markedly constant over long periods of time. Almost all illnesses are reflected, to some degree, in the blood and thereby cause alterations in the profile. At the time of writing, blood samples are being studied at two hospitals in Toronto in order to try to establish a correlation between changes in blood profile and specific illnesses. In a preliminary test twenty-four blood samples were tested by a hematologist over a period of several days and then by a biophysicist in fifteen minutes using a physical technique equivalent to the one described above. The biophysicist and the hematologist agreed on which donors were healthy and which were sick in twenty-three of the cases. The possibility has been suggested that in the future your blood profile may form a part of your medical record to be used for the rapid diagnosis of illnesses.

Toward a cure for leukemia

The spread of leukemia may be arrested by administering to the patient a powerful cycle-acting drug that preferentially destroys cells, such as the cancerous cells, which are in the process of growing and dividing. Unfortunately, these drugs kill not only the leukemia cells but **all** cells in the process of growing and dividing, including the blood-forming stem cells found in the bone marrow. Therefore, although the leukemia may be eradicated, the patient will die unless he can be given a successful bone-marrow transplant. The problem associated with transplants is to overcome the body's natural tendency to reject foreign cells. The mechanism whereby the body rejects anything foreign to it is known as the **immune** response. It is the body's major defense mechanism for fighting infection by bacteria and viruses. The foreign object called an **antigen** activates cells in the body to produce a specific **antibody** which combines with the antigen to render it inactive.[1] This problem can be overcome in the following manner. A small test dose of bone marrow from the donor is injected into the recipient prior to the main bone-marrow transplant. The immune response cells in the body begin their job by starting to grow and divide. About four days after the injection they are most active in division. At this time the powerful cycle-acting drug is administered which now destroys the immune response cells as well as the leukemia cells and the blood-forming stem cells. If the main bone-marrow transplant is now attempted, the body will not reject it, since the immune response cells have been destroyed. But a new problem arises. The injected bone marrow contains immune response cells of its own which will act to reject the patient's cells and ultimately cause his death. This problem may be solved by separating out the immune response cells from the blood-forming stem cells in the donor's bone marrow and injecting only the latter cells into the patient.

If the cells in a bone-marrow cell suspension are separated according to their volumes, some sort of profile is obtained (see Fig. 11.3). Studies of irradiated mice that were injected with bone-marrow cells from different small portions of such a distribution have shown that only the bone-marrow cells of large volume are blood-forming stem cells. By mixing samples of bone-marrow cells from different small portions of the distribution with foreign cells and providing conditions for incubation, the immune response cells can be identified. The result is indicated schematically in Fig. 11.3. The cells of small volume are responsible for initiating the rejection mechanism. Therefore, only the bone-marrow cells of large volume are injected into the patient. The results of the first such transplant carried out in a Toronto hospital give considerable hope that a cure for leukemia may not be far off.

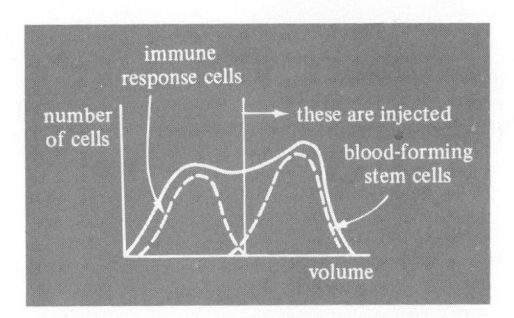

Fig. 11.3. The distribution of bone-marrow cells showing separately the immune response cells and the blood-forming stem cells.

---

[1] R. S. Speirs, "How Cells Attack Antigens," *Scientific American*, February, 1964. Available as *Scientific American Offprint 176* (San Francisco: W. H. Freeman and Co., Publishers).

11.4 THE FERMI ELECTRON GAS

We now enquire more closely into the nature of the electron gas that flows through a metal and examine some of its properties. Electrons are **fermions**[2] and therefore obey the Pauli exclusion principle.[3] If the electron gas were treated as a classical gas, the particles would have an average kinetic energy of $(3/2)kT$ where k is Boltzmann's constant and T is the temperature. At room temperature this average kinetic energy is about 0.025 eV. We shall see in the next chapter that the allowed energy states for the free electrons in a metal form bands which are several eV wide in energy. If all the conduction electrons in a metal were crowded into energy states lying within $\simeq 2 \times 0.025 = 0.05$ eV of the lowest energy state in the band, there would be more than 1000 electrons in each allowed quantum state. This clearly violates the Pauli exclusion principle and the **Fermi–Dirac distribution**[4] must be used.

The Fermi–Dirac distribution can be written as

$$\frac{N_i}{N} = \frac{1}{Z\left[\exp\left(\frac{E_i}{kT}\right) + 1\right]}$$

where N is the total number of particles, N_i is the number of particles having energy E_i (note that $N_i/N \leq 1$), and Z is the partition function.[5] For the large numbers of electrons ($\simeq 10^{20}$ or more) and very small separations between energy states ($\simeq 10^{-20}$ eV) in a typical metal, it is more convenient to consider a continuous distribution of electron energies and write the distribution as

$$n(E)\,dE = \frac{g(E)\,dE}{\exp\left(\frac{E - E_F}{kT}\right) + 1}$$

where $n(E)\,dE$ is the number of electrons per unit volume with energies between E and $E + dE$, $g(E)\,dE$ is the number of quantum states per unit volume with energy between E and $E + dE$ and E_F is a constant called the **Fermi energy** (see Problem 12).

The solution of the Schroedinger equation[6] for a gas of free electrons confined to a cubic volume of metal yields electronic energy states of energy

$$E = \frac{h^2}{32m_e a^2}(n_1^2 + n_2^2 + n_3^2)$$

---

[2] *MWTP*, Section 20.6.
[3] *MWTP*, Section 18.7.
[4] *MWTP*, Section 20.6.
[5] *MWTP*, Section 19.4.
[6] *MWTP*, Section 18.6.

where the **quantum numbers** n_1, n_2, n_3 can take integral values

$$n_i = 0, 1, 2 \ldots, \qquad i = 1, 2, 3;$$

h is Plancks' constant, m_e is the electron mass, and $2a$ is the length of the side of the cube. Three quantum numbers appear since the electron momentum can be written in terms of three components along three mutually perpendicular directions in the metal. There are actually two unique quantum states associated with each set of integers n_1, n_2, n_3—one for electron spin $+\frac{1}{2}$ and one for electron spin $-\frac{1}{2}$.

We can determine the number of quantum states per unit volume with energies between E and $E + dE$ by referring to Fig. 11.4, in which the values of n_1, n_2, and n_3 are plotted in **quantum number space**. Each point P represents a particular set of integer values of n_1, n_2, and n_3 and therefore represents two quantum states (with opposite spin). The volume of the spherical shell lying between r and $r + dr$ is $4\pi r^2\, dr$. The number of quantum states $g(r)\, dr$ between r and $r + dr$ is then

$$g(r)\, dr = 2 \times 4\pi r^2\, dr \times \tfrac{1}{8}$$
$$= \pi r^2\, dr$$

where the factor $\frac{1}{8}$ is introduced since negative values of the n's do not give different quantum states. That is, in counting states we must confine our counting to the all-positive octant of quantum number space.

Since $r^2 = n_1^2 + n_2^2 + n_3^2$, we can write

$$E = \frac{h^2}{32 m_e a^2} r^2$$

or

$$r = \frac{4a}{h}(2m_e E)^{1/2}$$

and

$$dr = \frac{2a}{h}(2m_e)^{1/2} E^{-1/2}\, dE.$$

Therefore,

$$g(E)\, dE = \pi \frac{(32 m_e a^2 E)}{h^2} \frac{2a}{h}(2m_e)^{1/2} E^{-1/2}\, dE$$
$$= \frac{64\pi a^3}{h^3} m_e^{3/2} (2E)^{1/2}\, dE$$

is the density of states in a cube of side $2a$ and volume $(2a)^3$ in real (x, y, z) space. Therefore, the density of states per unit volume is

$$g(E)\, dE = \frac{8\pi m_e^{3/2}}{h^3}(2E)^{1/2}\, dE.$$

Note that the cube dimensions do not appear in this expression which is, indeed, a general expression suitable for all geometries.

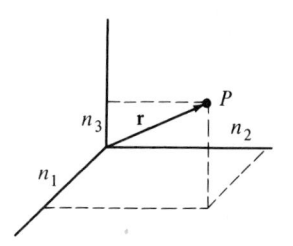

Fig. 11.4. Each point P in quantum number space represents a particular set of integer values of n_1, n_2, and n_3 and represents a pair of quantum states. All points in a spherical shell between r and $r + dr$ have approximately the same energy.

SEC. 11.4 THE FERMI ELECTRON GAS

The number of electrons in the electron gas having energies between E and $E + dE$ is then

$$n(E)\,dE = \frac{8\pi m_e^{3/2}(2E)^{1/2}\,dE}{h^3\left[\exp\left(\dfrac{E - E_F}{kT}\right) + 1\right]}.$$

First we shall examine the behavior of the term in brackets in the denominator. At $T = 0°$ K the exponential term is 0 for $E < E_F$ and is infinite for $E > E_F$. The reciprocal of the term in brackets is therefore unity for all energies less than the Fermi energy and zero for all energies greater than the Fermi energy. The behavior of the function

$$f(E) = \frac{n(E)}{g(E)} = \frac{1}{\exp\left(\dfrac{E - E_F}{kT}\right) + 1}$$

is shown in Fig. 11.5 for $T = 0°$K and for $T > 0°$K. As the temperature increases from $0°$ K, the number of particles per state changes from 1 to 0 more and more gradually with increase in the energy of the state.

The Fermi energy is determined by assigning electrons to the allowed energy states one by one in order of increasing energy until the last state is filled. This final state has energy $E = E_F$, by definition. If n is the density of electrons, then

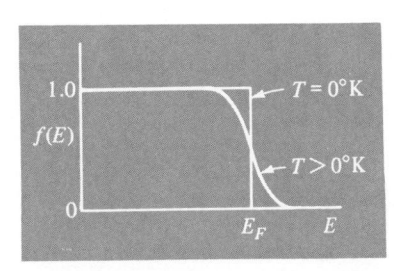

Fig. 11.5. The Fermi–Dirac distribution $f(E)$ for $T = 0°$K and $T > 0°$K.

$$n = \int_0^{E_F} g(E)\,dE$$

$$= \frac{8(2)^{1/2}\pi m_e^{3/2}}{h^3} \int_0^{E_F} E^{1/2}\,dE$$

$$= \frac{16(2)^{1/2}\pi m_e^{3/2}}{3h^3} E_F^{3/2}$$

and

$$E_F = \frac{h^2}{8m_e}\left(\frac{3n}{\pi}\right)^{2/3}.$$

Inserting numerical values for the constants into this expression gives

$$E_F = 3.66 \times 10^{-19} n^{2/3} \text{ eV}.$$

Example. Estimate the Fermi energy for copper.

Solution. The electron configuration[7] for a copper atom in the ground state is

$$1s^2 2s^2 2p^6 3s^2 3p^6 3d^{10} 4s.$$

Since each copper atom has a single electron in the $4s$ state outside closed shells, it is reasonable to assume that each copper atom contributes one electron to the

---

[7] *MWTP*, Section 18.7.

electron gas in the metal. The number of copper atoms per unit volume is

$$\frac{N_0 \rho}{W}$$

where $N_0 = 6.02 \times 10^{26}$ atoms·(kg·mole)$^{-1}$ is Avogadro's number, $\rho = 8.94 \times 10^3$ kg·m^{-3} is the density of copper, and $W = 63.5$ kg·(kg-mole)$^{-1}$ is the atomic weight of copper. Therefore,

$$n = \frac{6.02 \times 10^{26} \times 8.94 \times 10^3}{63.5}$$
$$= 8.48 \times 10^{28} \text{ atoms·m}^{-3}$$
$$= 8.48 \times 10^{28} \text{ electrons·m}^{-3}$$

and

$$E_F = 3.66 \times 10^{-19}(8.48 \times 10^{28})^{2/3}$$
$$= 7.06 \text{ eV}.$$

A piece of copper metal at 0°K would have electrons with energies up to 7.06 eV but no higher; classically, one would expect all energies to be zero.

The electron energy distribution

The number of electrons having energies between E and $E + dE$ is

$$n(E)\,dE = \frac{8\pi m_e^{3/2}(2E)^{1/2}\,dE}{h^3 \left[\exp\left(\dfrac{E - E_F}{kT}\right) + 1\right]}$$

$$= \left(\frac{3n}{2}\right) \frac{E_F^{-3/2} E^{1/2}\,dE}{\left[\exp\left(\dfrac{E - E_F}{kT}\right) + 1\right]}.$$

The function $n(E)$ is plotted in Fig. 11.6 for $T = 0°$ K and $T = 1000°$K. As T increases from 0°K, the shape of the curve in the region of the Fermi energy departs only slowly from the shape for $T = 0°$K.

The average energy $\bar{E}_0$ per electron at $T = 0°$K is found by calculating the total energy per unit volume E_0 and dividing by the electron density. That is,

$$\bar{E}_0 = \frac{E_0}{n} = \frac{1}{n} \int_0^{E_F} E n(E)\,dE$$
$$= \frac{1}{n}\left(\frac{3n}{2}\right) E_F^{-3/2} \int_0^{E_F} E^{3/2}\,dE$$
$$= \frac{3}{5} E_F.$$

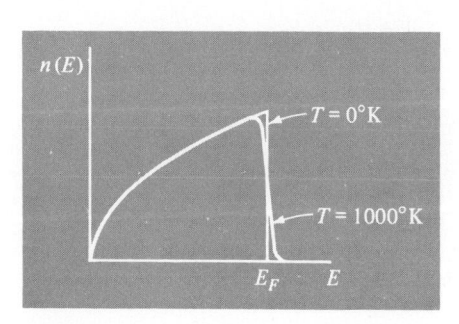

Fig. 11.6. The distribution of electron energies in a metal at $T = 0°$K and $T = 1000°$K.

Therefore, the average energy of an electron in a metal at 0°K is a few eV (4.23 eV for copper). An ideal gas whose molecules have an average kinetic energy of 4.23 eV has a temperature of 33,000°K. This means that if electrons behaved classically, copper would have to have a temperature of 33,000°K in order for the free electrons to have the same average energy as they actually have at 0°K.

11.5 TEMPERATURE VARIATION OF ELECTRICAL CONDUCTIVITY

The calculation of the conductivity of pure metals at high temperatures can be estimated reasonably by treating the lattice vibrations according to the Einstein model[8] of individual oscillators of frequency $v_E = k\theta_E/h$ with θ_E the Einstein temperature. For $T > \theta_E$ the energy of the oscillator[9] may be equated to kT. That is,

$$\tfrac{1}{2} M d^2 (2\pi v_E)^2 = kT$$

where M is the mass of the positive ions and d is the amplitude of their oscillations. Solving for d^2 yields

$$d^2 = \frac{kT}{2\pi^2 M v_E^2}.$$

The average time τ between collisions may be written as

$$\tau \simeq \frac{\lambda}{v_F}$$

where λ is the mean free path of the electrons and v_F is the **Fermi speed** of the electrons defined by

$$\tfrac{1}{2} m_e v_F^2 = E_F.$$

The mean free path can be estimated (see Problem 14) as

$$\lambda \simeq \frac{1}{nd^2}$$

where n is the density of positive ions. Therefore (see Section 11.2)

$$\begin{aligned}
\sigma &= \frac{n e^2 \tau}{2 m_e} \\
&\simeq \frac{n e^2}{2 m_e} \frac{1}{n d^2 v_F} \\
&= \frac{n e^2}{2 m_e} \frac{2\pi^2 M v_E^2}{n k T v_F} \\
&= \frac{M}{m_e} \frac{\pi^2 e^2 v_E^2}{v_F k T} = \frac{M}{m_e} \frac{\pi^2 k (e\theta_E)^2}{v_F h^2 T}.
\end{aligned}$$

[8] *MWTP*, Section 22.5.
[9] *MWTP*, Section 11.4 and 11.6.

Substitution of approximate values of the parameters gives conductivities of the correct order of magnitude. Furthermore, this result predicts that the resistivity of a pure metal should be proportional to the absolute temperature at sufficiently high temperatures, a prediction that is verified by experiments.

For low temperatures, the Debye model[10] of the lattice vibrations predicts that the conductivity becomes proportional to T^{-5}.

The resistivity of a metal containing impurities is well represented by

$$\rho = \rho_T + \rho_i$$

where ρ_T is the resistivity resulting from thermal motions of the lattice and ρ_i is the resistivity resulting from the presence of impurities. For small impurity concentrations, ρ_i is independent of temperature.

11.6 ELECTROMOTIVE FORCE[11]

If a difference of potential should exist in a conductor, the free electrons would immediately begin to move in order to reduce the difference of potential. As a result the energy associated with establishing the difference in potential is converted into kinetic energy of electron motion. The electrons in turn lose this energy to the crystal lattice through collisions. The production of an electric current in a conductor can only be sustained, however, if some external agent continuously supplies the required electrostatic energy. A device that continuously transforms some other form of energy (chemical, mechanical, thermal, etc.) into electrical energy is known as a source of **electromotive force** or **emf**. A source of emf in a closed conducting electrical circuit is capable of maintaining a continuous current in that circuit. A source of emf of 1 volt does 1 joule of work per coulomb of charge during the energy transformation process.

The energy supplied by the source of emf reappears as heat in the conductor. If the potential difference across the conductor maintained by the source of emf is V volts, the work done on a charge of dq coulombs falling through this potential difference is Vdq joules. If the charge takes dt sec to fall through the potential difference V, the work done per second will be Vdq/dt. Since this work reappears as heat, the rate of heat production is

$$\text{Heat energy} \cdot \text{sec}^{-1} = V\frac{dq}{dt} = VI = I^2R$$

where we have used Ohm's law in the last substitution.

Now **power** is the rate of doing work. Since the heat energy dissipated per second is just equal to the work done per second in moving the elec-

---

[10]*MWTP*, Section 22.5.
[11]See also Section 8.2.

SEC. 11.6　ELECTROMOTIVE FORCE

trons through the conductor, the electrical power P supplied to the circuit and subsequently lost or dissipated as heat is

$$P = VI = I^2R.$$

Since the appearance of heat is associated with the fact that a conductor shows resistance to current flow, we usually speak of power being dissipated in the resistance or of heat being produced in the resistance. The unit of power is the watt (W), named after James Watt. The power dissipation is 1 watt when one joule of electrical energy is lost per second.

A source of emf is characterized by the magnitude $\mathcal{E}$ of the emf it produces. The source of emf is capable of maintaining a certain difference of potential V in a closed electrical circuit. Both $\mathcal{E}$ and V are measured in the same units, volts. We enquire now about the relation between them.

The electric field within a conductor may be written as

$$\mathbf{E} = \mathbf{E}_1 + \mathbf{E}_2$$

where $\mathbf{E}_1$ is the electrostatic field due to any charge distribution within the conductor and $\mathbf{E}_2$ is the field produced by the source of emf. Integrating $\mathbf{E}$ around the circuit gives

$$\oint \mathbf{E} \cdot d\mathbf{l} = \oint \mathbf{E}_1 \cdot d\mathbf{l} + \oint \mathbf{E}_2 \cdot d\mathbf{l}$$
$$= \oint \mathbf{E}_2 \cdot d\mathbf{l}$$

since the first integral is zero. We define the **current density j** by the equation

$$\mathbf{j} = ne\mathbf{v}_d$$

where n is the number of electrons per unit volume and $\mathbf{v}_d$ is the drift velocity. In Section 11.2 we developed the following relation between the drift velocity and the field

$$\mathbf{v}_d = -\frac{e\tau}{2m_e}\mathbf{E}.$$

Substituting into the expression for the current density gives

$$\mathbf{j} = -\frac{ne^2\tau}{2m_e}\mathbf{E} = -\frac{\mathbf{E}}{\rho}$$

where ρ is the resistivity. We note now that

$$\mathbf{E}_2 \cdot d\mathbf{l} \equiv \mathbf{E} \cdot d\mathbf{l}$$

since $\mathbf{E}_2$ is the field that produces the current. Substituting for $\mathbf{E}_2$ in terms of the current density yields

$$\mathbf{E} \cdot d\mathbf{l} = -\rho\mathbf{j} \cdot d\mathbf{l} = \rho \, dl \, \frac{I}{A}$$
$$= I \, dR$$

where $d\mathbf{l}$ is assumed to be parallel to $\mathbf{E}$ and, therefore, antiparallel to $\mathbf{j}$. The line integral becomes

$$\oint \mathbf{E} \cdot d\mathbf{l} = I \oint dR = IR.$$

But the line integral is just the total work done by the source of emf per unit charge to move the charge around the circuit; this, by definition, is the emf of the source. Therefore,

$$\mathcal{E} = IR.$$

Applying this equation to the circuit shown in Fig. 11.7 gives

$$\mathcal{E} = I(R_1 + R_2 + R_3).$$

Ohm's law, on the other hand, need not be applied to a whole circuit but can be applied to individual elements in a circuit. The potential difference between the ends of the conducting element of resistance R_1 is $V_1 = IR_1$. The usual terminology is to say that there is a **potential drop** or **voltage drop** $V_1 = IR_1$ across the resistance R_1. In the circuit of Fig. 11.7 there are three potential drops,

$$V_1 = IR_1, \qquad V_2 = IR_2, \qquad V_3 = IR_3.$$

The total potential drop across all three elements is

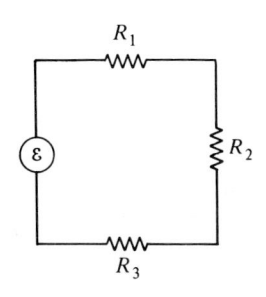

$$\begin{aligned} V &= V_1 + V_2 + V_3 \\ &= I(R_1 + R_2 + R_3) \\ &= \mathcal{E}. \end{aligned}$$

Fig. 11.7. A circuit containing a source of emf and three resistors.

Thus, the emf is equal to the sum of the potential drops around the circuit. This is a general relation true for all circuits regardless of the total number of resistive elements and the total number of sources of emf.

Example. Deduce the equivalent resistance of three resistors that satisfy Ohm's law and that are connected in series. Specialize the result to the case of resistors of equal resistance.

Solution. In Fig. 11.8 we see resistors R_1, R_2, and R_3 connected in series. The current is the same in each resistor. The voltage drop V_i across resistor R_i is

$$V_i = IR_i$$

so that the total voltage drop V across the three resistors is

$$V = V_1 + V_2 + V_3 = I(R_1 + R_2 + R_3).$$

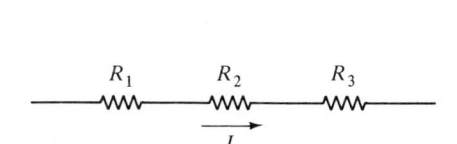

Fig. 11.8. Resistors R_1, R_2, and R_3 connected in series.

Now if R is a resistance equivalent to that of the three resistors, then

$$V = IR.$$

It follows that

$$R = R_1 + R_2 + R_3.$$

In general, when resistors are connected in series, the equivalent resistance is equal to the sum of the resistances of the individual resistors. If $R_1 = R_2 = R_3$, then

$$R = 3R_1.$$

Example. Deduce the equivalent resistance of three resistors that satisfy Ohm's law and that are connected in parallel. Specialize the result to the case of resistors of equal resistance.

Solution. In Fig. 11.9 we see resistors R_1, R_2, and R_3 connected in parallel. The current subdivides among the three resistors as indicated. In this case the voltage drop across each resistor is the same. Therefore, the current I_i through resistor R_i is

$$I_i = \frac{V}{R_i}$$

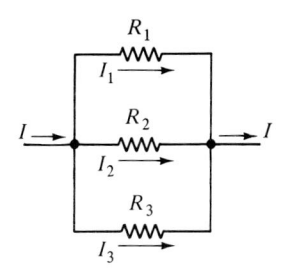

Fig. 11.9. Resistors R_1, R_2, and R_3 connected in parallel.

so that the total current I is

$$I = I_1 + I_2 + I_3 = V\left(\frac{1}{R_1} + \frac{1}{R_2} + \frac{1}{R_3}\right).$$

Now if R is a resistance equivalent to that of the three resistors, then

$$I = \frac{V}{R}.$$

It follows that

$$\frac{1}{R} = \frac{1}{R_1} + \frac{1}{R_2} + \frac{1}{R_3}.$$

In general, when resistors are connected in parallel, the reciprocal of the equivalent resistance is equal to the sum of the reciprocals of the resistances of the individual resistors. If $R_1 = R_2 = R_3$, then

$$R = \frac{R_1}{3}.$$

Some typical sources of emf are the electrolytic cell, the thermocouple, and the electric generator. A battery is merely a collection of several cells grouped together. A source of emf such as a battery offers some resistance to the flow of current due to the materials that comprise the battery. This **internal resistance** produces a potential drop when the battery is connected into a circuit and effectively reduces the potential difference between the terminals of the battery when a current is flowing to a value less than the emf of the battery. The reduction is very small if the internal resistance is very small compared to the resistance in the circuit (see Problem 19).

Example. An electric circuit consists of a source of emf with internal resistance r and an external resistance R (called a **load**) in which we wish to dissipate as much of the available electrical energy as possible. What value should R have?

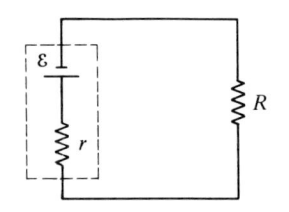

Fig. 11.10. A source of emf connected to an external resistance R.

Solution. The circuit is shown in Fig. 11.10. The current in the circuit is

$$I = \frac{\mathcal{E}}{r+R}.$$

The power P dissipated in the resistor R is

$$P = I^2 R = \frac{\mathcal{E}^2 R}{(r+R)^2}.$$

The value of R for which P is a maximum is obtained by setting

$$\frac{dP}{dR} = 0.$$

That is,

$$\frac{dP}{dR} = \mathcal{E}^2 \frac{d}{dR}\left[\frac{R}{(r+R)^2}\right] = \mathcal{E}^2 \left[\frac{1}{(r+R)^2} - \frac{2R}{(r+R)^3}\right] = 0$$

so that

$$r + R = 2R$$

or

$$R = r.$$

The condition for **maximum power transfer** from the source to the external resistor is obtained when the resistance of the load is equal to the internal resistance of the source. It is generally true that maximum power transfer between electrical circuits occurs when the output resistance of the one circuit is equal to the input resistance of the other. When this condition is satisfied, the circuits are said to be **matched**.

11.7 THE FLOW OF CURRENT IN CIRCUITS CONTAINING RESISTANCE, CAPACITANCE, AND INDUCTANCE

A circuit with resistance and capacitance

Let us consider the circuit shown in Fig. 11.11. When the switch is placed in position 1, the capacitor C begins to accumulate charge, the process continuing until current flow ceases and the potential drop across C is equal to the emf $\mathcal{E}$ (internal resistance of the source of emf is neglected for simplicity).

Let us suppose that at some time t after the closing of the switch, the capacitor has accumulated charge q and the current flowing is I. The potential drop across a capacitor of capacitance C containing charge q is q/C. We showed in Section 11.6 that the emf is equal to the sum of potential drops around the circuit. Therefore, we write

$$\mathcal{E} = IR + \frac{q}{C}$$

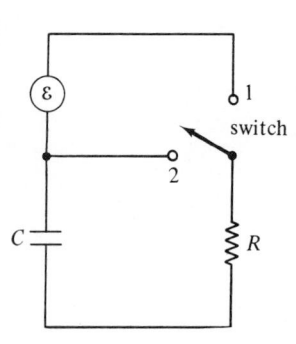

Fig. 11.11. A circuit containing capacitance and resistance.

for this circuit. When current ceases to flow, the potential drop or potential difference across the capacitor is

$$V_0 = \varepsilon = \frac{q_0}{C}$$

where q_0 is the final charge on the capacitor. Substituting into the equation for ε gives

$$\frac{q_0}{C} = R\frac{dq}{dt} + \frac{q}{C}.$$

This equation can be rearranged to give

$$\frac{dq}{q_0 - q} = \frac{dt}{RC}.$$

Upon integration we obtain

$$\ln(q_0 - q) = -\frac{t}{RC} + \text{constant}.$$

The constant of integration is evaluated by noting that $q = 0$ when $t = 0$. Therefore,

$$\text{constant} = \ln q_0$$

and

$$\ln(q_0 - q) = \ln q_0 - \frac{t}{RC},$$

$$\frac{q_0 - q}{q_0} = \exp\left(-\frac{t}{RC}\right),$$

$$q = q_0\left[1 - \exp\left(-\frac{t}{RC}\right)\right],$$

or

$$V = V_0\left[1 - \exp\left(-\frac{t}{RC}\right)\right].$$

The potential of the capacitor increases with time as shown in Fig. 11.12. The time required for the charge (or voltage) of the capacitor to build up to $(1 - 1/e) = 0.632$ of the final value is $t = RC$. The product RC is known as the **time constant** of the resistor-capacitor circuit; it is measured in seconds.

Let us imagine now that the switch in Fig. 11.11 is moved to the 2 position. The source of emf is removed from the circuit and current begins to flow as the capacitor discharges through the

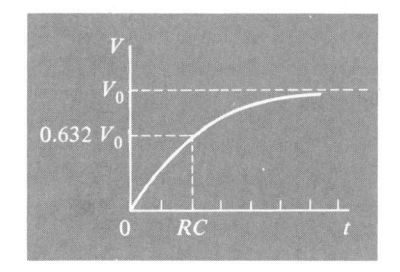

Fig. 11.12. Variation of potential on a capacitor being charged through a resistor.

resistor. Therefore,

$$0 = IR + \frac{q}{C}$$

for any time t so that

$$\frac{dq}{dt} = -\frac{q}{RC},$$

$$\frac{dq}{q} = -\frac{dt}{RC}$$

or

$$\ln q = -\frac{t}{RC} + \text{constant}.$$

When $t = 0$, $q = q_0$ so that

$$\text{constant} = \ln q_0.$$

Therefore,

$$q = q_0 \exp\left(-\frac{t}{RC}\right)$$

or

$$V = V_0 \exp\left(-\frac{t}{RC}\right).$$

The variation of the potential with time is shown in Fig. 11.13. After $t = RC$ sec, the potential on the capacitor has decreased to $e^{-1} = 0.368$ of the initial value. Indeed, the time constant is usually defined in terms of the discharge of a capacitor as "the time required for the voltage to decrease to e^{-1} of the initial value."

A circuit with resistance and inductance

When switch 1 in Fig. 11.14 is closed, a current begins to flow. The current does not reach its maximum value immediately, however, since there is a back emf $L(dI/dt)$ produced in the inductance. We now write

$$\mathcal{E} - L\frac{dI}{dt} = IR$$

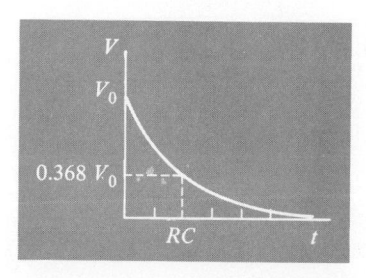

Fig. 11.13. Decrease of potential of a capacitor discharging through a resistor.

where the back emf is introduced with a negative sign since it opposes the emf $\mathcal{E}$ of the source. The solution to the above equation is easily seen to be

$$I = I_0\left[1 - \exp\left(-\frac{R}{L}t\right)\right]$$

where $I_0 = \mathcal{E}/R$ is the final current attained in the circuit. The current in an LR circuit increases with time exactly as does the potential across the capacitor in a CR circuit (Fig. 11.12). The time constant for the LR circuit is L/R sec.

We leave it as an exercise for the reader to show that when switch 2 in Fig. 11.14 is closed and switch 1 is opened, the current decreases according to

$$I = I_0 \exp\left(-\frac{R}{L}t\right)$$

(see Problem 30).

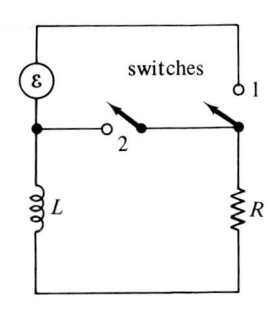

Fig. 11.14. A circuit containing resistance and inductance.

A circuit with resistance, inductance, and capacitance

When the switch in Fig. 11.15 is put in the 1 position, the capacitor is charged through the inductor and resistor. At any time during the charging process the following relation holds

$$\mathcal{E} - L\frac{dI}{dt} = IR + \frac{q}{C}.$$

Substituting $I = dq/dt$ and rearranging gives

$$\frac{d^2q}{dt^2} + \frac{R}{L}\frac{dq}{dt} + \frac{q}{LC} = \frac{\mathcal{E}}{L}.$$

The general behavior of the solution of this equation depends on the magnitude of R/L as compared to $1/LC$; three possibilities exist. The details of the solution are given in the shaded area.

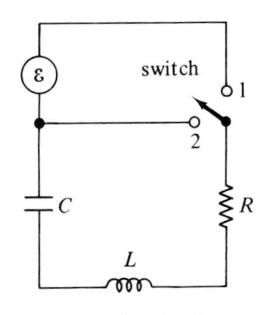

Fig. 11.15. A circuit containing inductance, capacitance, and resistance.

1. **Critically damped charging**

   If $R^2/L^2 = 4/LC$, then

   $$q = q_0\left[1 - \left(1 + \frac{R}{2L}t\right)\exp\left(-\frac{R}{2L}t\right)\right].$$

   This behavior is plotted in curve a of Fig. 11.16. For this case the charge on the capacitor reaches its equilibrium value q_0 in the shortest possible time without overshooting this value.

2. **Overdamped charging**

   If $R^2/L^2 > 4/LC$, the charge on the capacitor steadily approaches its equilibrium value q_0 without overshooting this value but in a time which is longer than for the case of critical damping. A typical overdamped charging curve is given as curve b of Fig. 11.16.

3. **Underdamped charging**

   If $R^2/L^2 < 4/LC$, the charge on the capacitor will oscillate sinusoi-

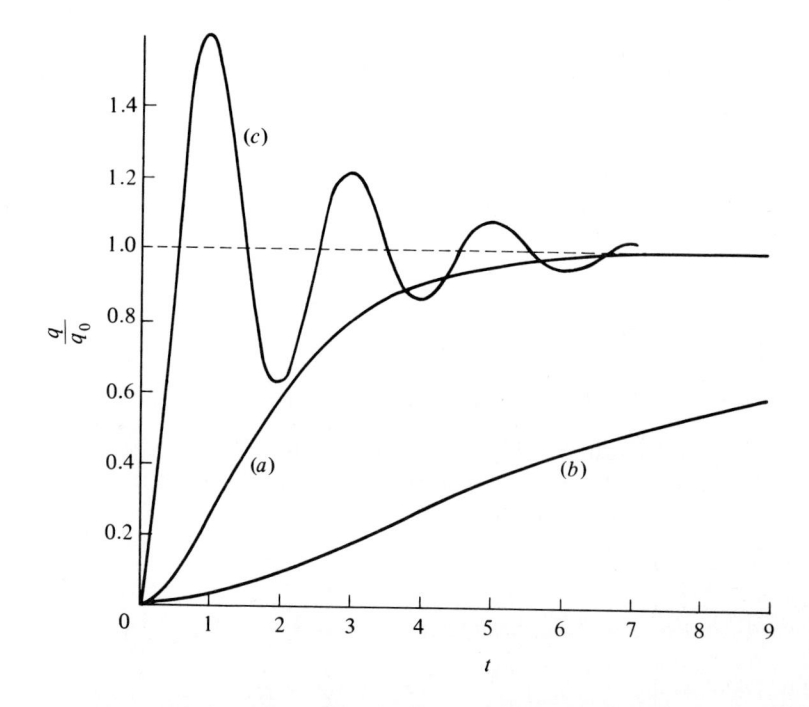

Fig. 11.16. Charging curves for an *LRC* circuit showing (a) critical damping, (b) overdamping, (c) underdamping.

dally with time with an amplitude that decays exponentially with time as the charge approaches the equilibrium value q_0. Curve c of Fig. 11.16 provides a typical example of underdamped charging. This result can be compared with the case of an *LC* circuit containing no resistance (see Section 8.8). For the latter case the sinusoidal oscillations continue undamped indefinitely. The frequency of oscillation is seen, by comparison of the equations for q with and without resistance present, to be

$$\omega = \frac{\left(\frac{4}{LC} - \frac{R^2}{L^2}\right)^{1/2}}{2}$$

$$= \frac{(4LC - R^2C^2)^{1/2}}{2LC} \text{ rad} \cdot \text{sec}^{-1}.$$

Solution of the equation for an *LCR* circuit

In order to solve the equation

$$\frac{d^2q}{dt^2} + b\frac{dq}{dt} + cq = f$$

SEC. 11.7 — THE FLOW OF CURRENT IN CIRCUITS

where $b = R/L$, $c = 1/LC$, and $f = \mathcal{E}/L$ we introduce a new variable

$$\alpha = q - f/c.$$

It is readily seen that in terms of the variable α the equation becomes

$$\frac{d^2\alpha}{dt^2} + b\frac{d\alpha}{dt} + c\alpha = 0,$$

which is a homogeneous linear differential equation with constant coefficients.[12] The equation may be written in the form

$$(D^2 + bD + c)\alpha = 0$$

where

$$D^2 = \frac{d^2}{dt^2} \quad \text{and} \quad D = \frac{d}{dt}.$$

The roots of the characteristic equation

$$D^2 + bD + c = 0$$

are

$$\frac{-b \pm (b^2 - 4c)^{1/2}}{2}.$$

Three cases may be distinguished.

(1) $b^2 - 4c = 0$

The two roots of the characteristic equation are equal so that the solution of the differential equation is of the form

$$\alpha = (m + nt)\exp\left(-\frac{bt}{2}\right)$$

where m and n are constants. Substituting for α yields

$$q = \frac{f}{c} + (m + nt)\exp\left(-\frac{bt}{2}\right).$$

Now

$$\frac{f}{c} = \frac{\mathcal{E}}{L}LC = \mathcal{E}C = q_0,$$

the final charge on the capacitor when equilibrium is reached. Therefore,

$$q = q_0 + (m + nt)\exp\left(-\frac{bt}{2}\right),$$

The initial conditions at time $t = 0$ are

$$q = 0 \quad \text{and} \quad \frac{dq}{dt} = 0.$$

Therefore,

$$q = 0 = q_0 + m, \quad t = 0.$$

---

[12]Consult any text on differential equations.

Since

$$\frac{dq}{dt} = n \exp\left(-\frac{bt}{2}\right) - \frac{b}{2}(m + nt) \exp\left(-\frac{bt}{2}\right),$$

at time $t = 0$

$$\frac{dq}{dt} = 0 = n - \frac{bm}{2}.$$

Therefore,

$$m = -q_0 \quad \text{and} \quad n = \frac{bm}{2} = -\frac{q_0}{2}\frac{R}{L}$$

so that

$$q = q_0\left[1 - \left(1 + \frac{R}{2L}t\right)\exp\left(-\frac{R}{2L}t\right)\right].$$

(2) $b^2 - 4c > 0$

There are two distinct real roots to the characteristic equation. The solution to the differential equation then takes the standard form

$$q = q_0 + m \exp\left\{\left[\frac{-b + (b^2 - 4c)^{1/2}}{2}\right]t\right\}$$
$$+ n \exp\left\{\left[\frac{-b - (b^2 - 4c)^{1/2}}{2}\right]t\right\}.$$

The constants m and n can be evaluated as above.

(3) $b^2 - 4c < 0$

The roots of the characteristic equation are imaginary or complex. The solution to the differential equation then is of the form

$$q = q_0 + m \exp\left(-\frac{bt}{2}\right)\cos\left[\frac{(4c - b^2)^{1/2}}{2}t + \delta\right]$$

where m and δ are constants which can be evaluated as before.

Discharge of a capacitor through resistance and inductance

If the switch of Fig. 11.15 is now placed in the 2 position, the capacitor will begin to discharge from its steady-state potential q_0/C toward zero potential. Again, the three possible modes of behavior (critical damping, overdamping, and underdamping) will occur dependent on the numerical values of L, C, and R. We leave it as an exercise for the reader to work out the appropriate expressions for the charge as a function of time (see Problem 29).

QUESTIONS AND PROBLEMS

1. Using conservation of charge, deduce the following general expression:

$$\nabla \cdot \mathbf{j} = -\frac{\partial \rho}{\partial t}.$$

   Hint: Apply the divergence theorem.

2. A resistor consists of two coaxial conducting cylindrical shells of radii r_a and r_b ($r_b > r_a$) and length L. Calculate the resistance between the shells if the intervening medium has resistivity ρ.

3. By substitution in the expression derived for the high temperature electrical conductivity of a pure metal in Section 11.5 show that it predicts values of the correct order of magnitude.

4. Discuss the manner in which living cells attack antigens.

5. Section 11.3 describes one method of separation of cells according to their physical characteristics. Suggest other simple physical principles that might form the bases of other practical methods.

6. Human blood is classified into four types (A, B, AB, O) depending on the presence of certain "A" and "B" antibodies in the blood cells. How is the blood type of a patient determined?

7. Derive the expression

$$\Delta R \simeq \frac{3\rho V}{2\pi^2 b^4}$$

   given in Section 11.3 for the change in resistance of a section of fluid due to the presence of a spherical cell of volume V.

*8. Certain types of algae are shaped like tetrahedrons. Derive an expression for the change in resistance one might expect if these cells were passed through the device described in Section 11.3. For the calculation, assume the orientation for the tetrahedron shown in Fig. 11.17.

*9. The Schroedinger equation for an electron confined to a cubic volume of metal is

$$-\frac{\hbar^2}{2m_e}\left(\frac{\partial^2 \psi}{\partial x^2} + \frac{\partial^2 \psi}{\partial y^2} + \frac{\partial^2 \psi}{\partial z^2}\right) = E\psi$$

   where m_e is the mass, E is the energy, and ψ is the wave function of the electron.

   (a) Show that

$$\psi = A \sin k_x(x+a) \sin k_y(y+a) \times \sin k_z(z+a)$$

   is a solution to the equation provided that

$$E = \frac{\hbar^2}{2m_e}(k_x^2 + k_y^2 + k_z^2).$$

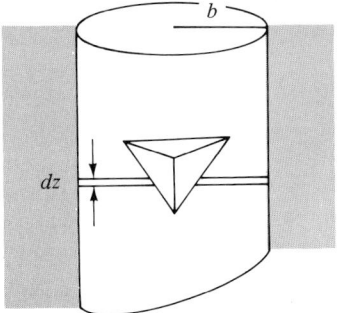

Fig. 11.17. A cell of tetrahedral shape suspended in a fluid flowing through a cylindrical orifice of radius b.

(b) A mathematical way of saying that the electron be confined to the volume of the metal is by insisting that ψ (and thus the probability of finding the electron) vanish on all the boundaries (that is, at $x = \pm a$, $y = \pm a$, $z = \pm a$). Show that

$$k_x = \frac{\pi n_x}{2a}, \quad n_x = 1, 2, 3, \ldots.$$

(c) Use the results of (a) and (b) to conclude that the possible energy levels are given by

$$E = \frac{h^2}{32m_e a^2}(n_x^2 + n_y^2 + n_z^2)$$

where n_x, n_y, and n_z are positive integers.

(d) Show that for given quantum numbers the energy of the electrons is proportional to $v^{-2/3}$ where v is the volume.

10. Calculate the Fermi energy for aluminum, gold, and silver.

11. Calculate the pressure exerted by the Fermi gas in copper metal at 0°K.

*12. Show that for a continuous distribution of electron energies the Fermi–Dirac distribution can be written as

$$n(E)\, dE = \frac{g(E)\, dE}{\exp\left(\dfrac{E - E_F}{kT}\right) + 1}.$$

13. The variation of the resistance of a metal with temperature provides the basis of a useful thermometer, the platinum resistance thermometer. Why is platinum the most suitable metal for this purpose?

14. Show that the mean free path of a conduction electron in a metal is given by

$$\lambda \simeq \frac{1}{nd^2}$$

where d is the amplitude of oscillation of the positive ions and n is the density of positive ions.

15. In copper each atom contributes one electron to the conduction process. Calculate the average time between collisions and the mean free path of copper at 20°C. At 40°K, $\tau = 2 \times 10^{-9}$ sec. Calculate the mean free path. What is the resistivity?

16. Describe the operation of each of the following sources of emf: the electrolytic cell, the thermocouple, the electric motor.

17. Suppose that m electrolytic cells of emf $\mathcal{E}$ and internal resistance r are connected in series to form a battery as shown in Fig. 11.18. Deduce the total emf and internal resistance of the battery. If the battery is connected to a load R, calculate the current that flows through the load.

18. Suppose that m electrolytic cells of emf $\mathcal{E}$ and internal resistance r, are connected in parallel to form a battery as shown in Fig. 11.19. Deduce the total emf and internal resistance of the battery. If the battery is connected to a load R, calculate the current that flows through the load.

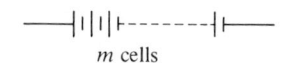

Fig. 11.18. A collection of m electrolytic cells connected in series.

19. An electrolytic cell of emf $\mathcal{E}$ and internal resistance r is connected to a load

R. Deduce the terminal voltage of the battery. What reduction in voltage occurs if $r = 2\,\Omega$ and $R = 1000\,\Omega$?

20. Write a short essay on the construction of and uses for voltage-controlled resistors and capacitors.

21. A $500\,\Omega$ resistor is rated at 5 W. Calculate the maximum allowable current.

22. What is the power output of a cyclotron producing a $5\,\mu A$ beam of 2.0 MeV protons? If the beam has a diameter of 1.0 mm, calculate the current density.

23. Gustav Robert Kirchhoff stated two laws that are useful for calculating currents in circuits. State **Kirchhoff's laws** and their physical significance.

24. Apply Kirchhoff's laws to the circuit shown in Fig. 11.20 to calculate the currents flowing in the resistors.

25. Calculate the power produced by each battery and the power dissipated by each resistor in the circuit shown in Fig. 11.21.

26. Twelve resistors R are arranged in a cubic structure. Calculate the total resistance across a body diagonal of the cube.

27. Show that the quantities RC and L/R have the units of time.

28. Show that the time constant for a parallel-plate capacitor containing a dielectric of resistivity ρ and dielectric constant K is independent of the area and separation of the plates.

29. Deduce the appropriate expressions to describe the charge as a function of time for the discharge of a capacitor through a resistance and inductance.

30. (a) In the study of the build up and decay of current in an LR circuit, why is the circuit shown in Fig. 11.14 discussed rather than a circuit equivalent to that shown in Fig. 11.11 for a CR circuit?
    (b) If switch 2 is closed before switch 1 is opened, what effect would this have, in practice, on the source of emf?
    (c) Deduce the expression for the decrease of the current.

31. Calculate the number of time constants required for the energy stored in the capacitor in Fig. 11.11 to reach one-half its equilibrium value.

32. Show that the total energy dissipated by the resistor of the CR circuit shown in Fig. 11.11 is equal to the energy initially stored in the capacitor.

33. A circuit consists of a resistor R, a capacitor C, and a source of emf $\mathcal{E}$. Assume that $R = 1.0\,M\Omega$, $C = 1.0\,\mu F$, and $\mathcal{E} = 10\,V$. Calculate
    (a) the rate at which the charge on C is increasing,
    (b) the rate at which energy is being dissipated in R,
    (c) the rate at which energy is being delivered by the source of emf 1.0 sec after the connection is made.

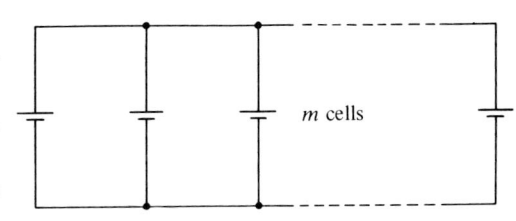

Fig. 11.19. A collection of m electrolytic cells connected in parallel.

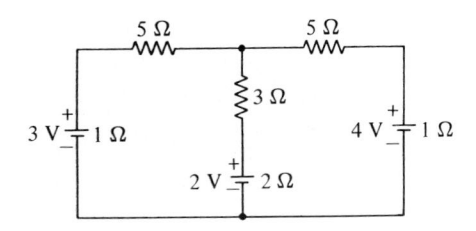

Fig. 11.20. A circuit that may be solved easily by the application of Kirchhoff's laws.

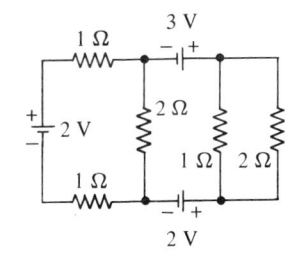

Fig. 11.21. A simple circuit containing batteries and resistors.

JOHN BARDEEN

12 Band theory of solids: conductors, semiconductors, and insulators

12.1 INTRODUCTION

Solids consist of atoms packed closely together. For such a circumstance, two important physical questions can be asked.

1. What is the origin of the forces holding the atoms together?
2. What are the allowed energies of the electrons in the solid?

In answer to the first question we can state that although several types of **binding** have been identified, such as ionic, covalent, van der Waals, and metallic,[1] the electrostatic forces between charged particles are the basis for all forms of binding. The primary distinction between the various types of binding relates to the manner in which the electrons are distributed around the atoms in the solid. Almost all solids are crystalline in form;[2] that is, they are formed of atoms in a periodic symmetrical array. (Concrete, glass, and plastics are important exceptions.) Stable crystalline lattices are formed because the attractive electrostatic forces between electrons and nuclei are opposed by repulsive electron-electron forces.

We know that isolated atoms possess well-defined sets of allowed energy levels for their electrons.[3] When atoms come together to form a solid, the interactions between them cause the energy levels of the individual atoms to be displaced so that a large number of closely spaced levels result. The resulting **energy bands** are of extreme importance in the determination of the properties of solids.

12.2 METALLIC CRYSTALS

Binding occurs in some metals because of the interaction of the positive ion cores with the conduction electrons. We shall study in detail the behavior of lithium atoms in lithium metal. Lithium provides a particularly suitable example for illustration since each contributing lithium atom possesses only three electrons and we therefore need to take into account only the $1s$ and $2s$ electron energy levels[4] in discussing the metallic structure.

Lithium atoms are arranged in a body-centered cubic structure[5] when in the crystalline state. The nearest neighbor distance is 3.0×10^{-10} m ($\equiv 0.30$ nm). In such a crystal lattice an electron cannot be farther than about 0.20 nm from the nearest lithium nucleus. The distribution of charge

---

[1] *MWTP*, Section 21.5.
[2] *MWTP*, Chapter 21.
[3] *MWTP*, Chapter 18.
[4] *MWTP*, Section 18.7.
[5] *MWTP*, Section 21.3.

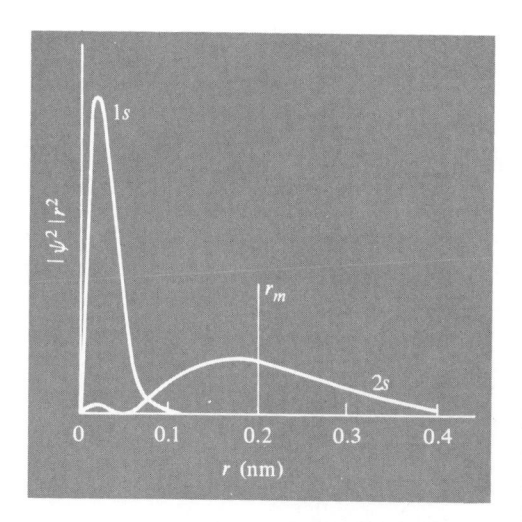

Fig. 12.1. 1s and 2s probability densities in a lithium atom; r_m is the maximum distance that an electron can be from the nearest lithium nucleus in lithium metal.

density associated with the wave functions for 1s and 2s electrons in a lithium atom is given in Fig. 12.1. The two 1s electrons are close to the nucleus while the single 2s electron can be very far away. When lithium atoms are brought together into a metallic solid, the 2s (valence) electron can no longer be considered to belong to a single lithium atom since the positive "cores" (nucleus and 1s electrons) of neighboring lithium atoms are separated by only 0.30 nm. That is, the 2s wave functions of the free atoms have a greater spatial extent than the separation of neighboring cores. As a result the valence electrons belong to the crystal as a whole rather than to individual atoms. Since no valence electron ever gets so far from a lithium nucleus in the metal as it does from the nucleus in a free atom, its potential energy in the metal is much reduced over the potential energy in a free atom. The electron kinetic energies, which are also less than for an isolated atom, must be consistent with the Pauli exclusion principle, since the electrons form a Fermi gas (see Section 11.4). For lithium metal, conduction electron kinetic energies range from 0 to 4.7 eV. The average kinetic energy of these electrons is dependent on the separation of the atoms in the metal; it would decrease if the separation were larger and increase if it were smaller.

A schematic diagram of the total energy as a function of internuclear separation in lithium metal is given in Fig. 12.2. As the separation decreases, the kinetic energy of a conduction electron increases but the potential energy decreases. The total energy reaches a minimum value at separation r_0 as determined by the detailed manner in which the kinetic and potential energies vary with internuclear separation. In a stable crystal lattice nearest neighbors are a distance r_0 apart.

The relative importance of kinetic and potential energy in metallic binding varies from metal to metal. In some metals ion core repulsion becomes significant and in metals like the transition metals with incomplete d shells, covalent binding involving inner electron shells is important. The basic binding process in all metals is as described for lithium but the details of the calculations vary considerably and are quite complicated.

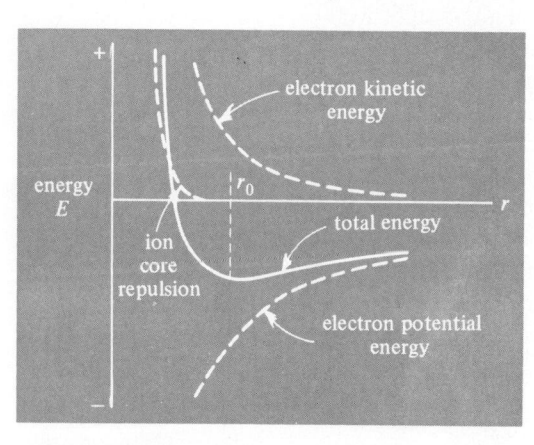

Fig. 12.2. Energy as a function of internuclear separation r in lithium metal.

12.3 BAND THEORY

As an introduction to energy bands in solids, let us first examine the low-lying energy levels available to six hydrogen atoms evenly spaced in a row. As the atoms are brought more closely together, the $1d$ and $2s$ levels each begin to split up into six separate levels as shown in Fig. 12.3. Bringing atoms together leaves the total number of quantum states unchanged; this is a general principle. The $1s$ and $2s$ states now define separate bands with a **forbidden region** between them. The bands for separation r_0 are given in Fig. 12.3. It might be noted that for $r \ll r_0$ the bands overlap.

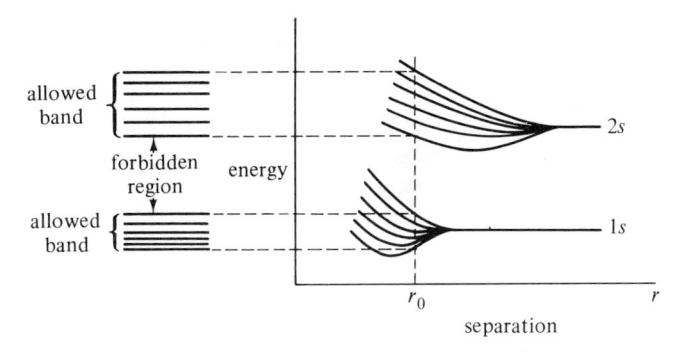

Fig. 12.3. Energy levels for six evenly spaced hydrogen atoms as a function of internuclear separation. Energy bands for separation r_0 are shown on the left.

In solids containing 10^{20} atoms or more, the energy levels in a given band have separations of the order of 10^{-19} eV or less. As a general rule, the width of a band increases as the binding energy of the electrons within that band decreases. This is illustrated in Fig. 12.4. In some solids the low-lying energy bands are separate while in others they overlap; this is also illustrated in Fig. 12.4. The widths of bands are at most a few tens of eV and often less. Therefore, the separations between adjacent electron energy states are typically 10^{-19} eV or less. Since energy separations as small as 10^{-19} eV are not measurable, it is logical to treat a band as if it consisted of a continuous distribution of energies. The widths of the bands are dependent only on the equilibrium separation of the atoms in the solid and not on the number of atoms in the material. Therefore, the forbidden regions, such as those shown in Fig. 12.3 and Fig. 12.4, remain forbidden regardless of the number of atoms in the solid.

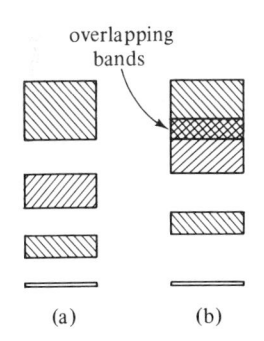

Fig. 12.4. (a) Energy bands in solids usually increase in width as the binding energy decreases. (b) In some solids some of the bands may overlap.

Metals

Metallic binding occurs when the number of valence electrons per atom is small. The elements on the left-hand side of the periodic table[6] therefore form metals. In Fig. 12.5 we show the energy levels of sodium atoms as a function of internuclear distance. The equilibrium internuclear separation in metallic sodium is 0.367 nm. The electron configuration for a sodium atom is $1s^2\ 2s^2\ 2p^6\ 3s$; there is a single electron in the $3s$ shell. Since the $3s$ state can contain two electrons, the $3s$ band is only half-full. The $1s$, $2s$, and $2p$ bands are all very narrow and the corresponding electrons are tightly bound to the sodium nuclei. The $3s$ electrons in the metallic crystal form a free electron gas and sodium is a good conductor of electricity as are all crystalline solids that have energy bands that are only partly filled.

12.4 CONDUCTORS AND INSULATORS

Metals, which are the best conductors of electricity, have conductivities that are more than 10^{24} times larger than the conductivities of the best insulators. No other property of solids has such an enormous range of values. In addition, the conductivity of a pure crystal of metal is greater than the conductivity of an impure crystal while poor conductors (insulators and semiconductors) have larger conductivities when impure than when pure. In this section we shall discover that the difference between conductors and insulators can be explained on the basis of the band theory of electron energy levels.

We have seen that the energies of electrons in solids can only have values that lie within allowed energy bands. In all solids the low-lying bands are completely filled. Electrons in a filled band can carry no electric current. We can see this by considering the application of an external field. If a current were to occur in a filled band under the action of the applied field, an increase in energy of the electrons would result. However, all the levels in the band are filled, and the action of the external field does not create new levels, since the number of these is determined solely by the number of atoms in the solid. Therefore, the only way in which the electrons in a filled band can gain energy is by being excited to an unfilled band.

The mean free path for electrons in a solid is of the order of 10^{-8} m. Even in an electric field as large as 10^4 V · m^{-1}, an electron in a solid would gain only 10^{-4} eV between collisions. Since the spacing between filled energy bands and unfilled bands in a solid is usually several eV or

[6]*MWTP*, Section 18.7. See also the Periodic Table in Appendix E in this book.

more, the probability for electrons to be excited to an unfilled band from a filled band under the action of an electric field is negligible.

Thermal excitation is a possible mechanism for lifting electrons from low-energy levels to higher-energy levels. The ratio of the number of electrons occupying a level at energy E_2 to the number occupying a level at energy $E_1 < E_2$ at a temperature T is, according to the Maxwell–Boltzmann distribution,[7]

$$\frac{\exp\left(-\frac{E_2}{kT}\right)}{\exp\left(-\frac{E_1}{kT}\right)} = \exp\left[-\frac{(E_2 - E_1)}{kT}\right].$$

This factor then corresponds to the fraction of the electrons thermally excited from energy E_1 to energy E_2 at temperature T. For $E_2 - E_1 \simeq 1$ eV and $T = 300°$K, the factor is $\simeq 10^{-16}$ so that thermal excitation does not lead to any significant conduction in insulators. (It can be important in semiconductors as we shall see in the next section.)

If a solid has a partially filled band, it is a good conductor of electricity for there are many levels of higher energy available for the electrons in the unfilled band. (Recall that levels in a band are $\simeq 10^{-19}$ eV apart while the electrons gain 10^{-4} eV or less between collisions and that bands, especially at high energies, are several eV wide.) Such a partially filled band is called the **conduction band**. If a solid does not have a partially filled band, it cannot be a good conductor.

The band structure of sodium is given in Fig. 12.5. The 1s, 2s, and 2p levels in a sodium atom are filled while there is only one electron in the 3s level. Therefore, in sodium metal the 1s, 2s, and 2p bands are filled while the 3s band is only partially filled and therefore sodium is a good conductor of electricity. Note also that the 3s and 3p bands in sodium overlap so that even more levels are available to the electrons in the partially filled band.

When potassium (K) and chlorine (Cl) unite to form a potassium chloride (KCl) crystal, the binding is ionic. The chlorine atoms pick up the single valence electrons from the potassium atoms

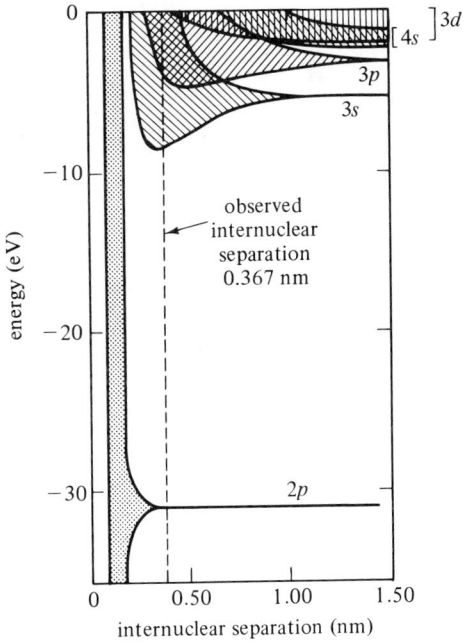

Fig. 12.5. Formation of bands in metallic sodium. The 2s band is at -63.4 eV and the 1s band is at -1041 eV.

[7] *MWTP*, Section 19.4.

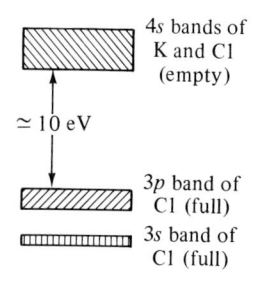

Fig. 12.6. KCl energy bands. The 3s and 3p bands of K are at much lower energy than those of Cl.

and each of the K⁺ and Cl⁻ ions in the crystal have closed electron shells corresponding to the rare gas argon (Ar). The band structure of KCl is shown in Fig. 12.6. There are just enough electrons to fill the 3s and 3p bands and KCl is an insulator.

We can make the generalization that all ionic crystals with rare-gas ions (such as LiF, NaCl, KCl, RbF, CsI, NaF) are insulators, while all alkalis (Li, Na, K, Rb, Cs) form metallic solids and are conductors. We must be careful, however, in making further generalizations. Some solids that might be expected to have only filled bands actually have a partially filled band because one of the filled bands overlaps with a normally empty band. The combined band then has more states available than there are electrons and the solid is a conductor rather than an insulator. The only additional generalization that we can make is that atoms with 1, 2, or 3 electrons outside a rare-gas shell produce metals.

There is one additional effect that occurs in some solids with covalent binding (e.g., diamond, silicon, germanium, gray tin). In these solids the bands split into two groups and some cross over (rather than merely overlap). We shall discuss these solids in more detail in the next section.

12.5 SEMICONDUCTORS

The energy bands in silicon as a function of the nearest neighbor distance are shown in Fig. 12.7; the lower-energy, filled bands are omitted from the diagram. Both the 3s and 3p states split into two groups as the internuclear separation decreases. The empty region between crossed bands is known as the **energy gap**. The splitting of the bands is associated with the nature of the wave functions for electrons in these states and will not be discussed further here; it is a quantum-mechanical effect. The magnitude of the energy gap at the observed internuclear separation varies considerably from substance to substance; for diamond it is 7 eV, for silicon 1.09 eV, for germanium 0.72 eV, and for gray tin 0.2 eV. Each band originating from the splitting of the 3p band can accommodate 3 electrons per atom and each band originating from the splitting of the 3s band can accommodate 1 electron per atom. The atoms in these solids each contribute 4 valence electrons so that the band below the energy gap is filled and the band above is empty. The energy gap in diamond is large enough that electron excitation to the unfilled band is negligible and diamond is a good insulator. For the other three, however, the energy gap is small enough for thermal excitation to boost an appreciable number of electrons to the unfilled band, even at room temperature; these solids are known as **intrinsic semiconductors**.

Intrinsic semiconductors

In an intrinsic semiconductor at $T = 0°K$ all electrons are in the **valence** (filled) band and no electrons are in the conduction band. As the temperature rises from 0°K, electrons begin to be thermally excited into the conduction band. The number of vacancies or **holes** left in the valence band is equal to the number of electrons in the conduction band. Holes carry positive charge but otherwise behave in a manner analagous to electrons in solids. A mass m_h may be associated with a hole; m_h need not in general equal m_e. The concentration of electrons n_e in the conduction band is

$$n_e(E) = g_e(E)f(E)$$

where $g_e(E)$ is the density of states in the conduction band and $f(E)$ is the Fermi distribution function (see Section 11.4). The concentration of holes n_h in the valence band is

$$n_h(E) = g_h(E)[1 - f(E)]$$

where $g_h(E)$ is the density of states in the valence band. To a first approximation

$$g_e(E) = \frac{4\pi(2m_e)^{3/2}}{h^3}(E - E_g)^{1/2}$$

$$g_h(E) = \frac{4\pi(2m_h)^{3/2}}{h^3}(-E)^{1/2}$$

where the zero of energy is taken at the top of the valence band and E_g is the gap energy.

Therefore, the number of electrons in the conduction band per unit volume is

$$n_e = \int_{E_g}^{\infty} g_e(E)f(E)\,dE$$
$$= 2\left(\frac{2\pi m_e kT}{h^2}\right)^{3/2} \exp\left(\frac{E_F - E_g}{kT}\right)$$

(see Problem 3). Similarly, the number of holes in the valence band per unit volume is

$$n_h = \int_{-\infty}^{0} g_h(E)[1 - f(E)]\,dE$$
$$= 2\left(\frac{2\pi m_h kT}{h^2}\right)^{3/2} \exp\left(\frac{-E_F}{kT}\right)$$

(see Problem 4).

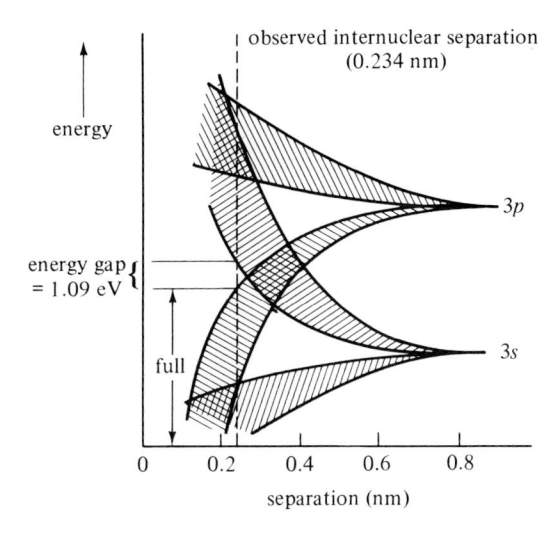

Fig. 12.7. Energy bands in silicon. At 0.234 nm, the observed internuclear separation, the bands have crossed leaving an energy gap of 1.09 eV.

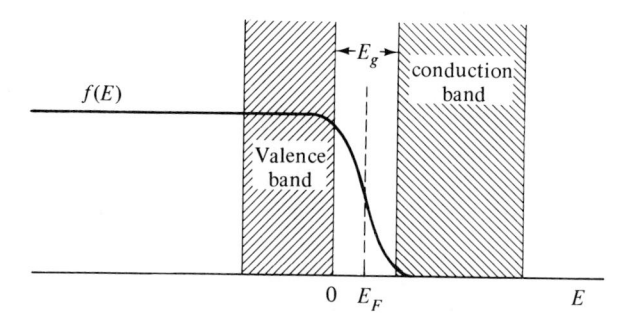

Fig. 12.8. The Fermi level occurs at $E_g/2$ in an intrinsic semiconductor.

For an intrinsic semiconductor the number of electrons in the conduction band is equal to the number of holes in the valence band. It therefore follows that

$$E_F = \frac{E_g}{2} + kT \ln\left(\frac{m_h}{m_e}\right)^{3/4}$$

(see Problem 5). For the special case $m_e = m_h$,

$$E_F = \frac{E_g}{2}.$$

That is, the Fermi level is at the mid point of the energy gap (see Fig. 12.8) and

$$n_e = n_h = 2\left(\frac{2\pi m_e kT}{h^2}\right)^{3/2} \exp\left(-\frac{E_g}{2kT}\right).$$

We saw in Section 11.2 that the conductivity of a conductor is proportional to the density of free electrons. When an electric field is set up in a semiconductor, the electrons in both the valence band and the conduction band move in the direction opposite to the field. However, it is often more convenient to consider the holes in the valence band which move in the direction of the field as the carriers of current in the valence band. Therefore, both electrons in the conduction band and holes in the valence band give rise to current. The conductivity of an intrinsic semiconductor is given by

$$\sigma \propto T^{3/2} \exp\left(-\frac{E_g}{2kT}\right)$$

for the case $m_e = m_h$. The exponential term is most sensitive to changes in T so that a measurement of the temperature variation of the conductivity allows the energy gap to be readily determined.

The sensitivity of the conductivity of an intrinsic semiconductor to variations in T leads to use of these materials in the measurement of temperature and in electronic circuits as temperature compensating devices.

Impurity semiconductors

Semiconductors can have their electrical properties altered by intentionally incorporating impurity atoms into the crystal. Impurity atoms are called **donors** if they provide electrons in energy levels lying just below the conduction band or **acceptors** if they provide vacant energy levels to which electrons from the valence band can be easily excited. Germanium or silicon with donor impurities is called *n*-type (negative charge carrier type) germanium or silicon, while acceptor impurities produce *p*-type (positive charge carrier type) germanium or silicon (see Fig. 12.9). The reason for this terminology will become apparent shortly.

Typical donors in germanium and silicon are phosphorus (P), arsenic (As), and antimony (Sb); these are all elements from group V of the peri-

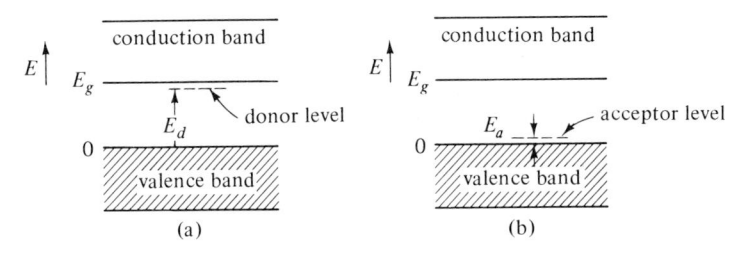

Fig. 12.9. (a) A schematic energy level diagram for an *n*-type semiconductor. (b) A schematic energy level diagram for a *p*-type semiconductor.

odic table and have 5 valence electrons. Typical acceptors are boron (B), aluminum (Al), gallium (Ga), and indium (In); these are all elements from group III of the periodic table and have 3 valence electrons. For example, when an arsenic atom is substituted for a germanium atom in a germanium crystal, only 4 of its valence electrons are required to maintain the covalent bonds with germanium atoms in the lattice. At sufficiently high temperatures ($\simeq 20°$K for germanium), the extra electron is thermally excited into the conduction band and can move throughout the crystal. The low temperature required for thermal excitation reflects the fact that the extra electron is only very weakly bound to the arsenic atom. The reason for this is that the electron is situated essentially in a dielectric medium and its potential energy is reduced accordingly. The experimental values for the binding energy of the extra valence electron for arsenic, phosphorus, and antimony donors in germanium are 0.0127, 0.0120, and 0.0097 eV, respectively. This binding energy is equal to $E_g - E_d$ where E_d is the energy of the donor level [see Fig. 12.9(a)]. A similar effect occurs when, for example, an aluminum atom is substituted into the crystal lattice. Now, however, the acceptor atom has one too few electrons to complete the covalent bonds and a hole is created. This hole is free to move for the same reason that the extra electron is free to move in *n*-type germanium. Experimental values for the binding energy of the hole (charge $+e$) to the acceptor atom (charge $-e$) for boron, aluminum, gallium, and iodine acceptors in germanium are 0.0104, 0.0102, 0.0108, and 0.0112 eV, respectively. These energies are equal to E_a, the energy of the acceptor level [see Fig. 12.9(b)].

The position of the Fermi level for germanium containing donors or acceptors now depends on the concentration of donors or acceptors. Denoting n_d as the concentration of donor atoms, we find that the position of the Fermi level approaches the value E_g as n_d becomes large or as T becomes small, while the Fermi level approaches the value $E_g/2$ (that for intrinsic semiconductors) as n_d becomes small or T becomes large. When the Fermi level lies more than a few kT below E_g (which is the usual case), essentially all the donors are ionized and the number of electrons in the conduction band is the same as the number of donor atoms. The conductivity then is given by

$$\sigma \propto n_d.$$

Semiconductors with donor impurities are called *n*-type because there are many more electrons in the conduction band than there are holes in the valence band, and current flow is due mainly to the movement of negative charge.

If acceptor impurities are added to germanium, the Fermi level is depressed below the value $E_g/2$. At room temperature essentially all acceptors are ionized, there are many more holes in the valence band than there are electrons in the conduction band, and the current is due mainly to the movement of holes (positive charge carriers). For this reason semiconductors with acceptor impurities are called *p*-type. The conductivity of a *p*-type semiconductor is

$$\sigma \propto n_a$$

where n_a is the concentration of acceptor atoms.

The conductivity of both *p*-type and *n*-type germanium or silicon depends on the concentration of impurity atoms. Concentrations of donors or acceptors are ordinarily only a few parts per million. Great care must be exercised in the preparation of *n*- and *p*-type (**extrinsic**) semiconductors to ensure that the crystal is as free from unintentional impurities as possible before injection of the desired donor or acceptor atoms. This process is called **doping**.

12.6 THE HALL EFFECT

In 1879 Edwin Hall first observed that when a conductor is placed in a magnetic field that is perpendicular to the direction of the flow of current in the conductor, a voltage appears across the sample in the direction perpendicular to both the current and the magnetic field. The **Hall effect**, as it came to be known, provides the most convincing experimental evidence in support of the concept of positively charged current carriers (holes) in crystals. The **Hall emf** arises from the deflection of the moving charges caused by the magnetic field. An accumulation of charge builds up on the faces of the sample until the associated electric field is sufficient to cancel the effect of the magnetic field. This is illustrated in Fig. 12.10. Electrons are assumed to travel in the y direction with speed v_y. A magnetic field of magnitude B_x is established in the x direction. The Lorentz force experienced by the electrons is

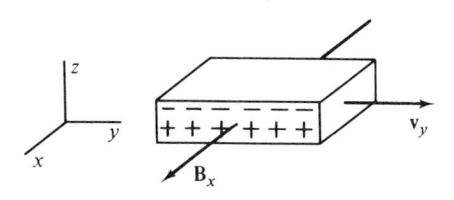

Fig. 12.10. Illustration of the Hall effect for electrons.

$$\mathbf{F} = -e(\mathbf{E} + \mathbf{v} \times \mathbf{B}).$$

The Hall electric field in the z direction is given by the condition

$$F_z = 0 = -e(E_z - v_y B_x).$$

That is,

$$E_z = v_y B_x = \frac{j_y B_x}{n(-e)}$$

where j_y is the current density and n is the concentration of electrons. The **Hall coefficient** R_H is

$$R_H = \frac{E_z}{j_y B_x} = -\frac{1}{ne}.$$

Note that the Hall coefficient contains only quantities that can be measured experimentally, and that it is negative for electrons.

Some observed values of R_H are listed in Table 12.1. The calculated values are deduced directly from the concentration of valence electrons. For the monovalent metals, lithium (Li), sodium (Na), potassium (K),

Table 12.1 Hall coefficients

Metal	Observed value $(m^3 \cdot C^{-1}) \times 10^{11}$	Calculated value $(m^3 \cdot C^{-1}) \times 10^{11}$
Lithium (Li)	−17.0	−13.1
Sodium (Na)	−25.0	−24.4
Potassium (K)	−42	−47
Cesium (Cs)	−78	−73
Copper (Cu)	− 5.5	− 7.4
Beryllium (Be)	+24.4	− 2.5
Zinc (Zn)	+ 3.3	− 4.6
Cadmium (Cd)	+ 6.0	− 6.5

cesium (Cs), and copper (Cu), the agreement between observed and calculated values is reasonable. For the divalent metals, beryllium (Be), zinc (Zn), and cadmium (Cd), the sign of the observed coefficient is opposite to that calculated for electrons, implying that the current is carried by positive charges in these metals. Band theory provides the explanation—we have seen that vacant states near the top of an otherwise filled band behave like carriers of positive charge.

12.7 EFFECTIVE MASS

In both semiconductors and metals electrical conductivity occurs as a result of the transport of charge by mobile carriers. A description of the charge carriers in terms of a free electron gas within the solid can at best be only qualitatively correct. In fact, the electrons must be described in terms of waves that interact with the periodic lattice of the solid. Even in a perfect crystal lattice in which there are no collisions between the charge carriers and the atoms of the crystal, acceleration of charge carriers by an external field causes a continuous transfer of momentum between the charge carriers and the crystal lattice. This momentum transfer can drastically alter the behavior of the mobile carriers. Account may be taken of this effect by ascribing a hypothetical value to the electron mass, called the **effective mass** m^*. If the electron loses momentum to the lattice, the effective mass is greater than the true electron mass; if the electron gains momentum from the lattice, the effective mass is less than the true electron mass. The effect might be described by writing $F + F' = m_e a$, where $F = eE$ is the force of the external field on the electron charge e, and F' is the effective force on the electron due to momentum transfer with the lattice. In this equation m_e is the electron mass and a the resultant acceleration. In terms of the effective mass concept this equation is replaced by $F = m^* a$ where m^* is chosen to give the same acceleration in both equations.

In Section 6.4 we introduced the **cyclotron frequency** of a charged particle moving in a magnetic field. In terms of the effective mass of an electron the cyclotron frequency of an electron in a solid is given by

$$f = \frac{eB}{2\pi m^*}.$$

By measuring the frequency f, we can deduce the effective mass m^*. A technique known as **cyclotron resonance**[8] has been developed for this purpose.

[8] A. F. Kip, "Cyclotron Resonance in Solids," *Contemporary Physics*, **1** (June, 1960), 355.

12.8 p-n JUNCTIONS; TRANSISTORS

It is possible to produce *p*- and *n*-type regions in a single semiconductor crystal. The boundary between the *p*- and *n*-type regions is called a **p-n junction**. In Fig. 12.11 we show the valence and conduction bands in a semiconductor containing a *p-n* junction. The Fermi level is constant throughout the semiconductor when there is equilibrium (no net current flow). In equilibrium there are very few electrons in the conduction band in the *p*-type region while there is a relatively large number in the conduction band in the *n*-type region. Therefore, the flow of electrons from the *p*-type to the *n*-type region must necessarily be small even though such electrons are accelerated in the junction region so that such electron movement is favored. The current flow from the *n*-type to the *p*-type region is also small since electron flow in this direction is not favored; there is an energy barrier that the electrons must surmount. In equilibrium these two currents are equal in magnitude but opposite in direction so that there is no net current. Exactly similar statements can be made about hole currents.

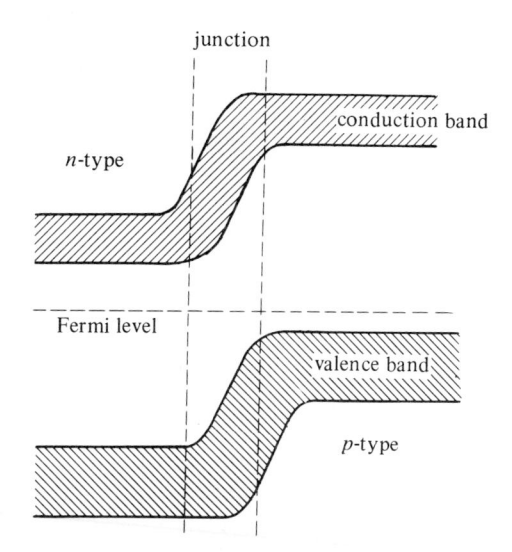

Fig. 12.11. Valence and conduction bands through a *p-n* junction.

The width of the junction depends on the concentration of donor and acceptor atoms in the semiconductor. For large concentrations, the junction is narrow; for small concentrations it is wide. From an alternative point of view we could say that the width of the junction is determined by the amount of excess charge available on each side of the junction. This charge sets up the barrier across the junction which inhibits electron flow from *n*-type to *p*-type. The junction width adjusts itself until the barrier is the right height to produce zero net current.

An additional potential can be applied across the junction externally (from a source of emf). If the barrier across the junction is increased, the junction is said to have **reverse bias**; if the barrier is decreased, the junction has **forward bias**. When the junction is reverse biased, electron flow from *n*-type to *p*-type is even further inhibited. Consequently, we might expect current flow from *p*-type to *n*-type to be increased; however, this type of current was already favored and cannot increase since electrons in the conduction band will still arrive at the barrier at the same rate and will continue to be accelerated as before. The result is that there is only a small net electron flow from *p*-type to *n*-type under reverse bias.

When the junction is forward biased, the barrier is reduced. This does not significantly affect current flow from *p*-type to *n*-type for the reasons

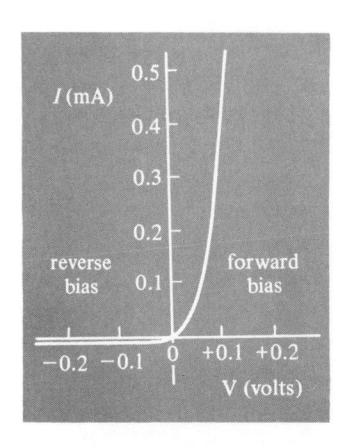

Fig. 12.12. Current vs voltage for a p-n junction.

just given. However, the reduction of the barrier greatly increases flow from n-type to p-type because of the very large numbers of electrons available. A plot of current flow I as a function of bias voltage V for a typical p-n junction is shown in Fig. 12.12. If I_0 is the current flow from n-type to p-type and from p-type to n-type in equilibrium, the maximum current from p-type to n-type under reverse bias is I_0. In Fig. 12.12 we see that the current becomes constant at a small negative value (current is positive in the forward direction, by convention). Typical values of I_0 for a p-n junction are of the order of 10^{-5} A.

If V is the magnitude of the forward bias across the junction, the energy difference between the bottom of the conduction band in the n-type region and the bottom of the conduction band in the p-type region is reduced by eV electron volts under forward bias. We might then expect the increase in the number of electrons traveling from n-type to p-type to be proportional to $\exp(eV/kT)$. Thus the current under forward bias should increase exponentially with bias voltage; this is indicated in Fig. 12.12 and is found to be true to a very good approximation in practice. The current as a function of voltage (including both reverse and forward bias) can be written as

$$I = I_0 \left[\exp\left(\frac{eV}{kT}\right) - 1 \right].$$

Precisely similar statements can be made for the hole current.

Devices that permit current to flow readily in one direction only find wide application in electrical and electronic circuits.[9] The p-n **junction diode** is such a device and operates on the principles just described.

The bipolar junction transistor

The bipolar junction transistor is a single semiconductor crystal containing two p-n junctions. The invention of the transistor and its subsequent development in a variety of forms for a multitude of applications is certainly one of the two or three most significant technological events of this century. For their contributions relating to the invention and development of the transistor, John Bardeen, Walter H. Brattain, and William B. Shockley were awarded the Nobel prize in physics in 1956. Transistors can be made in extremely small sizes and can operate at minute power levels. They are much smaller than vacuum tubes, more rugged and

---

[9] J. J. Brophy, *Basic Electronics for Scientists* (New York: McGraw-Hill Book Company, 1966). This is one of many possible references.

reliable, and their use has revolutionized the electronics and communications industries. Their use in computers and airborne communication and control equipment has been especially significant.

Bipolar junction transistors come in two basic types: *n-p-n* and *p-n-p*. We shall discuss the operation of an *n-p-n* transistor here; the operation of a *p-n-p* transistor is identical in principle. An *n-p-n* transistor consists of two *p-n* junctions separated by a very thin *p*-type region (approximately 2×10^{-5} m thick). The central *p*-type region is called the **base**, while the *n*-type regions are called the **emitter** and the **collector**. Bias voltages are applied so that the emitter junction is forward biased and the collector junction is reverse biased. Fig. 12.13 shows a simplified band diagram for an *n-p-n* transistor under normal bias conditions.

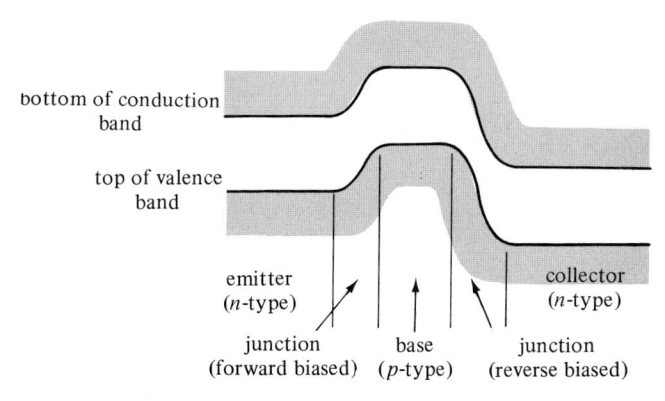

Fig. 12.13. Simplified band diagram for an *n-p-n* transistor with normal bias conditions.

The donor concentration in the emitter is made much larger than the acceptor concentration in the base. Therefore, the current through the junction consists mostly of electrons flowing to the right rather than holes to the left. This is important since the base is so thin that the electrons flowing through the emitter-base junction keep right on going, experience the accelerating field in the reverse-biased collector-base junction, and flow into the collector with scarcely any reduction in current. The collector current I_c is then only slightly less than the emitter current I_e. When the voltage across the emitter-base junction is varied according to some input signal, the emitter current I_e varies according to the relation

$$I_e = I_0 \left[\exp\left(\frac{eV}{kT}\right) - 1 \right]$$

developed for a *p-n* junction earlier in this section. The variations in the emitter current produce corresponding changes in the collector current.

234 BAND THEORY OF SOLIDS CHAP. 12

Typical curves of I_c as a function of collector voltage with I_e as a parameter are shown in Fig. 12.14. For sufficiently large collector voltage, the I_c curves have nearly zero slope. This means that a small change in I_e with a resultant small change in I_c produces a very large change in V_c. The power gain of a transistor in a typical circuit is of the order of $10^4 - 10^5$. For this reason transistors find wide application in amplifiers in electronic circuits. Also, the fact that the transistor exists in the two complementary *n-p-n* and *p-n-p* configurations greatly increases the variety of circuits in which transistors can be employed.

12.9 SUPERCONDUCTORS[10,11,12,13,14]

In 1911 Kamerlingh Onnes, in the course of his investigations of electrical resistance, found that the resistance of a sample of mercury dropped from 0.08 Ω at about 4°K to less than 3×10^{-6} Ω at about 3°K—**superconductivity** had been discovered. Work carried out during the subsequent 60 years has shown that superconductivity is a very widespread phenomenon among conducting solids. It was, however, not until 1957 that a microscopic theory was formulated. The **BCS theory** of John Bardeen, Leon Cooper, and Robert Schrieffer has proven to be rather successful in explaining the wealth of experimental facts concerning superconductivity. In this section we shall state some of the experimental results and try to present some insight into the mechanism responsible for this remarkable phenomenon.

The outstanding electromagnetic property of a superconductor is obviously its superconductivity. If a superconductor is an element in a circuit carrying a steady current, the voltage drop across the superconductor is zero within the accuracy of measurement. If a superconductor is in the form of a loop, a current can be induced in it;

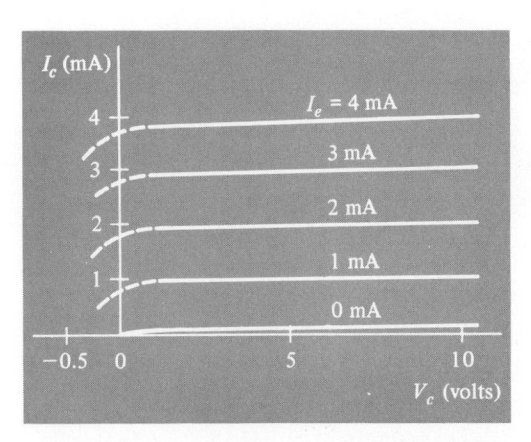

Fig. 12.14. Collector current as a function of collector voltage for an *n-p-n* transistor.

---

[10] *MWTP*, Section 23.9.
[11] B. T. Matthais, "Superconductivity," *Scientific American*, November, 1957. Available as *Scientific American Offprint 227* (San Francisco: W. H. Freeman and Co., Publishers).
[12] G. A. Saunders, "The Electron Pair Theory of Superconductivity," *Contemporary Physics*, 7 (February, 1966), 192.
[13] A. W. B. Taylor, "The Microscopic Theory of Superconductivity," *Contemporary Physics*, 9 (November, 1968), 549.
[14] T. H. Geballe, "New Superconductors," *Scientific American*, November, 1971.

such currents have been observed to persist without measurable decay for years. These two observations, although related, are not equivalent.

A second striking electromagnetic property is the **Meissner effect**, the complete exclusion of any magnetic field from the interior of a superconductor.

Table 12.2 Transition temperatures of some superconductors

Metal	Transition temperature (°K)
Titanium (Ti)	0.53
Osmium (Os)	0.71
Aluminum (Al)	1.18
Tin (Sn)	3.72
Vanadium (V)	5.13
Lead (Pb)	7.18

The best conductors at room temperature, such as copper, silver, and gold, are apparently not superconductors. Some of the metals that are superconductors are listed in Table 12.2, along with the temperatures at which they make the transition from the normal to the superconducting state.

The transition temperature T_c for a given metal depends on the isotopic mass M of the nuclei in the metal through the relation

$$T_c \propto M^{-1/2};$$

this is the **isotope effect**.

The application of an external magnetic field alters the temperature at which the transition occurs. To a first approximation the onset of superconductivity occurs at a temperature

$$T = T_c\left[1 - \frac{B}{B_0}\right]^{1/2}$$

where B_0 is the **critical field**. That is, superconductivity does not occur in a bulk sample at any temperature if it is placed in a magnetic field $B > B_0$.

The existence of an **energy gap** separating the superconducting state from the normal state constitutes another important experimental observation. The gap width $\Delta(T)$ increases from 0 at $T = T_c$ to a maximum value $\Delta(0)$ at 0°K. For most superconductors $\Delta(0)$ is related to T_c by

$$\frac{2\Delta(0)}{kT_c} \simeq 3.5.$$

The temperature dependence of $\Delta(T)$ is approximately given by

$$\Delta(T) = \Delta(0) \tanh \frac{T_c \Delta(T)}{T \Delta(0)}.$$

The experimental results point clearly to the fact that in the transition of a metal to the superconducting state the lattice and its properties are essentially unchanged whereas some of the properties of the conduction electrons are radically changed. Early theoretical attempts suggested that if the phenomenon was to be explained at all, the electrons would have to be treated according to the laws of quantum mechanics. Furthermore, the early theoretical work suggested that a theory which assumed that the conduction electrons were noninteracting particles could never explain the infinite conductivity. Therefore, the simplest model that seemed capable of explaining superconductivity was that of a gas of electrons interacting with each other through a two-particle interaction and governed in their motion by the laws of quantum mechanics. This was the model to which Bardeen, Cooper, and Schrieffer addressed themselves.

The interaction that couples pairs of electrons together in the BCS theory is an electron-electron attraction that takes place through their common coupling to the thermal vibrations of the lattice. If this coupling exceeds the normal electron-electron Coulomb repulsion, a superconducting state is possible. The stronger the coupling of the electrons to the lattice vibrations, the stronger the electron-electron attraction. This explains why very good conductors at room temperature do not become superconductors at low temperatures. The clue pointing to this interaction mechanism was the isotope effect which shows that the critical temperature depends only on the mass of the nuclei and not on the configuration or number of electrons. The state of lowest energy in the BCS model is that for which coupled pairs of electrons, known as **Cooper pairs**, have opposite spin orientations and oppositely directed linear momenta; the net momentum of each pair is identical. That is, the superconducting state is a quantum state on a macroscopic scale—the motions of all the electrons are locked together. A finite amount of energy, equal to the gap energy, must be supplied to excite the electron distribution as a whole to the higher-energy normal state. The density of available electron states as predicted by the BCS theory for a superconductor at 0°K is shown in Fig. 12.15. If the superconducting state has a net momentum (see Fig. 12.16), the associated

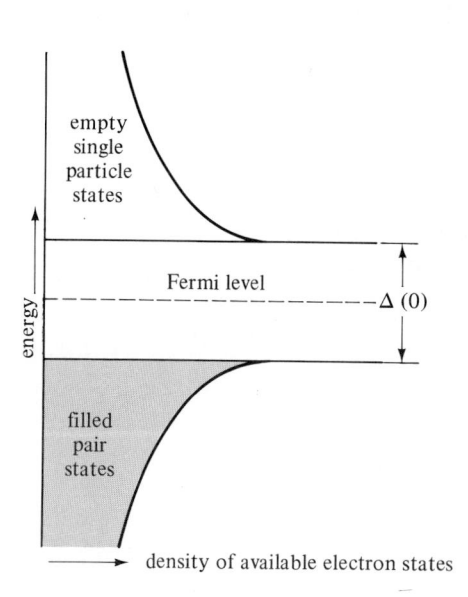

Fig. 12.15. Density of available electron states in a superconductor at 0°K as predicted by the BCS theory.

current cannot be attenuated by a large number of microscopic interactions as in a normal metal.

Superconductivity is a macroscopic quantum phenomenon. The macroscopic quantum state is represented by assigning a macroscopic number of electrons to a single wave function[15] Ψ. Since by definition the electrons are in precisely the same state, the equation of motion for the macrostate is identical to the equation of motion for any electron in the state. The first experiments to illustrate the consequences of a macroscopic quantum state were performed using a hollow superconducting cylinder. When a cylinder becomes superconducting in a magnetic field, it traps the flux enclosed by it. The remarkable fact exposed by the experiments was that the trapped magnetic flux Φ_B appeared only in discrete units of $h/2e$.

In 1962 Brian Josephson derived equations to describe two weakly coupled quantum macrostates. Devices consisting of two superconductors nearly in contact and called **Josephson junctions** have been constructed and have served to verify the predictions of Josephson's equations. The phenomena associated with weakly coupled superconductors include some striking illustrations of the quantum nature of the ordering process in superconductors[16,17]

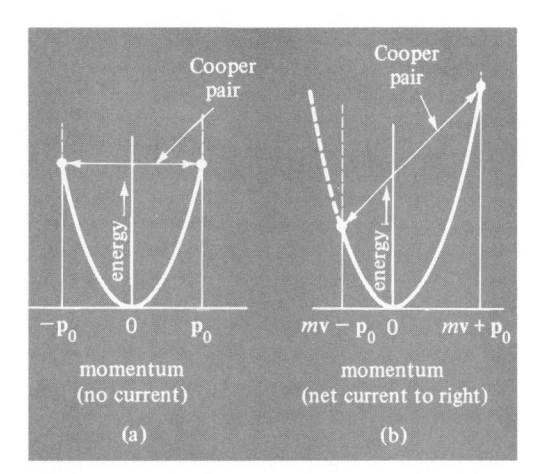

Fig. 12.16. The effect of an electric field on a one-dimensional superconductor. (a) No field applied. (b) A field is applied. The solid lines represent filled states and the dashed lines represent empty states.

QUESTIONS AND PROBLEMS

1. Why are the 3s and 3p bands of potassium at much lower energy than those of chlorine?

2. Some solids that might be expected to have only filled bands actually have an unfilled band because one of the filled bands overlaps with a normally empty band. Give several examples of this type of conductor.

*3. Show that the conduction electron concentration in a semiconductor is

$$n_e = 2\left(\frac{2\pi m_e kT}{h^2}\right)^{3/2} \exp\left(\frac{E_F - E_g}{kT}\right)$$

as stated in Section 12.5.

[15] *MWTP*, Section 18.6.
[16] B. W. Petley, "The Josephson Effects," *Contemporary Physics*, **10** (March, 1969), 139.
[17] D. N. Langenberg, D. J. Scalapino, and B. N. Taylor, "The Josephson Effects," *Scientific American*, May, 1966.

*4. Show that the hole concentration in a semiconductor is

$$n_h = 2\left(\frac{2\pi m_h kT}{h^2}\right)^{3/2} \exp\left(\frac{-E_F}{kT}\right)$$

as stated in Section 12.5.

5. Show that for an intrinsic semiconductor

$$E_F = \frac{E_g}{2} + kT \ln\left(\frac{m_h}{m_e}\right)^{3/4}$$

6. Show that the product $n_e n_h$ is independent of the position of the Fermi level in an intrinsic semiconductor.

7. For a two-dimensional intrinsic semiconductor the density of states for an electron is given by

$$g_e(E) = \frac{4\pi m_e}{h^2}$$

and for a hole by

$$g_h(E) = \frac{4\pi m_h}{h^2}.$$

(a) Denoting the bottom of the conduction band by E_c show that the density of electrons n_e in that band is

$$n_e = \frac{4\pi m_e kT}{h^2} \exp\left(-\frac{E_c - E_F}{kT}\right).$$

(b) Derive an analogous expression for the hole density and use it to show that

$$E_F = \frac{E_c + E_v}{2} + \frac{kT}{2} \ln\left(\frac{m_h}{m_e}\right)$$

where E_F is the Fermi energy and E_v the energy of the top of the valence band.

8. Using a modified form of the Bohr theory calculate the donor ionization energy in indium antimonide, the radius of the ground-state orbit, and the minimum donor concentration at which there will be an appreciable overlap effect between the orbits of adjacent impurity atoms. The dielectric constant of indium antimonide is 17 and the electron effective mass is 0.014 m_e.

9. The simplest model for an n-type semiconductor consists of a conduction band below which there are n_d donor levels per cubic centimeter at an energy E_d above the top of the valence band. The influence of the valence band is neglected. At $T = 0$ all donor levels are filled with electrons. Deduce an expression for the energy of the Fermi level for this model assuming that E_F lies more than a few kT below the bottom of the conduction band.

*10. Suppose one wanted to do cyclotron resonance experiments with copper ($E_F = 7.0\,\text{eV}$, $\sigma = 6.0 \times 10^5$ ohm$^{-1}\cdot$cm^{-1}) using an external magnetic field of 1.0 Wb$\cdot$m^{-2} and at a frequency of 1.0×10^{12} Hz.
(a) Show that the radius of an electron orbit is much larger than the **skin**

depth (see Section 14.5). Use the fact that free electrons travel at the Fermi speed v_F.

(b) Show that since the electrons experience the magnetic field only for a small part of their orbit (while near the surface) resonance will occur only at frequencies such that

$$B = \frac{\omega m^*}{en}$$

where n is an integer.

11. Write an essay on the scientific work of John Bardeen.

12. Justify the following statement: The transistor is one of the two or three most significant technological events of this century. Comment on other developments that might qualify for this honor.

13. A semiconducting specimen has length 0.15 m, width 0.080 m and thickness 0.010 m. A current of 12 mA flows lengthwise through the specimen and a magnetic field 0.70 Wb·m^{-2} is applied perpendicular to both the length and width of the specimen. Calculate the Hall voltage across the width of the specimen given that the Hall coefficient is 3.84×10^{-4} m^3·C^{-1}.

14. Draw a potential diagram for a **Zener diode**. Explain how these devices work and how they can be used as voltage regulators.

15. Explain the mechanism of a **field-effect transistor**. How does this differ from that of the bipolar junction transistor discussed in Section 12.8?

16. Draw a potential diagram for a **PIN diode**. What properties do such devices have which make them useful in radar circuitry?

17. The **de Haas–van Alphen effect** has been used to study the anisotropy of the Fermi energy in metals. Discuss this effect.

18. Electronic devices based on the Josephson effect offer the ultimate in precise electrical measurements.[18] Discuss this statement.

19. Write an essay on the technological implications of superconductivity.[19,20,21,22]

20. The range of applications of superconductivity in technology is limited in a very fundamental way by the low temperatures at which all known mate-

[18] J. Clarke, "*Electronics with Superconducting Junctions,*" *Physics Today*, August, 1971.

[19] T. A. Buchhold, "Applications of Superconductivity," *Scientific American*, March, 1960. Available as *Scientific American Offprint 270* (San Francisco: W. H. Freeman and Co., Publishers).

[20] J. E. Kunzler and M. Tanenbaum, "Superconducting Magnets," *Scientific American*, June, 1962. Available as *Scientific American Offprint 279* (San Francisco: W. H. Freeman and Co., Publishers).

[21] J. K. Hulm, D. J. Kasun, and E. Mullan, "Superconducting Magnets," *Physics Today*, August, 1971.

[22] D. P. Snowden, "Superconductors for Power Transmission," *Scientific American*, April, 1972.

rials become superconductors. Discuss some aspects of the problem of attaining high-temperature superconductors.[23,24,25]

21. In order to explain the Meissner effect F. and H. London postulated as a fundamental equation in a superconductor

$$\nabla \times \left(\frac{m_e \mathbf{j}}{n_s e^2}\right) = -\mathbf{B}$$

where n_s is the density of electrons in the superconducting state. Show that this postulate leads to the equation

$$\frac{m_e}{n_s \mu_0 e^2} \nabla^2 \mathbf{B} = \mathbf{B}.$$

This equation has a solution of the form

$$\mathbf{B} = \mathbf{B}_0 \exp\left(\frac{-x}{\lambda_L}\right)$$

where the parameter λ_L is the **penetration depth** of the magnetic field $\mathbf{B}$ into the superconductor and x is measured from the edge of the superconductor. Derive an expression for λ_L. Calculate the penetration depth for a superconductor for which $n_s = 10^{29}$ m^{-3}. Hint: for the purposes of this problem you may assume the equations derived for free space are applicable in the superconductor.

22. A superconducting solenoid was constructed using niobium zirconium (Nb$_3$Zr) wire. The coil was 10 cm in diameter, consisted of 984 turns of 0.95 cm diameter wire, and was 25 cm long. (a) Use the infinite length approximation to calculate the inductance of the coil. (b) The current in the coil was observed to decay by 1 part per billion per hour. Calculate the conductivity of the wire.

[23] V. L. Ginzburg, "The Problem of High-Temperature Superconductivity," *Contemporary Physics*, 9 (July 1968), 355.
[24] W. A. Little, "Superconductivity at Room Temperature," *Scientific American*, February, 1965.
[25] B. T. Matthias, "The Search for High-Temperature Superconductors," *Physics Today*, August, 1971.

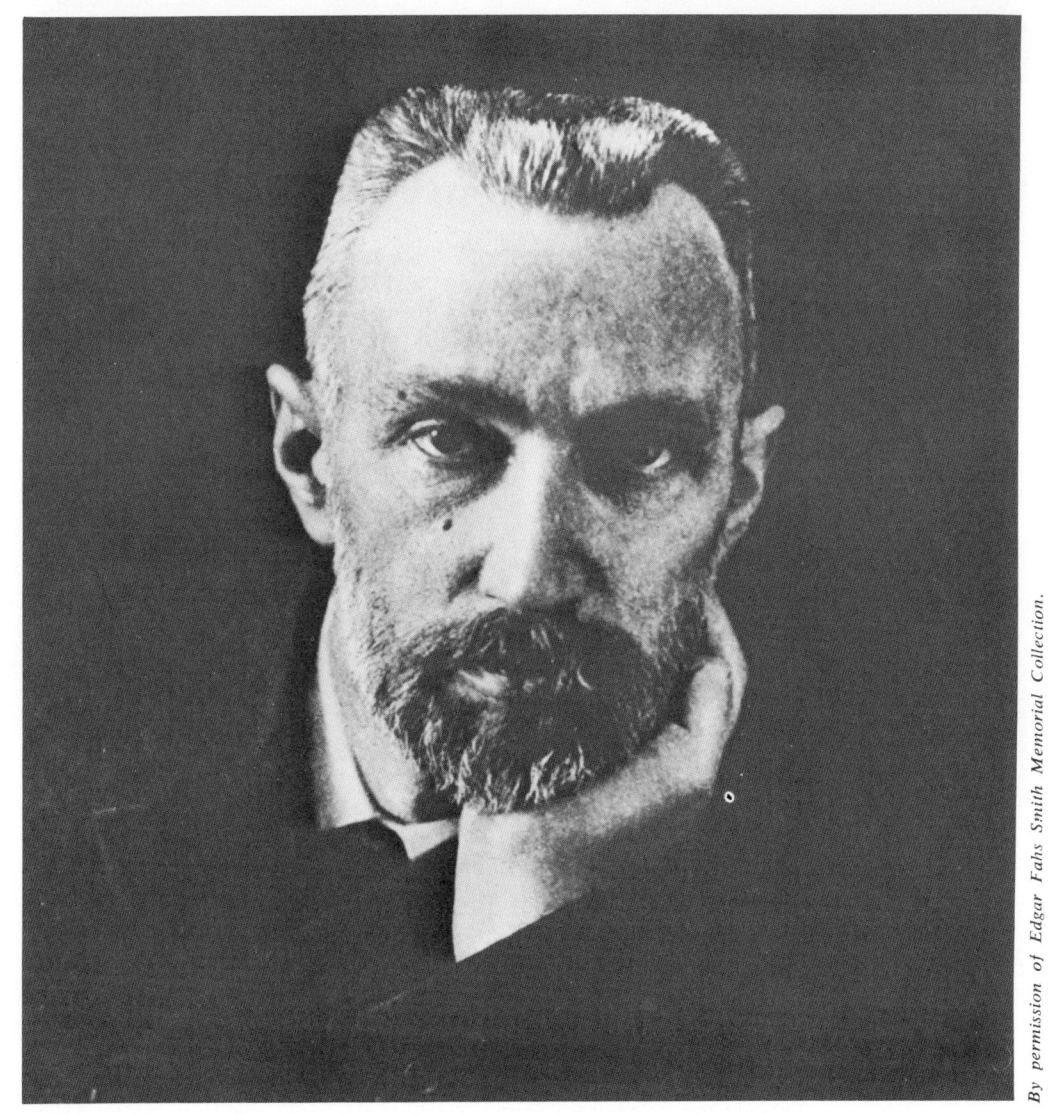

PIERRE CURIE

13 Magnetic properties of matter

13.1 INTRODUCTION

So far in our discussion we have been concerned with the magnetic field only in the absence of material media. If there is a medium present, **B** in the medium is, in general, changed from what it would be if there were no medium present. The field may be either increased or decreased. Materials are classified into three main types (**diamagnetic, paramagnetic, ferromagnetic**) depending on the manner in which they interact with externally imposed magnetic fields. The origin of magnetic effects in matter is in the motion of atomic electrons relative to the nuclei in the material. This should not surprise us since we know that the motion of charged particles produces magnetic fields. The magnetic fields due to electron motion in matter will interact with externally imposed magnetic fields to produce a net field that is dependent on the nature of the magnetic field produced by the atoms of the substance. Almost all of this chapter is devoted to a discussion of the atomic contributions to magnetism in matter.

13.2 MAGNETIC PARAMETERS

Magnetic fields are associated with current flow. It is convenient at this point to introduce a new field vector **H** called the **magnetic field intensity** which is to be identified with some current configuration. When **H** is applied to a volume of empty space, it gives rise to a magnetic field **B** where

$$\mathbf{B} = \mu_0 \mathbf{H}.$$

However, if there is a magnetic material present in the volume of space, there may be a change in the magnetic field due to the response of the atomic electrons in the solid to the presence of the field vector **H**. This response is referred to as **magnetic polarization** or **magnetization**.

We define the magnetization **M** as the magnetic moment per unit volume in the magnetic medium (recall the definition of the electric polarization in Section 10.2). Assuming that the magnetization is linearly dependent on the field intensity **H**, we write

$$\mathbf{M} = \chi_m \mathbf{H}$$

where the constant of proportionality χ_m called the **magnetic susceptibility** is a dimensionless quantity characteristic of the medium. **M** and **H** have the same dimensions so that we can quite properly write the expression

$$\mathbf{B} = \mu_0(\mathbf{H} + \mathbf{M})$$

for the field produced by a magnetic intensity **H** in a medium of magnetization **M**. Substituting for **M** in terms of **H** we obtain

$$\mathbf{B} = \mu_0(\mathbf{H} + \chi_m \mathbf{H})$$
$$= \mu_0(1 + \chi_m)\mathbf{H}.$$

We now define a constant μ called the **magnetic permeability** to be

$$\mu = \mu_0(1 + \chi_m).$$

In empty space $\chi_m = 0$ and $\mu = \mu_0$; this is the origin of the name "the permeability of free space" given to the constant μ_0 in Section 7.1. For a diamagnetic material $\mu/\mu_0 < 1$, for a paramagnetic material $\mu/\mu_0 > 1$, and for a ferromagnetic material $\mu/\mu_0 \gg 1$. The relation between **B** and **H** can now be written simply as

$$\mathbf{B} = \mu \mathbf{H}.$$

It is instructive for the reader to compare the magnetic parameters just given with the corresponding electric parameters of Section 10.3.

In isotropic media, **B**, **H**, and **M** all have the same direction and μ and χ_m are scalar quantities. This is not true in nonisotropic media such as single crystals. Many crystalline materials, however, actually consist of very large numbers of very small, randomly oriented crystals (metals are one example). These media behave on the macroscopic scale as if they were isotropic.

It should be noted that for most of the commonly used magnetic materials the relation between **B** and **H** is nonlinear. By way of contrast it might be pointed out that for most of the commonly used dielectric materials the relation between **D** and **E** is linear.

Stored energy

In Section 8.7 we showed that the energy density u_B in a magnetic field in free space is

$$u_B = \frac{1}{2} \frac{B^2}{\mu_0}.$$

This can be written in terms of H as

$$u_B = \frac{1}{2} \mu_0 H^2.$$

It is straightforward to show by similar arguments that the energy density in a magnetic field in matter is

$$u_B = \frac{1}{2} \frac{B^2}{\mu} = \frac{1}{2} \mu H^2$$

(see Problem 1).

Example. Show that $H = nI$ within a toroidal solenoid of n turns per unit length, carrying a current I and filled with a magnetic material of permeability μ.

Solution. In Section 7.3 we showed that, if $r \gg d$ where r is the radius of the toroid and d is the diameter of an individual turn, then

$$B = \mu_0 nI$$

within such a toroidal solenoid assuming that there was no magnetic material within it. For the present case we need simply replace μ_0 by μ to give

$$B = \mu nI.$$

Therefore,

$$H = \frac{B}{\mu} = nI.$$

Boundary conditions at magnetic surfaces

We consider now the interface $ABCD$ between two magnetic media of permeabilities μ_1 and μ_2, respectively, as shown in Fig. 13.1. We first apply Gauss' law to the pillbox to obtain

$$\int_S \mathbf{B} \cdot d\mathbf{S} = 0.$$

If we assume the pillbox to have negligible dimensions perpendicular to the interface, the integral reduces to

$$-B_1(n)S + B_2(n)S = 0$$

where $B_1(n)$ and $B_2(n)$ are the components of $\mathbf{B}_1$ and $\mathbf{B}_2$, respectively, normal to the interface and S is the area of one of the ends of the "pillbox." That is,

$$B_1(n) = B_2(n)$$

and the **normal component of the magnetic field is continuous across the boundary between two magnetic media**.

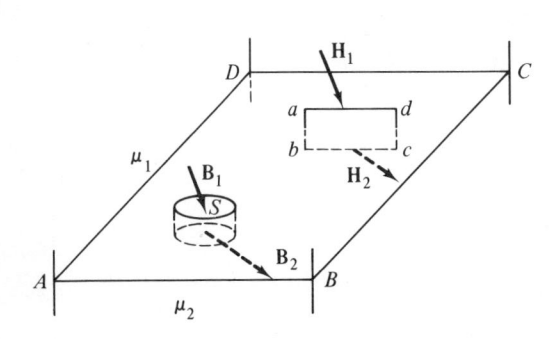

Fig. 13.1. The boundary conditions on **B** and **H** at the interface between two magnetic media are determined by application of Gauss' law to the "pillbox" and by evaluation of the line integral along the path $abcda$.

We assume that no current is flowing through the closed loop $abcda$ which has sides ab and cd of negligible dimensions. Therefore, the line integral of **H** around this path is zero and

$$\oint \mathbf{H} \cdot d\mathbf{l} = 0.$$

Letting $bc = da = l$, we have

$$-H_1(t)l + H_2(t)l = 0$$

SEC. 13.3 DIAMAGNETISM 245

or
$$H_1(t) = H_2(t)$$

where $H_1(t)$ and $H_2(t)$ are the tangential components of $\mathbf{H}_1$ and $\mathbf{H}_2$, respectively, at the interface. Therefore, the **tangential component of H is continuous across the boundary between two magnetic media**.

13.3 DIAMAGNETISM

The application of an external magnetic field produces a small negative magnetic moment in all matter. This is the diamagnetic contribution to magnetization and is characterized by a negative susceptibility. The external magnetic field exerts a force on the orbital electrons in the atoms of the material according the relation

$$\mathbf{F}_{ext} = -e\mathbf{v} \times \mathbf{B}.$$

This force affects the electron orbits in such a manner as to produce a dipole moment that opposes the external field. This effect is, in fact, an example of electromagnetic induction. As the external field rises from an initial zero value, back emf's are induced in all atoms in the material since the orbiting electrons can be considered to be closed conducting circuits. The induced fields, according to Lenz's law, are in a direction so as to oppose the increase in external field.

Since there is no resistance to current flow in such an orbital circuit, the induced fields will persist until the external field is removed. The diamagnetic effect is independent of the orientation of the atoms in the material which means that thermal vibrations do not affect the diamagnetic susceptibility. Therefore, the susceptibility is a constant for a given material, independent of temperature.

Although all materials display diamagnetism, some materials have additional responses to the presence of a magnetic field because they possess permanent magnetic dipole moments. The diamagnetic behavior of these latter materials is masked by the more prominent paramagnetic and ferromagnetic effects.

The Larmor precession and induced magnetic moment
of an atom in an external magnetic field

Suppose that the motion of an electron in an atom is governed by an instantaneous force $\mathbf{F}$. In the presence of an external magnetic field $\mathbf{B}$ the equation of motion of the electron is

$$m_e \frac{d^2\mathbf{r}}{dt^2} = \mathbf{F} - e\mathbf{v} \times \mathbf{B}$$

where **r** describes the instantaneous position of the electron. This equation is valid for a set of axes at rest relative to the observer. It is useful to transform the equation of motion to a set of axes rotating with angular frequency **ω** about the direction of **B**. The transformation to rotating axes is discussed in Appendix B. It follows that, provided the force on the electron due to the external magnetic field is small compared with the force **F** holding it within the atom, then

$$m_e \frac{\partial^2 \mathbf{r}}{\partial t^2} = \mathbf{F} - e\mathbf{v} \times \mathbf{B} - 2m_e \boldsymbol{\omega} \times \frac{\partial \mathbf{r}}{\partial t}$$
$$= \mathbf{F} - e\mathbf{v} \times \mathbf{B} + 2m_e \mathbf{v} \times \boldsymbol{\omega}$$

where $\partial^2 \mathbf{r}/\partial t^2$ and $\partial \mathbf{r}/\partial t$ are the acceleration and velocity, respectively, of the electron in the rotating frame of reference (see Problem 4). If we choose

$$\boldsymbol{\omega} = \left(\frac{e}{2m_e}\right)\mathbf{B} = \boldsymbol{\omega}_L,$$

this equation of motion becomes

$$m \frac{\partial^2 \mathbf{r}}{\partial t^2} = \mathbf{F};$$

$\boldsymbol{\omega}_L$ is called the **Larmor precessional frequency**. That is, to an observer rotating with angular frequency $\boldsymbol{\omega}_L$, the motion of the electron appears the same as it would to a stationary observer in the absence of the external field. In the presence of the external field the electron has precessional motion about **B** at frequency $\boldsymbol{\omega}_L$ in addition to its motion in zero field.

As a consequence of the Larmor precession, the electron acquires a component of angular momentum $L\hbar$ about the direction of **B** given by

$$l\hbar = m_e \overline{a^2} \boldsymbol{\omega}_L = m_e \overline{a^2} \left(\frac{e}{2m_e}\right) \mathbf{B}$$

where $\overline{a^2}$ is the mean square distance of the electron from an axis parallel to B through the nucleus. It may be shown that

$$\overline{a^2} = \frac{2}{3} r^2$$

where the electron is at a fixed distance r from the nucleus (see Problem 5). We have seen (Section 7.6) that the magnetic moment $\boldsymbol{\mu}$ of an orbiting electron is

$$\boldsymbol{\mu} = -\left(\frac{e}{2m_e}\right) l\hbar.$$

Therefore, it follows that

$$\boldsymbol{\mu} = -\left(\frac{e}{2m_e}\right) m_e \left(\frac{2}{3} r^2\right) \left(\frac{e}{2m_e}\right) \mathbf{B}$$
$$= -\frac{e^2}{6m_e} r^2 \mathbf{B}.$$

On summing over all electrons in the atom and including the possibility of orbits for which r is not constant, we obtain

$$\mathbf{\mu} = -\left(\frac{e^2}{6m_e}\right)(\sum \overline{r^2})\mathbf{B}.$$

For a sample containing n atoms per unit volume, the induced magnetic moment $\mathbf{M}$ per unit volume is

$$\mathbf{M} = n\mathbf{\mu} = -\frac{ne^2}{6m_e}(\sum \overline{r^2})\mu_0 \mathbf{H}$$

and the diamagnetic susceptibility is

$$\chi_m = \frac{M}{H} = -\frac{n\mu_0 e^2}{6m_e}\sum \overline{r^2}.$$

In the MKS system the diamagnetic volume susceptibility is of the order of -10^{-5} and is virtually independent of temperature since $\sum \overline{r^2}$ is insensitive to temperature.

13.4 PARAMAGNETISM

Almost all free atoms have permanent magnetic dipole moments (see Section 7.6). These moments arise from the orbital electron motion and from the intrinsic electron spin. In any atom the inner closed shells have no net magnetic moment since all quantum states are filled. Electrons in unfilled shells, however, will in general produce a resultant magnetic moment.

In molecules it so happens that the forces responsible for chemical binding strongly affect the arrangement of the individual magnetic moments of the various electrons in the molecule. As a result the stable state of a molecule is nearly always one in which there is no permanent magnetic moment. Of the common gases only oxygen (O_2) and nitrous oxide (NO) are paramagnetic.

In the solid state the binding forces also tend to produce situations in which there are no permanent magnetic moments. As an example, we consider a crystal of sodium chloride (NaCl). Its structure is ionic; the transfer of an electron from each sodium atom to each chlorine atom assures an inert rare gas configuration for the electrons of each ion. There is zero resultant angular momentum and NaCl is diamagnetic.

Paramagnetism does occur, however, in some solids as, for example, in the salts of the transition group ions. The significance of the transition group elements for paramagnetism is as follows. These elements occur in positions in the periodic table where **inner** electron shells are only partially filled. The electrons associated with these inner shells generally possess magnetic moments. Since these electrons do not participate in the chemical binding, many of the resultant solids possess permanent magnetic moments.

Neglecting interactions between permanent magnetic dipole moments, it can readily be shown (see Problem 9) that the average component of the dipole moment in the direction of an applied field is

$$\overline{\mu \cos \theta} = \frac{\mu^2 \mu_0 H}{3kT}.$$

The derivation is identical to that carried out in Section 10.4 for polar molecules and is applicable only for small H or high T. If n is the number of magnetic dipoles per unit volume, the magnetization is

$$M = n\overline{\mu \cos \theta} = \frac{n\mu^2 \mu_0 H}{3kT}$$

and the susceptibility is

$$\chi_m = \frac{M}{H} = \frac{n\mu^2 \mu_0}{3kT}.$$

The fact that the magnetic susceptibility is inversely proportional to the temperature was discovered by Pierre Curie and is known as the **Curie law**. It is valid for solutions of paramagnetic salts but is only an approximation for solids, where account must be taken of the interaction of the dipole moments in the material. The form of the Curie law appropriate for solids (called the **Curie–Weiss law**) is

$$\chi_m = \frac{\text{const}}{T - T_c}$$

where T_c is a constant known as the **Curie temperature**. The Curie temperature is so low for most paramagnetic substances that the Curie law is a good approximation at room temperature.

We can determine the order of magnitude of the susceptibility of a paramagnetic substance by putting $\mu = \beta$ (the Bohr magneton) in the Curie law and assuming a value for n of about 10^{29} atoms $\cdot$ m^{-3}. At room temperature (300°K) the susceptibility is

$$\chi_m \simeq \frac{10^{29} \times (0.927 \times 10^{-23})^2 \times (4\pi \times 10^{-7})}{3 \times (1.38 \times 10^{-23}) \times (300)}$$
$$\simeq +10^{-3}.$$

That is, the paramagnetism is considerably greater than the diamagnetism that is also present.

Quantum theory of paramagnetism

We consider a collection of noninteracting electron spins in a magnetic field $\mathbf{B}_0$. There are only two possible orientations of the magnetic moment vector relative to the applied field so that the solution of the Schroedinger equation[1] yields two states of energy

$$\mu B_0 m_s = 2\beta B_0 m_s$$

[1] *MWTP*, Section 18.6.

SEC. 13.4　　　　　　　　　　PARAMAGNETISM

(see Section 7.6) where the quantum number $m_s = +\frac{1}{2}$ or $-\frac{1}{2}$ corresponding to the electron spin being antiparallel or parallel to the magnetic field, respectively. There are two energy levels separated by an energy $2\beta B_0$ (see Fig. 13.2). The thermal equilibrium populations per unit volume of the two levels, as given by the Maxwell–Boltzmann distribution, are

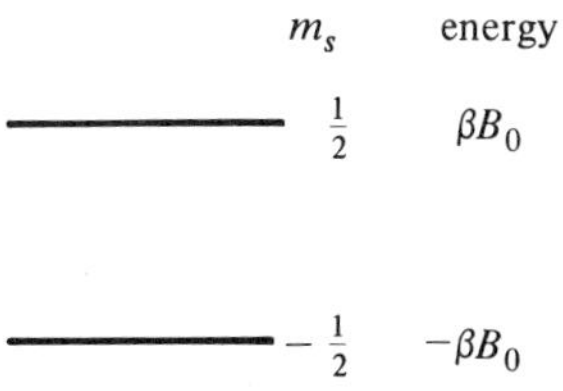

$$\frac{n_{-1/2}}{n} = \frac{\exp\left(\frac{\beta B_0}{kT}\right)}{\exp\left(\frac{\beta B_0}{kT}\right) + \exp\left(\frac{-\beta B_0}{kT}\right)}$$

$$\frac{n_{1/2}}{n} = \frac{\exp\left(-\frac{\beta B_0}{kT}\right)}{\exp\left(\frac{\beta B_0}{kT}\right) + \exp\left(\frac{-\beta B_0}{kT}\right)}.$$

Fig. 13.2.　Energy-level scheme for an electron with only spin angular momentum.

In these expressions $n_{-1/2}$ and $n_{1/2}$ are the populations of the lower and upper states, respectively, and $n = n_{-1/2} + n_{1/2}$ is the total number of electron spins per unit volume. The resultant magnetization is

$$M = \beta(n_{-1/2} - n_{1/2}) = n\beta \frac{\exp\left(\frac{\beta B_0}{kT}\right) - \exp\left(-\frac{\beta B_0}{kT}\right)}{\exp\left(\frac{\beta B_0}{kT}\right) + \exp\left(-\frac{\beta B_0}{kT}\right)}$$

$$\simeq \frac{n\beta^2 B_0}{kT}$$

for $\beta B_0 \ll kT$. Since $B_0 = \mu_0 H_0$,

$$M = \frac{n\beta^2 \mu_0 H_0}{kT}$$

and the paramagnetic susceptibility is

$$\chi_m = \frac{n\mu_0 \beta^2}{kT}.$$

We now consider a collection of noninteracting atoms or ions with magnetic moments

$$\boldsymbol{\mu} = -g_J \beta \mathbf{J}$$

where g_J is the Landé g-factor and $\mathbf{J}\hbar$ is the total angular momentum. It follows (see Problem 14) that the magnetization is

$$M = \frac{nJ(J+1)g_J^2 \beta^2 B_0}{3kT}$$

provided that $g_J J \beta B_0 \ll kT$. Therefore, the paramagnetic susceptibility is

$$\chi_m = \frac{n\mu_0 J(J+1)g_J^2 \beta^2}{3kT} = \frac{n\mu_0 \mu_{\text{eff}}^2 \beta^2}{3kT}$$

where

$$\mu_{\text{eff}} = g_J[J(J+1)]^{1/2}$$

is the **effective Bohr magneton number**.

The above relation gives a good description of the paramagnetism of solids containing **rare-earth ions**. Some examples are given in Table 13.1. The $4f$ shell in the rare-earth transition elements is building up to its full complement of 14 electrons.

Table 13.1 Paramagnetism due to some rare-earth ions

Ion	Electron configuration (outer shells)	J	L	S	μ_{eff} (calc)	μ_{eff} (expt)
Ce^{3+}	$4f^1\ 5s^2\ 5p^6$	$\frac{5}{2}$	3	$\frac{1}{2}$	2.54	2.4
Pr^{3+}	$4f^2\ 5s^2\ 5p^6$	4	5	1	3.58	3.5
Gd^{3+}	$4f^7\ 5s^2\ 5p^6$	$\frac{7}{2}$	0	$\frac{7}{2}$	7.94	8.0
Ho^{3+}	$4f^{10}\ 5s^2\ 5p^6$	8	6	2	10.60	10.4
Yb^{3+}	$4f^{13}\ 5s^2\ 5p^6$	$\frac{7}{2}$	3	$\frac{1}{2}$	4.54	4.5

In Table 13.2 we list similar results for some **iron-group ions**. The $3d$ shell in the iron-group transition elements is building up to its full complement of 10 electrons. Note that in this case $g_J[J(J+1)]^{1/2}$ does not agree with the experimental results. Therefore, the additional column $g_S[S(S+1)]^{1/2} = 2[S(S+1)]^{1/2}$ has been included; this column gives reasonable agreement with the experimental results. The conclusion is that the iron-group ions behave as if only their spin angular momentum is effective in producing the paramagnetism. This is a consequence of the electric field at the ion that results from the crystal lattice in which it resides. The phenomenon is referred to as **quenching of the orbital angular momentum**. One may now ask why such fields do not affect the rare-earth ions included in Table 13.1. We leave this puzzle as an exercise for the reader (see Question 12).

Table 13.2 Paramagnetism due to some iron-group ions

Ion	Electron configuration ($3d$ shell only)	J	L	S	μ_{eff}(calc) $g_J[J(J+1)]^{1/2}$	$2[S(S+1)]^{1/2}$	μ_{eff} (expt)
Ti^{3+}	$3d^1$	$\frac{3}{2}$	2	$\frac{1}{2}$	1.55	1.73	1.8
V^{3+}	$3d^2$	2	3	1	1.63	2.83	2.8
Fe^{3+}	$3d^5$	$\frac{5}{2}$	0	$\frac{5}{2}$	5.92	5.92	5.9
Co^{2+}	$3d^7$	$\frac{9}{2}$	3	$\frac{3}{2}$	6.63	3.87	4.8
Cu^{2+}	$3d^9$	$\frac{5}{2}$	2	$\frac{1}{2}$	3.55	1.73	1.9

Pauli paramagnetism

In almost all metals the valence electrons are detached from the individual atoms and become conduction electrons. The bound electrons are attached to positive ions that have closed electron shells and are therefore diamagnetic. The conduction electrons give rise to both a diamagnetic and a paramagnetic effect. Wolfgang Pauli pointed out that in order to calculate the paramagnetic susceptibility of the conduction electrons, the Fermi–Dirac distribution must be used rather than the Maxwell–Boltzmann distribution (see Question 15). We will present only a semiquantitative discussion of the so-called **Pauli paramagnetism** of the conduction electrons.

Consider first a sample of paramagnetic molecules of magnetic moment μ. The probability that a molecule will line up parallel to an external magnetic field of magnitude B exceeds the probability that the molecule will line up antiparallel to the field by an amount $\simeq \mu B/kT$. Therefore, for n molecules per unit volume the net magnetic moment $\simeq n\mu^2 B/kT$, which is essentially the result that we obtained earlier. Almost all electrons in a metal, however, have zero probability of changing their orientation when the magnetic field is applied, since the required energy states are already filled. In fact, it is only those electrons with energies within $\simeq kT$ of the Fermi energy E_F (see Section 11.4) that can change their orientations. That is, only the fraction $\simeq kT/E_F = T/T_F$ (with T_F the Fermi temperature of the metal) of the electrons should be included in a calculation of the paramagnetic susceptibility of the conduction electrons. Therefore,

$$\chi_m \simeq \left(\frac{n\mu^2 \mu_0}{kT}\right)\left(\frac{T}{T_F}\right) = \left(\frac{n\mu_0 \mu^2}{kT_F}\right).$$

Since T_F is of the order of $3 \times 10^{4} °K$, the paramagnetic susceptibility of a metal is $\simeq +10^{-5}$ and independent of temperature. This contrasts sharply with a nonmetallic paramagnetic substance for which the susceptibility at room temperature $\simeq +10^{-3}$ and varies inversely with the temperature.

An accurate calculation based on the free electron gas model of Section 11.4 yields

$$\chi_m = \frac{3n\mu_0 \mu^2}{2kT_F}.$$

for the paramagnetic susceptibility of a metal. The effect of the magnetic field on the translational motion of the electrons in a metal gives an additional diamagnetic contribution to the susceptibility equal to $-\frac{1}{3}$ of the paramagnetic contribution. The total susceptibility of a free electron gas is

$$\chi_m = \frac{n^2 \mu \mu_0^2}{kT_F}.$$

Nuclear paramagnetism

We have seen that many atomic nuclei possess magnetic dipole moments (Section 7.6). These moments are smaller than the magnetic moment of the electron by a factor $\simeq 10^{-3}$. Therefore, the susceptibility of a nuclear paramagnetic system will be less than that of an electronic paramagnetic system by a factor $\simeq 10^{-6}$, assuming equal numbers of atomic and nuclear spins. The magnetization is given by

$$M = \frac{nI(I+1)g_N^2\beta_N^2 B_0}{3kT}$$

where $I\hbar$ is the nuclear spin angular momentum, g_N is the nuclear g-factor, and β_N is the nuclear magneton. In discussions of nuclear paramagnetism it is more common to write $\mathbf{\mu} = \gamma\hbar\mathbf{I}$ than $\mathbf{\mu} = g_N\beta_N\mathbf{I}$. The parameter γ is called the **magnetogyric ratio**. Therefore, the magnetization is usually written as

$$M = \frac{nI(I+1)\gamma^2\hbar^2 B_0}{3kT}$$

and the susceptibility as

$$\chi_m = \frac{n\mu_0 I(I+1)\gamma^2\hbar^2}{3kT}.$$

13.5 FERROMAGNETISM

A ferromagnetic solid is characterized by a very high value of the susceptibility and by residual magnetization. Both these effects can be explained by a single phenomenon, the spontaneous magnetization of small regions called **domains** in the solid. We shall consider first the magnetization within a single domain.

The three common ferromagnetic elements (iron, cobalt, and nickel) each have partially filled 3d shells plus two 4s electrons. The electron configuration for iron, for example, is

$$1s^2 2s^2 2p^6 3s^2 3p^6 3d^6 4s^2;$$

cobalt and nickel have 7 and 8 3d electrons, respectively. Normally, we should expect that the 6 3d electrons in iron would be divided into two groups of three—three with spin in one direction and three with spin in the opposite direction. However, measurements of the spectrum of radiation emitted by excited iron atoms show that five of the 3d electrons have spin in one direction and only one electron has opposite spin. The reason for this is quantum-mechanical in nature and involves **exchange energy**. It happens that the energy of the iron atom is lowest when the electron spins in the partially filled 3d shell are as described.

Iron atoms each have a magnetic moment of 4β (orbital moments cancel). When iron atoms come together into a single small region or domain of a solid crystal, all the magnetic moments are aligned in the same direction. This is also attributed to the exchange interaction. It has been shown that the energy of interaction of atoms i and j having spins $\mathbf{S}_i$ and $\mathbf{S}_j$ contains an exchange term

$$E_{\text{ex}} = -2J_{ij}\mathbf{S}_i \cdot \mathbf{S}_j$$

where J_{ij} is a measure of the strength of the interaction and is related to the overlap of the charge distributions of the two atoms. Although this quantum-mechanical result has no classical analogue, it is electrostatic in origin. The $3d$ electrons in one atom move slightly away from the $3d$ electrons in a neighboring atom if the spins are aligned. This in turn reduces the electrostatic repulsion energy. Thus the energy of the system is lower for the magnetized state than for the unmagnetized state, and the system spontaneously magnetizes when the solid is formed.

This effect can be described also in terms of the band theory introduced in Chapter 12. Figure 13.3(a) shows the $3d$ bands in unmagnetized iron and Fig. 13.3(b) shows the bands in the magnetized state. Unmagnetized iron would have three electrons per atom in each of the $m_s = +\frac{1}{2}$ and $m_s = -\frac{1}{2}$ bands. However, the configuration of Fig. 13.3(b) is preferred and an iron domain is always magnetized. Putting five electrons into one spin state lowers the energy by the exchange interaction. The Fermi energy in the filled band increases, however, since the maximum electron energy is higher. In iron the increase in the average electron kinetic energy, which is $(3/5)E_F$ (see Section 11.4), is less than the exchange energy so that the magnetized state is stable.

The lowering of the energy due to exchange effects predominates over the increase in Fermi energy in only a few of the elements and alloys with unfilled bands. The relative magnitude of the effects is strongly dependent

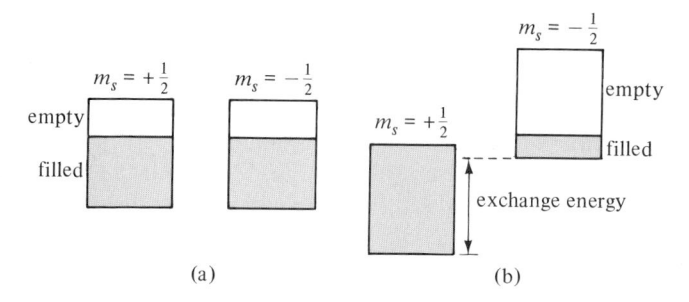

Fig. 13.3. (a) The $3d$ energy band in unmagnetized iron separated according to the two spin states. (b) The $3d$ energy band for magnetized iron. The magnetized state is preferred since the total energy is lower.

on the magnitude of the atomic separation in the solid. Some alloys (for example, copper-manganese) have atomic separactions in the iron-cobalt-nickel range and are ferromagnetic although neither copper nor manganese is ferromagnetic itself.

The energy difference between the magnetized and unmagnetized state is only a few tenths of an electron volt. Thus we should expect that increasing the temperature would eventually destroy spontaneous magnetization in a ferromagnet. The critical temperature for the loss of spontaneous magnetization is the Curie temperature which is 1043°K, 1400°K, and 631°K for iron, cobalt, and nickel, respectively. A ferromagnetic susceptibility can be calculated in a manner similar to the paramagnetic susceptibility. Now, however, the approximation of low field or high temperature cannot be made and, of course, the interaction between moments is included. The Curie temperature is defined in a similar manner for both paramagnetic and ferromagnetic materials.

Domains

In a ferromagnetic solid moments are aligned normally only within very small regions called **ferromagnetic domains**, which typically have linear dimensions of about 10^{-5} m. If the solid is a single crystal, the energy is lowest if the solid forms four domains as shown in Fig. 13.4(a). This configuration produces zero magnetization macroscopically, although each domain is magnetized; the domain magnetization corresponds to a field of about 1 Wb·m^{-2}. Since there is very little external field, there is very little energy in the field. This reduction in energy is partially offset by the energy required to create a domain wall, and by the **anisotropy energy**, which is connected with the energy difference that arises when the crystal is magnetized in different directions with respect to the crystal axes. The minimization of the energy due to these three effects determines the actual domain configuration in a given ferromagnetic material.

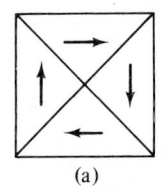

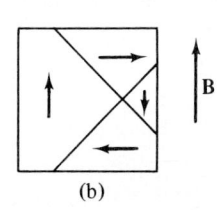

Fig. 13.4. (a) Domains in an unmagnetized single crystal. (b) Partially magnetized single crystal.

When an external magnetic field is applied to a ferromagnetic solid, the domain walls move,[2] as indicated in Fig. 13.4(b) for the single crystal; this produces a net magnetization on a macroscopic scale. For very large values of the magnetic field intensity **H**, the domains all rotate into the direction of **H** and the magnetization of the solid becomes a maximum equal to the magnetization of an individual domain. In a polycrystalline metal the domains are randomly oriented. As **H** increases from zero in

[2]*Ferromagnetic Domain Wall Motion, Ealing Film Loop P80-2033/1* (Ealing Films, Cambridge, Mass.).

such a specimen, the domain walls move and a net magnetic field **B** results. This is illustrated in Fig. 13.5. At point X domain wall movement ceases and further magnetization occurs via rotation of domains into the direction of **H**. The saturation value of the magnetization occurs at point Y, beyond which there is no increase in the field **B** with increasing **H**.

Since solids are not perfect structures, we should expect a polycrystalline solid to contain various imperfections such as interstitial atoms, strains, precipitated colloidal particles, and vacancies. These imperfections retard the motion of domain walls. Therefore, when **H** is returned toward zero for the solid of Fig. 13.4, the curve of B vs H is not simply retraced because some domain wall motion is irreversible. Some permanent field **B** remains even when $\mathbf{H} = 0$. If **H** is varied from zero to large values in opposite directions, the B vs H curve shown in Fig. 13.6 results. This curve is called a **hysteresis loop**. The value B_1 of the field when $H = 0$ is known as the **remanence** and the value H_1 (in the reverse direction) required to return the field to zero is known as the **coercive force**. Relatively pure metals composed of large crystals without significant impurities are used for transformers and rotating electric machinery in order to keep hysteresis energy losses to a minimum. On the other hand, for permanent magnets, a large remanence and coercive force are desired. For these magnets, alloys composed of small crystals with many impurities are employed.

Magnetic domains that are cylindrical in shape and are known as **magnetic bubbles**[3] show considerable promise as the basis for large-volume storage elements in electronic computers.

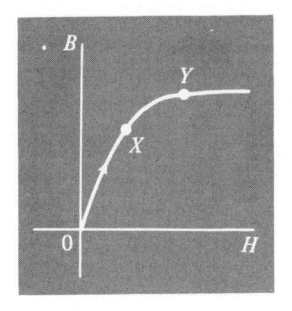

Fig. 13.5. B vs H for a polycrystalline ferromagnetic solid.

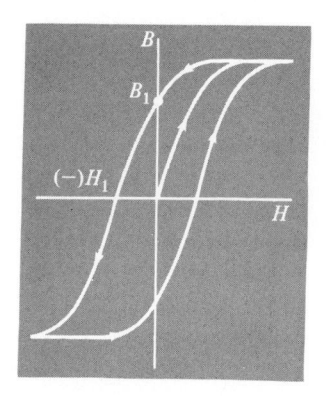

Fig. 13.6. Hysteresis loop for a ferromagnetic solid.

13.6 ANTIFERROMAGNETISM AND FERRIMAGNETISM

In some solids the exchange interaction causes the magnetic moments of neighboring atoms to be aligned antiparallel. Such a solid is termed **antiferromagnetic**. It has a positive susceptibility that decreases as the temperature is lowered below the Curie temperature. The antiparallel alignment occurs very often when a negative ion, commonly oxygen, is positioned between two positive ions possessing magnetic moments. The electron spins in an oxygen (O^{2-}) ion are paired with those of the positive ions on each side, which results in the positive ions being aligned antiparallel (see Fig. 13.7). NiO, FeF_2, and MnS are examples of antiferromagnetic compounds.

Some solids, in which the antiferromagnetic exchange just described is operative, contain ions having two or more different magnitudes of magne-

---

[3] A. H. Bobeck and H. E. D. Scovil, "Magnetic Bubbles," *Scientific American*, June, 1971.

tic moments. Instead of having a vanishingly small magnetization, these solids exhibit spontaneous magnetization as do ferromagnetic materials. They are called **ferrimagnetic**. The basic character of ferrimagnetic materials can be illustrated by considering magnetite which has the chemical formula Fe_3O_4. Magnetite forms a complex crystal structure with eight molecules per unit cell. Two-thirds of the iron ions in the crystal are triply charged (Fe^{3+}) and one-third of the iron ions are doubly charged (Fe^{2+}), while all oxygen ions are doubly charged (O^{2-}). The iron ions are found in two different positions in the crystal lattice where they have either four oxygen ions as their nearest neighbors (tetrahedral or T sites) or six oxygen ions as their nearest neighbors (octahedral or O sites). Half of the Fe^{3+} ions are at T sites and half at O sites while all Fe^{2+} ions are at O sites. The ions interact through the exchange interaction which turns out to be much stronger between O and T sites than between two O or two T sites. Therefore, the Fe^{3+} ions give zero net magnetic moment while each Fe^{2+} ion contributes a net magnetic moment (of 4β) to the magnetization. Through chemical modification of the basic Fe_3O_4 structure, ferrimagnetic solids with almost any desired magnetic properties can be produced.

Ferrimagnetic materials (or ferrites) are widely used as storage elements in computers. They have very small electrical conductivity, unlike ferromagnetic materials. Therefore, their response is essentially magnetic and they respond only very weakly to electric signals. The magnetic response of ferrites to fast-changing currents in a computer is not appreciably slowed down by unwanted electrical effects. Ferrites are thus ideal for use at high frequencies where response times of 10^{-6} sec or less are required.

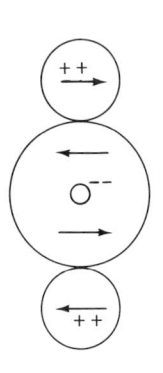

Fig. 13.7. Antiferromagnetism occurs when the magnetic moments of positive ions are aligned antiparallel due to the exchange interaction with electrons in an O^{2-} ion.

13.7 ROCK MAGNETISM AND CONTINENTAL DRIFT

The hypothesis of **continental drift**[4,5,6] is an old one. As early as 1620 Francis Bacon had suggested that the Western Hemisphere had once been joined to Europe and Asia. It was not, however, until the late 1960's that completely convincing experimental confirmation was available. Many lines of evidence now favor the idea that the present continents were once assembled either into two great land masses—Gondwanaland in the south and Laurasia in the north—or into one single great land mass named

[4] J. T. Wilson, "Continental Drift," *Scientific American*, April, 1963. Available as *Scientific American Offprint 868* (San Francisco: W. H. Freeman and Co., Publishers).
[5] P. M. Hurley, "The Confirmation of Continental Drift," *Scientific American*, April, 1968. Available as *Scientific American Offprint 874* (San Francisco: W. H. Freeman and Co., Publishers).
[6] R. S. Dietz and J. C. Holden, "The Breakup of Pangaea," *Scientific American*, October, 1970. Available as *Scientific American Offprint 892* (San Francisco: W. H. Freeman and Co., Publishers).

Pangaea. Measurements of rock magnetism have provided important contributions to these studies.

Iron-bearing rocks become weakly magnetized at their time of formation—during cooling in the case of lavas and during deposition in the case of sediments. The permanent magnetic fields in the rocks are aligned with the direction of the earth's magnetic field at the time and place of their formation. Unless this magnetism is disturbed by reheating or by physical distortion, it is retained as a permanent record of the direction of the earth's magnetic field at the time the rock was formed. Measurements of the magnetism in rock samples of all ages from the various continents have been taken. From these measurements it has been possible to reconstruct the variation in position of the geomagnetic poles during the earth's history. It was found that for each continent the position of the north magnetic pole followed a different path backward in time. This result was interpreted to mean that the continents had, in fact, moved independently relative to the present position of the magnetic pole. Assuming that the magnetic pole had not moved very far from the earth's axis of rotation, which itself had not changed its position relative to the principal mass of the earth, it followed that the continents must have moved over the surface of the earth.

The study of rock magnetism has also revealed a distinct pattern of magnetic field reversals over the past 3.6 million years.[7] This observation has particular significance with respect to the oceanic ridges. In the early 1960's it was proposed that these were created by rising currents of material which then spread outward to form new ocean floors. Studies of the magnetism in rocks of the ocean floors revealed that on either side of a ridge the rocks were magnetized in a stripelike pattern with adjacent stripes having opposite polarity. In fact, the pattern of the magnetization was observed to be symmetric on either side of a ridge. The past history of the earth's magnetic field was laid out horizontally in the magnetism of the rocks of the ocean floor going away from the ridge in both directions. It is thought that new, hot material is continually rising from the ridges, that it becomes magnetized in the direction of the earth's field as it cools, and that it is then pushed outward carrying with it the history of magnetic field reversals. The dates of the reversals have been determined from radioactive dating work and, therefore, the distance between stripes gives the rate of spreading of the ocean floor.

Although this result does not unequivocally require continental drift, the directions and rates of motion for both ocean-floor spreading and continental drift are entirely compatible.

An example of part of Gondwanaland, tentatively reconstructed, is

---

[7] A. Cox, G. B. Dalrymple, and R. R. Doell, "Reversals of the Earth's Magnetic Field," *Scientific American*, February, 1967.

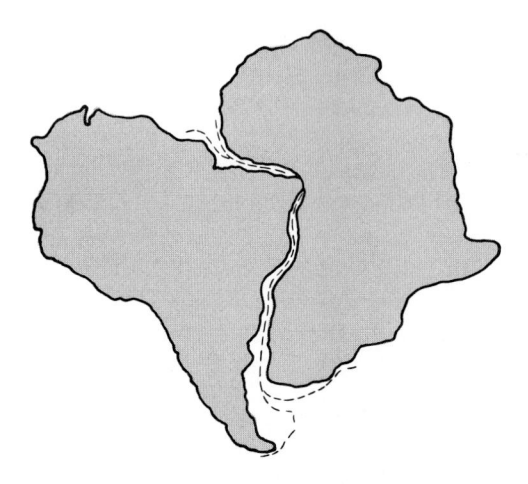

Fig. 13.8. An illustration of the way in which South America and Africa may have fitted together some 200 million years ago.

shown in Fig. 13.8. The dashed line indicates the fit if made along the continental slope at the 500 fathom contour line.

QUESTIONS AND PROBLEMS

1. A **Rowland ring** consists of a toroidal coil wound on an iron ring. For a current of 2.0 A flowing in such a coil constructed to have 10 turns per centimeter, B is measured to be 1.0 Wb·m^{-2}. Calculate the ratio of the magnetic permeability for the iron core to the magnetic permeability of free space for these particular operating conditions.

2. Show by arguments similar to those given in Section 8.5 that the energy density in a magnetic field in matter is

$$u_B = \frac{1}{2}\frac{B^2}{\mu} = \frac{1}{2}\mu H^2.$$

3. A superconductor exhibits perfect diamagnetism. Comment on this statement.

*4. Show that the equation of motion of an atomic electron in the presence of an external magnetic field **B** in a frame of reference rotating with frequency **ω** about the direction of **B** is given approximately by

$$m_e \frac{\partial^2 \mathbf{r}}{\partial t^2} = \mathbf{F} - e\mathbf{v} \times \mathbf{B} + 2m_e \mathbf{v} \times \boldsymbol{\omega}.$$

5. Show that the mean square distance $\overline{a^2}$ of an atomic electron from an axis through the nucleus is

$$\overline{a^2} = \frac{2}{3}r^2$$

where the electron is at a constant distance r from the nucleus.

6. Estimate the diamagnetic susceptibility of 1.0 m^3 of argon gas at 273°K and atmospheric pressure.

7. The diamagnetic susceptibility of one kg-mole of helium gas is -2.4×10^{-8}. Deduce the mean square distance of each electron from the nucleus in the helium atom. Compare your answer to a_0^2 where $a_0 = 0.0529$ nm is the radius of the first Bohr orbit in the hydrogen atom.[8]

8. The electronic wavefunction of a hydrogen atom in its ground state is

$$\Psi = (\pi a_0^3)^{-1/2} \exp\left(\frac{-r}{a_0}\right)$$

[8] *MWTP*, Section 18.6.

where $a_0 = \hbar^2/m_e^2 = 0.0529$ nm. Calculate the diamagnetic susceptibility of atomic hydrogen using the statistical interpretation of the wave function to obtain a value for $\overline{r^2}$.

9. Show that the average component of the magnetic dipole moment in the direction of an applied magnetic field in a paramagnetic substance is

$$\overline{\mu \cos \theta} = \frac{\mu^2 \mu_0 H}{3kT}.$$

10. Check the calculated μ_{eff} values for the rare-earth ions listed in Table 13.1.

11. Check the calculated μ_{eff} values for the iron-group ions listed in Table 13.2.

12. Why is quenching of orbital angular momentum important for iron-group ions (Table 13.2) but not for rare-earth ions (Table 13.1)?

13. The quantum numbers J, L, S describing the lowest energy state of an atom or ion may be deduced from **Hund's rules**.[9] State these rules and show how they apply to the rare-earth ions listed in Table 13.1.

14. Show that the magnetization per unit volume for a collection of noninteracting ions with magnetic moments

$$\boldsymbol{\mu} = -g_J \beta \mathbf{J}$$

is

$$M = \frac{nJ(J+1)g_J^2 \beta^2 B_0}{3kT}$$

provided that $g_J J \beta B_0 \ll kT$.

15. Why is it necessary to employ the Fermi–Dirac distribution for the calculation of the paramagnetic susceptibility of the conduction electrons in a metal, whereas the Maxwell–Boltzmann distribution is applicable to the calculation of the paramagnetic susceptibility of a sample of oxygen gas molecules? Give a physical explanation.

*16. If a magnetic field **B** is applied to a metal, the conduction electron spins must rearrange themselves so that the Fermi level of those whose moments are parallel to the field is the same as that of those electrons whose moments are antiparallel to the field (see Fig. 13.9). The concentration n_P of electrons with moments parallel to the field is

$$n_P = \tfrac{1}{2} \int_{-\mu B}^{E_F} f(E) g(E + \mu B) \, dE$$

where $f(E)$ is the Fermi–Dirac distribution, $g(E)$ is the density of states, and the factor $\tfrac{1}{2}$ has been introduced to account for only spins parallel to the field.

(a) Show that for $kT \ll E_F$ ($f(E) \simeq 1$, $E < E_F$)

$$n_P \simeq \frac{1}{2} \int_0^{E_F} f(E) g(E) \, dE + \frac{1}{2} \mu B g(E_F).$$

[9] G. Herzberg, *Atomic Spectra and Atomic Structure* (New York: Dover Publications, 1944).

(b) Derive an analogous expression for the concentration n_A of electrons with moments antiparallel to the field.

(c) Show that the magnetization $M = \mu(n_P - n_A)$ is given by

$$M \simeq \mu^2 B g(E_F).$$

(d) Show that, for a Fermi gas,

$$g(E_F) = \frac{3n}{2kT_F}$$

and hence

$$\chi = \frac{M}{H} = \frac{3n\mu_0\mu^2}{2kT_F}$$

as stated in Section 13.4.

(e) Why are electrons in the valence bands neglected?

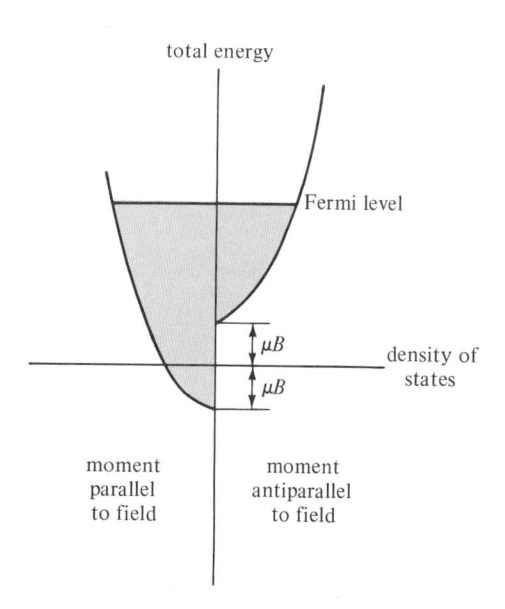

Fig. 13.9. The free electron gas model of Pauli paramagnetism.

17. Even though a permanent horseshoe magnet is not in use, it should have its poles connected with a soft iron **keeper**. Why is this so?

18. When constructing transformers to be used for alternating current circuits, why is it desirable to choose a metal for the core with a relatively small hysterisis loop?

19. Pierre Weiss postulated that an internal interaction in a paramagnetic material caused the magnetic moments to line up parallel to one another. This interaction he described by an exchange field $\mathbf{H}_E$ and further assumed that

$$\mathbf{H}_E = \lambda \mathbf{M}$$

where λ is known as the **molecular field coefficient**. Use the Curie law and the relation $\mathbf{H}_i = \mathbf{H}_0 + \mathbf{H}_E$ where $\mathbf{H}_i$ is the actual field experienced by a magnetic moment and $\mathbf{H}_0$ is the external field, to show that

$$\chi = \frac{C}{T - T_c}, \quad T_c = \lambda C$$

where C is the Curie constant. At T_c there is a singularity and spontaneous magnetization occurs.

20. How are the moments oriented in a **canted antiferromagnet** such as hematite?
21. Discuss evidence (other than that presented in Section 13.6) in support of the hypothesis of continental drift.
22. Comment on the botanical evidence for and against continental drift.[10]
23. Describe several methods employed in the measurements of the magnetic remanence in rocks.[11]

[10] N. W. Radforth, "The Ancient Flora and Continental Drift," in *Continental Drift*, G. D. Garland [ed.] (Toronto: University of Toronto Press, 1966).

[11] F. D. Stacey, *Physics of the Earth* (New York: John Wiley & Sons, Inc., 1969).

PAVEL CERENKOV

From MODERN MEN OF SCIENCE, Volume I, by McGraw-Hill Encyclopedia of Science and Technology Editors. 1966. Used with permission of McGraw-Hill Book Company.

14

Electromagnetic waves in matter: classical effects

14.1 INTRODUCTION

When an electromagnetic wave interacts with matter, a variety of possible effects may occur. For example, some or all of the wave may be reflected or absorbed; scattering may take place due to the interaction of the wave with electrons, both bound and free; free electrons and holes may be created in association with the disappearance of some of the energy in the beam. The relative magnitudes of these effects are strongly frequency dependent. Moreover, we shall discover that in order to explain some of the interactions that occur, we must assume that the energy carried by the wave is quantized or localized into pulses of electromagnetic waves called photons. In this chapter we shall consider those effects that can be described classically using continuous wave theory, and in the next chapter we shall investigate some interactions in which the quantum nature of both electromagnetic radiation and matter must be taken into account.

14.2 THE GENERAL FORM OF MAXWELL'S EQUATIONS

We saw in Section 10.3 that Gauss' law for a dielectric medium is written

$$\int_S \mathbf{D} \cdot d\mathbf{S} = \int_v \rho \, dv.$$

Maxwell's first equation in differential form then becomes

$$\nabla \cdot \mathbf{D} = \rho.$$

Ampère's law for free space was given in Section 9.4 as

$$\oint \mathbf{B} \cdot d\mathbf{l} = \mu_0 \int_S \left(\mathbf{j} + \epsilon_0 \frac{\partial \mathbf{E}}{\partial t} \right) \cdot d\mathbf{S}.$$

However, in free space

$$\mathbf{B} = \mu_0 \mathbf{H} \quad \text{and} \quad \mathbf{D} = \epsilon_0 \mathbf{E}$$

so that Ampère's law can also be written as

$$\oint \mathbf{H} \cdot d\mathbf{l} = \int_S \left(\mathbf{j} + \frac{\partial \mathbf{D}}{\partial t} \right) \cdot d\mathbf{S}.$$

Since **H** and **D** are both field vectors that are associated with the source(s) of the electromagnetic field rather than with the space (empty or not) in which the field is located, this latter form of Ampère's law is a general one, true in all situations. Maxwell's second equation in differential form becomes

$$\nabla \times \mathbf{H} = \mathbf{j} + \frac{\partial \mathbf{D}}{\partial t}.$$

SEC. 14.3 PLANE WAVES USING COMPLEX NOTATION

The third and fourth equations can be written in terms of **H** merely by substituting $\mu \mathbf{H}$ for **B**.

Maxwell's four equations in their general form are then

$$\nabla \cdot \mathbf{D} = \rho$$

$$\nabla \times \mathbf{H} = \mathbf{j} + \frac{\partial \mathbf{D}}{\partial t}$$

$$\nabla \times \mathbf{E} = -\mu \frac{\partial \mathbf{H}}{\partial t}$$

$$\nabla \cdot \mathbf{B} = 0.$$

14.3 PLANE WAVES USING COMPLEX NOTATION

The use of complex algebra (see Appendix C) simplifies the discussion of many problems involving waves. We shall show how a sinusoidal plane wave can be written in exponential form using complex notation.

Plane sinusoidal waves in free space

The free space solutions for the wave equations for **E** and **B** were given in Section 9.7 as

$$\mathbf{E} = \mathbf{E}_0 \sin 2\pi \left(vt - \frac{x}{\lambda} \right)$$

$$\mathbf{B} = \mathbf{B}_0 \sin 2\pi \left(vt - \frac{x}{\lambda} \right).$$

We note that

$$\mathbf{E} = \mathbf{E}_0 \exp \left\{ i \left[2\pi \left(vt - \frac{x}{\lambda} \right) \right] \right\}$$

can be written as

$$\mathbf{E} = \mathbf{E}_0 \left[\cos 2\pi \left(vt - \frac{x}{\lambda} \right) + i \sin 2\pi \left(vt - \frac{x}{\lambda} \right) \right].$$

Both the real and imaginary part represent quantities which vary sinusoidally with time with frequency v. The two sinusoidal variations are identical, with the exception of a phase difference of $\pi/2$ radians. Either the real or imaginary part is suitable for describing a sinusoidal wave; the imaginary part corresponds to the solution quoted above to the wave equation for **E**. The exponential expression for a plane sinusoidal wave traveling in the $+x$ direction is

$$\mathbf{E} = \mathbf{E}_0 \exp \left\{ i \left[2\pi \left(vt - \frac{x}{\lambda} \right) \right] \right\}$$

$$= \mathbf{E}_0 \exp \left[i(\omega t - kx) \right]$$

where $\omega = 2\pi\nu$ is the angular frequency and $k = 2\pi/\lambda$ is the **propagation constant**.[1]

The exponential expression for a plane sinusoidal wave traveling in an arbitrary direction is

$$\mathbf{E} = \mathbf{E}_0 \exp[i(\omega t - \mathbf{k} \cdot \mathbf{r})]$$

where the **propagation vector**[2] is

$$\mathbf{k} = k_x \mathbf{i} + k_y \mathbf{j} + k_z \hat{\mathbf{k}}.$$

(Note that $\hat{\mathbf{k}}$ is a unit vector in the z direction; it should not be interpreted as a unit vector in the $\mathbf{k}$ direction. The use of k for both the unit vector in the z direction and the propagation constant is standard and, hopefully, will not lead to confusion.) Since

$$\mathbf{r} = x\mathbf{i} + y\mathbf{j} + z\hat{\mathbf{k}},$$

the scalar product of $\mathbf{r}$ and $\mathbf{k}$ is

$$\mathbf{k} \cdot \mathbf{r} = k_x x + k_y y + k_z z.$$

The quantity $\mathbf{k} \cdot \mathbf{r}$ is constant for all points on a plane of constant phase (see Problem 1).

Substitution of the exponential expression for $\mathbf{E}$ into the (free space) wave equation for $\mathbf{E}$

$$\nabla^2 \mathbf{E} = \epsilon_0 \mu_0 \frac{\partial^2 \mathbf{E}}{\partial t^2}$$

shows that the exponential is a solution provided that

$$k = \omega(\epsilon_0 \mu_0)^{1/2}$$

(see Problem 2). The wave speed was shown in Section 9.7 to be

$$c = \nu\lambda = (\epsilon_0 \mu_0)^{-1/2}.$$

Now

$$c = \nu\lambda = \frac{\omega}{2\pi}\frac{2\pi}{k} = \frac{\omega}{k} = (\epsilon_0 \mu_0)^{-1/2}.$$

Therefore,

$$k = \omega(\epsilon_0 \mu_0)^{1/2}$$

which is identical to the above result as it must be. Thus the exponential solution is consistent with the solution given in Section 9.7.

14.4 PLANE WAVES IN ISOTROPIC DIELECTRICS

The wave equations for $\mathbf{E}$ and $\mathbf{H}$ in an isotropic dielectric (nonconducting) medium of permittivity ϵ and permeability μ are (see Problem 4)

---

[1] *MWTP*, Section 15.3.
[2] The quantity $\mathbf{k}$ is often referred to as the **wave vector**.

SEC. 14.4　　　PLANE WAVES IN ISOTROPIC DIELECTRICS

$$\nabla^2 \mathbf{E} = \epsilon\mu \frac{\partial^2 \mathbf{E}}{\partial t^2}$$

and

$$\nabla^2 \mathbf{H} = \epsilon\mu \frac{\partial^2 \mathbf{H}}{\partial t^2}.$$

These equations have as solutions the plane sinusoidal waves given by

$$\mathbf{E} = \mathbf{E}_0 \exp[i(\omega t - \mathbf{k} \cdot \mathbf{r})]$$
$$\mathbf{H} = \mathbf{H}_0 \exp[i(\omega t - \mathbf{k} \cdot \mathbf{r})]$$

provided that the wave speed is

$$v = \frac{\omega}{k} = (\epsilon\mu)^{-1/2}.$$

The wave speed in a material medium is less than the wave speed in free space since

$$\frac{v}{c} = \left(\frac{\epsilon_0 \mu_0}{\epsilon\mu}\right)^{1/2}$$

which is less than one since $\epsilon > \epsilon_0$ and $\mu > \mu_0$. The ratio of the speed in free space to the speed in a medium is called the **index of refraction** n of the medium; that is,

$$n = \frac{c}{v} = \left(\frac{\epsilon\mu}{\epsilon_0 \mu_0}\right)^{1/2}.$$

In practice, the values of n, ϵ, and μ are frequency dependent. Therefore, this relation holds only if n, ϵ, and μ are all measured for electromagnetic waves of the same frequency.

The variation of the index of refraction with frequency gives rise to the phenomenon of **dispersion**. We are all familiar with what happens when a parallel beam of white light passes through a glass prism—a spectrum of colors is formed. This is an example of dispersion and it results from the fact that the index of refraction of glass is greater for high-frequency visible radiation (blue) than for low-frequency visible radiation (red). In Fig. 14.1 we show a graph of the index of refraction of a typical glass as a function of frequency for frequencies in the visible portion of the electromagnetic spectrum (Section 9.7). The graph shows us that the index of refraction of glass is about 1.5 over the entire range, and that the dispersion, $\Delta n/\Delta v$, gives an increase in n of about 0.006 per 10^{14} Hz increase in v.

Taking the curl of the equation for **H** we have (see Section 9.7)

$$\nabla \times \mathbf{H} = \nabla \times \mathbf{H}_0 \exp[i(\omega t - \mathbf{k} \cdot \mathbf{r})]$$
$$= \mathbf{H}_0 \times \nabla \exp[i(\omega t - \mathbf{k} \cdot \mathbf{r})]$$

since

$$\nabla \times \mathbf{H}_0 = 0.$$

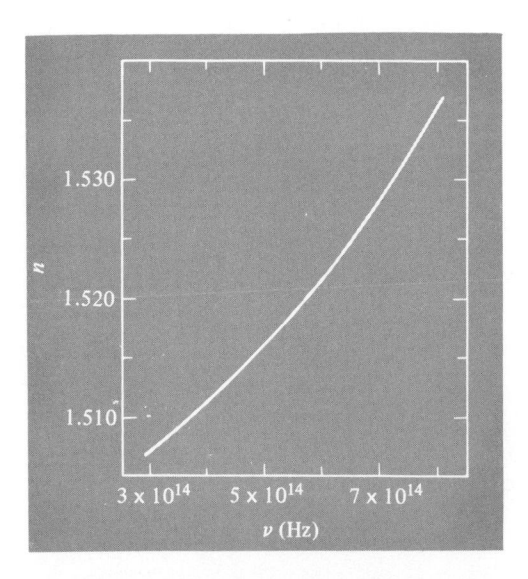

Fig. 14.1. The index of refraction of a typical glass as a function of frequency for frequencies in the visible portion of the electromagnetic spectrum.

Now

$$\nabla \{\exp [i(\omega t - \mathbf{k} \cdot \mathbf{r})]\}$$
$$= \left(\mathbf{i}\frac{\partial}{\partial x} + \mathbf{j}\frac{\partial}{\partial y} + \hat{\mathbf{k}}\frac{\partial}{\partial z}\right)\{\exp [i(\omega t - k_x x - k_y y - k_z z)]\}$$
$$= \exp i(\omega t - \mathbf{k} \cdot \mathbf{r})[\mathbf{i}(-ik_x) + \mathbf{j}(-ik_y) + \hat{\mathbf{k}}(-ik_z)]$$
$$= (-i) \exp [i(\omega t - \mathbf{k} \cdot \mathbf{r})]\mathbf{k}.$$

Therefore, using Maxwell's second equation and noting that $\mathbf{j} = 0$ in a nonconductor,

$$\nabla \times \mathbf{H} = \mathbf{H}_0 \times \mathbf{k}(-i) \exp [i(\omega t - \mathbf{k} \cdot \mathbf{r})]$$
$$= \epsilon \frac{\partial \mathbf{E}}{\partial t}.$$

Since $\mathbf{k}$ is in the direction of wave propagation, we see that $\mathbf{H}_0$ and $\partial \mathbf{E}/\partial t$ are both perpendicular to the direction of propagation. It follows that $\mathbf{H}$ and $\mathbf{E}$ are both perpendicular to $\mathbf{k}$ and (as in free space) the electromagnetic wave in the isotropic dielectric medium is transverse.

Substituting for $\mathbf{E}$ in the last equation, we obtain

$$\mathbf{H}_0 \times \mathbf{k}(-i) \exp [i(\omega t - \mathbf{k} \cdot \mathbf{r})] = \epsilon \mathbf{E}_0 \exp [i(\omega t - \mathbf{k} \cdot \mathbf{r})](i\omega).$$

The relation between the magnitudes of $\mathbf{E}_0$ and $\mathbf{H}_0$ (and, therefore, $\mathbf{E}$ and $\mathbf{H}$) is

$$H_0 k = \epsilon E_0 \omega.$$

Therefore,

$$\frac{E}{H} = \frac{E_0}{H_0} = \frac{k}{\omega \epsilon} = \frac{1}{v\epsilon}$$
$$= \left(\frac{\mu}{\epsilon}\right)^{1/2}.$$

The ratio of E to H is often called the **characteristic impedance** Z of the medium. The impedance determines the rate at which energy is radiated into the medium by the source. It turns out that the wave speed $v = (\epsilon \mu)^{-1/2}$ and the impedance $Z = (\mu/\epsilon)^{1/2}$ are the two natural parameters to describe traveling waves in a given medium.

Energy flow

The energy transported per unit time in the direction of an electromagnetic wave in free space was given in Section 9.7 by the vector

$$\frac{1}{\mu_0}\mathbf{E} \times \mathbf{B}.$$

This could also have been written as

$$\mathbf{E} \times \mathbf{H}.$$

This latter expression also applies for a dielectric medium where now $\mathbf{B} = \mu\mathbf{H}$. The vector $\mathbf{S} = \mathbf{E} \times \mathbf{H}$, called the **Poynting vector**, gives the instantaneous value of the rate of energy flow in the medium. The mean rate of energy flow is found by averaging $\mathbf{E} \times \mathbf{H}$ over one period of the electromagnetic wave.

Reflection and refraction

Light undergoes both reflection and refraction at the interface between two media of different indices of refraction. These familiar phenomena should be describable in terms of electromagnetic theory. We shall consider two nonconducting and nonmagnetic media. The approximation $\mu = \mu_0$ is a good one since we have seen in Chapter 13 that almost all materials are only very weakly magnetic with susceptibilities much less than unity, which means permeabilities very close to μ_0 since

$$\mu = \mu_0(1 + \chi_m).$$

Let us assume that the interface between the two media, of permittivities ϵ_1 and ϵ_2, lies in the xz plane, as indicated in Fig. 14.2. We specify the direction of the incident wave by $\mathbf{k}_1$, the direction of the refracted wave by $\mathbf{k}_2$, and the direction of the reflected wave by $\mathbf{k}_3$. Both a reflected and refracted wave are required in order to satisfy the boundary conditions at the interface. We assume that $\mathbf{k}_1$ lies in the yz plane and write the following equations for the waves:

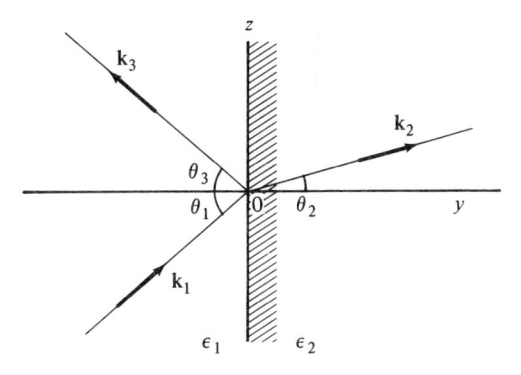

Fig. 14.2. A wave $\mathbf{k}_1$ incident on the boundary between two dielectrics gives rise to a refracted wave $\mathbf{k}_2$ and a reflected wave $\mathbf{k}_3$.

incident $\quad \mathbf{E}_1 = \mathbf{E}_{01} \exp[i(\omega_1 t - \mathbf{k}_1 \cdot \mathbf{r})]$

refracted $\quad \mathbf{E}_2 = \mathbf{E}_{02} \exp[i(\omega_2 t - \mathbf{k}_2 \cdot \mathbf{r} + \alpha)]$

reflected $\quad \mathbf{E}_3 = \mathbf{E}_{03} \exp[i(\omega_3 t - \mathbf{k}_3 \cdot \mathbf{r} + \beta)]$

where α and β allow for possible phase shifts of the refracted and reflected waves.

We showed in Section 10.3 that the tangential component of the electric field must be continuous across the interface between two dielectric media. Therefore, we write

$$E_1^t + E_3^t = E_2^t$$

for all points in the xz plane, where the superscript t indicates the tangential component. Since $y = 0$ on the xz plane, we have

$$E^t_{01} \exp i(\omega_1 t - k_{1x}x - k_{1z}z) + E^t_{03} \exp i(\omega_3 t - k_{3x}x - k_{3z}z + \beta)$$
$$= E^t_{02} \exp i(\omega_2 t - k_{2x}x - k_{2z}z + \alpha).$$

Since this equation must hold for all time t, it follows that

$$\omega_1 = \omega_2 = \omega_3.$$

Therefore, the incident, reflected, and refracted waves all have the same frequency. For the equality to hold at all points on the interface we must have

$$k_{1x} = k_{2x} = k_{3x}$$
$$k_{1z} = k_{2z} = k_{3z}.$$

Since $\mathbf{k}_1$ is in the yz plane, $k_{1x} = 0$ which shows that the vectors $\mathbf{k}_1$, $\mathbf{k}_2$, and $\mathbf{k}_3$ all lie in the yz plane.

Now $\mathbf{k}_1$ and $\mathbf{k}_3$ have the same magnitude since they describe waves in the same medium. Therefore, from

$$k_{1z} = k_{3z}$$

we have

$$k_1 \sin \theta_1 = k_3 \sin \theta_3$$
$$= k_1 \sin \theta_3$$

or

$$\sin \theta_1 = \sin \theta_3;$$

the angles of incidence and reflection are equal.

By writing

$$k_{1z} = k_{2z}$$

in the form

$$k_1 \sin \theta_1 = k_2 \sin \theta_2$$

we obtain, by substitution for k_1 and k_2 from

$$k = \omega(\epsilon \mu_0)^{1/2},$$
$$\omega_1 (\epsilon_1 \mu_0)^{1/2} \sin \theta_1 = \omega_2 (\epsilon_2 \mu_0)^{1/2} \sin \theta_2$$

or

$$\frac{\sin \theta_1}{v_1} = \frac{\sin \theta_2}{v_2}$$

$$\frac{c}{v_1} \sin \theta_1 = \frac{c}{v_2} \sin \theta_2$$

and

$$n_1 \sin \theta_1 = n_2 \sin \theta_2$$

which is **Snell's law of refraction**. The electromagnetic theory does indeed predict the phenomena of reflection and refraction for electromagnetic waves.

The refracted wave $\mathbf{k}_2$ also lies in the yz plane Therefore, we can write
$$k_2^2 = k_{2y}^2 + k_{2z}^2$$
or
$$k_{2y}^2 = k_2^2 - k_{2z}^2.$$
But
$$\frac{k_1}{n_1} = \frac{k_2}{n_2}$$
and
$$k_{2z} = k_{1z} = k_1 \sin \theta_1.$$
Therefore,
$$k_{2y}^2 = k_1^2 \left[\left(\frac{n_2}{n_1}\right)^2 - \sin^2 \theta_1 \right].$$
When $n_2 < n_1$, it is possible to have
$$\frac{n_2}{n_1} < \sin \theta_1$$
for sufficiently large θ_1. In this event k_{2y} is the square root of a negative number. We conclude that there is no refracted ray but rather that **total internal reflection** occurs. Although total reflection occurs, the incident wave actually penetrates into the second medium but its amplitude falls off exponentially with distance from the boundary (see Problem 14); the direction of propagation is along the boundary. Such an exponentially-damped wave is known as an **evanescent wave**.

Fresnel's equations

The relations between the amplitudes and phases of plane waves reflected and refracted at boundaries between different dielectric media can be derived by applying the boundary conditions satisfied by **E** and **H** (see Sections 10.3, 13.2). We will consider the special case illustrated in Fig. 14.3 in which **E** of the incident wave is taken to be perpendicular to the yz plane. In this case the electric field and the tangential component of the electric field are one and the same. Since the tangential component of **E** is continuous across the boundary, therefore
$$E_1 + E_3 = E_2.$$
The vectors **H** lie in the yz plane. Since the tangential component of **H** is also continuous across the boundary, therefore
$$H_1 \cos \theta_1 - H_3 \cos \theta_3 = H_2 \cos \theta_2.$$

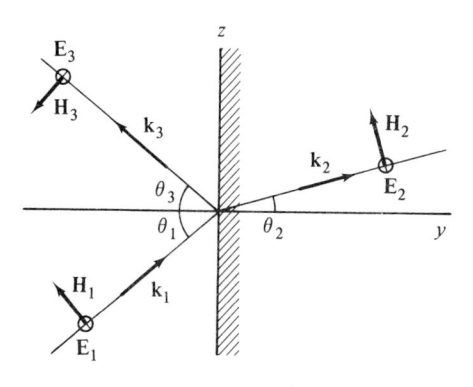

Fig. 14.3. Definition of field vectors and propagation vectors for the derivation of one of Fresnel's equations.

Noting that $H/E = (\epsilon/\mu)^{1/2}$, we can write this equation as

$$\left(\frac{\epsilon_1}{\mu_1}\right)^{1/2} E_1 \cos\theta_1 - \left(\frac{\epsilon_3}{\mu_3}\right)^{1/2} E_3 \cos\theta_3$$
$$= \left(\frac{\epsilon_2}{\mu_2}\right)^{1/2} E_2 \cos\theta_2.$$

Assuming that both media are nonmagnetic so that $\mu_1 = \mu_2 = \mu_3 = \mu_0$ and noting that n is proportional to $\epsilon^{1/2}$ it follows that

$$n_1(E_1 \cos\theta_1 - E_3 \cos\theta_3) = n_2 E_2 \cos\theta_2.$$

Further, using the facts that $\theta_1 = \theta_3$ and $n_1 \sin\theta_1 = n_2 \sin\theta_2$ derived above, we obtain

$$(E_1 - E_3) \tan\theta_2 = E_2 \tan\theta_1 = (E_1 + E_3) \tan\theta_1.$$

Therefore,

$$\frac{E_3}{E_1} = \frac{\tan\theta_2 - \tan\theta_1}{\tan\theta_2 + \tan\theta_1} = \frac{\sin(\theta_2 - \theta_1)}{\sin(\theta_2 + \theta_1)}.$$

This equation gives the amplitude of the reflected wave and is one of Fresnel's equations. From this equation we see that if $n_1 > n_2$ so that $\theta_2 > \theta_1$, then E_1 and E_3 have the same sign; that is, **there is no phase change upon reflection at a dense-rare interface.** However, if $n_1 < n_2$ so that $\theta_2 < \theta_1$, then E_1 and E_3 are of opposite sign; that is, **there is a phase change of π rad upon reflection at a rare-dense interface.** Alternately, we can say that $\beta = 0$ for $n_1 > n_2$ and π for $n_1 < n_2$ in the equation for the reflected wave given on page 267.

It is easily seen that the refracted wave is always in phase with the incident wave so that $\alpha = 0$ in the equation for the refracted wave given on page 267.

The derivation of the second of Fresnel's equations in which **H** of the incident wave is taken to be perpendicular to the yz plane is left as an exercise for the reader (see Problem 16).

14.5 THE FIZEAU EXPERIMENT

The Fizeau experiment, first performed in 1851, provides a measurement of the speed of light traveling in water which itself is moving with respect to the laboratory. According to the prediction[3] of the special theory of relativity, the speed v of the light in the moving water as measured by an observer in the laboratory is

[3] *MWTP*, Section 5.2.

$$v = \frac{\frac{c}{n} + u}{1 + \frac{u}{nc}}$$

where c is the speed of light in vacuum, n is the index of refraction of water, and u is the speed of the water relative to the laboratory. Since $u \ll c$, it follows from the binomial theorem that

$$v \simeq \left(\frac{c}{n} + u\right)\left(1 - \frac{u}{nc}\right)$$
$$\simeq \frac{c}{n} + u\left(1 - \frac{1}{n^2}\right).$$

The experimental arrangement is indicated in Fig. 14.4. Water flows through the tube at a speed u of the order of 10 m · sec^{-1}. Light from the source is divided into two beams by the half-silvered mirror A. One beam then traverses the path $ABCDAR$ through the water to the receiver R; the second beam traverses the path $ADCBAR$. The relative time delay for the light to traverse the two paths is measured by interferometric techniques.[4] The experimental results are in agreement with the prediction of the special theory of relativity.

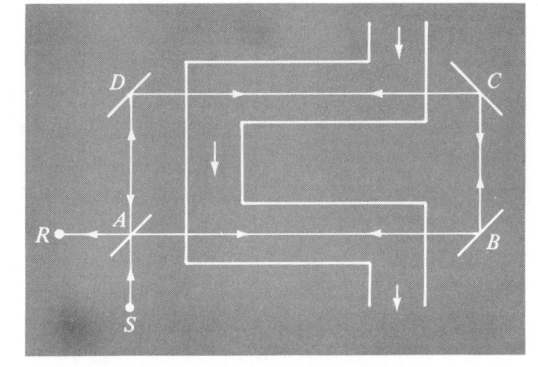

Fig. 14.4. Schematic diagram of the apparatus for the Fizeau experiment.

14.6 PLANE WAVES IN CONDUCTING MEDIA

In a medium with a finite conductivity σ Maxwell's second equation becomes

$$\nabla \times \mathbf{H} = \mathbf{j} + \frac{\partial \mathbf{D}}{\partial t} = \sigma \mathbf{E} + \epsilon \frac{\partial \mathbf{E}}{\partial t}.$$

Eliminating $\mathbf{H}$ from Maxwell's equations as in Section 9.7 yields the equation

$$\nabla^2 \mathbf{E} = \epsilon\mu \frac{\partial^2 \mathbf{E}}{\partial t^2} + \mu\sigma \frac{\partial \mathbf{E}}{\partial t}.$$

This wave equation represents a damped wave motion where the amplitude decreases as the wave moves along because of the presence of the $\partial \mathbf{E}/\partial t$ term. If we consider a wave traveling along the x axis, the solution to the wave equation is (see Problem 20)

$$\mathbf{E} = \mathbf{E}_0 \exp\left[i\left(\omega t - \frac{x}{\delta}\right)\right] \exp\left(-\frac{x}{\delta}\right)$$

[4] See Chapters 17 and 18.

where

$$\delta = \left(\frac{2}{\mu\sigma\omega}\right)^{1/2}$$

is known as the **skin depth**. As the wave progresses along the x axis its amplitude decreases in accordance with the term $\exp(-x/\delta)$. The skin depth decreases with increase in frequency for a given conductor. For example, at 1 MHz for copper δ is only 7×10^{-3} cm while at 60 Hz it is 0.9 cm.

The index of refraction of a conducting medium is a complex quantity of the form

$$n = n_1 - in_2.$$

The real part n_1 determines the wave speed in the conductor while the imaginary part n_2 is related to the skin depth and determines the rate at which the amplitude of the wave decays. Therefore, a complex index of refraction indicates that absorption of energy from the wave occurs. Calculations of the intensities of the transmitted and reflected waves are rather complex and only the results are quoted here. Metals should be almost perfect reflectors of electromagnetic radiation for all frequencies up to those of visible light. The fact that copper is strongly colored shows that this is not true at visible frequencies. The explanation lies in the fact that the conductivity decreases as the frequency increases and that electrons bound into the crystal structure begin to play a role at optical frequencies.

14.7 DISPERSION

According to electromagnetic wave theory, the index of refraction of a medium should be given by (see Section 14.4)

$$n = \left(\frac{\epsilon\mu}{\epsilon_0\mu_0}\right)^{1/2}.$$

For most media $\mu \simeq \mu_0$ so that

$$n = \left(\frac{\epsilon}{\epsilon_0}\right)^{1/2} = K^{1/2}$$

to a very good approximation where K is the dielectric constant. Dielectric constants are usually measured at low frequencies while the refractive index is usually measured at optical frequencies. Under these conditions very poor agreement is obtained with the theoretical predictions except for the case of simple, nonpolar gases (for example, air, nitrogen, oxygen, argon, carbon dioxide).

The variation of the refractive index with wavelength for radiation in the optical region has been discussed in Section 14.4 and is known as **dis-**

SEC. 14.7   DISPERSION

persion. In most cases the value of n increases as the wavelength decreases; this is known as **normal dispersion**. The reverse case, where the value of n decreases with decreasing wavelength, occurs only for frequencies near which the material absorbs electromagnetic radiation, that is, near **absorption lines** characteristic of the material. This effect is known as **anomalous dispersion**. Both types of dispersion can be explained on the basis of a classical theory involving the assumption that atoms contain electrons that vibrate at certain natural frequencies and that the alternating electric field of an electromagnetic wave sets the electrons into forced oscillation.

We shall illustrate this approach by considering a gas of dielectric constant K subject to an alternating electric field

$$\mathbf{E} = \mathbf{E}_0 \exp(i\omega t).$$

We omit the position dependence of the field since we assume that the wavelength of the electromagnetic wave is much larger than the size of an atom. This limits the simple theory to wavelengths greater than about 10^{-8} m, which corresponds to soft x-rays. When an electron suffers a displacement $\mathbf{s}$ due to the action of the field $\mathbf{E}$, a restoring force[5] $-m_e\omega_0^2\mathbf{s}$ acts to return the electron to its equilibrium position; the frequency $\omega_0 = 2\pi v_0$ is associated with the natural frequency v_0 of oscillation of the electron of mass m_e. Damping[6] also occurs due to collisions, radiation of energy, etc; this may be represented by a term

$$-m_e\gamma \frac{d\mathbf{s}}{dt}$$

where γ is a constant. The motion of the electron is then determined from the equation

$$\mathbf{F} = m_e \mathbf{a} = m_e \frac{d^2\mathbf{s}}{dt^2}$$

where

$$\mathbf{F} = -e\mathbf{E} - m_e\gamma \frac{d\mathbf{s}}{dt} - m_e\omega_0^2\mathbf{s}.$$

Therefore,

$$m_e\left(\frac{d^2\mathbf{s}}{dt^2} + \gamma \frac{d\mathbf{s}}{dt} + \omega_0^2\mathbf{s}\right) = -e\mathbf{E}_0 \exp(i\omega t).$$

The left-hand side of this equation has the characteristic form of the equation for damped oscillations. However, the presence of the external driving force of frequency ω prevents the oscillations from dying out. The response to the external driving force is, however, dependent on the relative magni-

---

[5] *MWTP*, Section 11.3.
[6] *MWTP*, Section 11.8.

tudes of the frequencies ω and ω_0. The solution to the equation of motion is, to a good approximation,

$$\mathbf{s} = -\frac{e\mathbf{E}}{m_e[(\omega_0^2 - \omega^2) + i\gamma\omega]};$$

the details are given in the shaded area.

Solution of the equation for forced oscillation

The equation

$$m_e\left(\frac{d^2\mathbf{s}}{dt^2} + \gamma\frac{d\mathbf{s}}{dt} + \omega_0^2\mathbf{s}\right) = -e\mathbf{E}\exp(i\omega t)$$

is a second-order linear differential equation with constant coefficients. The solution is the sum of a particular integral and a complementary function. The complementary function is the solution of the equation

$$\frac{d^2\mathbf{s}}{dt^2} + \frac{d\mathbf{s}}{dt} + \omega_0^2\mathbf{s} = 0$$

which is

$$\exp\left(-\frac{\gamma t}{2}\right)\left\{\mathbf{A}\cos\left[\left(\omega_0^2 - \frac{\gamma^2}{4}\right)^{1/2} t\right] + \mathbf{B}\sin\left[\left(\omega_0^2 - \frac{\gamma^2}{4}\right)^{1/2} t\right]\right\}.$$

The particular integral is any solution to the original equation which involves no arbitrary constant. Let us try the solution

$$\mathbf{s} = \beta\mathbf{E}$$

where β is a constant. Now

$$\frac{d\mathbf{s}}{dt} = i\omega\beta\mathbf{E}$$

and

$$\frac{d^2\mathbf{s}}{dt^2} = -\omega^2\beta\mathbf{E}$$

so that, upon substitution in the original equation, we obtain

$$m_e(-\omega^2\beta\mathbf{E} + i\gamma\omega\beta\mathbf{E} + \omega_0^2\beta\mathbf{E}) = -e\mathbf{E}.$$

Thus,

$$\mathbf{s} = \beta\mathbf{E}$$

is a solution if

$$\beta = -\frac{e}{m_e[(\omega_0^2 - \omega^2) + i\gamma\omega]}$$

and the particular integral is

$$-\frac{e\mathbf{E}}{m_e[(\omega_0^2 - \omega^2) + i\gamma\omega]}.$$

SEC. 14.7 DISPERSION

The complete solution to the differential equation is then

$$s = -\frac{eE}{m_e[(\omega_0^2 - \omega^2) + i\gamma\omega]} + \exp\left(-\frac{\gamma t}{2}\right)\left\{A \cos\left[\left(\omega_0^2 - \frac{\gamma^2}{4}\right)^{1/2} t\right] + B \sin\left[\left(\omega_0^2 - \frac{\gamma^2}{4}\right)^{1/2} t\right]\right\}.$$

The terms involving **A** and **B** are as often positive as negative when evaluated for a large number of electrons in the gas, since they depend on the initial conditions in individual atoms. Therefore, these terms average to zero over the volume of the gas, and

$$s = -\frac{eE}{m_e[(\omega_0^2 - \omega^2) + i\gamma\omega]}$$

is a suitable expression for the displacement of an individual electron.

Because of the displacement **s** of the electron, an instantaneous dipole moment $-e\mathbf{s}$ is created. The gas acquires an electric polarization

$$\mathbf{P} = -n_0 e\mathbf{s}$$
$$= \frac{n_0 e^2}{m_e[(\omega_0^2 - \omega^2) + i\gamma\omega]} \mathbf{E}$$

where n_0 is the number of gas molecules per unit volume. However, we saw in Section 10.4 that

$$\mathbf{P} = n_0 \alpha \mathbf{E}_l$$

in general, where α is the molecular polarizability and $\mathbf{E}_l$ is the local field, which may be different from the external field. The relation between the dielectric constant K and molecular polarizability was given in Section 10.4 as

$$\frac{K-1}{K+2} = \frac{n_0 \alpha}{3\epsilon_0}.$$

Since $K = n^2$, we have

$$\frac{K-1}{K+2} = \frac{n^2-1}{n^2+2} = \frac{n_0 \alpha}{3\epsilon_0}$$
$$= \frac{n_0 e^2}{3m_e \epsilon_0} \frac{1}{[(\omega_0^2 - \omega^2) + i\gamma\omega]}$$

using the value for α suggested above. Both the dielectric constant and the index of refraction are complex. For gases at low pressures, we can write the approximate expressions,

$$K_1 = n_1^2 - n_2^2 \simeq n_1^2 = 1 + \frac{n_0 e^2}{2m_e \epsilon_0 \omega}\left[\frac{\omega_0 - \omega}{(\omega_0 - \omega)^2 + \gamma^2/4}\right]$$

$$K_2 = 2n_1 n_2 \simeq 2n_2 = \frac{n_0 e^2}{2m_e \epsilon_0 \omega}\left[\frac{\gamma/2}{(\omega_0 - \omega)^2 + \gamma^2/4}\right]$$

for the real and imaginary parts of the complex dielectric constant as defined by

$$K = K_1 - iK_2$$

(see Problem 22), and the real and imaginary parts of the complex refractive index as defined by

$$n = n_1 - in_2.$$

The real part n_1 of the refractive index is called the **dispersive component**; its behavior near the resonant frequency ω_0 is shown in Fig. 14.5. Note that n_1 decreases in magnitude with increasing frequency (decreasing wavelength) in the vicinity of an absorption line giving anomalous dispersion as stated earlier.

The imaginary part n_2 of the refractive index is called the **absorptive component**; its behavior near the resonant frequency ω_0 is shown in Fig. 14.6. The magnitude of n_2 at frequency ω is a measure of the energy that a gas will absorb from an incident electromagnetic wave of frequency ω. The value of n_2 is maximum at $\omega = \omega_0$ and falls to half its maximum value at frequencies $\omega = \omega_0 \pm \gamma/2$. The quantity

$$\Delta \nu = \frac{|\omega - \omega_0|}{2\pi} = \frac{\gamma}{4\pi}$$

is the **half-width** of the absorption line in Hz; $\gamma/2$ is the half-width in rad·sec^{-1}.

Usually a gas will absorb at a number of discrete frequencies. The value of n_2 is negligible everywhere except in the vicinity of an absorption line. Therefore, $n \simeq n_1$ except near an absorption line. The behavior of n_1 as a function of frequency over a wide frequency range is shown in Fig. 14.7. The index of refraction attains constant values between resonances but the limiting value on the high-frequency side is smaller than on the low-frequency side. Finally, the index of refraction approaches the value of unity above all resonances. However, the index is slightly less than unity at high frequencies since the electrons can be considered to be free when interacting with the high-frequency wave. Measured values of the index of refraction in the x-ray region are indeed less than one.

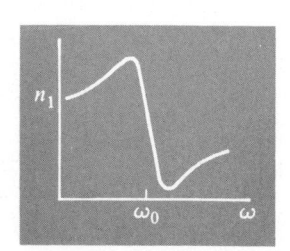

Fig. 14.5. Variation of the dispersive component of the refractive index near the resonant frequency ω_0.

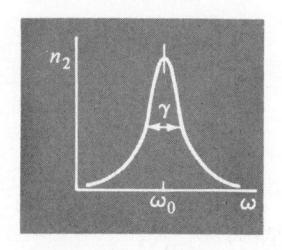

Fig. 14.6. Variation of the absorptive component of the refractive index near the resonant frequency ω_0.

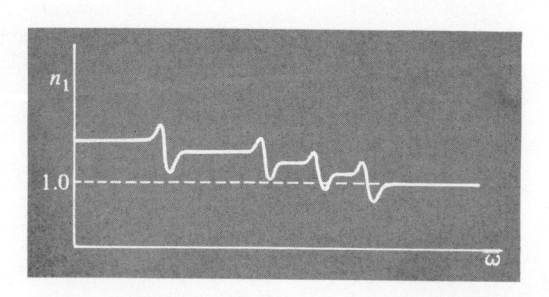

Fig. 14.7. Variation of the index of refraction of a gas with frequency.

14.8 CERENKOV RADIATION

In 1958 the Nobel prize in physics was awarded to three Russian scientists for their extensive studies of the radiation generated by charged particles moving through a material medium. Pavel Cerenkov carried out the basic investigations from 1934 to 1938. Ilya Frank and Igor Tamm in 1937 provided the correct interpretation of Cerenkov's experiment.

When a charged particle moves in a medium with a uniform speed u greater than the speed $v(\omega)$ of an electromagnetic wave in the medium, an electromagnetic wave of frequency ω is generated through interaction of the charged particle with the molecules of the medium. This wave (called **Cerenkov radiation**[7]) propagates in the direction $\theta(\omega)$ where

$$\cos \theta(\omega) = \frac{v(\omega)}{u}.$$

Cerenkov radiation is an electromagnetic shock-wave phenomenon and is closely analogous to bow waves formed when a vessel moves through water at a speed greater than the speed of the surface waves on the water and to the sonic boom created by an object moving through air at a speed greater than the speed of sound in air (see Fig. 14.8).

The Cerenkov effect can be described in terms of a theory[8] in which the electromagnetic field generated by a moving charged particle is expressed as a superposition of plane monochromatic waves. In general, both ordinary waves and evanescent waves are present in the plane waves; this representation of the field is called the **angular spectrum**. An evanescent wave (see Section 14.5) has an amplitude that decreases exponentially with distance in a direction perpendicular to the direction of propagation. The evanescent waves decay exponentially while the ordinary waves propagate with constant amplitude and carry off energy. For the case of uniform motion, only evanescent waves are present and no radiation of energy occurs. However, when the speed of the charged particle exceeds the speed of propagation of an electromagnetic wave in the medium, ordinary waves are generated, giving rise to Cerenkov radiation. In this model the ordinary waves are generated by interaction of the vacuum field of the charged particle with the molecules of the dielectric medium.

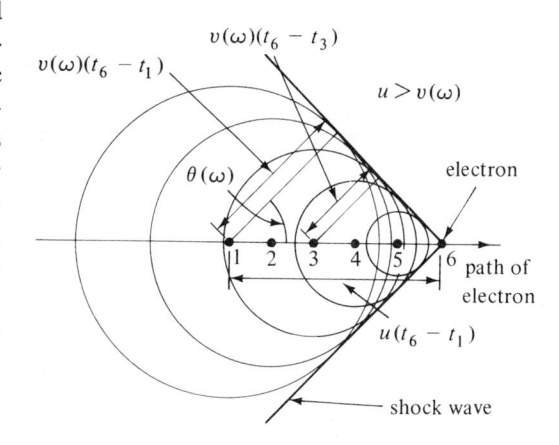

Fig. 14.8. The generation of Cerenkov radiation. The shock wave is formed through the superposition of spherical waves originating along the path of the electron. The constituent waves are shown at a particular instant of time.

The study of solvated electrons by pulse radiolysis

A **solvated electron**[9] is an electron that is released into solution and becomes bound to the solvent molecules. The importance of solvated elec-

---

[7] J. V. Jelley, "Cerenkov Radiation: Its Origin, Properties and Applications," *Contemporary Physics*, **3** (October 1961), 45.

[8] R. Asby and E. Wolf, "Theory of Cerenkovian Effects," *The Physics Teacher*, **9** (April 1971), 207.

[9] J. L. Dye, "The Solvated Electron," *Scientific American*, February, 1967.

trons to biology is that they are thought to be involved with the sequence of chemical reactions that leads to radiation damage in living tissue. Solvated electrons can be produced in a wide variety of solvents, including water, by a burst of high-energy ionizing radiation. In the experiment described in this section the ionization is produced by energetic electrons from a **linear accelerator** or **linac**.[10]

The minimum energy required to ionize a molecule of water is 13 eV. Therefore, an electron of energy $\simeq$ 10 MeV can ionize a very large number of water molecules. The effect of the interaction on the water molecule can be described in the following symbolic manner:

$$H_2O = H_2O^+ + e^-.$$

The electron ejected from the water molecule loses its excess energy within 10^{-13} sec to the surrounding water molecules. Meanwhile, the H_2O^+ ion can react with an H_2O molecule to form a hydroxyl radical (OH) and a hydronium ion (H_3O^+). The electron may then be recaptured by the H_3O^+ ion to produce a hydrogen atom and an H_2O molecule or it can be trapped by surrounding H_2O molecules and become a solvated electron as indicated schematically in Fig. 14.9. The presence of solvated electrons causes water to take on a blue color; that is, solvated electrons give rise to absorption in the red portion of the visible spectrum. The absorption line occurs at a wavelength of approximately 600 nm. The lifetimes of solvated electrons are extremely short, particularly in concentrated solutions of biological interest.

Fig. 14.9. A solvated electron.

Pulse radiolysis may be used to study the decay of solvated electrons following a burst of ionizing radiation. The principle of the technique is illustrated in Fig. 14.10. A short pulse of energetic electrons is passed through the sample thereby creating solvated electrons. A pulse of analyzing light of wavelength $\simeq$ 600 nm is passed through the sample Δt sec after the pulse of electrons; the absorption of the light is measured. The whole procedure is repeated for a series of delay times of the light pulse. In this manner a plot of absorption as a function of the time following the pulse of ionizing radiation can be obtained (see Fig. 14.11). If the decay curve is exponential in form, it can be characterized by a **rate constant** which is just the time for the absorption to fall to e^{-1} of its initial value.

Because of the nature of the operation of a linac a single burst of duration $\simeq$ 36 nsec actually consists of about 100 pulses each of 10 psec duration separated by 350 psec [see Fig. 14.12(a)].[11] During each pulse there is a build-up of solvated electrons; during the time between pulses the

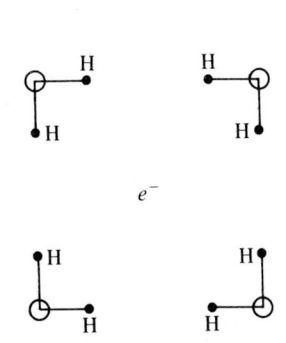

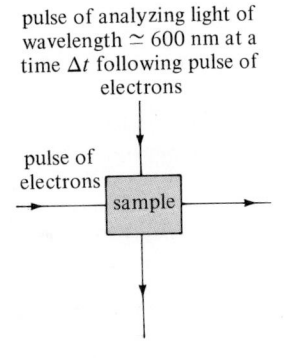

Fig. 14.10. Schematic representation of a pulse radiolysis experiment.

---

[10]W. Panofsky, "The Linear Accelerator," *Scientific American*, October, 1964. Available as *Scientific American Offprint 234* (San Francisco: W. H. Freeman and Co., Publishers).

[11]The method of pulse radiolysis described in this paragraph was developed by J. W. Hunt at the University of Toronto in 1965.

SEC. 14.8   CERENKOV RADIATION   279

solvated electrons decay away. In Fig. 14.12(b) we assume the rate constant for the decay to be $\simeq 100$ psec. Cerenkov radiation provides a convenient source of analyzing light. In a typical case the electrons leave the linac drift tube traveling at a speed of 0.99992 c (40 MeV electrons); the speed of light in air is 0.99971 c. Therefore, the electrons as they leave the vacuum chamber of the linac are traveling faster than the speed of light in air and Cerenkov radiation is produced. By means of a system of mirrors, as shown in Fig. 14.13, the Cerenkov radiation can be delayed by an amount Δt_1 relative to the pulse of electrons and used to monitor the decay of the solvated electrons [see Fig. 14.12(c)]. The light of wavelength $\simeq 600$ nm transmitted by the sample during the pulse of Cerenkov radiation is accu-

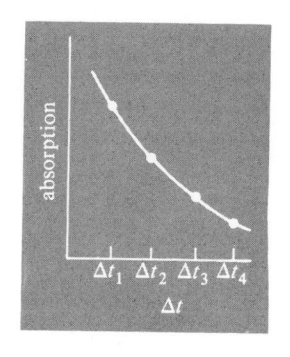

Fig. 14.11. Absorption due to solvated electrons as a function of the time following a pulse of ionizing radiation.

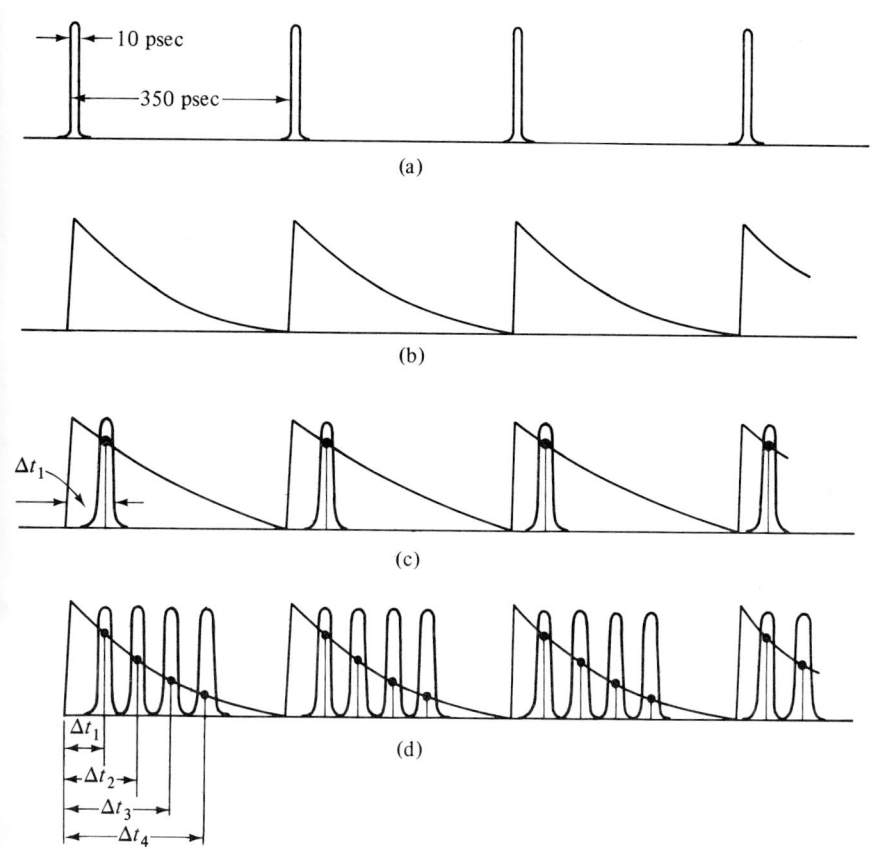

Fig. 14.12. (a) Pulses from the linac. (b) Build-up and decay of solvated electrons. (c) Pulses of Cerenkov radiation delayed by an amount Δt_1. (d) Repetition of the experiment for different delay times $\Delta t_1, \Delta t_2, \Delta t_3, \Delta t_4, \ldots$ permits the decay curve to be traced out.

mulated during the 100 pulses to give a measurable signal. Movement of the mirrors allows the Cerenkov radiation to be delayed by different amounts $\Delta t_1, \Delta t_2, \Delta t_3, \Delta t_4, \ldots$ [see Fig. 14.12(d)]. Each Δt_i gives one point on the decay curve shown in Fig. 14.11.

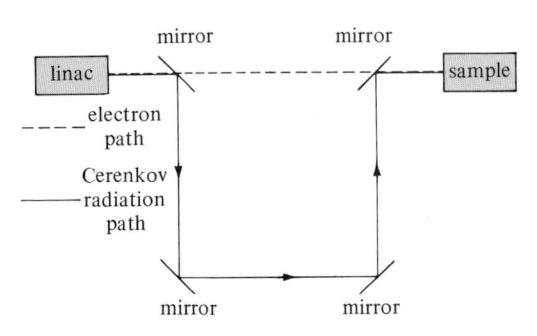

Fig. 14.13. A system of mirrors is used to delay the pulse of Cerenkov radiation relative to the pulse of electrons.

14.9 POLARIZATION

In Section 9.7 we showed that electromagnetic waves may be linearly (or plane) polarized. Using complex notation for a plane harmonic wave, we have

$$\mathbf{E} = \mathbf{E}_0 \exp[i(\omega t - \mathbf{k} \cdot \mathbf{r})]$$
$$\mathbf{H} = \mathbf{H}_0 \exp[i(\omega t - \mathbf{k} \cdot \mathbf{r})].$$

Linear or plane polarization occurs if $\mathbf{E}_0$ and $\mathbf{H}_0$ are constant real vectors. The direction of the electric vector is taken as the direction of polarization.

Let us now consider the special case of two linearly polarized waves having the same amplitude, polarized at right angles to each other, and having a phase difference of $\pi/2$. We choose coordinate axes such that the direction of polarization of one wave is along the x axis and the direction of the second is along the y axis; both waves travel, therefore, in the z direction. The electric field of the composite wave can be written as

$$\mathbf{E} = \mathbf{i}E_0 \exp[i(\omega t - kz)] + \mathbf{j}E_0 \exp\left[i\left(\omega t - kz \pm \frac{\pi}{2}\right)\right]$$

where

$$\mathbf{k} \cdot \mathbf{r} = k_z z = kz.$$

Since

$$\exp\left(i\frac{\pi}{2}\right) = \cos\frac{\pi}{2} + i\sin\frac{\pi}{2} = i,$$

we obtain

$$\mathbf{E} = E_0(\mathbf{i} \pm i\mathbf{j}) \exp(\omega t - kz).$$

The electric vector is constant in magnitude but rotates in space with angular frequency ω. The wave is said to be **circularly polarized** [see Fig. 14.14(a)]. When the complex term is $(\mathbf{i} + i\mathbf{j})$, the electric vector rotates counterclockwise when viewed against the direction of propagation and the wave is said to be **left** circularly polarized. When the complex term is $(\mathbf{i} - i\mathbf{j})$, the electric vector rotates clockwise when viewed against the direction of propagation and the wave is said to be **right** circularly polarized.

If the two component fields do not have the same amplitude, the resultant electric field vector at some given point in space not only rotates but also changes its magnitude in such a way that the tip of the vector moves along an elliptical path. The wave is now said to be **elliptically** polarized [see Fig. 14.14(b)].

A **complex vector amplitude** $\mathcal{E}_0$, defined as

$$\mathcal{E}_0 = \mathbf{i}E_0 + i\mathbf{j}E'_0$$

where E_0 and E'_0 are the amplitudes of the component waves, is often used. The composite wave is

$$\mathbf{E} = \mathcal{E}_0 \exp\left[i(\omega t - kz)\right].$$

The type of polarization now depends on the nature of $\mathcal{E}_0$. The following cases occur:

1. $\mathcal{E}_0$ real—linear polarization;
2. $\mathcal{E}_0$ complex, $E_0 \neq E'_0$—elliptic polarization;
3. $\mathcal{E}_0$ complex, $E_0 = E'_0$—circular polarization.

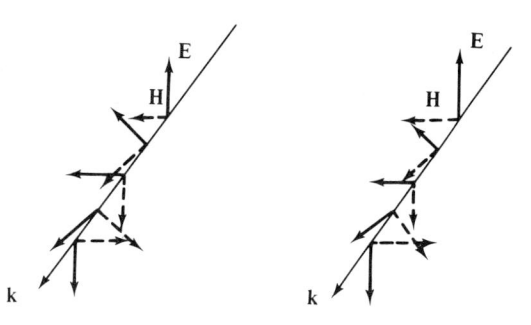

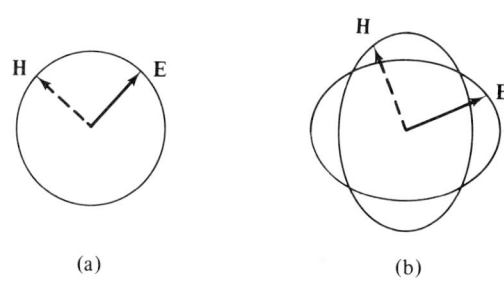

Fig. 14.14. (a) Fields in a circularly polarized wave. (b) Fields in an elliptically polarized wave.

Matrix representation of polarization

A more general way of writing the complex amplitude $\mathcal{E}_0$ is

$$\mathcal{E}_0 = \mathbf{i}\mathcal{E}_{0x} + \mathbf{j}\mathcal{E}_{0y}$$

where $\mathcal{E}_{0x}$ and $\mathcal{E}_{0y}$ may be complex. This pair of complex amplitudes can be conveniently expressed as a matrix[12] of the form

$$\begin{bmatrix} \mathcal{E}_{0x} \\ \mathcal{E}_{0y} \end{bmatrix} = \begin{bmatrix} E_0 \exp(i\theta_x) \\ E_0 \exp(i\theta_y) \end{bmatrix}.$$

The values of $\mathcal{E}_{0x}$ and $\mathcal{E}_{0y}$ are usually **normalized** so that

$$\mathcal{E}_{0x}^2 + \mathcal{E}_{0y}^2 = 1.$$

Therefore, a plane wave linearly polarized in the x direction is represented by the matrix

$$\begin{bmatrix} 1 \\ 0 \end{bmatrix},$$

---

[12] A short introduction to matrices is given in Appendix D.

and a plane wave linearly polarized in the y direction by

$$\begin{bmatrix} 0 \\ 1 \end{bmatrix}.$$

The matrix

$$(2)^{-1/2} \begin{bmatrix} 1 \\ 1 \end{bmatrix}$$

represents a wave linearly polarized at 45° with respect to the x or y axis. A circularly polarized wave is represented by

$$\begin{bmatrix} 1 \\ \pm 1 \end{bmatrix},$$

the plus sign indicating left circular polarization.

Optical elements such as lenses, mirrors, and polarizers can be represented by 2×2 matrices of the form

$$\begin{bmatrix} \alpha & \beta \\ \gamma & \delta \end{bmatrix}.$$

If light incident upon an optical element is represented by the matrix

$$\begin{bmatrix} A \\ B \end{bmatrix},$$

light emerging from the optical element is represented by the matrix

$$\begin{bmatrix} \alpha & \beta \\ \gamma & \delta \end{bmatrix} \begin{bmatrix} A \\ B \end{bmatrix} = \begin{bmatrix} \alpha A + \beta B \\ \gamma A + \delta B \end{bmatrix}.$$

The matrices representing linear and circular polarizers are listed below:

linear polarizer		circular polarizer	
$\begin{bmatrix} 1 & 0 \\ 0 & 0 \end{bmatrix}$,	$\begin{bmatrix} 0 & 0 \\ 0 & 1 \end{bmatrix}$	$\begin{bmatrix} 1 & i \\ -i & 1 \end{bmatrix}$,	$\begin{bmatrix} 1 & -i \\ i & 1 \end{bmatrix}$
x direction	y direction	right	left

For example, if a beam of light linearly polarized in the x direction is incident upon a right circular polarizer, the emergent beam is given by

$$\begin{bmatrix} 1 & i \\ -i & 1 \end{bmatrix} \begin{bmatrix} 1 \\ 0 \end{bmatrix} = \begin{bmatrix} 1 \\ -i \end{bmatrix}.$$

The emergent light is right circularly polarized.

Optical activity and molecular dissymmetry

Some substances are able to rotate the plane of polarization of an incident plane-polarized light wave.[13] For a specified medium the amount

[13] S. F. Mason, "Optical Activity and Molecular Assymmetry," *Contemporary Physics*, **9** (May 1968), 239.

of rotation depends on the distance traveled through it and the wavelength of the light. The latter effect is known as **rotary dispersion**. The rotation may be either right-handed (**dextrorotatory**) or left-handed (**laevorotatory**). Right-handed rotation means that if we look against the oncoming light the plane of vibration is rotated in the clockwise sense. Quartz crystals, turpentine, and sugar solutions provide examples of materials that exhibit optical activity. In Fig. 14.15 we show the rotary dispersion for light traveling through a 1.0 mm piece of quartz. Note that violet light is rotated nearly four times as much as red light.

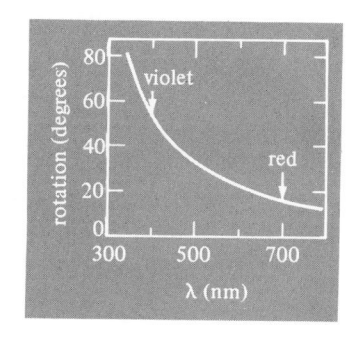

Fig. 14.15. Rotary dispersion by a 1.0 mm piece of quartz.

Optically active molecules have a dissymmetric molecular structure. This means that they do not possess a center of inversion, a plane of symmetry, or even a rotation-reflection symmetry axis.[14] The tartaric acid molecule (see Fig. 14.16) is an example of a **dissymmetric molecule**. The two forms are related structurally as an object and its mirror image. A dissymmetric molecule may be thought of as approximating to a conducting helix, with a natural dipole frequency, which can absorb energy from a plane-polarized electromagnetic wave of appropriate frequency. A linearly polarized wave is a superposition of equal amounts of right and left circularly polarized radiation (see Problem 34). The interaction of the wave with the molecule induces alternating electric and magnetic dipole moments in the molecule, both directed along the axis of the helix. The sense of winding of the helix determines whether the interaction of the molecule with the right or left circularly polarized component is favored. The differential interaction imposes new phase relations on the circular components reradiated from the helix. Therefore, the plane of polarization of the resultant transmitted light is rotated, the direction of rotation depending on the sense in which the helix is wound.

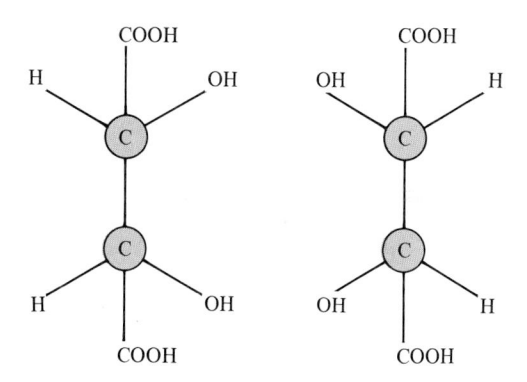

Fig. 14.16. The left- and right-handed forms of the tartaric acid molecule.

In a liquid the axes of the helices are randomly oriented, but the optical activity does not vanish since it is the sense of winding of the helix relative to its own axis rather than the orientation of that axis with respect to the direction of the light beam that determines the sense of rotation of the plane of polarization. As a result the liquid has different indices of refraction for right and left circularly polarized light. Therefore, as the linearly polarized light travels through the liquid, the phase of one circularly

[14] *MWTP*, Section 21.2.

polarized component gets ahead of the other. It follows that the plane of linear polarization is rotated, the direction of rotation being the same as the direction of rotation of the circularly polarized component with the smaller index of refraction.

The general basis for optical activity can be demonstrated at the macroscopic level by replacing dissymmetric molecules by copper helices of macroscopic dimensions and using electromagnetic waves from the microwave region.[15]

It is interesting to note that helical organic molecules produced by living organisms are single-handed. For example, all DNA molecules (see Section 3.5) are right-handed helices. This fact may well constitute a fundamental clue to the sorting out of the evolution of life on the earth!

14.10 UNPOLARIZED RADIATION

A single atom in a gas may be stimulated to emit electromagnetic radiation as the result of a collision. The duration of the emission is characterized by a **mean decay time** τ. The polarization of the emitted radiation is constant during this time interval. If at a later time the atom is again stimulated to emit radiation by a second collision, the state of polarization of the radiation bears no relation to that emitted following the first excitation assuming collisions to occur in a random manner. If we now consider a collection of atoms, an observer will see an electromagnetic wave that is a superposition of the waves emitted by the individual radiating atoms. It is easy to see that the state of polarization of the total electromagnetic wave remains constant during a time interval short compared to τ. After an interval of many mean lives, the atoms that were initially radiating are either no longer radiating or have been reexcited to radiate several times. The state of polarization of the total wave changes in a random manner with a time scale of the order of τ. Therefore, if the measurement process requires a time long compared to τ, no polarization will be detected. This is illustrated in Fig. 14.17. The individual contributions to the total electric vector are arranged randomly in the plane perpendicular to the direction of propagation of the radiation and the phase relations between them vary randomly in time. Such radiation is said to be **unpolarized**. It is often convenient to describe unpolarized radiation as consisting of two orthogonal components equal in magnitude but having a rapidly and randomly varying phase relation (see Fig. 14.18). These two descriptions of unpolarized radiation are equivalent (see Problem 31).

Since typical mean decay times for atoms emitting visible radiation are of the order of 10^{-10} sec, most common sources of visible light are, for all practical purposes, sources of unpolarized radiation.

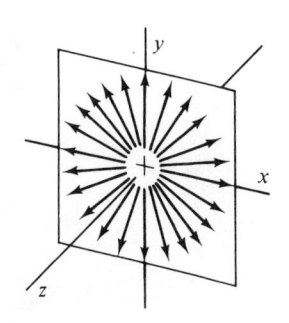

Fig. 14.17. The components of the electric vector in an unpolarized wave are arranged symmetrically about the direction of propagation.

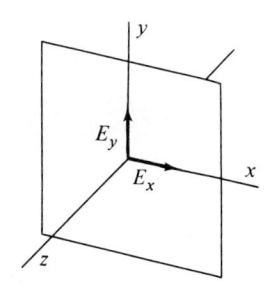

Fig. 14.18. Unpolarized radiation may be described by two orthogonal components of equal magnitude and having a randomly varying phase relation.

[15] G. E. Foxcroft and C. J. Crumper, *School Science Review*, **164** (November 1966), 28.

Polarization by reflection

There is a variety of techniques available for producing polarized light from an unpolarized beam. As an example we consider the polarization of a beam of light by reflection. A beam of unpolarized radiation is incident upon a reflecting surface making an angle θ with the normal to the surface. The radiation may be considered to consist of a component $\mathbf{E}_{\|}$ in the plane of incidence and a component $\mathbf{E}_{\perp}$ perpendicular to the plane of incidence (see Fig. 14.19). The two components are not equally reflected. This is illustrated in Fig. 14.20 for reflection from glass. Note that for $\theta = 56°$, only the component $\mathbf{E}_{\perp}$ is reflected—this is called the **polarizing angle** θ_p or the **Brewster angle** after Sir David Brewster. In general,

$$\tan \theta_p = \frac{n_2}{n_1}$$

where n_1 and n_2 are the refractive indices of the first and second medium, respectively. This result follows from Fresnel's equations (Problem 16). **For light incident upon a reflecting surface at the Brewster angle, the reflected light is plane polarized with its electric vector perpendicular to the plane of incidence.** The transmitted light is a mixture of $\mathbf{E}_{\|}$, all of which is transmitted, and $\mathbf{E}_{\perp}$ of which about 0.85 is transmitted. By using a pile of plates rather than just one, most of $\mathbf{E}_{\perp}$ can be removed from the transmitted beam as a result of repeated reflections. The result is indicated in Fig. 14.21.

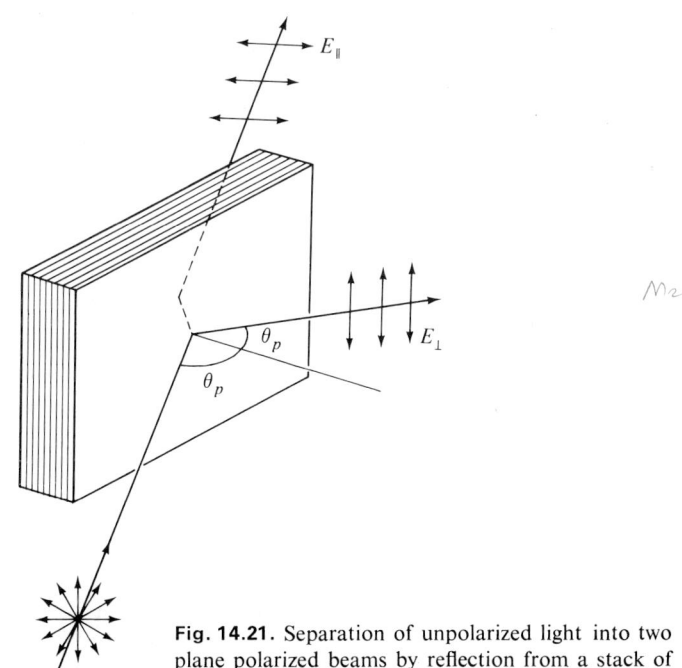

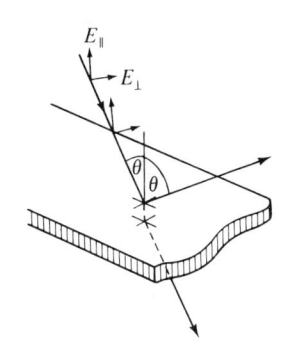

Fig. 14.19. A beam of unpolarized light incident upon a reflecting surface.

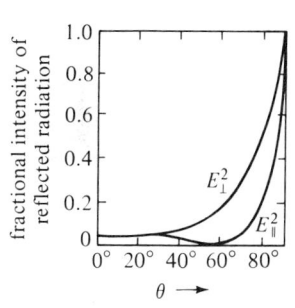

Fig. 14.20. Variation of the reflected intensity with θ for the components parallel and perpendicular to the plane of incidence.

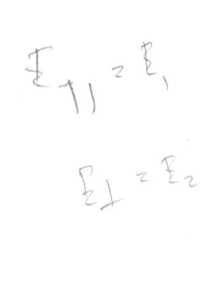

Fig. 14.21. Separation of unpolarized light into two plane polarized beams by reflection from a stack of glass plates.

14.11 DOUBLE REFRACTION

Crystals are generally electrically anisotropic. This means that the electric polarization in the crystal produced by an external electric field depends on the relative directions of the electric field and the crystal lattice. The speed of propagation of an electromagnetic wave in a crystal is a function of the direction of propagation and of the state of polarization[16] of the wave. All crystals, with the exception of those with cubic symmetry, are said to be **birefringent** or **doubly refracting** since there are two possible values for the phase velocity for a given direction of propagation of an electromagnetic wave in the crystal. Cubic crystals behave as do isotropic dielectric media.

In general, the dielectric constant of a crystal has three **principal** values, K_1, K_2, and K_3, associated with three **principal axes**[17] of the crystal. For each principal dielectric constant, there is a corresponding **principal index of refraction** n_1, n_2, or n_3. If an electromagnetic wave moves in the direction of one of the principal axes, say the x axis, then two possible phase velocities exist: $cK_2^{-1/2}$ if the electric vector is in the y direction; $cK_3^{-1/2}$ if the electric vector is in the z direction. Thus, the two waves are polarized at right angles to each other. The same behavior can be shown to be true for any direction of the propagation in a crystal—there are two phase velocities corresponding to two mutually perpendicular polarizations.

Crystals may posses one or two **optic axes**, directions in the crystal for which the two values of the phase velocity are equal. If a crystal possesses two optic axes, it is said to be **biaxial**; if it possesses only one optic axis, it is **uniaxial**. A biaxial crystal has its three principal indices of refraction all different while a uniaxial crystal has two of its principal indices of refraction equal. The two equal indices in a uniaxial crystal are called the **ordinary** index (n_0) while the third index is called the **extraordinary** index (n_e). The extraordinary index of refraction is specified for a direction perpendicular to the optic axis. A crystal is said to be **positive** if $n_e > n_0$ and **negative** if $n_e < n_0$. The indices of refraction of some common crystals are listed in Table 14.1.

Refraction at a crystal boundary is treated exactly as in Section 14.5. Now however, there are two refracted waves in the crystal as indicated in Fig. 14.22. If we denote the propagation vector of the wave in air by $\mathbf{k}_0$ and those in the crystal by $\mathbf{k}_1$ and $\mathbf{k}_2$, we find that

$$k_0 \sin \theta = k_1 \sin \theta_1 = k_2 \sin \theta_2$$

where θ is the angle of incidence and θ_1 and θ_2 are the angles of refraction.

---

[16] The word "polarization" is used in two different senses in this section. When used with the adjective "electric" it refers to charge separation in the crystal under the action of an external electric field. Otherwise, it refers to the orientation of the electric field vector in the electromagnetic wave, as discussed in Section 14.9.

[17] C. Kittel, W. D. Knight, and M. A. Ruderman, *Mechanics* (New York: McGraw-Hill Book Company, 1965).

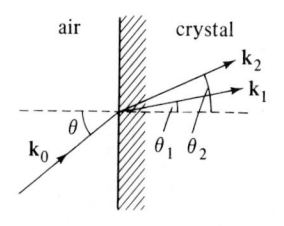

Fig. 14.22. Double refraction at a crystal boundary.

Table 14.1 Indices of refraction of some common crystals

isotropic	Fluorite Sodium chloride Diamond	1.392 1.544 2.417		
		n_o	n_e	
uniaxial (positive)	Ice Quartz Zircon	1.309 1.544 1.923	1.310 1.553 1.968	
		n_o	n_e	
uniaxial (negative)	Sodium nitrate Calcite Tourmaline	1.587 1.658 1.669	1.336 1.486 1.638	
		n_1	n_2	n_3
biaxial	Gypsum Mica Topaz	1.520 1.552 1.619	1.523 1.582 1.620	1.530 1.588 1.627

Now however, k_1 and k_2 are not constant, in general, since their magnitudes vary with the directions of $\mathbf{k}_1$ and $\mathbf{k}_2$. Thus Snell's law is not valid in general for crystals. An exception occurs for the ordinary wave in a uniaxial crystal. For this wave $k_1 = k_2 =$ constant for all directions in the crystal and

$$\frac{\sin \theta}{\sin \theta_1} = n_o,$$

the ordinary index of refraction.

In Fig. 14.23 we show the result of the interaction of a narrow beam of light with a slab cut from a uniaxial crystal. The beam is incident perpendicular to the face of the crystal. Two beams emerge from the crystal, one having traveled straight through (the ordinary beam) and the other having been displaced parallel to the direction of the incident beam (the extraordinary beam). The amount of displacement experienced by the extraordinary beam is determined by the nature of the crystal and the orientation of the optic axis relative to the face of the crystal. If the crystal is rotated about an axis parallel to the incident beam, the spot produced by the ordinary beam on the screen will remain fixed, while that produced by the extraordinary beam revolves in a circle about the fixed spot.

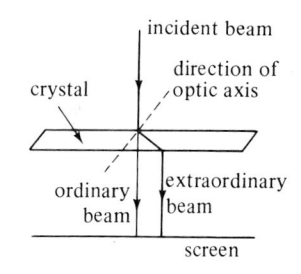

Fig. 14.23. A narrow beam of light can be split into two beams by a doubly refracting crystal.

QUESTIONS AND PROBLEMS

1. Show that the quantity $\mathbf{k} \cdot \mathbf{r}$ in the equation for a plane sinusoidal wave is a constant for all points on a plane of constant phase.

2. Show that

$$\mathbf{E} = \mathbf{E}_0 \exp[i(\omega t - \mathbf{k} \cdot \mathbf{r})]$$

is a solution of the free space wave equation for $\mathbf{E}$ provided that $k = \omega(\epsilon_0 \mu_0)^{1/2}$.

3. Evaluate the characteristic impedance of free space in ohms.

4. Starting from Maxwell's equations for a homogeneous isotropic linear medium deduce the wave equations for $\mathbf{E}$ and $\mathbf{H}$.

*5. In 1650 P. Fermat formulated the remarkable principle that light will travel between two points along the path that minimizes the travel time. This has come to be called the **principle of least time** or **Fermat's principle**. Show that Snell's law of refraction is a direct consequence of this principle. Hint: For arbitrary points A and C as shown in Fig. 14.24, find the point B (and thus θ_1 and θ_2) which minimizes the total travel time.

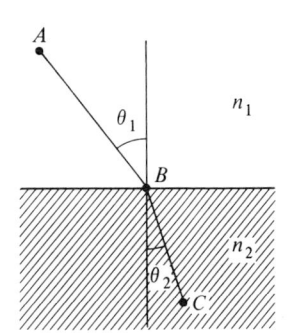

Fig. 14.24. The refraction of an electromagnetic wave at the boundary of two dielectrics.

6. An underwater object appears to be closer than it actually is when viewed from above the surface. Calculate the apparent depth d_a of a fish when it is actually 2.0 m below the surface. Use Fig. 14.25 and assume that θ_1 and θ_2 are both small angles ($n_{\text{air}} = 1.00$, $n_{\text{water}} = 1.33$).

7. A **light pipe** used for bending light around corners consists of a solid plastic tube of index of refraction 1.60. Calculate the critical angle for total internal reflection of light in such a pipe.

8. In a Fizeau experiment water flows through the apparatus at a speed of 25 m·sec^{-1}. Calculate the difference between the speed of light traveling through the moving water and the speed of light in stationary water. Compare the result with that predicted by the Galilean transformation.

9. (a) Consider a light ray incident at an angle θ on one face of a prism as in Fig. 14.26. Derive an expression for the index of refraction n of the prism in terms of the **angle of minimum deviation** δ_m and the angle φ of the prism. Hint: the deviation of a ray of light as it passes through a prism is a minimum when the ray passes symmetrically through the prism.

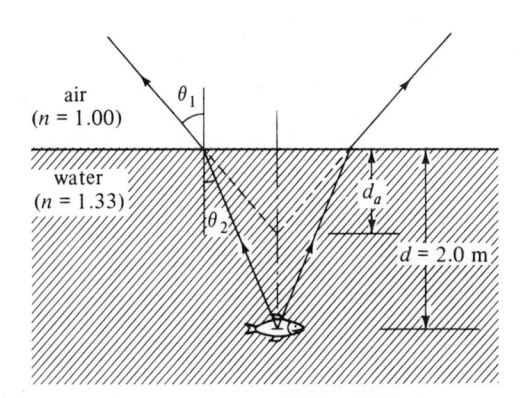

Fig. 14.25. A fish swimming in water appears to be closer to the surface than it actually is.

(b) For a prism for which φ is small obtain an expression for the dispersion $(\delta_b - \delta_r)$ of the spectrum assuming that each ray passes through the prism approximately at the angle of minimum deviation. The subscripts b and r denote blue and red, respectively.

(c) Use Fig. 14.1 to estimate the dispersion of a glass prism with $\varphi = 10°$.

*10. Show that the wave equation for the electric field in an inhomogeneous dielectric (K is a function of $\mathbf{r}$) is

$$\nabla^2 \mathbf{E} - \frac{K}{c^2}\frac{\partial^2 \mathbf{E}}{\partial t^2} = -\nabla\left(\frac{\nabla K}{K} \cdot \mathbf{E}\right).$$

*11. Radio waves of certain frequencies are reflected back to us by the ionized gas in the earth's upper atmosphere. Suppose that a radio wave is emitted from the surface of the earth ($n = 1.00$) at an angle θ_i (see Fig. 14.27). As it proceeds upward it encounters increasingly more ions and thus a decreasing index of refraction.
   (a) Show that the radius of curvature of the path, $R = (d\theta/dl)^{-1}$, is given by
   $$1/R = -(1/n)(dn/dl) \tan \theta.$$
   Use the fact that $n \sin \theta$ is constant along the path.
   (b) Calculate the radius of curvature if the index of refraction depends on height z as
   $$n = \frac{C}{z + C}$$
   where C is a constant.

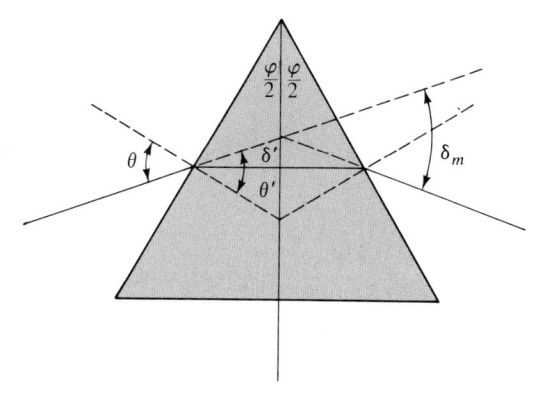

Fig. 14.26. The deviation by a prism is a minimum when a ray passes through the prism symmetrically.

12. **Whistlers** are a form of evanescent wave encountered in the earth's magnetic field. Write a short essay on whistlers.

13. A point source of light is situated 15.0 cm below the surface of a pond. The light illuminates a circular patch on the surface. Deduce the diameter of the patch.

14. Show that, when total reflection occurs, the incident wave actually penetrates into the second medium but its amplitude falls off exponentially with distance.

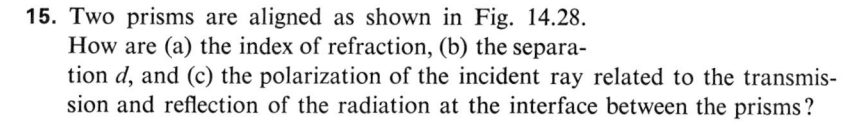

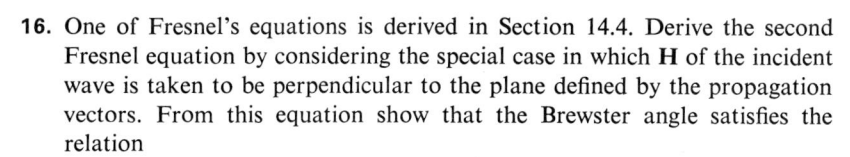

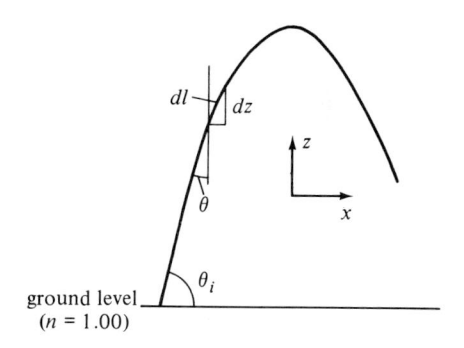

Fig. 14.27. Radio waves transmitted from the earth may be reflected by ions in the earth's upper atmosphere.

15. Two prisms are aligned as shown in Fig. 14.28. How are (a) the index of refraction, (b) the separation d, and (c) the polarization of the incident ray related to the transmission and reflection of the radiation at the interface between the prisms?

16. One of Fresnel's equations is derived in Section 14.4. Derive the second Fresnel equation by considering the special case in which **H** of the incident wave is taken to be perpendicular to the plane defined by the propagation vectors. From this equation show that the Brewster angle satisfies the relation
$$\tan \theta_p = \frac{n_2}{n_1}$$
where n_1 and n_2 are the refractive indices of the first and second medium, respectively.

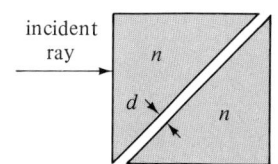

Fig. 14.28. Two closely spaced prisms.

290   ELECTROMAGNETIC WAVES IN MATTER: CLASSICAL EFFECTS   CHAP. 14

17. The **reflectance** R is defined as

$$R \equiv \frac{\text{reflected power}}{\text{incident power}} = \frac{E_r^2}{E_i^2}$$

where E_r and E_i are the reflected and incident amplitudes, respectively. Determine the reflectance of water ($n = 1.33$) for plane-polarized light whose electric vector is in the plane of incidence for several angles of incidence. Repeat the calculations for light whose electric vector is perpendicular to the plane of incidence. Plot your results.

18. Starting from Fresnel's equations show that for normal incidence on a boundary between two dielectrics

$$R = \left(\frac{n_1 - n_2}{n_1 + n_2}\right)^2$$

and

$$T = \frac{4 n_1 n_2}{(n_1 + n_2)^2}$$

where R is the reflectance (see previous problem) and $T = 1 - R$ is the **transmittance**.

19. Calculate the skin depth of sea water ($\sigma = 5.0$ ohm$^{-1}\cdot$m^{-1}, $K_m = 1.00$) for waves of frequency ν. Electromagnetic disturbances occur in the vicinity of the earth on many time scales. The oceans tend to damp these disturbances. Calculate the frequencies of waves that can penetrate 1.0 mm, 1.0 cm, and 1.0 m into the sea without appreciable power loss (say, less than 10%).

20. Show that

$$\mathbf{E} = \mathbf{E}_0 \exp\left[i\left(\omega t - \frac{x}{\delta}\right)\right] \exp\left(-\frac{x}{\delta}\right)$$

is an approximate solution of the wave equation

$$\nabla^2 \mathbf{E} = \epsilon\mu \frac{\partial^2 \mathbf{E}}{\partial t^2} + \mu\sigma \frac{\partial \mathbf{E}}{\partial t}$$

for a good conductor.

21. Calculate the effective resistance per unit length of a wire of radius a and conductivity σ carrying electromagnetic waves of angular frequency ω.

22. For gases at low pressures, show that to a good approximation the real and imaginary parts of the complex dielectric constant are given by

$$K_1 = 1 + \frac{n_0 e^2}{2 m_e \omega \epsilon_0}\left[\frac{\omega_0 - \omega}{(\omega_0 - \omega)^2 + \frac{\gamma^2}{4}}\right]$$

$$K_2 = \frac{n_0 e^2}{2 m_e \omega \epsilon_0}\left[\frac{\frac{\gamma}{2}}{(\omega_0 - \omega)^2 + \frac{\gamma^2}{4}}\right].$$

*23. Show that in the case of anomalous dispersion for gases the maximum and minimum values of n_1 occur at the positions where n_2 reaches one-half its maximum value. Assume $\gamma/\omega_0 \ll 1$.

24. In Section 14.6 it was stated that the measured values of the index of refraction of a gas in the x-ray region are less than unity. Does this mean that the speed of x-rays in a gas is greater than the speed of x-rays in a vacuum?

25. The Cerenkov effect can be used to measure the speed of high-energy particles. Suppose a beam of electrons is passing through water ($n = 1.33$). Calculate the speed and energy of electrons in the beam if the radiation is being emitted predominantly at an angle of 30° to the direction of travel.

26. Pulse radiolysis is a technique that is applicable to the study of physicochemical reactions in the pico-to nanosecond range. **Flash photolysis** is a technique for studying reactions in the microsecond range. Describe this latter technique.

27. Two states of polarization represented by the complex vector amplitudes $\mathcal{E}_1$ and $\mathcal{E}_2$ are said to be **orthogonal** if $\mathcal{E}_1 \mathcal{E}_2^* = 0$ (the asterisk denotes complex conjugate). Show that the matrices

$$\begin{bmatrix} 2 \\ -i \end{bmatrix} \text{ and } \begin{bmatrix} 1 \\ 2i \end{bmatrix}$$

represent a pair of orthogonal states of elliptic polarization. Show these states in a diagram.

28. A **Nicol prism** is a device for utilizing double refraction to produce polarized light. Explain the principle of the device.

29. What is a **quarter-wave plate**? Deduce a matrix to represent such a device.

30. What is a **half-wave plate**? Deduce a matrix to represent such a device.

31. Two descriptions of unpolarized radiation are given in Section 14.10. Show that they are equivalent.

32. What is **polaroid**? Describe what you would see if you looked at a source of unpolarized light through a polaroid disc as you rotated the disc about its axis. Describe what you would see if you looked at the source through two polaroid discs as you rotated one of them about its axis with respect to the other.

33. Why do polaroid sunglasses reduce glare for highway driving?

34. Show that a linearly polarized electromagnetic wave is a superposition of equal amounts of right and left circularly polarized radiation.

35. Suppose that a beam of linearly polarized light passes through a sugar solution, is reflected by a mirror, and travels back through the solution. After the two traversals is the plane of polarization the same as that of the incident light or rotated through twice the angle resulting from a single traversal?

36. Two prisms made of right handed and left handed quartz are combined as shown in Fig. 14.29. A beam of unpolarized light is incident normally onto the system. Describe the emergent radiation.

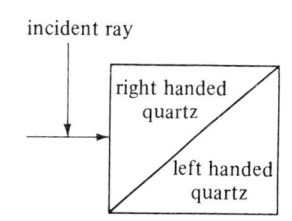

Fig. 14.29. Light incident normally on a combined pair of quartz prisms, one made of right-handed quartz and the other of left-handed quartz.

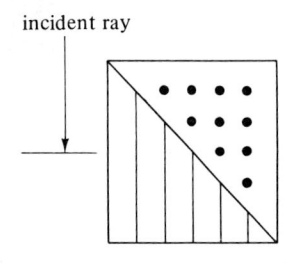

Fig. 14.30. Light incident on a combined pair of calcite prisms cut to have their optic axes perpendicular to each other.

37. A narrow beam of light is incident on a slab cut from a uniaxial crystal. The angle between the beam and the face of the crystal is 90°. Describe what will occur if (a) the optic axis of the crystal is parallel to the direction of the incident radiation, and (b) the optic axis of the crystal is parallel to the face of the crystal onto which the radiation falls.

38. Two prisms made of calcite and cut to have their optic axes perpendicular to one another are combined as shown in Fig. 14.30. The directions of the optic axes are indicated. A beam of unpolarized light is incident normally onto the system. Describe the emergent radiation.

ALBERT EINSTEIN

Excerpted from ATOMIC PHYSICS TODAY, by Otto R. Frisch, Basic Books, Inc., Publishers, New York, 1961.

15

Electromagnetic waves in matter: quantum effects

15.1 INTRODUCTION

In the early years of this century it became apparent that some of the interactions of light with matter could be explained only by assuming that the energy carried by a beam of light of frequency v is concentrated into discrete packages, called **photons**,[1] of energy hv. We now accept the view that the energy carried by any electromagnetic radiation is localized in photons whose paths through space are governed by a wave equation. In many of the interactions of electromagnetic radiation with matter we can ignore the quantum character of the radiation and treat the radiation as a classical electromagnetic wave. On the other hand, there are many interactions in which the quantum nature of the radiation must be taken into account. This **wave-particle duality** causes no difficulty so long as we realize that the photon theory merely adds quantum properties to the already familiar wave properties of electromagnetic radiation. According to the correspondence principle we can always apply the classical theory in situations involving large numbers of photons. No confusion arises so long as we do not attempt to apply wave and quantum concepts simultaneously.

15.2 THE INTERACTION OF PHOTONS WITH MATTER

When electromagnetic radiation penetrates into a medium, it may pass through unhindered, it may be partially or totally absorbed, or it may be scattered; after scattering, absorption may or may not occur. The relative magnitudes of these effects depend on the nature of the medium involved as well as on the frequency of the electromagnetic radiation. We have already discussed absorption in Sections 14.5 and 14.6. We were able to describe absorption, in general, in terms of a complex index of refraction of atoms in the medium. Absorption in a conductor was seen to be very rapid; it is associated with the presence of conduction electrons and will be discussed more fully in Section 15.3.

If we wish to enquire more closely into the nature of the absorption of electromagnetic radiation by an atom, we must take account of the quantum nature of both the atom and the radiation. The three principal processes by which high-energy photons give up their energy to matter are the **photoelectric effect**,[2] the **Compton effect**, and **pair production**. The threshold for the photoelectric effect is in the energy region of 2 to 5 eV which corresponds to wavelengths of 620 to 248 nm (in the visible and ultraviolet regions of the spectrum). The threshold for pair production is 1.022 MeV so that this process occurs only for high-energy x-rays and γ-rays. The Compton effect is operable at all frequencies and results in a change in wavelength $\Delta\lambda$ of

[1] *MWTP*, Chapter 18.
[2] *MWTP*, Section 18.3.

the scattered photon given by

$$\Delta\lambda = 2.43 \times 10^{-3}(1 - \cos\theta) \text{ nm}$$

where θ is the angle of scattering. This wavelength shift is independent of frequency. For $\theta = 180°$, $\Delta\lambda = 4.86 \times 10^{-3}$ nm, the maximum value. Now $\lambda \geq 400$ nm in the visible region, so that

$$\frac{\Delta\lambda}{\lambda} \leq 1.2 \times 10^{-5}.$$

This shift in wavelength is less than the usual natural width of an atomic absorption line (see Problem 3). Therefore, the Compton effect, although operable, is not observable at visible wavelengths.

For wavelengths in the visible region, absorption and scattering are described in terms of the interaction of the electromagnetic wave with free electrons or with the bound electrons of an atom (or atomic lattice). The motion of electrons is quantized[3] so that only certain energies or frequencies are permitted. We shall turn our attention first to a discussion of absorption and scattering at optical frequencies.

15.3 OPTICAL ABSORPTION

Photons are removed from a beam of light by collisions that produce either excited states or free electrons and holes. If a beam of initial intensity I_0 is incident upon some absorbing medium, the intensity I of the beam after traversing a thickness x of the medium is observed to satisfy the relation

$$I = I_0 \exp(-\mu x)$$

where μ, the (optical) **absorption coefficient**, is frequency dependent. This expression may be derived as follows (see Fig. 15.1). Light of initial intensity I_0 and frequency v is absorbed in passing through a medium. The intensity I of light incident on a slab of thickness dx and perpendicular to the incident light beam is reduced by an amount dI on passing through the slab. The number of photons n per unit volume in the beam is proportional to the intensity of the beam. Assuming that for a homogeneous medium the probability of absorption of a specific photon in the slab is independent of the number of photons present, it follows that the fractional change dI/I in the intensity is independent of the position x in the medium and proportional only to the thickness dx causing the absorption. Therefore,

$$\frac{dI}{I} = -\mu\, dx$$

where, for a given frequency, μ is a constant characteristic of the medium.

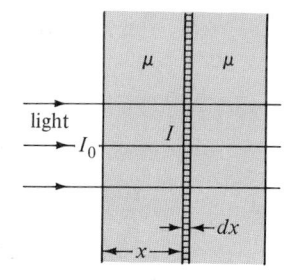

Fig. 15.1. Diagram for the calculation of the absorption of light by a medium.

[3] *MWTP*, Sections 18.4 and 18.6.

The minus sign is used since the intensity is reduced by the absorption. The intensity I of the beam after it has penetrated a thickness x of the medium is given by

$$\int_{I_0}^{I} \frac{dI}{I} = -\int_0^x \mu\, dx$$

or

$$\ln\left(\frac{I}{I_0}\right) = -\mu x$$

or

$$I = I_0 \exp(-\mu x).$$

The absorption coefficient μ is dependent upon the frequency (or energy) of the photons in the beam. There are usually several processes that contribute to absorption at a given frequency so that the absorption coefficient is really a sum of several coefficients, each referring to a specific absorption process.

Absorption and scattering by atoms and molecules

Both the electric and magnetic fields in an electromagnetic wave interact with the electrons in an atom. The electric force on the electron has magnitude euB where u is the electron speed. We saw in Section 14.4 that in a medium

$$\frac{E}{H} = \left(\frac{\mu}{\epsilon}\right)^{1/2}.$$

Since $B = \mu H$, we have

$$\frac{E}{B} = \frac{1}{(\mu\epsilon)^{1/2}} = v$$

where v is the speed of the electromagnetic wave in the medium. The ratio of the magnetic to the electric force on the electron is then

$$\frac{F_m}{F_e} = \frac{euB}{eE} = \frac{u}{v}$$

which is much less than unity for electrons in an atom (or molecule). Therefore, we can neglect the magnetic interaction.

We saw in Section 14.6 that absorption is a maximum at certain natural frequencies of the atoms (or molecules). These natural frequencies correspond to the energy differences between certain of the allowed states. If a photon has an energy equal to the difference in energy between two appropriate quantum states, the photon may be absorbed, leaving the atom (or molecule) in an **excited state**. Not all possible excitations are allowed;[4] only those satisfying the **selection rules** of quantum mechanics

[4] *MWTP*, Sections 18.6 and 18.8.

are permitted. The electrons act as forced electric dipoles under the influence of the electric field in the wave. An electron so excited may immediately return to the ground state by reemitting the energy just absorbed. The radiation is electric dipole in nature, as discussed in Section 9.8. Alternatively, the energy may be reemitted at different frequencies if the excited electron passes through intermediate energy states during its return to the ground state. In either case, the direction of emission is random so that the intensity of the transmitted wave is reduced by this **scattering** process (see Fig. 15.2).

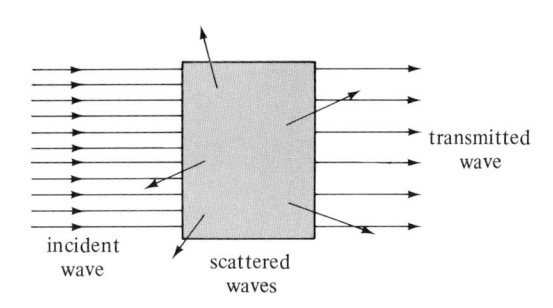

Fig. 15.2. Scattering of an electromagnetic wave incident upon a medium reduces the intensity of the transmitted wave.

Scattering is most pronounced at those frequencies corresponding to the emission frequencies of the atom or molecule as just described. However, scattering may still be significant at frequencies other than those of the emission spectrum. This occurs because any system will respond to an external driving force even if the frequency of the external force does not correspond to a natural frequency of the system. The response to the forced oscillations, however, is very small far from a natural resonance, but it increases as the frequency of the external force approaches a resonance frequency. The blue color of the sky is due to such an effect. Molecules that make up the atmosphere have emission frequencies in the ultraviolet region. Therefore, their response to an electromagnetic wave decreases as the frequency of the wave decreases. The response to waves with frequencies in the blue region is greater than the response to waves with lower frequencies and the scattered radiation is predominantly blue. It can be shown that the scattered intensity is proportional to the fourth power of the driving frequency. This is referred to as **Lord Rayleigh's blue sky law.**

Absorption by solids

A photon incident upon an insulator can excite an electron from the valence band to the conduction band if its energy is sufficiently large (see Fig. 15.3). The electron and hole so produced are free to move independently through the crystal; however, they experience a Coulomb attraction for one another. Therefore, it is possible for stable, bound electron-hole states to be formed. These states are analogous to the stationary states of the Bohr model of the hydrogen atom (see Problem 5). The photon energy necessary to create an electron-hole pair or **exciton** starting from a filled valence band is less than the

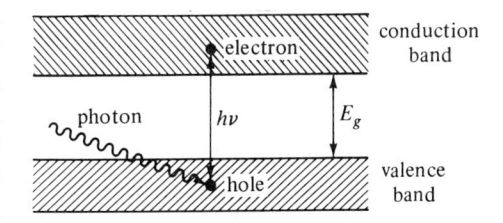

Fig. 15.3. When an electron is excited from the conduction band to the valence band, a hole is created in the valence band.

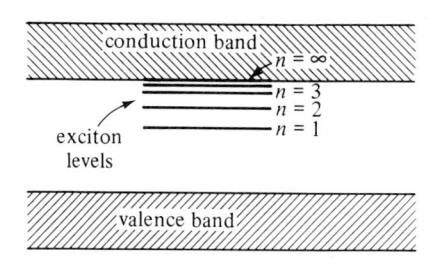

Fig. 15.4. Illustration of exciton levels.

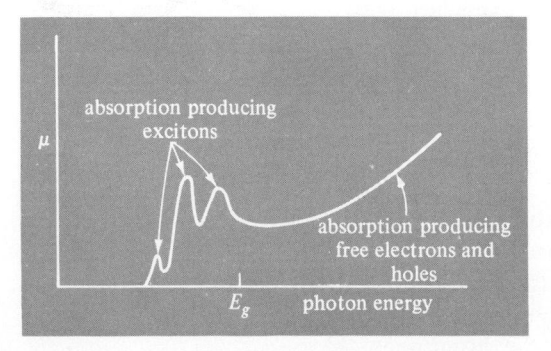

Fig. 15.5. Variation of the absorption coefficient at optical and ultraviolet frequencies for an insulator.

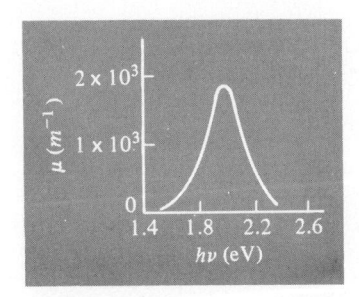

Fig. 15.6. The F-center absorption band of KBr at 20°C.

gap energy E_g (see Fig. 15.4). An exciton is an electrically neutral excited state of a crystal and therefore cannot contribute to the electrical conductivity. An exciton can travel through a crystal giving up its energy of formation when the electron drops back into the hole state from which it came.

No absorption of light by an insulator can occur below the lowest exciton energy (usually from 0.75 to 0.90 of the gap energy). As the photon energy increases one or more excitons may be produced. Finally, the photon energy becomes large enough to produce a free electron and a hole. The variation of the absorption coefficient μ of a pure insulator with photon energy is shown schematically in Fig. 15.5. The threshold for exciton production may be in the visible or ultraviolet region of the spectrum, the latter being most common.

Absorption at much lower energies is possible if the solid has chemical impurities or imperfections that produce energy levels in the gap between the valence and conduction bands. For example, an excess of potassium atoms in a KCl crystal produces K^+ ions and an equivalent number of Cl^- holes plus electrons. The region of a Cl^- **vacancy** has an excess positive charge and so traps the electron. An electron bound to a negative-ion center is called an **F-center**; the name comes from the German word Farbe which means color. Another way of describing an F-center is to say that an electron replaces a Cl^- ion in the KCl crystal lattice. A photon can excite the electron of the F-center to a higher-energy state. The optical absorption due to F-centers is much weaker than that producing electron-hole pairs, since there are many more electrons in the valence band than there are F-centers in a crystal. The F-center absorption band of KBr at 20°C is shown in Fig. 15.6. A schematic energy-level diagram for F-center absorption in KCl is shown in Fig. 15.7.

The F-center just described is an example of a **color center** in a solid. The solid appears colored because F-center absorption removes light from an incident beam at selected frequencies and transmits light of other frequencies. Other types of color centers can occur in solids due to chemical impurities or to vacancies. Since exciton absorption is usually in the ultraviolet region, a solid without imperfections is transparent. The absorption coefficient of glass at optical frequencies is less than $10^{-2}\ m^{-1}$, for example. If photon energies in the visible region of the electromagnetic spectrum are not large enough to excite electrons to higher-energy states, no absorption can occur and the solid is transparent.

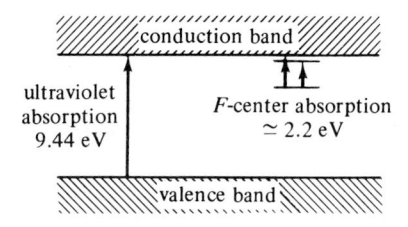

Fig. 15.7. Energy-level diagram for F-center absorption in KCl.

Luminescence

The term **luminescence** is used to denote the absorption of energy by matter and its reemission as visible or near visible radiation. If the energy is reemitted on the average within 10^{-8} sec of the excitation,[5] the process is called **fluorescence**; if the energy is reemitted more than 10^{-8} sec after the excitation, the process is called **phosphorescence**. Crystalline luminescent solids are called **phosphors**; delay periods may be anywhere from microseconds to hours.

Photoconductivity

The increase in the electrical conductivity of an insulating crystal due to light incident upon it is known as **photoconductivity**. The simplest possible model of a photoconductor assumes that electron-hole pairs are produced throughout the crystal by an external source of light, that recombination occurs by the direct annihilation of electrons with holes, and that the hole current may be neglected in comparison with the electron current. The rate of change dn/dt of the electron concentration is given by

$$\frac{dn}{dt} = n_p - An^2$$

where n_p is the number of photons absorbed by the sample per unit volume per second and An^2 is the recombination rate. A steady state is reached when

$$\frac{dn}{dt} = 0 \quad \text{or} \quad n_p = An^2.$$

The steady-state electron concentration is

$$n_0 = \left(\frac{n_p}{A}\right)^{1/2}$$

---

[5] The lifetime of an atomic state for a permitted electric dipole transition at optical frequencies is of the order of 10^{-8} sec.

and the associated conductivity σ is

$$\sigma = \frac{n_0 e v_d}{E} = \left(\frac{n_p}{A}\right)^{1/2} \frac{e v_d}{E}$$

where v_d is the drift speed of the electrons and E the magnitude of the applied electric field. This relation predicts that for a given potential across the crystal the photocurrent will vary as the one-half power of the intensity of the incident light.

If the light source is switched off, then

$$\frac{dn}{dt} = -An^2$$

which upon solution yields

$$n = \frac{n_0}{1 + An_0 t}.$$

Therefore, the time t_0 required for the electron concentration to drop to one-half of its steady-state value (that is, the **response time**) is

$$t_0 = \frac{1}{(n_p A)^{1/2}} = \frac{n_0}{n_p}.$$

Note that, for a fixed level of illumination, the response time of a photoconductor is predicted on this simple model to be proportional to the photoconductivity; that is, sensitive photoconductors have long response times.

15.4 ABSORPTION OF ULTRAVIOLET RADIATION, X-RAYS, AND γ-RAYS

The photoelectric effect

When light (visible or ultraviolet) is incident upon a metal, electrons may be emitted from the surface. This is called the **photoelectric effect**.[6] We have seen in Chapters 11 and 12 that there are free electrons in a metal and that these electrons move essentially without hindrance throughout the volume of the metal. Since electrons do not leave the surface of a metal in large numbers, there must be a barrier of some sort that prevents electrons from escaping easily from the metal. Additional energy must be supplied to the electrons in some way before an appreciable number will penetrate the surface barrier. Heating a metal to high temperature in a vacuum results in **thermionic** emission of electrons; the additional electron energy is provided by thermal excitation ($\frac{1}{2}kT$ per degree of freedom[7]).

[6] *MWTP*, Section 18.3.
[7] *MWTP*, Section 17.5.

SEC. 15.4 ABSORPTION OF ULTRAVIOLET RADIATION, X-RAYS, AND γ-RAYS

In the photoelectric effect the additional energy is provided by a photon from the incident light.

We characterize a solid by a **work function** ϕ where ϕ is the minimum energy required to free an electron from the solid. The free electrons in metals have a distribution of energies from zero to a maximum of up to several eV. The work function refers to the additional energy required to remove the most energetic electron from the conduction band (see Fig. 15.8). A photon of energy $h\nu$ will produce photoelectrons of maximum kinetic energy K_{max} where

$$h\nu = K_{max} + \phi;$$

this is the photoelectric equation first proposed by Albert Einstein. In the photoelectric process an individual photon disappears, all its energy being transferred to the electron. The threshold energy E_0 for photoelectric emission is given by

$$E_0 = h\nu_0 = \phi$$

or

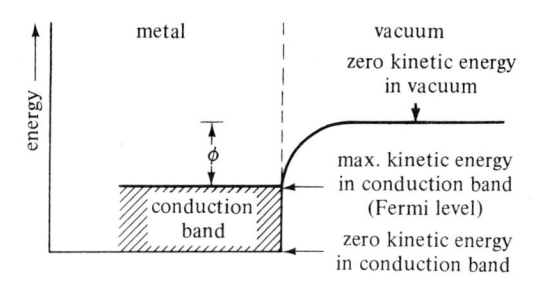

$$\nu_0 = \frac{\phi}{h}.$$

Fig. 15.8. Energy relations for electrons near a metal surface.

The work functions and threshold frequencies for several metals are given in Table 15.1.

If photons incident upon a medium have sufficient energy, electrons may be freed from atoms through an **atomic photoelectric effect**. In this process the photon disappears and an atomic electron receives sufficient energy to escape from the atom, the photon supplying energy at least as great as the binding energy of that electron in the atom. The process is shown schematically in Fig. 15.9. The atomic photoelectric effect is the predominant mode of absorption of x-rays and low energy γ-rays in a medium.

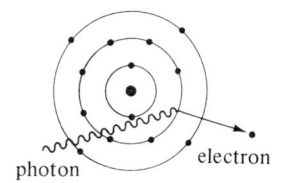

Fig. 15.9. The atomic photoelectric effect.

Table 15.1 Photoelectric properties of some metals

Metal	Work function ϕ eV	Threshold frequency ν_0 sec^{-1}
Sodium	2.28	5.50×10^{14} (green)
Calcium	2.71	6.53×10^{14} (blue)
Uranium	3.63	8.75×10^{14} (ultraviolet)
Tantalum	4.12	9.93×10^{14} (ultraviolet)
Tungsten	4.50	1.08×10^{15} (ultraviolet)
Nickel	5.01	1.21×10^{15} (ultraviolet)

The complete absorption of an x-ray photon in the photoelectric effect does not violate energy-momentum conservation since the atomic electron involved is bound to the massive nucleus by electrostatic forces. Therefore, the nucleus is able to participate in the absorption process by recoiling in such a way as to conserve momentum. But the nuclear mass is very large so that the nucleus is able to take up the required momentum without absorbing significant kinetic energy. The probability of photoelectric absorption of an x-ray photon by a particular electron is expected to decrease with increasing photon energy as the nucleus must absorb more and more momentum. However, once the photon energy is smaller than the binding energy of the electron, the probability of absorption by that electron drops rapidly to zero. The probability of photoelectric absorption varies with the energy of the x-ray photons as indicated in Fig. 15.10 for lead. **Absorption edges** appear at energies equal to the binding energies of the electrons in each of the subshells of the atom. We label K and L absorption edges; they occur at the minimum energies required to eject an electron from the atomic K and L shells,[8] respectively.

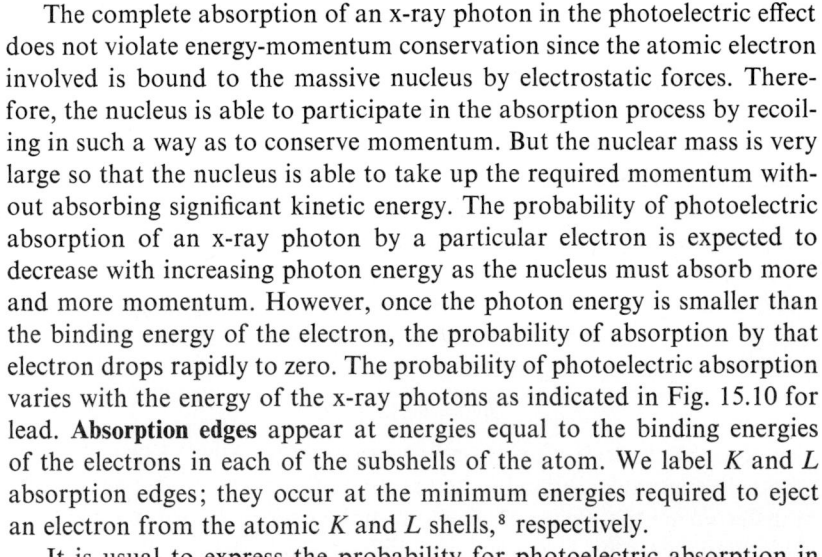

It is usual to express the probability for photoelectric absorption in terms of a **photoelectric cross section per atom**, σ_{PE}^a, defined by

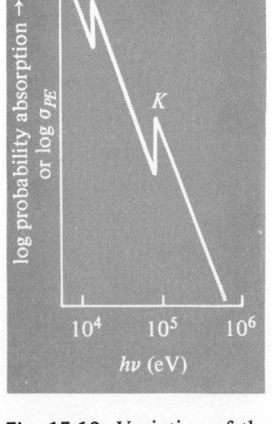

$$\sigma_{PE}^a = \frac{\bar{N}_P}{\bar{I}}$$

Fig. 15.10. Variation of the probability of photoelectric x-ray absorption with photon energy.

where $\bar{N}_P$ is the average number of quanta absorbed per sec per atom from a beam containing $\bar{I}$ quanta per sec per m^2. By using sophisticated quantumelectrodynamical calculations it has been shown that for photoelectric absorption by K shell electrons

$$\sigma_{PE}^a \propto Z^5 (h\nu)^{-7/2}$$

for $h\nu >$ binding energy of K electrons, but $h\nu < m_e c^2$. The strong dependence on Z reflects the fact that the strength of the coupling to the nucleus increases rapidly with increasing Z and so the nucleus finds it much easier to take up the necessary momentum.

The Compton effect

In some interactions of photons with electrons the original photon disappears but only a portion of the available energy appears as kinetic energy of the electron. The remainder of the energy appears as a new photon of lower energy than the original. This process was discovered by Arthur H. Compton and is named after him. The **Compton effect**, or **Compton scattering** as it is often called, is pictured in Fig. 15.11. A photon of energy $h\nu$ interacts with an electron that recoils at an angle ϕ with

[8] *MWTP*, Section 18.7.

SEC. 15.4 ABSORPTION OF ULTRAVIOLET RADIATION, X-RAYS, AND γ-RAYS

respect to the orginal photon direction. A new (or scattered) photon of energy $h\nu'$ moves off at an angle θ to the original photon direction. Since the recoil electron has acquired momentum, we must ascribe some momentum to the incoming photon. This is reasonable from the point of view of relativistic mechanics, for the kinetic energy of a particle,[9] is given by

$$K = (m - m_0)c^2$$

where m_0 is the rest mass of a particle and m is its relativistic mass. A photon has rest mass $m_0 = 0$ and a relativistic mass

$$m = \frac{K}{c^2} = \frac{h\nu}{c^2}.$$

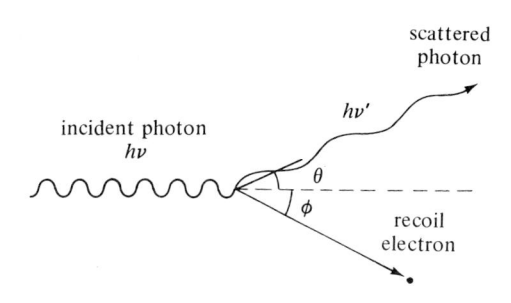

Fig. 15.11. The Compton effect.

The photon momentum is then

$$p = mc = \frac{h\nu}{c} = \frac{h}{\lambda}.$$

We can find the relation between the wavelengths of the original and scattered photons by applying conservation of energy and conservation of momentum. Conservation of momentum tells us that the momentum vectors of the two photons and the recoil electron must form a closed triangle as shown in Fig. 15.12, where $\mathbf{p}_0$ is the momentum of the incident photon, $\mathbf{p}_1$ is the momentum of the scattered photon, and $\mathbf{p}_e$ is the momentum of the recoil electron. Applying the law of cosines to the momentum triangle we obtain

$$p_e^2 = p_0^2 + p_1^2 - 2p_0 p_1 \cos \theta.$$

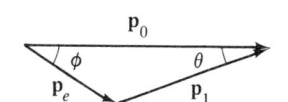

Fig. 15.12. Momentum triangle for the Compton effect.

Multiplying by c^2 we have

$$p_e^2 c^2 = p_0^2 c^2 + p_1^2 c^2 - 2p_0 c \, p_1 c \cos \theta$$
$$= K_0^2 + K_1^2 - 2K_0 K_1 \cos \theta,$$

where K_0 and K_1 are the energies of the incident and scattered photons, respectively.

The relativistic mass m of an electron and its total energy $E = mc^2$ are related to its momentum p_e by the expression[10]

$$m^2 c^4 = p_e^2 c^2 + m_e^2 c^4$$

where m_e is the rest mass of the electron. From conservation of energy we have

$$K_0 + m_e c^2 = K_1 + mc^2$$

[9] *MWTP*, Section 12.4.
[10] *MWTP*, Section 12.5.

so that
$$m^2c^4 = (K_0 - K_1 + m_e c^2)^2$$
$$= p_e^2 c^2 + m_e^2 c^4$$

from above. This last equation can be simplified to give
$$K_0 K_1 (1 - \cos\theta) = m_e c^2 (K_0 - K_1).$$

Now
$$K_0 = h\nu_0 = \frac{hc}{\lambda_0}, \quad K_1 = \frac{hc}{\lambda_1}$$

so that
$$\lambda_1 - \lambda_0 = \Delta\lambda = \frac{h}{m_e c}(1 - \cos\theta).$$

The change in wavelength $\Delta\lambda$ depends only on the angle θ. The value of the constant term is
$$\frac{h}{m_e c} = \frac{6.63 \times 10^{-34}}{9.11 \times 10^{-31} \times 3.00 \times 10^8} = 2.43 \times 10^{-12} \text{ m}$$
$$= 2.43 \times 10^{-3} \text{ nm}$$

as quoted in Section 15.2. The quantity $h/m_e c$ is known as the **Compton wavelength** of the electron. Absorption of radiation via the Compton effect is predominant for all materials in the $1-3$ MeV range and continues to be predominant in light elements up to 10 MeV or more.

For x-rays of wavelengths greater than $\simeq 2 \times 10^{-11}$ m, the photon energy becomes small compared to the rest mass of an electron and the scattering can be treated by classical electromagnetic theory.

The **scattering cross section per atom** σ_s^a for lead varies with the energy of the x-ray photons as shown in Fig. 15.13.

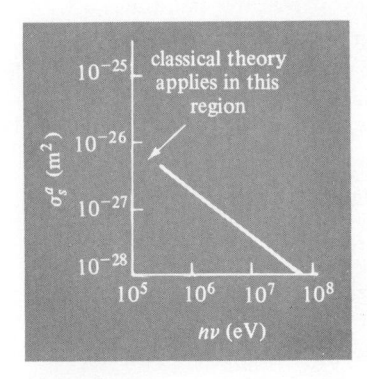

Fig. 15.13. The variation of σ_s^a with $h\nu$ for lead.

Pair production

Above 1.022 MeV photon energy **pair production** can occur. In this process a photon in the field of a charged particle disappears and a positron-electron pair is created. Interaction with the electric field of the nucleus is preferred for pair production, although interaction with orbital electrons gives a substantial contribution. The reaction is described by the equation
$$h\nu \rightarrow e^+ + e^- + 2K$$

SEC. 15.4   ABSORPTION OF ULTRAVIOLET RADIATION, X-RAYS, AND γ-RAYS

where K is the kinetic energy of each of the positron and electron. (The recoil of the nucleus or atom carries off a very small amount of the energy and momentum; this recoil is neglected in the equation just quoted.)

The rest mass of an electron is 0.511 MeV, as also is the rest mass of a positron. Therefore, a minimum energy of $2 \times 0.511 = 1.022$ MeV is required for pair production, giving the threshold quoted above. The probability for pair production increases slowly with energy above the threshold. The probability of pair production by photons of a given energy increases with the atomic number of the material since the number of positive charges on the nucleus and the number of orbital electrons are equal to the atomic number. Pair production is the predominant mode of photon absorption above about 5 MeV in heavy elements, such as lead. The **cross section per atom for pair production** σ_{PR}^a varies with the energy of the x-ray photons for lead as indicated in Fig. 15.14.

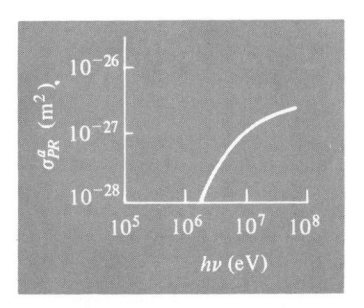

Fig. 15.14. The variation of σ_{PR}^a with $h\nu$ for lead.

The relative importance of the photoelectric effect, the Compton effect, and pair production depends on the atomic number of the element involved. The absorption coefficients for the three processes are shown in Fig. 15.15 for aluminum ($Z = 13$) and in Fig. 15.16 for lead ($Z = 82$). The absorption coefficient μ is the atomic cross section multiplied by the number of atoms·m^{-3} and has units m^{-1}. It is the reciprocal of the thickness of material required to attenuate a beam by a factor e. This thickness is called the **attenuation length**.

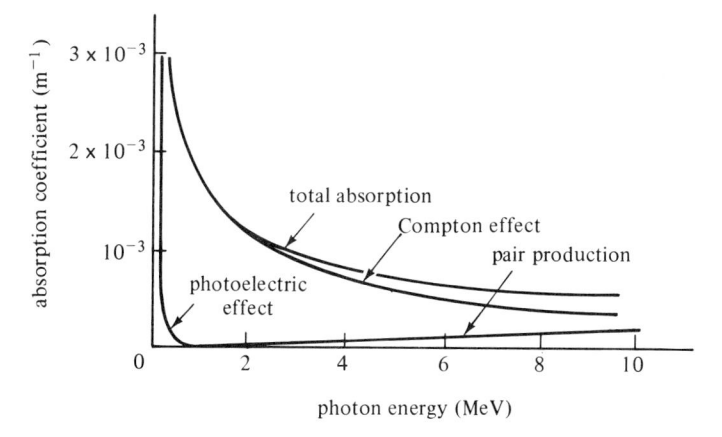

Fig. 15.15. Relative values of absorption coefficients for aluminum ($Z = 13$).

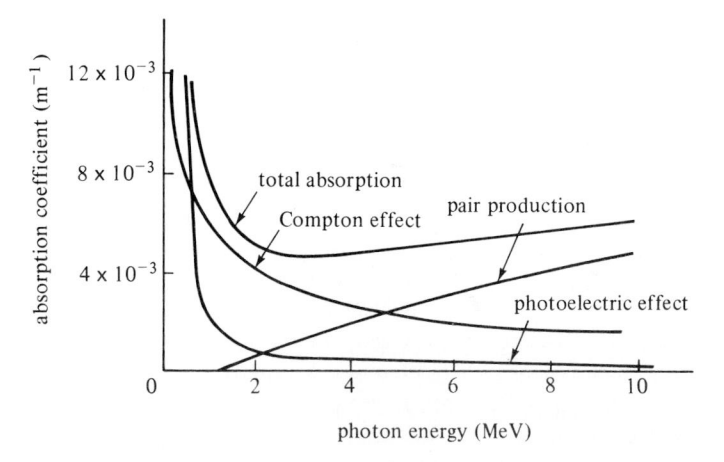

Fig. 15.16. Relative values of absorption coefficients for lead ($Z = 82$).

15.5 BIOLOGICAL INTERACTIONS

Life originated on earth in a process involving the photochemical synthesis of biologically important molecules from simple inorganic substances present in the atmosphere and the oceans[11]. Life is sustained today through the transformation of water and carbon dioxide into simple carbohydrates and oxygen in the presence of sunlight in a process called **photosynthesis**.[12,13] Life slowly and continually evolves through **mutations** or alterations of genetic material caused primarily by the effects of ionizing radiation.

The origin of life

At the time that life first appeared on the earth some 3.5×10^9 years ago the chief constituents of the atmosphere were hydrogen, ammonia, water vapor, and methane. As a result of the interaction of sunlight with these gases and with inorganic salts present in the oceans, life began. The first experimental confirmation of the process was provided in 1953

---

[11] G. Wald, "The Origin of Life," *Scientific American*, August, 1954. Available as *Scientific American Offprint 47* (San Francisco: W. H. Freeman and Co., Publishers).
[12] G. Wald, "Life and Light," Scientific American October, 1959. Available as *Scientific American Offprint 61* (San Francisco: W. H. Freeman and Co., Publishers).
[13] D. I. Arnon, "The Role of Light in Photosynthesis," *Scientific American*, November, 1960. Available as *Scientific American Offprint 75* (San Francisco: W. H. Freeman and Co., Publishers).

in experiments carried out by Harold Urey and Stanley Miller. They demonstrated that when these substances were exposed to ultraviolet radiation or to an electrical discharge, a variety of organic compounds were synthesized, including amino acids, cyanides, urea, aldehydes, acetic acid, and lactic acid. Since these early experiments, pyrimidines, purines, simple sugars, and hosts of other biologically significant compounds have been produced by photochemical and electrochemical processes.

The importance of the interaction of light with green plants

Photosynthesis is the process in which plants absorb photons of visible light from the sun's radiation and thereby convert carbon dioxide (CO_2) and water (H_2O) into oxygen (O_2) and the simple carbohydrate $C_6H_{12}O_6$. The fundamental process in photosynthesis is the absorption of radiant energy. In almost all green plants the pigments effective for photosynthesis[14] are **chlorophyll a** and **chlorophyll b**. The absorption spectra of these molecules are shown in Fig. 15.17. It is the presence of these absorption bands that is responsible for the green coloring of the plants. The excited chlorophyll molecules give up their excess energy in a biochemical reaction that converts **adenosine diphosphate** (ADP) molecules to **adenosine triphosphate** (ATP) molecules. The structure of the ATP molecule is depicted in Fig. 15.18. The arrow indicates the bond that is formed as a result of the partial deexcitation of a chlorophyll molecule. It is in phosphate bonds of this type that the energy absorbed by the chlorophyll molecules from sunlight is stored in a common form that is available for all of the metabolic processes of the cell. The deexcitation of a single chlorophyll molecule results in the formation of the order of ten such bonds. In terms of energy units used by the biologist, photons of wavelength 450 nm have a g-mole equivalent energy equal to 64 kcal·g-mole^{-1} (see Problem 22) whereas the phosphate bonds have a mole equivalent energy equal to 8 kcal·g-mole^{-1}.

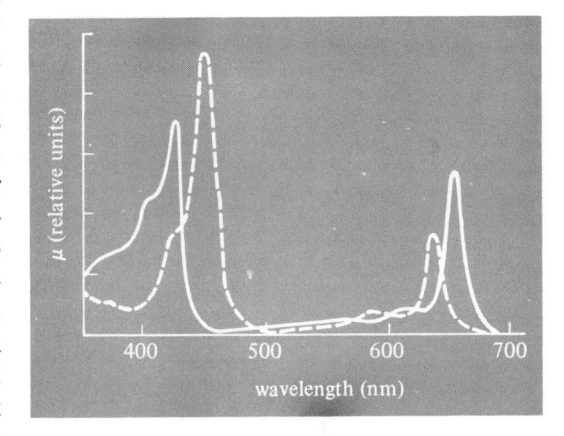

Fig. 15.17. The absorption spectra of chlorophyll a (solid line) and chlorophyll b (dashed line).

When the cell requires energy to carry out a biochemical reaction, an ATP molecule is converted to an ADP molecule through the removal of one phosphate group and with the release of the small amount of energy stored in the bond. The cycle continues throughout the life of the plant.

Storing energy in phosphate bonds has three main advantages:

[14] E. I. Rabinowitch and Govindjee, "The Role of Chlorophyll in Photosynthesis," *Scientific American*, July, 1965. Available as *Scientific American Offprint 1016* (San Francisco: W. H. Freeman and Co., Publishers).

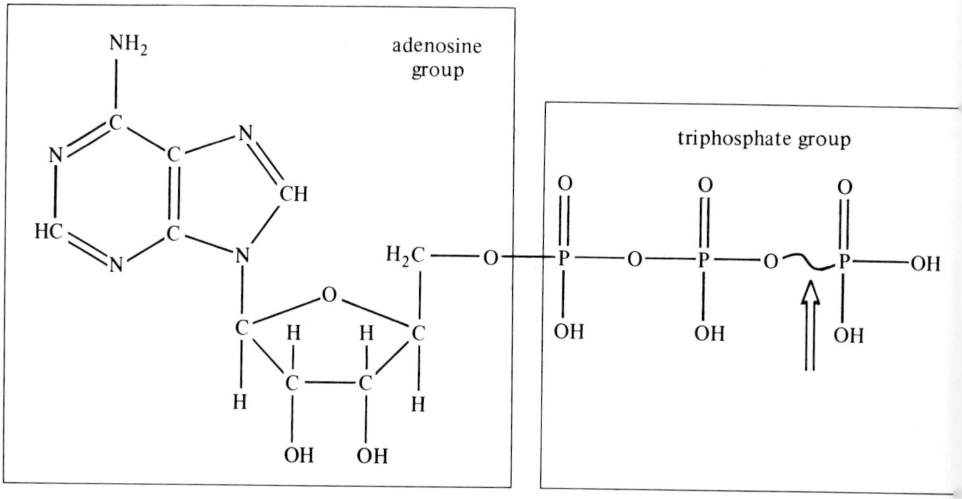

Fig. 15.18. The structure of the ATP molecule.

1. it provides a common unit of energy for later use in any biochemical reaction;
2. it prevents the sudden release of large amounts of energy that could damage the cell;
3. the release of small units of energy results in more efficient use of the available energy.

Mutations and biological evolution

The blueprints for the replication of cells are encoded in the sequences of bases that are attached to the backbone of sugar and phosphate groups in DNA molecules (see Section 3.5). An ordered set of three bases encodes an **amino acid** which is the basic building block, a sequence of approximately one thousand amino acids constitutes a single **gene**, and each DNA molecule is comprised of several thousand genes.[15,16,17] The genetic information of DNA is transferred to molecules of **ribonucleic acid** (**RNA**) which then serve as the protein templates for cell replication. Almost all

[15] F. H. C. Crick, "The Genetic Code," *Scientific American*, October, 1962. Available as *Scientific American Offprint 123* (San Francisco: W. H. Freeman and Co., Publishers).

[16] M. W. Nirenberg, "The Genetic Code II," *Scientific American*, March, 1963. Available as *Scientific American Offprint 153* (San Francisco: W. H. Freeman and Co., Publishers).

[17] F. H. C. Crick, "The Genetic Code III," *Scientific American*, October, 1966. Available as *Scientific American Offprint 1052* (San Francisco: W. H. Freeman and Co., Publishers).

the DNA is found in the **chromosomes** that are located within the nucleus of the cell. RNA molecules are synthesized in the nucleus after which many of them move to the **cytoplasm** in which most of the protein synthesis takes place. Nucleic acids both carry the hereditary information and initiate the cell replication process.

Experiments have indicated that cells may be duplicated $\simeq 10^8$ times before there is a 50% probability that even one gene in a daughter cell will be altered. These experiments suggest that cells are able to repair such damage[18] to their DNA as might be caused, for example, by ionizing radiation. This remarkable property of cells is clearly of importance in biological evolution. Repair mechanisms can make it possible for a species to maintain its genetic stability in an environment in which mutations occur at a relatively rapid rate. However, mutations are essential for evolution since they are the changes that allow variations among the individual members of a population. Therefore, if repair mechanisms were too efficient, no evolution would occur.

The repair mechanism in cells is affected by visible light in a process called **photoreactivation**. In 1948 it was observed that the number of actinomycetes (a type of soil bacteria) that survived large doses of ultraviolet radiation could be increased by a factor of several hundred thousand if the irradiated bacteria were subsequently exposed to an intense source of visible light. The effect of the ultraviolet is to create pyrimidine bases on a DNA strand. Thymine-thymine dimers (see Section 3.5) are the most readily formed of the three possible types of pyrimidine dimers. As a result, bacteria having DNA that is rich in thymine tend to be more sensitive to ultraviolet radiation than those for which the DNA is not rich in thymine. It has been shown that the presence of pyrimidine dimers blocks normal replication of DNA; bacteria possessing even a few such defects are unable to divide and form colonies.[19] Photoreactivation involves the action of an **enzyme** (an organic catalytic agent) that is selectively bound to the DNA molecule that has been irradiated with ultraviolet radiation. Visible light provides the source of energy required by the enzyme to allow it to cleave the pyrimidine dimer, thereby restoring the two bases to their original form.

It is interesting to speculate on the role played by the **ozone layer** of the earth's atmosphere in the evolutionary process. The radiation protection afforded by the atmospheric ozone (O_3) is indicated in Fig. 15.19 where the remarkable similarity between the absorption spectra of ozone

---

[18] P. C. Hanawalt and R. H. Haynes, "The Repair of DNA," *Scientific American*, February, 1967. Available as *Scientific American Offprint 1061* (San Francisco: W. H. Freeman and Co., Publishers).

[19] R. A. Deering, "Ultraviolet Radiation and Nucleic Acid," *Scientific American*, December, 1962. Available as *Scientific American Offprint 143* (San Francisco: W. H. Freeman and Co., Publishers).

and a nucleic acid is indicated. As indicated earlier, nucleic acids play an essential role.

It has been postulated that when the earth was young (before the ozone layer had formed) the rate of mutations was high and the probability of exact replication was very small. This gave rise to the widest range of variation with little competition because of the high nutrient concentrations present at that time. With the advent of green plants oxygen was released into the atmosphere. The action of ultraviolet radiation on the oxygen produced ozone; the appearance of the ozone layer caused a large reduction in the mutation rate which in turn signaled the beginning of exact replication. At this time evolutionary competition began. The successful organisms were those that were able to develop systems to utilize visible light both as an energy source and as a means to activate enzymes.

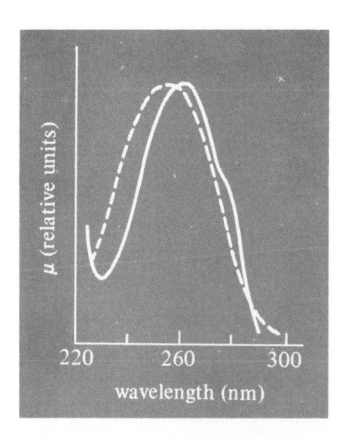

Fig. 15.19. The ultraviolet absorption spectra of ozone (dashed line) and a nucleic acid (solid line).

15.6 THE EINSTEIN COEFFICIENTS

One of the basic postulates of the Bohr theory[20] is that in the absence of any interaction with electromagnetic radiation an atomic system remains in its ground state. The probabilities of transitions taking place between stationary states were first discussed by Albert Einstein in 1917 before the advent of quantum mechanics. We will now present his hypothesis.

Let E_2 and E_1 represent the two stationary states of an atomic system where $E_2 > E_1$. Einstein considered three processes whereby the atomic system could change its state; these are indicated in Fig. 15.20. Note that in each process the total energy is conserved. In Fig. 15.20(a) a **spontaneous emission** process is indicated. The probability per unit time for such a process is designated A_{21}. This process occurs whether or not there is external radiation present. In Fig. 15.20(b) we show a **stimulated emission** process. This process requires the presence of external electromagnetic radiation at the characteristic frequency $v = (E_2 - E_1)/h$ of the atomic system. The probability per unit time for such a process is $\rho_v B_{21}$ where ρ_v is the energy density at frequency v; that is, the energy per unit volume in the electromagnetic field in the frequency range v to $v + dv$ is $\rho_v \, dv$. The induced photon is indistinguishable from the photon responsible for the process. In Fig. 15.20(c) we show an **absorption process**. This process also requires the presence of external radiation at the characteristic fre-

[20] *MWTP*, Section 18.4.

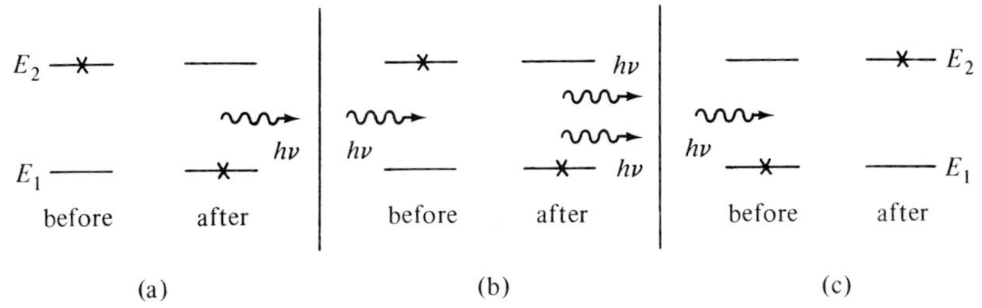

SEC. 15.6 THE EINSTEIN COEFFICIENTS

Fig. 15.20. Possible processes whereby a two-level atomic system can change its state: (a) a spontaneous emission process; (b) a stimulated emission process; (c) an absorption process.

quency v of the atomic system. The probability per unit time for an absorption process is $\rho_v B_{12}$.

Let us suppose that N weakly interacting, two-level atomic systems are in thermal equilibrium at temperature T in a cavity filled with blackbody radiation.[21] The populations of the energy levels then satisfy the Maxwell–Boltzmann distribution[22] so that

$$\frac{N_2}{N_1} = \exp\left(\frac{-hv}{kT}\right)$$

where $N = N_1 + N_2$. The condition for equilibrium is

$$N_2[A_{21} + \rho_v B_{21}] = N_1 \rho_v B_{12} = N_2 \rho_v B_{12} \exp\left(\frac{hv}{kT}\right).$$

This equation states that the number of atoms making transitions from state E_2 to state E_1 per second is just balanced by the number of atoms making transitions from state E_1 to state E_2. Therefore,

$$A_{21} = \rho_v \left[B_{12} \exp\left(\frac{hv}{kT}\right) - B_{21} \right]$$

or

$$\rho_v \, dv = \frac{A_{21} \, dv}{B_{21}\left[\left(\frac{B_{12}}{B_{21}}\right) \exp\left(\frac{hv}{kT}\right) - 1\right]}.$$

This equation may now be compared with Planck's blackbody radiation law

$$\rho_v \, dv = \frac{\left(\frac{8\pi h}{c^3}\right) v^3 \, dv}{\exp\left(\frac{hv}{kT}\right) - 1}.$$

In order for the two expressions for $\rho_v \, dv$ to be identical, it is necessary

[21] *MWTP*, Section 18.2.
[22] *MWTP*, Section 19.4.

that

$$B_{12} = B_{21} = B$$
$$\frac{A_{21}}{B_{21}} = \frac{A}{B} = \frac{8\pi h \nu^3}{c^3}.$$

Note that the coefficient of stimulated emission B_{21} is equal to the coefficient of absorption B_{12}, and that the importance of the coefficient of spontaneous emission A_{21} relative to the coefficient of stimulated emission increases as the cube of the frequency. This was Einstein's derivation of Planck's blackbody radiation law. The coefficients A and B are known as the **Einstein coefficients.**

From the relation between the Einstein coefficients we could show that, in an experiment carried out in the visible region of the electromagnetic spectrum using a conventional light source, $\rho_\nu B \ll A$ (see Problem 25); that is, stimulated emission may be neglected in comparison with spontaneous emission. On the other hand, in a typical experiment using radiofrequency radiation, $\rho_\nu B \gg A$; that is, spontaneous emission may be neglected in comparison with stimulated emission (see Problem 26).

We should emphasize that the above results may now be derived quite generally using quantum mechanics and that they apply equally well to both equilibrium and nonequilibrium interactions of radiation with matter.

QUESTIONS AND PROBLEMS

1. What thickness of absorbing material with absorption coefficient μ is necessary to reduce the intensity of a beam of radiation by a factor of 2?

*2. In Section 9.8 it was stated that the power radiated by an oscillating dipole is proportional to the fourth power of the frequency. The exact expression is

$$P = \frac{p_0^2}{12\pi\epsilon_0 c^3}\omega^4$$

where p_0 is the maximum dipole moment, P is the total power radiated, and ω is the angular frequency of oscillation. Show that the scattered intensity varies as the fourth power of the frequency when an electromagnetic wave

$$E_x = E_0 \cos \omega t$$

excites an electron bound to a molecule such that it exhibits a natural frequency ω_0. Assume that there is no damping and that $\omega_0 \gg \omega$.

3. The **natural line width** $\Delta \nu_n$ and the **spontaneous lifetime** τ_{sp} for an atom emitting light are given by

$$\Delta \nu_n = (\tau_{sp})^{-1} = \frac{e^2 \nu_0^2}{3\epsilon_0 c^3 m_e}.$$

Calculate Δv_n and τ_{sp} for an atom emitting radiation of wavelength 500 nm.

4. Compare the effectiveness of scattering by the atmosphere of red light of wavelength 650 nm to blue light of wavelength 450 nm.

*5. Calculate the ionization energy in eV of an exciton in germanium. Take $m_e^* = m_h^* = 0.10\, m_e$ and $K = 11.7$ for germanium. Would you expect to observe transitions from one exciton state to another? What wavelength light would be emitted by an $n = 2$ to $n = 1$ transition?

6. The **V-center** is another example of a possible color center in a solid. What is meant by the term V-center?

7. Write a short essay on the possible mechanism and evolutionary aspects of **bioluminescence**.[23]

8. Discuss a number of applications of the photoconductive effect.

9. Show that it is not possible to conserve both relativistic total energy and momentum if a free electron absorbs all the energy of a quantum.

10. State and comment on the fundamental difference between the photoelectric effect and the Compton effect.

11. A beam of monochromatic photons of wavelength 300 nm falls on sodium metal ($\phi = 2.46$ eV). Calculate the maximum kinetic energy of the photoelectrons. What is the maximum wavelength of photons that will still produce photoelectrons?

12. A beam of light of intensity 2.0×10^{-6} W·m^{-2} is incident normally on the surface of a block of sodium. Use arguments from classical physics to estimate the time required for an electron to gain sufficient energy to escape from the metal. The density of sodium is 970 kg·m^{-3}. Note that no time lag is observed experimentally.

13. Discuss **Thomson scattering** of x-rays.

14. A beam of 10 MeV photons is incident on a target. An electron scatters a photon through 30°. At what angle will the electron recoil? What will its energy be?

15. Show that the equation

$$K_0 K_1 (1 - \cos \theta) = m_e c^3 (K_0 - K_1)$$

mentioned in the discussion of Compton scattering follows from momentum-energy conservation.

*16. A beam of high-energy photons is incident on a target. At the same time it is observed that 15 keV electrons are coming out of the opposite side of the target along the direction of the photon path. Calculate the wavelength of the incident radiation. What are the energies of the photons and electrons that are scattered 45° by the target?

*17. A 1.5 MeV photon interacts with a lead-208 nucleus producing an electron and a position; each has the same energy and travels in the forward direction. Calculate the momentum and energy given to the recoiling nucleus.

[23] H. H. Seliger and W. D. McElroy, *Light: Physical and Biological Action* (New York: Academic Press, 1965).

*18. A beam of 1.0 GeV monochromatic photons is incident on a metal sheet producing electron-positron pairs traveling in the forward direction. Each particle produced moves at right angles to a uniform magnetic field **B**. Show that to a good approximation the energy of the photon is proportional to the sum of the radii of curvature of the particles.

19. Estimate the ratio of the attenuation lengths for 8 MeV x-rays in aluminum and lead.

20. Estimate the thickness of lead required to attenuate a beam of 1.0 MeV x-rays by a factor of 100.

21. The **mitochondrion** is the powerhouse of the cell. Discuss this statement.[24,25]

22. In Section 15.5 we stated that photons of wavelength 450 nm have a g-mole equivalent energy equal to 64 kcal. g-mole^{-1}. Justify this statement.

23. **Ribosomes** are the organelles that carry out the synthesis of proteins in living cells. Discuss the structure and functioning of the ribosomes.[26]

24. Discuss the effects of x-rays on human cells.[27]

*25. Suppose that a blackbody radiation source at a temperature of 2000°K is used for an experiment to study an atomic transition of wavelength 700 nm. Deduce the relative importance of spontaneous and stimulated emission processes in this experiment.

*26. An essentially monochromatic source of radiofrequency radiation of 30 MHz establishes a radiation field of $1.0 \times 10^3 \text{ W} \cdot \text{m}^{-3}$ within a sample. Calculate the relative importance of spontaneous and stimulated emission processes in this case.

*27. Show that

$$A_{21} + \rho_\nu B_{21} = A_{21}(1 + \bar{n})$$

where $\bar{n}$ is the average number of photons of frequency ν in the radiation field. Interpret this result for experiments in the radiofrequency region.

*28. Consider a system with two degenerate states E_1 and E_2 with $E_2 > E_1$.
   (a) If there are g_1 states of energy E_1 and g_2 of energy E_2, use an analogous development to that presented in Section 15.6 to show that

$$B_{12} = B_{21}\frac{g_2}{g_1} \quad \text{and} \quad \frac{A_{21}}{B_{21}} = \frac{8\pi h \nu^3}{c^3}.$$

   (b) Neglecting spontaneous emission, show that the absorption coefficient for light of frequency $h\nu = E_2 - E_1$ passing through n such systems

---

[24] D. E. Green, "The Mitochondrion," *Scientific American*, January, 1964.

[25] A. L. Lehninger, "How Cells Transform Energy," *Scientific American*, September, 1961. Available as *Scientific American Offprint 91* (San Francisco: W. H. Freeman and Co., Publishers).

[26] M. Monura, "Ribosomes," *Scientific American*, October, 1969.

[27] T. T. Puck, "Radiation and the Human Cell," *Scientific American*, April, 1960. Available as *Scientific American Offprint 71* (San Francisco: W. H. Freeman and Co., Publishers).

per unit volume is

$$\mu = \frac{c^2}{8\pi v^2 \Delta v}\left[n_1 \frac{g_2}{g_1} - n_2\right]\frac{1}{\tau_{sp}}$$

where Δv is the line width and τ_{sp} the spontaneous lifetime of the upper state. Use the fact that the intensity $I = \rho_v \Delta v c$ changes according to the relation

$$\frac{dI}{dx} = hv[n_2 \rho_v B_{21} - n_1 \rho_v B_{12}]$$

where x is the direction of propagation.

EDWARD M. PURCELL

16 Magnetic resonance

16.1 INTRODUCTION

In Section 7.6 we saw that all electrons, some nuclei, and some atoms possess magnetic moments associated with a spin or an orbital angular momentum, or both. In Section 13.4 we saw that, in general, when atoms form molecules or solids, the nature of the electrostatic binding is such that it leaves no resultant electronic magnetic moments. However, there are exceptions; molecules such as NO and O_2, metals, iron-group salts, and rare-earth group salts all exhibit electronic paramagnetism. Any solid that contains nuclei that possess magnetic moments shows nuclear paramagnetism.

When a substance possessing magnetic moments is placed in a static magnetic field, the moments experience torques which tend to align them. Although the direct observation of such an alignment is very difficult, it is relatively easy to observe the absorption of energy from a time-dependent magnetic field oscillating at a radio or microwave frequency using the techniques of **magnetic resonance**. The electronic or nuclear moments serve as minute magnetic probes that can be utilized to study local magnetic effects in condensed matter.

E. Zavoisky in Russia was the first to observe **electron paramagnetic resonance (EPR)**. In 1944 he carried out measurements on some iron-group salts. However, a group of physicists headed by Brebis Bleaney at Oxford University in England elucidated many of the basic principles that form the basis of our present understanding of EPR.[1]

Nuclear magnetic resonance (NMR)[2] in bulk samples was observed independently by two groups in the United States in 1945. Edward Purcell, Henry Torrey, and Robert Pound at Harvard University reported proton resonance absorption in solid paraffin; Felix Bloch, William Hansen, and Martin Packard of Stanford University performed proton resonance measurements in water. Although NMR was first used mainly by nuclear physicists for the accurate determination of nuclear moments, it has now become apparent that its applications are extremely numerous.

When nuclear quadrupole moments (see Section 3.9) in solids are located at sites of finite electric field gradient, they experience torques that tend to align them. This alignment may be detected through the absorption of magnetic dipole radiation from an appropriate radiofrequency magnetic field using the techniques of **nuclear quadrupole resonance (NQR)**.[3] The first observation of this effect was reported in 1950 by H. Dehmelt and H. Kruger in Germany. They measured the resonance of the

---

[1] R. A. Kamper, "Paramagnetic Resonance," *American Journal of Physics*, **28** (March 1960), 249.

[2] D. J. E. Ingram, "Nuclear Magnetic Resonance," *Contemporary Physics*, **7** (October 1965) 13 (November 1965) 103.

[3] J. C. Raich and R. H. Good, Jr. "Discussion of Quadrupole Precession," *American Journal of Physics*, **31** (May 1963), 356.

chlorine-35 nuclei in solid dichloroethylene. NQR is being used more and more as an analytical tool for solid-state physics and structural chemistry.

This chapter will be confined to a discussion of some of the principles of magnetic resonance that apply quite generally to all branches of the subject. Selected examples of the application of magnetic resonance will be considered. In all cases the discussion will be confined to the three branches of magnetic resonance mentioned above—nuclear magnetic resonance, nuclear quadrupole resonance, and electron paramagnetic resonance. Such techniques as **nuclear magnetic double resonance (NMDR)**, **electron nuclear double resonance (ENDOR)**, **ferromagnetic resonance (FMR)**, and **antiferromagnetic resonance (AFMR)** will not be specifically considered.

16.2 THE RESONANCE PHENOMENON

The description of the behavior of a system of noninteracting magnetic moments in a uniform magnetic field is a basic problem in magnetic resonance. Let us consider a nucleus with magnetic moment **μ** and spin angular momentum **I**$\hbar$ situated in a magnetic field **B**. The magnetic moment will experience a torque

$$\boldsymbol{\tau} = \boldsymbol{\mu} \times \mathbf{B}$$

and its equation of motion will be

$$\boldsymbol{\tau} = \boldsymbol{\mu} \times \mathbf{B} = \frac{d}{dt}(\mathbf{I}\hbar).$$

Since $\boldsymbol{\mu} = \gamma \mathbf{I}\hbar$ with γ the nuclear magnetogyric ratio (see Section 13.4), this equation may be written

$$\frac{d\boldsymbol{\mu}}{dt} = \gamma \boldsymbol{\mu} \times \mathbf{B}.$$

For a system of N noninteracting magnetic moments, the macroscopic magnetization **M** satisfies the equation of motion

$$\frac{d\mathbf{M}}{dt} = \gamma \mathbf{M} \times \mathbf{B}.$$

To solve this equation of motion it is useful to transform it to a frame of reference rotating with an angular frequency **ω** relative to the laboratory fixed frame (see Fig. 16.1 and Appendix B). In the rotating frame

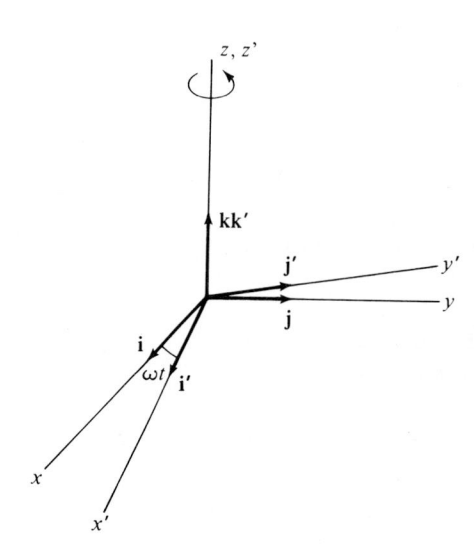

Fig. 16.1. The coordinate system $x'y'z'$ rotates at frequency ω with respect to the system xyz about the common z, z' axis. Note that x and x' coincide at $t = 0$.

SEC. 16.2   THE RESONANCE PHENOMENON

$$\frac{\partial \mathbf{M}}{\partial t} = \gamma \mathbf{M} \times \left(\mathbf{B} + \frac{\boldsymbol{\omega}}{\gamma}\right).$$

First we consider the system of magnetic moments to be situated in a static magnetic field pointing in the z direction; that is, we take $\mathbf{B} = B_0 \mathbf{k}$. If we choose

$$\boldsymbol{\omega} = -\gamma B_0 \mathbf{k},$$

then $\partial \mathbf{M}/\partial t = 0$ and $\mathbf{M}$ is a stationary vector in the rotating reference frame. Therefore, in the laboratory fixed frame, $\mathbf{M}$ precesses about the z axis with angular frequency $\omega_0 = \gamma B_0$, the **Larmor frequency**.

Now we consider the addition of a small time-varying magnetic field rotating in the xy plane. The total magnetic field is then

$$\mathbf{B} = B_0 \mathbf{k} + B_1[\cos \omega t \, \mathbf{i} - \sin \omega t \, \mathbf{j}]$$

where B_1 is the amplitude of the rotating component. We transform to a rotating reference frame by choosing

$$\boldsymbol{\omega} = -\omega \mathbf{k}$$

and taking the x axes in the two frames to coincide at $t = 0$. Therefore,

$$\frac{\partial \mathbf{M}}{\partial t} = \gamma \mathbf{M} \times \left[\left(B_0 - \frac{\omega}{\gamma}\right)\mathbf{k} + B_1 \mathbf{i}'\right]$$
$$= \gamma \mathbf{M} \times \mathbf{B}_{\text{eff}}$$

where $\mathbf{i}'$ is a unit vector in the x direction in the rotating frame of reference.

In physical terms this equation of motion states that in the rotating frame of reference $\mathbf{M}$ acts as if it experienced a static magnetic field $\mathbf{B}_{\text{eff}}$. Therefore, in the rotating frame $\mathbf{M}$ will precess in a cone of fixed angle about the direction of $\mathbf{B}_{\text{eff}}$ at an angular frequency $\gamma \mathbf{B}_{\text{eff}}$. For $\mathbf{M}$ aligned along the z axis at $t = 0$, the half-angle δ of the precessional cone is

$$\delta = \sin^{-1}\left(\frac{B_1}{B_{\text{eff}}}\right) = \sin^{-1}\left\{\frac{B_1}{\left[\left(B_0 - \frac{\omega}{\gamma}\right)^2 + B_1^2\right]^{1/2}}\right\}$$

(see Fig. 16.2). In a practical case $B_1 \ll B_0$ and $\delta \simeq 0$ unless $B_0 - \omega/\gamma$ becomes comparable to B_1; that is, unless $\omega \simeq \gamma B_0$. This is a **resonance phenomenon** since a rotating magnetic field which is small compared to the constant field B_0 can only appreciably reorient the magnetization if its frequency of rotation is approximately the Larmor frequency.

In the laboratory fixed frame the precessional motion about $\mathbf{B}_{\text{eff}}$ at frequency $\gamma \mathbf{B}_{\text{eff}}$ is superimposed upon a precession about the z axis at frequency γB_0.

Note that the motion of $\mathbf{M}$ is periodic. If $\mathbf{M}$ is initially aligned along the z axis, it periodically returns to that direction. Therefore, there is no net absorption of energy from the time-varying magnetic field since we

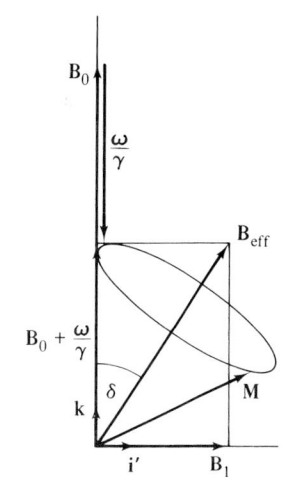

Fig. 16.2. The precession of $\mathbf{M}$ about the direction of $\mathbf{B}_{\text{eff}}$ in the rotating reference frame.

have neglected the interactions between the magnetic moments and the interactions of the magnetic moments with their environment.

In practice, magnetic resonance experiments make use of oscillating rather than rotating magnetic fields. A linearly polarized field of amplitude $2B_1$ is a superposition of two fields of amplitude B_1 rotating in opposite senses (see Section 14.8). If one of these components satisfies the resonance condition, the other component is off resonance by twice the Larmor frequency and will have a negligible effect on the system of magnetic moments.

Example. It is possible to rotate the magnetization **M** from its equilibrium direction (the z axis) into the xy plane by the application of a pulse of time-varying magnetic field at the Larmor frequency; such a pulse is called a **π/2 pulse** (see Fig. 16.3). Derive a formula for the length of a π/2 pulse.

Solution. We take the time-varying magnetic field as $B_1(\cos \omega_0 t \mathbf{i} - \sin \omega_0 t \mathbf{j})$. In the rotating frame of reference

$$\mathbf{B}_{\text{eff}} = B_1 \mathbf{i}'$$

and **M** precesses about the x' axis at a frequency

$$\omega_1 = \gamma B_1.$$

In a time t_ω, **M** precesses through an angle

$$\theta = \omega_1 t_\omega.$$

Therefore, for $\theta = \pi/2$

$$t_\omega = \frac{\pi}{2\gamma B_1}.$$

This is the length of a pulse required to rotate the magnetization from the z direction into the xy plane. Following the pulse, **M** precesses about the z axis in the laboratory frame of reference at the Larmor frequency.

The problem that we have just been discussing from the viewpoint of classical mechanics is also straightforward in quantum mechanics. The solution of the Schroedinger equation for a nuclear magnetic moment **μ** with associated spin angular momentum $I\hbar$ in a static magnetic field $B_0 \mathbf{k}$ yields $2I + 1$ states of energy

$$E_m = -\gamma \hbar B_0 m$$

where the quantum number m can take the values

$$m = I, I-1, \ldots, -(I-1), -I.$$

Each energy state corresponds to an allowed orientation of the magnetic moment in the external magnetic field. These we show in Fig. 16.4 for a magnetic moment **μ** with an associated spin angular momentum $3\hbar/2$ on the assumption that γ is positive. In Fig. 16.5 we show the corresponding energy states. The energy states are equally spaced and separated in

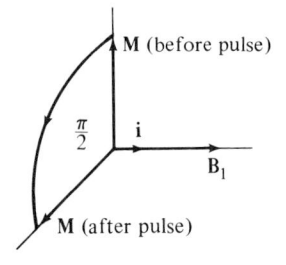

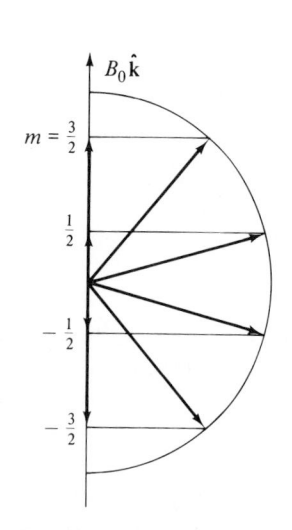

Fig. 16.3. A π/2 pulse.

Fig. 16.4. The allowed orientations of a magnetic moment **μ** with associated spin angular momentum $3\hbar/2$ in a static magnetic field $B_0 \mathbf{k}$.

angular frequency by

$$\omega = \frac{E_{m-1} - E_m}{\hbar} = \gamma B_0,$$

the classical Larmor frequency.

The same set of energy states holds for a system of N noninteracting magnetic moments and gives a good approximation to the energy levels of a system of interacting magnetic moments as long as the energy of interaction between magnetic moments is small compared to the energy of interaction of the individual magnetic moments with the magnetic field.

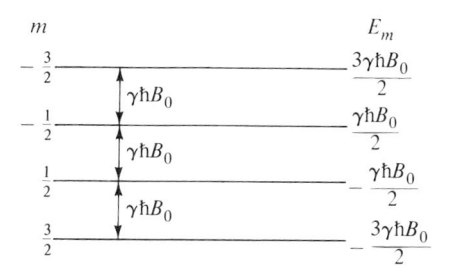

Fig. 16.5. The energy states for a magnetic moment with an associated spin angular momentum $3\hbar/2$ in a static magnetic field $B_0\mathbf{k}$.

16.3 ENERGY ABSORPTION AND SPIN-LATTICE RELAXATION

We now consider a macroscopic sample containing N nuclear magnetic moments with associated spin angular momentum quantum number $I = \frac{1}{2}$, for example, the protons in ice. In a magnetic field of 1.0 Wb·m^{-2} the proton resonance frequency is 42.6 MHz. If the sample is placed in an electromagnetic field varying at this frequency, energy will be exchanged between the magnetic moments (**the spin system**) and the radiation field. Only absorption and stimulated emission processes need be considered (see Section 15.6).

The energy levels of the spin system are shown in Fig. 16.6. The number of spins in the $m = -\frac{1}{2}$ state is n_-; the number in the $m = \frac{1}{2}$ state is n_+, where $n_+ + n_- = n$ is the total number of spins per unit volume. The application of a radiofrequency magnetic field of frequency $\gamma \hbar B_0$ will cause n_+ and n_- to change with time. The time rate of change of n_+ is given by

$$\frac{dn_+}{dt} = n_- W_{-\to +} - n_+ W_{+\to -}$$

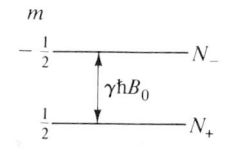

Fig. 16.6. The energy states for proton magnetic moments in a static magnetic field.

where $W_{-\to +}$ is the probability per second per magnetic moment of a stimulated emission process taking place and $W_{+\to -}$ the probability of an absorption process. Since $W_{-\to +} = W_{+\to -} \equiv W$,

$$\frac{dn_+}{dt} = W(n_- - n_+).$$

Similarly the time rate of change of n_- is given by

$$\frac{dn_-}{dt} = W(n_+ - n_-).$$

Combining these two equations gives

$$\frac{d(n_+ - n_-)}{dt} = -2W(n_+ - n_-)$$

or

$$\frac{d(\Delta n)}{dt} = -2W\Delta n$$

where $\Delta n = n_+ - n_-$ is the difference in population between the two states. The solution of this equation is

$$\Delta n = \Delta n_0 \exp(-2Wt)$$

where Δn_0 is the population difference at $t = 0$. The rate of absorption of energy from the radiofrequency field is

$$\frac{dE}{dt} = n_+ W\hbar\omega - n_- W\hbar\omega = \Delta n W\hbar\omega.$$

This result states that if a population difference exists at $t = 0$, there will be a net absorption of energy from the applied field until the populations of the two levels become equal. Since a resonance can be detected experimentally only through the absorption of energy from the applied field, this result suggests that the resonance signal will decrease exponentially with time.

Furthermore, the above result states that if no applied field is present (so that $W = 0$), the populations cannot change. However, this prediction is incorrect. If we place an unmagnetized sample into a static magnetic field, the sample becomes magnetized! Such a preferential alignment of spins requires that n_+ becomes greater than n_-. In this process the spin system must give up energy to other degrees of freedom of the total system. We call all of the degrees of freedom of the sample, with the exception of those associated with the spin system, the **lattice** and postulate the existence of a **spin-lattice coupling** whereby energy can be exchanged between the spin system and the lattice. This is shown schematically in Fig. 16.7. If the energy of interaction between the spins is much larger than the energy of interaction between the spins and the lattice, the spins and the lattice may individually be assumed to be in quasi-thermal equilibrium and describable by temperatures T_s and T, respectively. Energy will flow between the two systems until they come into equilibrium at a common temperature. In practice, the heat capacity[4] of the lattice is enormous relative to the heat capacity of the spin system and the lattice may be assumed to remain at a fixed temperature.

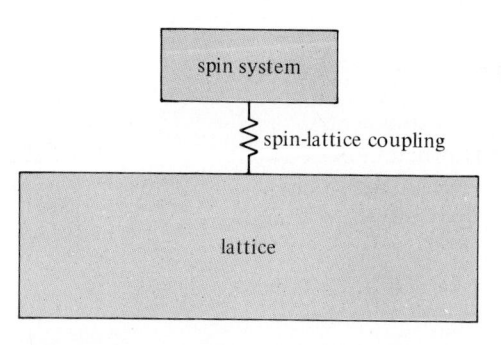

Fig. 16.7. Energy can be exchanged between the spin system and the lattice via spin-lattice coupling.

[4] *MWTP*, Section 17.6.

The time rate of change of n_+ due to the spin-lattice interaction is given by

$$\frac{dn_+}{dt} = n_- W_\downarrow - n_+ W_\uparrow$$

where $W_\downarrow$ is the probability per second per spin of a transition to the state $m = \frac{1}{2}$ and $W_\uparrow$ the probability of a transition to the state $m = -\frac{1}{2}$. Since we know from the Maxwell–Boltzmann distribution[5] that the equilibrium populations n_{+0} and n_{-0} must satisfy the equation

$$\frac{n_{-0}}{n_{+0}} = \exp\left(-\frac{\gamma \hbar B_0}{kT}\right),$$

it follows that

$$\frac{W_\uparrow}{W_\downarrow} = \exp\left(-\frac{\gamma \hbar B_0}{kT}\right)$$

(see Problem 5). We wish to draw special attention to the fact that $W_\uparrow \neq W_\downarrow$ although we previously saw that $W_{-\to+} = W_{+\to-}$. The explanation of what at first sight might appear to be a paradox is quite simple. A transition between the spin system and the lattice requires not only the existence of a spin-lattice coupling mechanism but also requires that the lattice is in an energy state that permits the transition; an interaction between the spin system and the applied field resulting in a transition requires only that an oscillation at the proper frequency is occurring. We illustrate this by considering a lattice with just two energy levels whose spacing is equal to that of the spin levels. In Fig. 16.8(a) conservation of

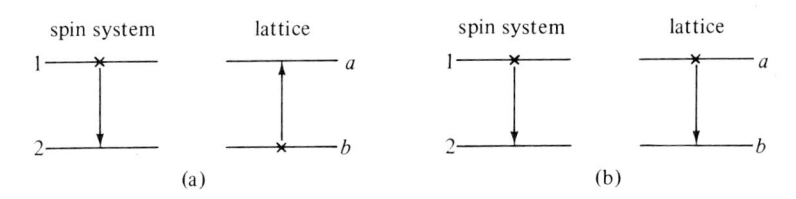

Fig. 16.8. (a) An allowed spin-lattice process. (b) A forbidden spin-lattice process.

energy is satisfied by the simultaneous transitions shown; in Fig. 16.8(b) conservation of energy is violated by the simultaneous transitions indicated. Therefore, Fig. 16.8(a) depicts an allowed spin-lattice process and Fig. 16.8(b) a forbidden spin-lattice process.

Similarly, the time rate of change of n_- due to the spin-lattice interaction is given by

$$\frac{dn_-}{dt} = n_+ W_\uparrow - n_- W_\downarrow.$$

[5] *MWTP*, Section 19.4.

Combining the equations for dn_+/dt and dn_-/dt gives

$$\frac{d(n_+ - n_-)}{dt} = 2(n_- W_\downarrow - n_+ W_\uparrow).$$

This equation may also be written as

$$\frac{d(\Delta n)}{dt} = \frac{\Delta n_0 - \Delta n}{T_1}$$

where

$$\Delta n_0 = n\left(\frac{W_\downarrow - W_\uparrow}{W_\downarrow + W_\uparrow}\right) \quad \text{and} \quad \frac{1}{T_1} = (W_\downarrow + W_\uparrow)$$

(see Problem 6).

The solution of this equation is

$$\Delta n = \Delta n_0 + A \exp\left(-\frac{t}{T_1}\right)$$

where A is a constant.

By inspection of this result we see that Δn_0 is the population difference for the spin system in thermal equilibrium with the lattice (assuming the spin system and lattice to approach thermal equilibrium as $t \to \infty$) and T_1 is a characteristic time associated with the approach to equilibrium. The constant T_1 is called the **spin-lattice relaxation time**. If at $t = 0$ we put an unmagnetized sample into a static magnetic field, then it follows that

$$0 = \Delta n_0 + A$$

or

$$A = -\Delta n_0.$$

Therefore, at later times

$$\Delta n = \Delta n_0 \left[1 - \exp\left(-\frac{t}{T_1}\right)\right]$$

which states that T_1 is a measure of the time required to magnetize an initially unmagnetized sample.

Finally, we combine the equation derived above to describe the time rate of change of the population difference between the spin levels for the interaction of the spin system with the applied field with the equation that describes the interaction of the spin system with the lattice. This gives

$$\frac{d(\Delta n)}{dt} = -2W\Delta n + \frac{\Delta n_0 - \Delta n}{T_1}.$$

Once a steady state is reached $d(\Delta n)/dt = 0$ and energy is absorbed by the spin system from the applied field at the same rate as energy is exchanged between the spin system and the lattice. In the steady state

$$\Delta n = \frac{\Delta n_0}{1 + 2WT_1}.$$

SEC. 16.3 ENERGY ABSORPTION AND SPIN-LATTICE RELAXATION

If $2WT_1 \ll 1$, $\Delta n = \Delta n_0$, and the absorption of energy by the spin system from the applied field does not appreciably alter the populations of the spin levels from their thermal equilibrium values.

The transition probability W can be evaluated using quantum-mechanical techniques that are beyond the scope of this text. It turns out that

$$W = \frac{\pi}{2}\gamma^2 B_1^2 g(\omega)$$

where B_1 is the amplitude of a radio-frequency magnetic field which is rotating at frequency ω. The function $g(\omega)$ is the **line shape function**. Because of the microscopic magnetic field inhomogeneity ΔB_0 that results from the interactions between neighboring magnetic moments in a real sample, magnetic resonance absorption occurs over a small range of frequencies $\Delta\omega = \gamma\Delta B$. The function $g(\omega)$ describes the distribution of resonance frequencies; it has the dimensions of time. The actual form of $g(\omega)$ will depend on the details of the interaction mechanism responsible for coupling the spins.

If a number of interacting spins are initially in phase and are precessing about a static field $B_0\mathbf{k}$, they will quickly get out of phase as a result of variation of precession frequencies resulting from the microscopic magnetic field inhomogeneity. The characteristic time for this dephasing to occur is called the **spin-spin relaxation time** and it is designated as T_2. Clearly an intimate connection exists between the line shape function and the spin-spin relaxation time. The relation is taken to be

$$T_2 = \frac{1}{2}[g(\omega)]_{\max}$$

where $[g(\omega)]_{\max}$ is the maximum value of $g(\omega)$. Alternately, T_2 may be defined by

$$T_2 = \frac{1}{\Delta\omega}.$$

Substitution for W in the steady-state solution for Δn yields the population difference

$$\Delta n = \frac{\Delta n_0}{1 + \frac{1}{2}\gamma^2 B_1^2 T_1 g(\omega)}.$$

We see that the steady-state value of Δn is reduced from the thermal equilibrium value Δn_0 by the factor

$$\frac{1}{1 + \frac{1}{2}\gamma^2 B_1^2 T_1 g(\omega)},$$

which is called the **saturation factor**. The largest degree of saturation occurs at the maximum of $g(\omega)$ for which the saturation factor is

$$\frac{1}{1 + \gamma^2 B_1^2 T_1 T_2}.$$

The rate of energy absorption by the spin system is given by

$$\frac{dU}{dt} = \Delta n W \hbar \omega$$

$$= \frac{\Delta n_0 \gamma^2 B_1^2 g(\omega) \hbar \omega}{4[1 + \frac{1}{2}\gamma^2 B_1^2 T_1 g(\omega)]}.$$

Therefore, if $\frac{1}{2}\gamma^2 B_1^2 T_1 g(\omega) \ll 1$, a measurement of dU/dt in a magnetic resonance experiment yields the line shape function $g(\omega)$. If, however, this inequality is not satisfied, the degree of saturation varies with $g(\omega)$; the center of the absorption line will be more strongly affected than the edges of the line. This effect, known as **saturation broadening**, is illustrated in Fig. 16.9.

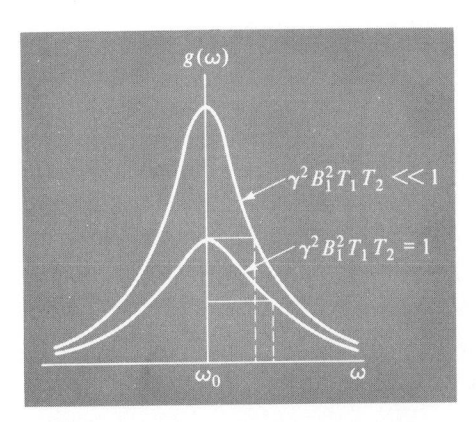

Fig. 16.9. An illustration of saturation broadening. The line width at half intensity is greater for $\gamma^2 B_1^2 T_1 T_2 = 1$ than for $\gamma^2 B_1^2 T_1 T_2 \ll 1$.

Example. For protons in water both T_1 and T_2 are about 4 sec. Given that the proton resonance frequency ν in a static magnetic field $B_0 = 1.0$ Wb·m^{-2} is 42.6 MHz, calculate the amplitude B_1 of the radio-frequency magnetic field for which the saturation factor will be unity. Comment on the result.

Solution. The saturation factor is $\gamma^2 B_1^2 T_1 T_2$ with the magnetogyric ratio γ given by $2\pi\nu/B_0$. Therefore,

$$\gamma^2 B_1^2 T_1 T_2 = 1$$

if

$$B_1 = \frac{1}{\gamma(T_1 T_2)^{1/2}} = \frac{B_0}{2\pi\nu(T_1 T_2)^{1/2}}$$

$$= \frac{1.0}{2\pi \times 42.6 \times 10^6 \times (4 \times 4)^{1/2}}$$

$$\simeq 10^{-9} \text{ Wb·m}^{-2}.$$

Since the signal amplitude is proportional to B_1, it would be impractical from an experimental point of view to reduce B_1 sufficiently to avoid saturation in this case.

16.4 BLOCH'S EQUATION

In 1946 Felix Bloch suggested a simple equation for the description of the magnetic properties of collections of nuclear magnetic moments in external magnetic fields. This equation, first derived from phenomenological arguments, has proved to be extremely useful and, for liquid samples at least, usually provides a correct quantitative description. The argument for obtaining Bloch's equation is based on the following points:

SEC. 16.4 BLOCH'S EQUATION

1. the equation of motion of a set of noninteracting magnetic moments in a homogeneous magnetic field is

$$\frac{d\mathbf{M}}{dt} = \gamma \mathbf{M} \times \mathbf{B};$$

2. in a static magnetic field $B_0\mathbf{k}$ the approach of the magnetization to its equilibrium value $M_z = M_0 = \chi_0 H_0 = \chi_0 B_0/\mu_0$ is often found experimentally to be describable by the equation

$$\frac{dM_z}{dt} = -\frac{(M_z - M_0)}{T_1};$$

3. if the magnetization is given a transverse component, the decay of this component toward zero is often found experimentally to be describable by the equations

$$\frac{dM_x}{dt} = -\frac{M_x}{T_2} \quad \text{and} \quad \frac{dM_y}{dt} = -\frac{M_y}{T_2};$$

4. in the presence of an applied magnetic field which is the sum of a static component and a radio-frequency component, the motion of the spins due to the relaxation processes can be superimposed on the motion of a set of noninteracting spins.

Therefore, Bloch's equation is

$$\frac{d\mathbf{M}}{dt} = \gamma \mathbf{M} \times \mathbf{B} - \frac{M_x\mathbf{i} + M_y\mathbf{j}}{T_2} - \frac{(M_z - M_0)\mathbf{k}}{T_1}.$$

Steady-state solution

As an example of the application of Bloch's equation we take

$$\mathbf{B} = B_0\mathbf{k} + B_1(\cos\omega t\,\mathbf{i} - \sin\omega t\,\mathbf{j})$$

and obtain the steady-state solution, that is, the solution after all transient effects arising from the switching on of the fields have decayed. To arrive at the solution we transform to a set of coordinate axes that rotate around $B_0\mathbf{k}$ at frequency ω. In this frame of reference, as we have seen in Section 16.2, there is an effective static magnetic field

$$\mathbf{B}_{\text{eff}} = \left(B_0 - \frac{\omega}{\gamma}\right)\mathbf{k} + B_1\mathbf{i}'$$

$$= \frac{\Delta\omega\mathbf{k} + \omega_1\mathbf{i}'}{\gamma}$$

where $\Delta\omega = \omega_0 - \omega$, with $\omega_0 = \gamma B_0$, and $\omega_1 = \gamma B_1$. Bloch's equation in the rotating frame of reference becomes

$$\frac{d\tilde{\mathbf{M}}}{dt} = \gamma(\tilde{\mathbf{M}} \times \mathbf{B}_{\text{eff}}) - \frac{\tilde{M}_x\mathbf{i}' + \tilde{M}_y\mathbf{j}'}{T_2} - \frac{(M_z - M_0)\mathbf{k}}{T_1}$$

where $\tilde{M}_x, \tilde{M}_y, \tilde{M}_z = M_z$ are the components of the magnetization in the rotating frame $(\mathbf{i}', \mathbf{j}', \mathbf{k}' = \mathbf{k})$ The steady-state condition implies that

$$\frac{d\tilde{M}_x}{dt} = \frac{d\tilde{M}_y}{dt} = \frac{dM_z}{dt} = 0.$$

The solution then yields (Problem 10)

$$\tilde{M}_x = \frac{\Delta\omega\gamma B_1 T_2^2}{1 + (T_2\Delta\omega)^2 + \gamma^2 B_1^2 T_1 T_2} M_0$$

$$\tilde{M}_y = \frac{\gamma B_1 T_2}{1 + (T_2\Delta\omega)^2 + \gamma^2 B_1^2 T_1 T_2} M_0$$

$$M_z = \frac{1 + (\Delta\omega T_2)^2}{1 + (T_2\Delta\omega)^2 + \gamma^2 B_1^2 T_1 T_2} M_0.$$

Let us consider the steady-state value of M_z. If $\gamma^2 B_1^2 T_1 T_2 \ll 1$, then $M_z \simeq M_0$; if $\gamma^2 B_1^2 T_1 T_2 \gg 1$, $M_z \simeq 0$. The factor $\gamma^2 B_1^2 T_1 T_2$ gives rise to saturation in agreement with the discussion in Section 16.3. Note that $\tilde{M}_x \neq \tilde{M}_y$. The reason for this asymmetry is that the directions $\mathbf{i}'$ and $\mathbf{j}'$ in the rotating frame of reference are not equivalent—the rotating magnetic field is directed along $\mathbf{i}'$.

The components of magnetization in the laboratory frame of reference are now easily deduced. For example,

$$M_x = \tilde{M}_x \cos\omega t\, \mathbf{i} + \tilde{M}_y \sin\omega t\, \mathbf{j}.$$

If the radio-frequency magnetic field is of the form

$$\mathbf{B} = 2B_1 \cos\omega t\, \mathbf{i},$$

the same solution results. The response of the spin system (that is, the establishment of a magnetization at the frequency of the stimulating radio-frequency field) consists of a component $\tilde{M}_x \cos\omega t\, \mathbf{i}$ which is in phase with the stimulating field and we can write

$$M_x = \chi(\omega) H_x$$

with $\chi(\omega) = \chi'(\omega) - i\chi''(\omega)$ where $\chi'(\omega)$ and $\chi''(\omega)$ are the real and imaginary parts of the radio-frequency susceptibility $\chi(\omega)$. It follows (Problem 11) that

$$\chi'(\omega) = \frac{\chi_0}{2}\omega_0 T_2 \frac{\Delta\omega T_2}{1 + (\Delta\omega T_2)^2},$$

$$\chi''(\omega) = \frac{\chi_0}{2}\omega_0 T_2 \frac{1}{1 + (\Delta\omega T_2)^2}$$

where χ_0 is the static magnetic susceptibility of the nuclear spin system. The components $\chi'(\omega)$ and $\chi''(\omega)$ are shown in Fig. 16.10.

The function

$$f_{T_2}(\Delta\omega) = \frac{T_2}{\pi} \frac{1}{1 + (\Delta\omega T_2)^2}$$

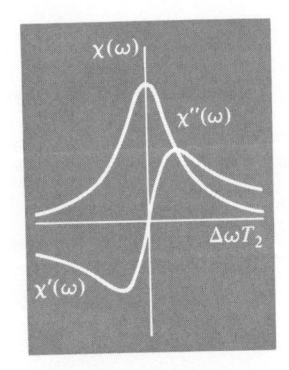

Fig. 16.10. The dispersive and absorptive components of the radio-frequency susceptibility in the vicinity of $\Delta\omega = 0$.

is a well-known mathematical form; it is called the **normalized Lorentzian form**. If Bloch's equation is valid, then

$$\chi''(\omega) = \frac{\pi}{2}\chi_0\omega_0 f_{T_2}(\Delta\omega).$$

Since the absorption of power by the spin system is proportional to $\chi''(\omega)$ (Problem 8), we see that Bloch's equation predicts magnetic resonance absorption line shapes that are Lorentzian in form. Note that the maximum value of $\chi''(\omega)$ is

$$\chi''_{max} = \frac{\chi_0}{2}\omega_0 T_2 = \frac{\chi_0}{2}\left(\frac{\omega_0}{\Delta\omega}\right).$$

Since in any experiment $\omega_0/\Delta\omega \gg 1$, it follows that $\chi''_{max} \gg \chi_0$; that is, a resonance experiment is characterized by its high sensitivity.

The simplest experimental arrangement for observing NMR absorption is shown schematically in Fig. 16.11. The magnetic field B_1 tends to tilt the direction of the macroscopic magnetization away from the B_0 direction as the radio frequency approaches the Larmor frequency of the nuclei in the sample. The energy absorbed in the process produces a drop in the amplitude of the output voltage of the source; the output amplitude of the source is amplified and fed to the vertical plates of the oscilloscope (see Section 5.3). In the arrangement shown the source frequency is kept constant and the applied field B_0 has superposed on it a low frequency (~ 100 Hz) modulation field $B_m = B_{m0} \cos \omega_m t$ of small amplitude. The signal used to produce the field sweep is fed to the horizontal plates of the oscilloscope. The NMR absorption signal is thereby displayed on the oscilloscope screen.

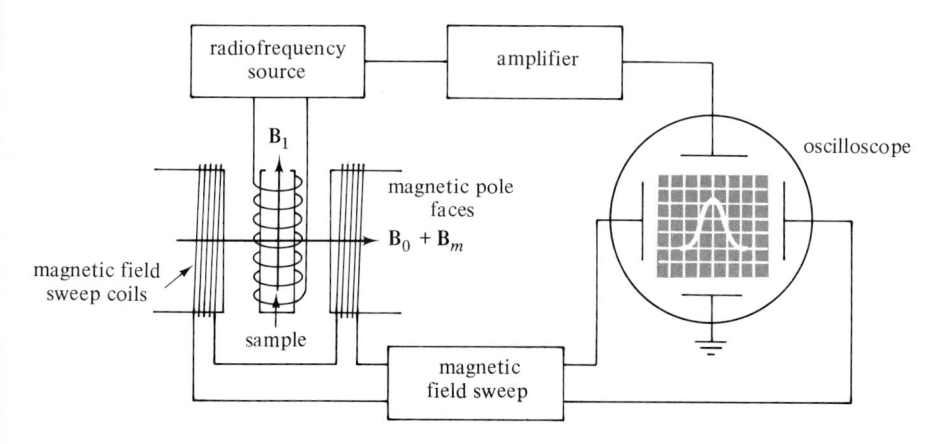

Fig. 16.11. A simple experimental arrangement for observing NMR absorption.

Response to a pulse

As a second example we suppose that a short burst of radio-frequency magnetic field is applied to a spin system at the resonance frequency in a static field $B_0\mathbf{k}$ for a time that is short compared to both T_1 and T_2. During the application of the pulse, Bloch's equation reduces to

$$\frac{d\tilde{\mathbf{M}}}{dt} = \gamma \tilde{\mathbf{M}} \times \mathbf{B}$$

which has as its solution in the rotating frame of reference

$$\tilde{M}_x(t) = 0$$
$$\tilde{M}_y(t) = M_0 \sin \gamma B_1 t$$
$$\tilde{M}_z(t) = M_z(t) = M_0 \cos \gamma B_1 t.$$

If the pulse is a $\pi/2$ pulse, at the end of the pulse ($t = t_w$)

$$\tilde{M}_x(t_w) = 0, \quad \tilde{M}_y(t_w) = M_0, \quad M_z(t_w) = 0.$$

Following the pulse, Bloch's equation in the rotating frame of reference becomes

$$\frac{d\tilde{\mathbf{M}}}{dt} = -\frac{\tilde{M}_x \mathbf{i}' + \tilde{M}_y \mathbf{j}'}{T_2} - \frac{(M_z - M_0)\mathbf{k}}{T_1}.$$

The solutions are

$$\tilde{M}_x = 0, \quad \tilde{M}_y = M_0 \exp\left(-\frac{t}{T_2}\right)$$

$$M_z = M_0 \left[1 - \exp\left(-\frac{t}{T_1}\right)\right]$$

where we neglect the difference between $t = 0$ and $t = t_w$.

In general, the **free precession** of the magnetization in the xy plane following a pulse can be expressed in the form

$$\tilde{M}_y = M_0 G_1(t)$$

where $G_1(t)$ is called the **free induction decay**. We see that Bloch's equation predicts that

$$G_1(t) = \exp\left(-\frac{t}{T_2}\right).$$

It is left as an exercise for the reader (Problem 13) to show that $G_1(t)$ is just the **Fourier transform** of the line shape function $f(\Delta\omega)$ for the function predicted by Bloch's equation. In fact, this is a general result.

16.5 CHEMICAL EXCHANGE

The primary reason that magnetic resonance is of interest to the chemist is that environmental influences, in general, produce several resonances in a sample at different frequencies. These resonances are

SEC. 16.5   CHEMICAL EXCHANGE

collectively referred to as the **magnetic resonance spectrum**. Frequently, each resonance of the spectrum can be treated by means of Bloch's equation as outlined in the previous section.

In this section we shall deal with the simplest example of a **rate process** which modifies the nuclear magnetic resonance spectrum. For example, the protons in a mixture of alcohol and water may be exchanged between water molecules and the hydroxyl position in the alcohol in a random statistical manner. The nature of the spectrum is determined by the exchange rate.

To begin the analysis it is useful to define a complex magnetization $\mathfrak{M}$ in terms of its components $\tilde{M}_x$ and $\tilde{M}_y$ in the rotating frame by

$$\mathfrak{M} = \tilde{M}_x + i\tilde{M}_y.$$

If the radiofrequency field is not too large, then $M_z \simeq M_0$ and the equation for $\mathfrak{M}$ is

$$\frac{d\mathfrak{M}}{dt} + \alpha \mathfrak{M} = -i\gamma B_1 M_0$$

with $\alpha = 1/T_2 - i\Delta\omega$ (see Problem 16). Let us label the two nuclear sites A and B. If no exchange of nuclei occurs, there will be two independent macroscopic moments with complex magnetization components $\mathfrak{M}$ given by

$$\frac{d\mathfrak{M}_A}{dt} + \alpha_A \mathfrak{M}_A = -i\gamma B_1 M_{0A}$$

$$\frac{d\mathfrak{M}_B}{dt} + \alpha_B \mathfrak{M}_B = -i\gamma B_1 M_{0B}.$$

To modify these equations for the present situation we must add terms to allow for exchange. We denote by τ_A the mean lifetime for a nucleus to be at an A site and by τ_B the mean lifetime for a nucleus to be at a B site. The fractional populations p_A and p_B of A and B sites, respectively, are related to τ_A and τ_B by

$$p_A = \frac{\tau_A}{\tau_A + \tau_B}, \qquad p_B = \frac{\tau_B}{\tau_A + \tau_B}.$$

Transitions from A sites to B sites are accompanied by a change of magnetization $\mathfrak{M}_A/\tau_A$ and those from B sites to A sites by a change $\mathfrak{M}_B/\tau_B$. Therefore, we write the modified Bloch equations

$$\frac{d\mathfrak{M}_A}{dt} + \alpha_A \mathfrak{M}_A = -i\gamma B_1 M_{0A} + \frac{\mathfrak{M}_B}{\tau_B} - \frac{\mathfrak{M}_A}{\tau_A}$$

$$\frac{d\mathfrak{M}_B}{dt} + \alpha_B \mathfrak{M}_B = -i\gamma B_1 M_{0B} + \frac{\mathfrak{M}_A}{\tau_A} - \frac{\mathfrak{M}_B}{\tau_B}.$$

The steady-state solution requires that

$$\frac{d\mathfrak{M}_A}{dt} = \frac{d\mathfrak{M}_B}{dt} = 0.$$

Noting that
$$M_{0A} = p_A M_0 \quad \text{and} \quad M_{0B} = p_B M_0,$$
we write for the total complex magnetization
$$\mathfrak{M} = \mathfrak{M}_A + \mathfrak{M}_B = -i\gamma B_1 M_0 \frac{\tau_A + \tau_B + \tau_A \tau_B(\alpha_A p_B + \alpha_B p_A)}{(1 + \alpha_A \tau_A)(1 + \alpha_B \tau_B) - 1}$$
(Problem 17).

In the limit that the lifetimes τ_A and τ_B are large relative to $(\omega_A - \omega_B)^{-1}$, the spectrum consists of distinct resonances in the vicinity of the frequencies ω_A and ω_B. This is the **slow exchange limit**. If ω is near ω_A, then $\mathfrak{M}_B \simeq 0$ and
$$\mathfrak{M}_A \simeq -i\gamma B_1 M_0 \frac{p_A \tau_A}{1 + \alpha_A \tau_A}.$$
The imaginary part is
$$\tilde{M}_{Ay} = -\gamma B_1 M_0 \frac{p_A T'_{2A}}{1 + (T'_{2A})^2 (\omega_A - \omega)^2}$$
(see Problem 18). This result shows that the exchange leads to a broadened resonance centered at ω_A with width described by the parameter
$$\frac{1}{T'_{2A}} = \frac{1}{T_{2A}} + \frac{1}{\tau_A}.$$
If T_{2A} is known, measurements of the width of the exchange-broadened resonance allow τ_A to be estimated.

In the **rapid exchange limit** the lifetimes τ_A and τ_B are sufficiently small that terms quadratic in τ_A and τ_B can be neglected. Then,
$$\mathfrak{M} = -i\gamma B_1 M_0 \frac{\tau_A + \tau_B}{\alpha_A \tau_A + \alpha_B \tau_B} = -\frac{i\gamma B_1 M_0}{p_A \alpha_A + p_B \alpha_B}.$$
The imaginary part is
$$\tilde{M}_y = -\gamma B_1 M_0 \frac{T'_2}{1 + (T'_2)^2 (p_A \omega_A + p_B \omega_B - \omega)^2}$$
(see Problem 19).

Therefore, a single resonance occurs centered at an average frequency $\bar{\omega}$ where
$$\bar{\omega} = p_A \omega_A + p_B \omega_B$$
and with a width described by the parameter
$$\frac{1}{T'_2} = \frac{p_A}{T_{2A}} + \frac{p_B}{T_{2B}}.$$
In the region between the two limiting cases
$$\tilde{M}_y = -\gamma B_1 M_0 \frac{\tau(\omega_A - \omega_B)^2}{[(\omega_A + \omega_B) - 2\omega]^2 + 4\tau^2(\omega_A - \omega)^2(\omega_B - \omega)^2}$$

assuming equal populations, equal lifetimes, and large spin-spin relaxation times T_{2A}, T_{2B} (see Problem 20). Using this formula, we show the magnetic resonance spectrum as a function of the exchange rate in Fig. 16.12. The two peaks coalesce into one broad resonance with a maximum at the average position for $\tau\Delta\omega \leq (2)^{1/2}$.

16.6 CHEMICAL SHIFTS AND ELECTRON-COUPLED SPIN INTERACTIONS

The magnetic resonance spectrum for a particular nucleus, for a liquid sample in a very homogeneous magnetic field, in general consists of several closely spaced resonances. The interpretation of such spectra provides the chemist and the biologist[6,7] with an invaluable tool for the structural analysis of organic molecules.

Chemical shift

Undoubtedly, the most important parameter for structural studies is the **chemical shift**. For each chemically distinct position of a particular type of nucleus in a molecule, the resonance frequency will be somewhat different. This displacement of the resonance for different chemical environments is referred to as a chemical shift.

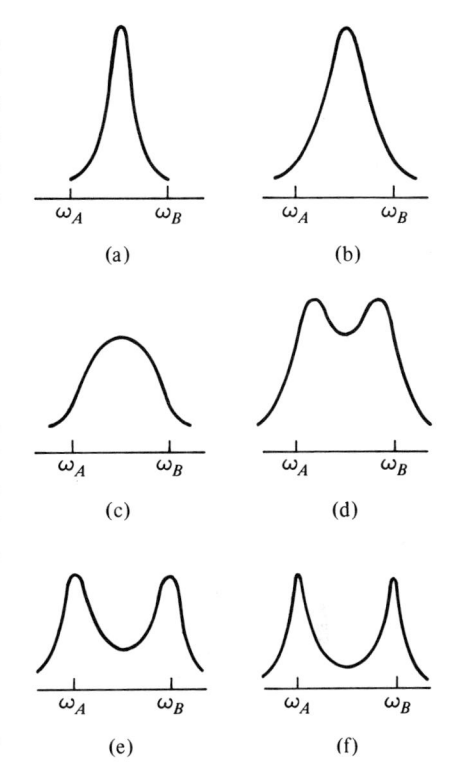

Fig. 16.12. The magnetic resonance spectrum as a function of exchange rate: (a) $\tau\Delta\omega = 0.5$, (b) $\tau\Delta\omega = 1$, (c) $\tau\Delta\omega = (2)^{1/2}$, (d) $\tau\Delta\omega = 2$, (e) $\tau\Delta\omega = 4$, (f) $\tau\Delta\omega = 10$.

A molecule placed in a magnetic field acquires a diamagnetic moment due to the induced orbital motion of its electrons (see Section 13.3). The moving electrons constitute a current which in turn sets up a secondary magnetic field at each nuclear site. The local magnetic field as a nuclear site may be written as

$$B_{\text{loc}} = B_0(1 - \sigma)$$

where B_0 is the external magnetic field and $-B_0\sigma$ is the secondary magnetic field due to the motion of the electrons. Note that the magnitude of the secondary field is proportional to the strength of the external field and that it acts to oppose the external field. The constant σ, which depends

---

[6]R. C. Ferguson and W. D. Phillips, "High Resolution Nuclear Magnetic Resonance Spectroscopy," *Science*, **157** (July 1967), 257.

[7]K. Wüthrich and R. G. Shulman, "Magnetic Resonance in Biology," *Physics Today*, April, 1970.

on the chemical environment, is known as the **screening constant**; for protons σ is about 10^{-5}. It is customary to express chemical shifts in terms of the dimensionless unit δ defined by

$$\delta = \frac{B - B_r}{B}$$

where B is the resonant field at fixed frequency for nuclei in a particular chemical environment, and B_r is the corresponding resonant field for the same type of nuclei in a reference environment. The proton chemical shifts of some simple gases are listed in Table 16.1. Note that methane (CH_4) has been taken as the reference substance.

Table 16.1 Proton chemical shifts of some simple gases

Gas	δ (parts per million)
Hydrogen (H_2)	− 4.2
Water (H_2O)	− 0.60
Methane (CH_4)	0
Ammonia (NH_3)	0.05
Hydrogen chloride (HCl)	0.45
Hydrogen bromide (HBr)	4.35
Hydrogen iodide (HI)	13.25

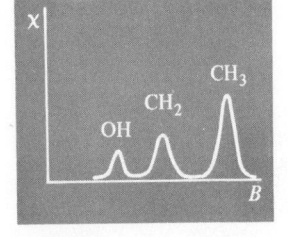

Fig. 16.13. The proton magnetic resonance spectrum of ethyl alcohol.

Studies of chemical shifts provide important clues to molecular structure. As a simple example we consider ethyl alcohol (CH_3CH_2OH). The proton magnetic resonance spectrum is shown in Fig. 16.13. In this molecule the methyl (CH_3), methylene (CH_2), and hydroxyl (CH) protons experience a different screening constant; that is, they are **inequivalent**. Since the intensities of the three resonances are approximately in the ratio 3:2:1, the identification of the different types of protons is immediate.

Electron-coupled spin interaction

A scalar interaction of the form

$$J_{ij}\mathbf{I}_i \cdot \mathbf{I}_j$$

can also occur between neighboring nuclei i and j in a molecule. The parameter J_{ij} is called the **spin-coupling constant**. This interaction is independent of the magnetic field. The interaction arises from an indirect coupling of the nuclear dipole moments via the electrons in the chemical bonds in the molecule. The interaction of one nucleus with a valence electron of its atom will preferentially result in an antiparallel arrangement of the nuclear and electronic spins. The two electron spins in a covalent

SEC. 16.6 CHEMICAL SHIFTS AND ELECTRON-COUPLED SPIN INTERACTIONS

bond must be antiparallel to each other. Therefore, it follows that the two nuclear spins will be preferentially antiparallel to each other.

As an example of the effect of an **electron-coupled spin interaction**, we consider the hydrogen deuteride (HD) molecule. In fact, the proton resonance spectrum will consist of three components as a result of this interaction. The deuteron has spin $I = 1$ which gives three possible spin states. In other words, because of the electron-coupled spin interaction, the proton spin can sense the three different orientations of the deuteron spin. Similarly, the deuteron resonance spectrum will consist of two components as a result of this interaction. The magnetic resonance spectra are shown in Fig. 16.14. From the measured separation of the component signals the spin-coupling constant J_{HD} has been deduced to be 43.5 Hz.

The method of solution of the general problem of analyzing a magnetic resonance spectrum in which both chemical shifts and electron-coupled spin interactions occur depends on the relative strengths of the interactions. If the chemical shift differences are large compared to the electron-coupled spin splittings, we can treat the latter adequately as a small correction to the spectrum due to chemical shifts. Let us return to the example of ethyl alcohol (CH_3CH_2OH). We noted that three peaks are expected on the basis of chemical-shift differences. The methyl protons interact with the methylene protons through the electron-spin coupled interaction. By its interaction with the two equivalent methylene protons, the methyl resonance is split into three components, with intensity ratios 1:2:1. That is, a given methyl proton may see both methylene protons with spins parallel to its own spin, may see both methylene protons with spins antiparallel to its own spin, or may see one methylene proton with spin parallel and one methylene proton with spin antiparallel to its own spin. Since in the latter case either methylene proton may be parallel to the methyl proton, the corresponding component will have twice the intensity of the other two components. Similarly, the methylene resonance is split into four components, with intensity ratios 1:3:3:1 by the coupling between methyl and methylene protons. Note that we have not considered any coupling to hydroxyl protons. In slightly acidified samples the hydroxyl protons undergo rapid chemical exchange with protons on H_2O molecules which effectively decouples the OH protons from the other protons in the molecule. The complete proton magnetic resonance spectrum of CH_3CH_2OH is given in Fig. 16.15.

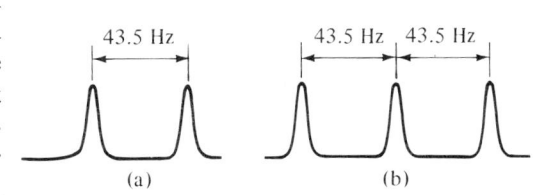

Fig. 16.14. (a) Deuteron magnetic resonance spectrum in HD. (b) Proton magnetic resonance spectrum in HD.

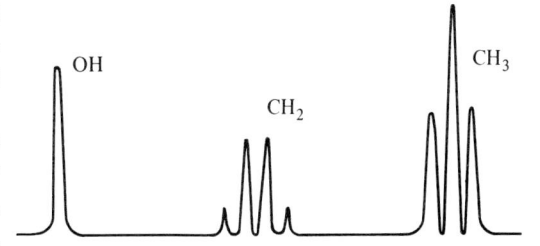

Fig. 16.15. The proton magnetic resonance spectrum of CH_3CH_2OH resulting from chemical shifts and electron-coupled spin interactions.

16.7 NUCLEAR MAGNETIC DIPOLAR INTERACTIONS IN SOLIDS

In solids the effects discussed in the last section are masked by the presence of the **dipole-dipole interaction** between neighboring nuclei. The magnitude of the magnetic field surrounding a nuclear dipole moment $\boldsymbol{\mu}$ is approximately $10^{-7}\,\mu/r^3$ where r is the distance from the dipole (see Section 7.5). For $r \cong 0.1$ nm, $10^{-7}\,\mu/r^3 \simeq 5 \times 10^{-4}$ Wb·m^{-2}. The energy of interaction of two nuclear dipole moments to a good approximation varies with the angle θ between the direction of the applied magnetic field and the vector joining the two dipoles as $(3\cos^2\theta - 1)$. Therefore, the effect of the dipole-dipole interaction between pairs of nuclei in a solid is to produce a broad resonance absorption. In a liquid sample the rapid and random motions of the molecules cause this interaction to average to zero.

The dipole-dipole interaction consists of two contributions. In Fig. 16.16 we picture two interacting nuclear dipole moments in an external magnetic field $B_0\mathbf{k}$. The magnetic moment $\boldsymbol{\mu}_1$ has been resolved into component vectors $\boldsymbol{\mu}_{1z}$ and $\boldsymbol{\mu}_{1xy}$ which are parallel and perpendicular, respectively, to the external field. The component $\boldsymbol{\mu}_{1xy}$ processes at its resonance frequency about $B_0\,\mathbf{k}$. Now, $\boldsymbol{\mu}_1$ sets up a magnetic field at the site of nucleus 2; this field consists of a static component due to $\boldsymbol{\mu}_{1z}$ and a time-varying component due to $\boldsymbol{\mu}_{1xy}$. The static component interacts with $\boldsymbol{\mu}_2$ independent of the type of nucleus concerned. The dynamic component will give rise to a resonant interaction with $\boldsymbol{\mu}_2$ if the two nuclei are alike; otherwise it will have a negligible effect.

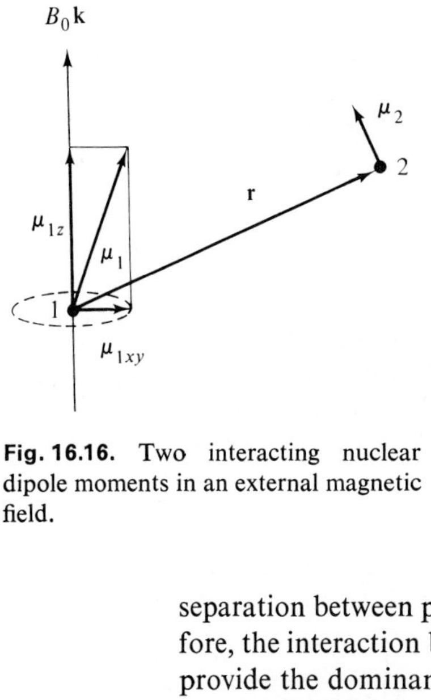

Fig. 16.16. Two interacting nuclear dipole moments in an external magnetic field.

As a simple example of the effect of the dipolar interaction we consider the proton magnetic resonance spectrum in a single crystal of gypsum (CaSO$_2$·2H$_2$O). The separation of the two protons in each water molecule is 0.16 nm, while the closest separation between protons on different water molecules is 0.30 nm. Therefore, the interaction between the protons in individual water molecules will provide the dominant contribution to the spectrum. The direction of the dipolar field $\mathbf{B}_d$ can be either parallel or antiparallel to the direction of the external field so that, for each direction of the proton-proton axis in the crystal, resonance absorption occurs at two frequencies given by[8]

$$\omega = \gamma(B_0 \pm B_d)$$

---

[8] Note that $\mathbf{B}_d$ is an effective magnetic field associated with this particular example; it is *not* the field due to an isolated magnetic dipole.

where

$$B_d = \frac{3}{2}\left(\frac{\mu_0}{2\pi}\right)\left(\frac{\mu}{r^3}\right)(3\cos^2\theta - 1),$$

the numerical factor $\frac{3}{2}$ resulting from a consideration of both static and dynamic effects. In fact, two orientations of the proton-proton axis occur in a gypsum crystal so that the spectrum may be expected to consist of two doublets. In Fig. 16.17 we show the observed spectrum for three different orientations of the crystal in the applied field. The angles are measured between the applied field and the normal to the (100) planes[9] in the crystal. For $\theta = 54°$ the two doublets are clearly distinguishable; for $\theta = 18°$ and $90°$ the separation of the two doublets is about the same and they coalesce to give a single doublet. Note that the spectrum is quite broad; this is due to the contributions from other protons in the crystal. From this spectrum the proton-proton separation can be determined.

As the temperature of a solid is increased various internal motions within the crystal begin to occur. The effect of these motions is to partially average out the dipolar interaction and to cause a narrowing of the observed resonance. This phenomenon is called **motional narrowing**. It is customary to discuss this type of alteration in the spectrum in terms of a quantity called the **experimental second moment** of the resonance. The second moment M_2 is defined by

$$M_2 = \int (\omega - \omega_0)^2 f(\omega - \omega_0) d(\omega - \omega_0)$$

where $f(\omega - \omega_0)$ is the line shape function and ω_0 is the frequency of maximum response. Two contributions to the second moment can be identified—an intramolecular contribution M_2'' due to nuclei within a single molecule and an intermolecular contribution M_2' due to nuclei on different molecules. The two parts can be separated experimentally by studying deuterated molecular forms (see Problem 29). In Fig. 16.18 we show the temperature dependence of the intramolecular contribution M_2'' for the proton resonance in a polycrystalline sample of benzene (C_6H_6). As we

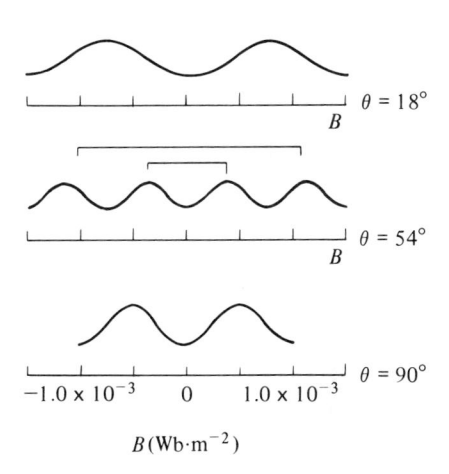

Fig. 16.17. The proton magnetic resonance spectrum of a single crystal of gypsum for three different orientations of the crystal in the applied field.

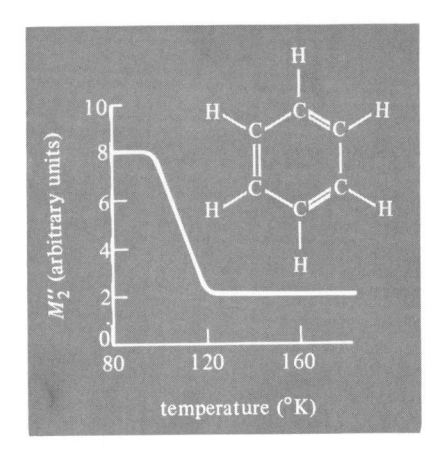

Fig. 16.18. The temperature dependence of the intramolecular contribution to the second moment of the proton resonance in benzene.

[9] *MWTP*, Section 21.4.

shall see, the rapid change in M_2'' occurring at about 100°K is due to the onset of rotation of the benzene molecules about their six-fold symmetry axes.

Theoretically, the intramolecular part $(M_2'')_{RL}$ of the second moment at low temperatures can be written

$$(M_2'')_{RL} = g(R)\overline{(1 - 3\cos^2\theta_{jk})^2}$$

where $g(R)$ is a function of the nearest neighbor proton-proton separation, the bar indicates an average over the random orientations of the crystallites (that is, small crystals) in the sample, and the subscript RL stands for rigid lattice and implies that the molecules are not rotating. When rotational motion begins, an average must first be taken over the rotational motion to obtain the contribution for a given crystallite orientation and then an average over crystallite orientations. That is,

$$(M_2'')_{\text{rot}} = g(R)\overline{(1 - 3\cos^2\theta_{jk})_{\text{AVG}}^2}$$

where AVG indicates an average over the rotational motion and the subscript rot implies that the molecules are rotating. By a quite general theorem

$$(1 - 3\cos^2\theta_{jk})_{AVG} = (1 - 3\cos^2\theta')\frac{(3\cos^2\gamma_{jk} - 1)}{2}$$

where the angles θ_{jk}, θ', and γ_{jk} are as shown in Fig. 16.19. This result applies to the rotation of any molecule about an axis of three-fold or higher symmetry independent of the details of the motion. Since the crystal axes are randomly oriented with respect to the external field, so are the rotation axes. Therefore

$$\overline{(1 - 3\cos^2\theta')^2} = \overline{(1 - 3\cos^2\theta_{jk})^2}$$

and

$$(M_2')_{\text{rot}} = (M_2'')_{RL}\left(\frac{3\cos^2\theta_{jk} - 1}{2}\right)^2.$$

Since γ_{jk} is the angle between the vector joining two nuclei in the molecule and the axis of rotation within the molecule, γ_{jk} is a constant independent of the rotation. For rotation about the six-fold axis of benzene, $\gamma_{jk} = \pi/2$ so that

$$(M_2'')_{\text{rot}} = \frac{(M_2'')_{RL}}{4}.$$

This is what is shown in the experimental results given in Fig. 16.18.

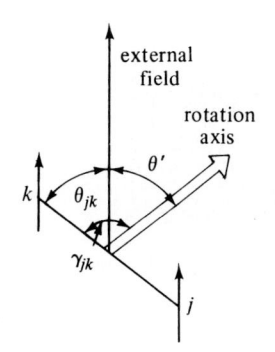

Fig. 16.19. Relevant angles for the description of the rotation of a molecule.

16.8 NUCLEAR SPIN-LATTICE RELAXATION

The first attempt to understand nuclear spin-lattice relaxation considered, as a relaxation mechanism, the modulation of the dipolar interaction resulting from motions of the lattice. We saw in Section 16.3 that the spin-lattice relaxation time is essentially the reciprocal of the probability per second that a spin will make a transition under the influence of the relaxation mechanism.

As a specific example we consider proton relaxation in hydrogen gas due to the intramolecular dipolar interaction rendered time dependent by the molecular collisions. Each proton in the molecule experiences a magnetic field due to the presence of the other proton. The frequency and phase of this magnetic field varies randomly with time due to the random nature of the collisions resulting from the thermal motions of the molecules. The probability that a spin will make a transition depends on the power available in the random local magnetic field at the resonance frequency for the spin; the total power available depends on the average of the square of the intramolecular dipolar field strength. A plot of the power available in the random local field as a function of frequency is known as the **power spectrum** of the interaction. In Fig. 16.20 we show schematically the power spectrum $J(\omega)$ of the intramolecular dipolar field for three gas densities ρ_1, ρ_2, and ρ_3 at a constant temperature. The area of the spectral density curve $\int J(\omega)\, d\omega$ depends only on the strength of the intramolecular dipolar field and is independent of gas density. For a particular gas density the power available in the random field will drop off at a frequency of the order of the mean collision frequency. Since the collision rate increases with the density for a fixed temperature, it follows that for the curves shown in Fig. 16.20 $\rho_1 > \rho_2 > \rho_3$. Let us suppose that the resonance frequency ω_0 is located as indicated in Fig. 16.20. This depicts the case if $\omega_0 \simeq 30$ MHz and the lowest density corresponds to a pressure of $\simeq 3$ atm at 300°K. Since

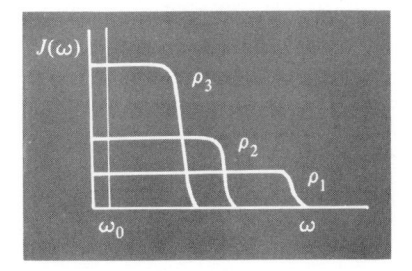

Fig. 16.20. The power spectrum of the intramolecular dipolar field in a sample of hydrogen gas for three densities at a constant temperature.

$$\frac{1}{T_1} \simeq J(\omega_0),$$

and since the mean collision frequency is proportional to the density for moderate densities, it follows that for the density range shown in Fig. 16.20, $J(\omega_0) \propto 1/\rho$ so that $T_1 \propto \rho$.

It is interesting to explore what happens as the density is further reduced. In Fig. 16.21 we give the power spectrum for densities $\rho_3 > \rho_4 > \rho_5$. The lowest density now corresponds to a pressure $\simeq 0.1$ atm at 300°K. For $\rho = \rho_4$ the resonance frequency of the nuclear spin system is roughly equal to the mean collision frequency and there is maximum transfer of energy between the spin system and the lattice; that is, T_1 goes through a minimum for $\rho = \rho_4$. For $\rho = \rho_5$ the coupling between the spin system

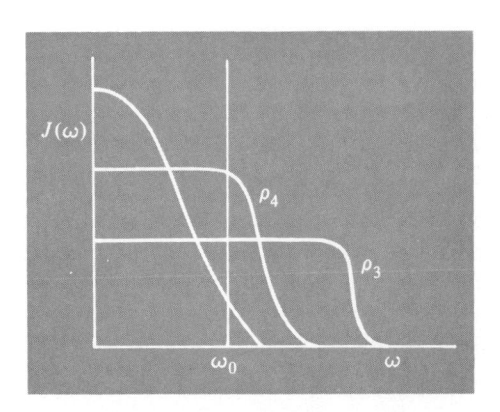

Fig. 16.21. The power spectrum as in Fig. 16.18 but for lower gas densities. Note that the frequency scale has been expanded.

and the lattice is again weaker and T_1 longer. The variation of T_1 with density for the five densities shown in Fig. 16.20 and 16.21 is shown in Fig. 16.22. Studies of T_1 in gases in the vicinity of the minimum provide important information concerning both the interactions within the free molecule and its interaction with other molecules.

We should note that other interaction mechanisms can also be important. For example, in hydrogen gas the **spin-rotation interaction mechanism** must also be considered. As a result of molecular rotation, a magnetic field is set up at the proton sites. This field is modulated by collisions and provides a second relaxation mechanism. In liquid hydrogen the average separation between molecules is greatly reduced and intermolecular magnetic fields become important. If the protons are replaced by deuterons that possess nuclear quadrupole moments (see Section 3.9), a time-dependent quadrupolar mechanism results.

16.9 ELECTRON PARAMAGNETIC RESONANCE IN SOLIDS

The general theory discussed in Sections 16.2, 16.3, and 16.4 is applicable to electron paramagnetic resonance (EPR) as well as to nuclear magnetic resonance (NMR). In practice, the experimental techniques used and the results obtained are quite different. In this section we will discuss only three isolated topics of particular significance for EPR.

Exchange narrowing

The width of an EPR line in a paramagnet with an exchange interaction among nearest neighbor electron spins is usually much narrower than expected on the basis of the dipolar interaction alone. This effect is called **exchange narrowing**. In Fig. 16.23 we illustrate the effect on an EPR line shape of an exchange interaction in addition to the usual dipolar interaction. The form of the exchange interaction between neighboring spins $\mathbf{S}_i$ and $\mathbf{S}_j$ is

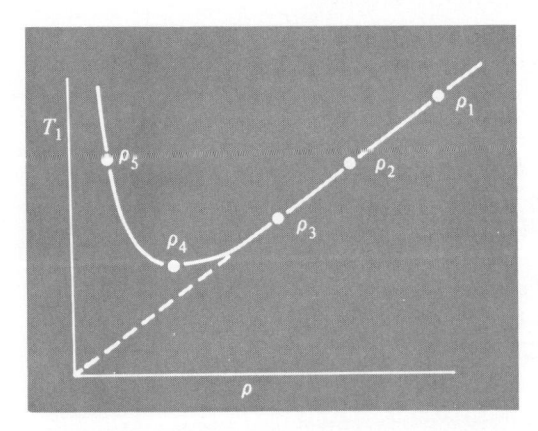

$$2J\mathbf{S}_i \cdot \mathbf{S}_j$$

Fig. 16.22. The variation of T_1 with ρ in hydrogen gas at 300°K.

where J is a measure of the strength of the interaction. Physically, this term may be thought of as flipping neighboring oppositely oriented spins at a rate J/h, thereby producing an effective migration of spin orientation through the lattice.

Crystal field theory

Crystal field theory is based upon the assumption that the paramagnetic ions reside in a potential whose sources are point charges (ions) or point dipoles (water molecules) lying entirely outside the paramagnetic ion. In order to obtain the energy states of the paramagnetic ion, we must add this potential to the other interactions. The potential must have the symmetry of the array of its sources and, indeed, the symmetry manifests itself by fixing the properties of the energy states.

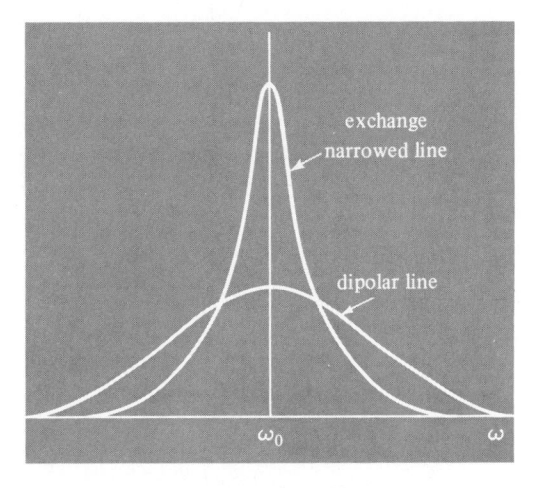

Fig. 16.23. Comparison of resonance line shapes with and without an exchange interaction.

The expression for the electrons of a paramagnetic ion in a crystal contains the following contributions:

1. kinetic energy of electrons;
2. Coulomb attraction between electrons and nuclei;
3. Coulomb repulsion between electrons;
4. spin angular momentum-orbital angular momentum coupling interaction;
5. crystal field interaction.

The relative magnitudes of the contributions (4) and (5) determine the manner in which the calculation is performed. Two distinct cases for which crystal field theory applies may be identified.

(a) Weak crystal field: (3) > (4) > (5)

This situation typifies the rare-earth ions (see Section 13.4) in ionic crystals. The paramagnetic electrons lie deep within the ion and are well shielded from the electric field of the crystal. We first neglect term (5) and solve the Schroedinger equation for the energy states of an isolated ion. This is the basic problem of atomic physics! The crystal field interaction is then added and it causes small perturbations of the free ion energy states.

(b) Medium crystal field: (3) > (5) > (4)

This situation typifies the iron-group ions (see Section 13.4) in ionic crystals. The electrons responsible for the paramagnetism are near the outer regions of the ion and experience stronger crystal field interactions than their own spin-orbit coupling interactions. We now begin with the energy states as they would be for an ion if no spin-orbit interaction existed. Next, the effect of the electric field of the crystal is considered. The effect of the crystal field is to quench the orbital angular momentum in the iron-group ions so that they exhibit spin-only properties. In fact, because of this interaction, magnetic resonance experiments in such systems are often referred to as **electron spin resonance** (ESR) experiments. Finally, the spin-orbit interaction is included; it has the effect of reinstating a small amount of the orbital angular momentum.

As an example of the medium crystal field case we consider an ion with a single paramagnetic electron in an energy state having $l = 2$. Since the quantum number m describing the electron can have values 2, 1, 0, -1, -2, the energy state has five-fold orbital degeneracy. We now enquire about the effect of a crystal electric field resulting from the six negative charges shown in Fig. 16.24. Clearly, the charge distribution for the ion must reflect the symmetry of the octahedral array of charges; the possible distributions or **orbitals** are shown in Fig. 16.25. It should be obvious that the orbitals (a), (b), and (c), called the $d\epsilon$ orbitals, all have the same electrostatic interaction energy with the array of charges. Although it is not so obvious, the orbitals (d) and (e), called the $d\gamma$ orbitals, have the same electrostatic interaction energy with the array of charges. As a consequence of the extension of the regions of high-charge density directly toward the negative charges for the $d\gamma$ orbitals, they lie higher in energy than the $d\epsilon$ orbitals. The effect of the octahedral crystal field is to par-

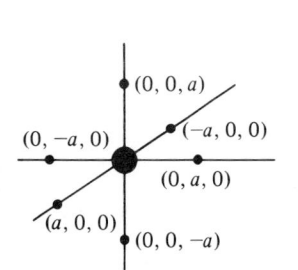

Fig. 16.24. An ion surrounded by an array of six negative charges.

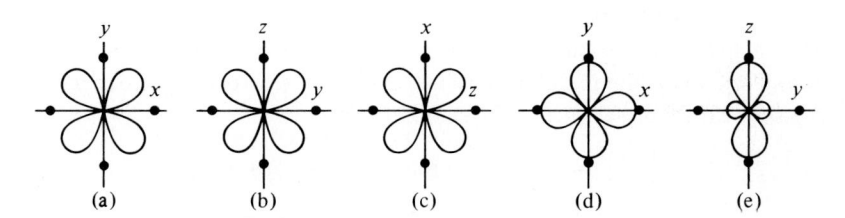

Fig. 16.25. Allowed charge distributions for the ion in the crystal electric field.

tially lift the five-fold degeneracy resulting in two energy states separated by an amount Δ (see Fig. 16.26). If we now consider a small symmetrical displacement of the charges on the z axis (a tetragonal distortion), a further lifting of the degeneracy results (see Fig. 16.26 and Problem 32). The $d\epsilon$ orbital is split into two energy states separated by an amount δ.

If the two possible orientations of the electron spin are taken into account, the degeneracy of each orbital energy level is doubled; that is, the ground-state level in the presence of the tetragonal distortion (Fig.

16.26) is a doublet. This degeneracy can be removed by the presence of a magnetic field and an EPR experiment can be performed using the resulting two energy levels. The predicted EPR spectrum would show spin-only properties and would be characterized by the electron spin g-factor, g_s.

Finally, the spin-orbit coupling interaction is considered. If the magnetic field is taken to be in the z direction, that is, along the tetragonal axis, the EPR experiment is characterized by the following g-factors:

$$g_z = g_s - 8\left(\frac{\lambda}{\Delta}\right) = g_\parallel$$

$$g_x = g_y = g_s - 2\left(\frac{\lambda}{\Delta}\right) = g_\perp$$

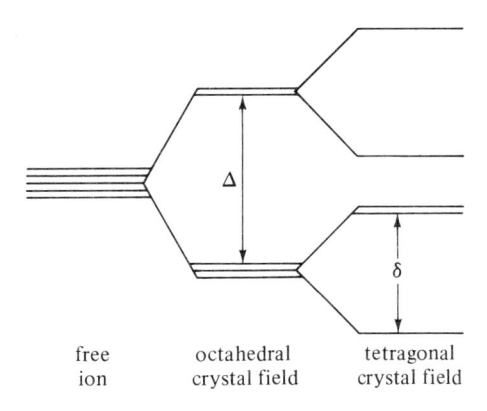

Fig. 16.26. The removal of orbital degeneracy by a crystal electric field.

where λ is the spin-orbit coupling constant and $\lambda/\Delta \ll 1$. The effect of the crystal electric field is to quench the orbital angular momentum; the effect of spin-orbit coupling is to reinstate a small amount of orbital angular momentum of order (λ/Δ).

Quadrupolar spin-lattice relaxation

The modulation of the crystal electric field by the lattice vibrations provides the dominant spin-lattice interaction for paramagnetic ions in crystals. Three principal processes have been identified by means of which the relaxation may proceed. For the **direct process**, the electron-spin transition is accompanied by the absorption or emission of a single quantum of lattice vibrational energy (called a **phonon**). The probability of such a process is proportional to the number of phonons in the lattice at the resonant frequency; therefore, for the direct process,

$$\frac{1}{T_1} \propto T$$

where T is the temperature of the lattice. The **Raman process** occurs as the result of the interaction of a phonon with an electron in which a change in the energy between the incoming and outgoing phonon is just sufficient to compensate for the change in energy associated with the electron-spin transition. At low temperatures the temperature dependence of T_1 is governed by the nature of the distribution of lattice phonons. For a Debye distribution[10]

$$\frac{1}{T_1} \propto T^7 \quad \text{or} \quad T^9.$$

[10] *MWTP*, Section 22.5.

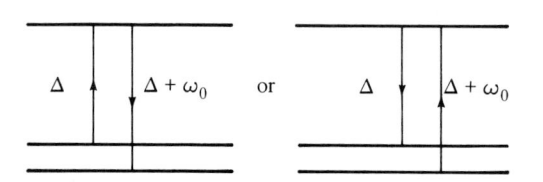

Fig. 16.27. An energy-level diagram for the lattice showing the energy states that participate in an Orbach process.

Finally, there is the **Orbach process** which is a two-stage process in which a phonon is actually absorbed at one frequency and emitted at another (see Fig. 16.27). This process is fundamentally different from the Raman process and gives rise to a temperature variation of T_1 of the form

$$\frac{1}{T_1} \propto \exp\left(-\frac{\Delta'}{kT}\right)$$

where Δ' is the energy of the absorbed phonon.

16.10 NUCLEAR QUADRUPOLE RESONANCE

Nuclei having spin angular momentum equal to or greater than $\hbar$ possess a nuclear quadrupole moment (see Section 3.9). The scalar electric quadrupole moment of a nucleus is defined as

$$eQ = \int_v \rho(3z'^2 - r^2)\,dv$$

where ρ is the density of charge within volume element dv at distance r from the origin and the z' direction is that of the spin axis of the nucleus. In the presence of a local electric field gradient the quadrupole moment experiences a torque. If the field gradient is axially symmetric, the axis of nuclear spin precesses about the symmetry axis of the field gradient. With reference to a system of principal axes the field gradient is defined by two quantities: (1) the **field gradient parameter** $q = q_{zz}$ is the maximum value of the field gradient; (2) **the asymmetry parameter**, defined by

$$\eta = \frac{q_{yy} - q_{xx}}{q_{zz}},$$

is a measure of the departure from axial symmetry of the electric field gradient. The components q_{ii} of the field gradient in terms of the electrostatic potential V are

$$q_{ii} = -\frac{\partial^2 V}{\partial i^2}, \qquad i = x, y, z$$

where the x, y, and z directions are labeled such that

$$|q_{zz}| > |q_{yy}| > |q_{xx}|.$$

The energy states for a nuclear quadrupole resonance experiment are found by solving the Schroedinger equation which describes the interaction of the nuclear quadrupole moment with the local electric field gradient. If the field gradient has axial symmetry, the energy states are

$$E_Q = \frac{eqQ}{4I(2I-1)}[3M^2 - I(I+1)]$$

where
$$M = I, I-1, \ldots, 0, \ldots, -I,$$
$I\hbar$ is the nuclear spin angular momentum, and eqQ is the **quadrupole coupling constant**. We note that the energy states are doubly degenerate with the exception of the state $M = 0$. Each state corresponds to an allowed orientation of the quadrupole in the field gradient. The energy states for a nucleus of spin $\tfrac{3}{2}$ in an axially symmetric electric field gradient are shown in Fig. 16.28(a) and for a nucleus of spin $\tfrac{5}{2}$ in Fig. 16.28(b).

Transitions between the energy states can be induced by applying a radio-frequency magnetic field at an appropriate frequency. Such a magnetic field interacts with the nuclear magnetic dipole moment to induce the transition. It is therefore somewhat unfortunate that experiments of this kind have come to be known as **nuclear quadrupole resonance** (NQR) experiments. Many different types of information are available from NQR experiments; we now discuss a number of examples.

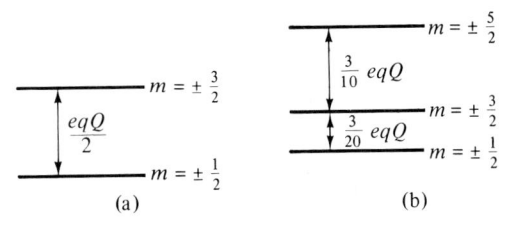

Fig. 16.28. Energy states for nuclei in axially-symmetric electric field gradients: (a) nuclear spin $\tfrac{3}{2}$, (b) nuclear spin $\tfrac{5}{2}$.

The ratio of nuclear quadrupole moments

Some nuclei have more than one isotope possessing a nuclear quadrupole moment, as for example chlorine-37 and chlorine-35. In a given solid the field gradient at a nuclear site is, to a good approximation, independent of the isotope. Therefore, a measurement of the ratio of NQR frequencies for two isotopes gives an accurate value of the ratio of nuclear quadrupole moments. For example, for chlorine,
$$\frac{Q(35)}{Q(37)} = 1.2688.$$

Quadrupole coupling constants and asymmetry parameters.

When the electric field gradient is not axially symmetric, the NQR frequencies depend on both the quadrupole coupling constant eqQ and the asymmetry parameter η. For a nucleus of spin $\tfrac{3}{2}$, there is a single resonance line of frequency
$$\omega = \frac{eqQ}{2\hbar}\left(1 + \frac{\eta^2}{3}\right)^{1/2}.$$

Therefore, we cannot obtain eqQ and η separately for such a nucleus from a NQR measurement. For a nucleus of spin $\tfrac{5}{2}$, there are two resonance lines whose frequencies ω' and ω'' are in the ratio

$$\frac{\omega'}{\omega''} = 2\left(1 - \frac{35}{27}\eta^2\right).$$

The ratio of resonance frequencies gives η and then either frequency will yield eqQ. A similar situation applies to nuclei with other values of spin. Note that we cannot obtain a direct measurement of the nuclear quadrupole moment from a NQR experiment.

Molecular structure

Since the NQR frequency is proportional to the field gradient parameter at the nuclear site, each inequivalent site in the unit cell of a crystalline solid will give rise to a separate resonance. The NQR spectrum of a substance therefore provides some information concerning crystal structure.

For example, let us consider the chlorine resonance in potassium hexachloroplatinate (K_2PtCl_6); the crystal structure is shown in Fig. 16.29. There are six chlorine nuclei per $PtCl_6$ molecule and four such molecules per unit cell. Each $PtCl_6$ molecule is located at the center of a cube defined by eight potassium atoms; the Pt-Cl axes are parallel to the cube faces. A single chlorine resonance line is observed indicating that all the chlorine nuclei in the unit cell occupy equivalent sites. In contrast, we quote the case of antimony trichloride ($SbCl_3$). The chlorine spectrum consists of two triplets. Each triplet results from a distorted pyramidal $SbCl_3$ molecule; the distortion causes the chlorine nuclear sites to be inequivalent. The presence of two chlorine triplets indicates that there are two inequivalent $SbCl_3$ molecules per unit cell.

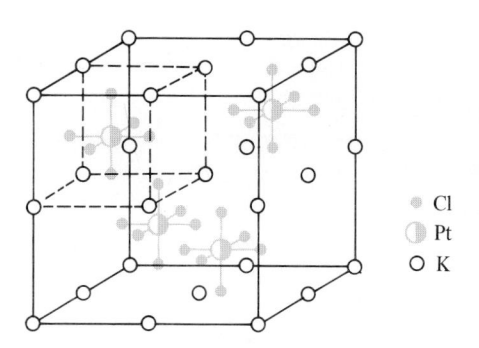

Fig. 16.29. A unit cell of potassium hexachloroplatinate (K_2PtCl_6).

Nature of chemical bonds

The electric field gradient at a nuclear site is determined by the distribution of charge within the vicinity of the nucleus. For a nucleus that forms a part of a tightly bonded molecule or ion within a solid, the electric field gradient is due predominantly to the charge distribution within the chemical bonds of the molecule or ion. Often we write

$$q_{\text{solid}} \simeq q_{\text{molecule}} \simeq f q_{\text{atom}}$$

where q_{solid}, the field gradient at the nuclear site within the solid, is taken to be approximately equal to q_{molecule}, the field gradient at the nuclear site within the free molecule or ion. The quantity q_{atom} is the field gradient at the nuclear site in a free atom. As a result of the redistribution of charge during the formation of the molecule or ion, the field gradient is modified; this alteration in the field gradient is described by the factor f. For example, let us consider the metal-ligand bonds in potassium hexachloroplatinate. The $PtCl_6$ ions form tightly bonded molecular units within the solid. More than 90% of the electric field gradient at a chlorine site comes from the charge distribution within the $PtCl_6$ ion. Therefore, $q_{solid} \simeq q_{molecule}$. In the formation of a $PtCl_6$ molecule charge migrates from the ligands to the central platinum atom. If a pure covalent bond were formed, there would be no charge migration and $q_{molecule} \simeq q_{atom}$; on the other hand, if a pure ionic bond were formed, the resulting chlorine ion would have a closed shell configuration and $q_{molecule} \simeq 0$. That is, $0 < f < 1$ as determined by the covalent character of the metal-ligand bond. In fact, an analysis of the NQR data suggests that the degree of covalency of the Pt—Cl bond is $\simeq 56\%$.

Structural phase transitions

NQR provides a simple and extremely sensitive probe for the detection of structural phase transitions. In Fig. 16.30 we see the temperature variation of the NQR frequencies of the bromine-79 nucleus in potassium hexabromoselenate (K_2SeBr_6). Above 240°K a single resonance line is observed, indicating that all chlorine nuclei are at equivalent sites. Below 240°K two resonance lines appear, the intensity of the one at the lower frequency being twice that of the other line. This indicates that a change in the structure to one of lower symmetry has occurred. It is suggested that the cubic cage defined by the potassium atoms in Fig. 16.28 becomes slightly distorted in one direction (the z direction) and that the $SeBr_6$ ions rotate through a small angle about their z axes (see Fig. 16.31). The bromine nuclei in the xy plane are no longer equivalent to those on the z axis. A second phase transition occurs at 221°K resulting in three resonance lines of equal intensity. This indicates that another change in structure has occurred to a structure of still lower symmetry. It is interesting to note that one of the lines is insensitive to this transition: that is, the one line exhibits no discontinuity of frequency or intensity at the transition point. The atoms responsible for this line suffer no sudden change in their local field gradient at this temperature (that is, during this phase transition).

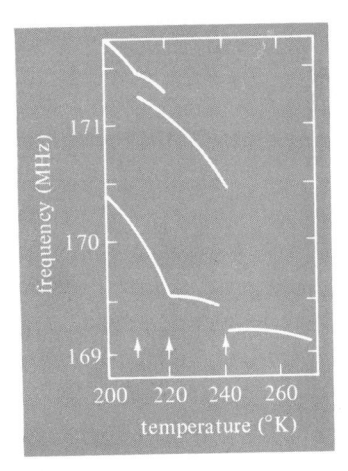

Fig. 16.30. The temperature variation of the bromine-79 NQR frequency in potassium hexabromoselenate.

16.11 THE PROTON MAGNETOMETER

The phenomenon of free precession discussed in Section 16.4 forms the basis of an accurate magnetometer for the measurement of the earth's magnetic field. Such instruments have found many applications including the measurement of magnetic fields in space from rockets[11] and as an aid on archaeological expeditions in search of artifacts from ancient civilizations.[12]

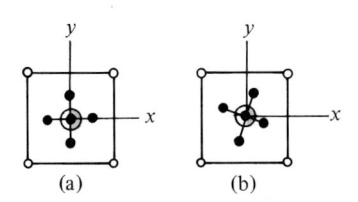

Fig. 16.31. (a) structure of K_2SeBr_6 above 240°K. (b) structure of K_2SeBr_6 below 240°K.

Protons are chosen as the nuclei to use for the magnetometer, primarily because of their large magnetogyric ratio; ordinary water provides a convenient sample containing a large number of protons. Typically, a 150 cm³ bottle of distilled water is used and 1000 turns of copper wire are wound around it. When a current of about 1 A is passed through the coil, a magnetic field of several hundred gauss (1 gauss = 10^{-4} Wb·m⁻²) is produced along the axis of the bottle. This polarizing field, oriented perpendicular to the lines of force of the earth's magnetic field, creates a macroscopic magnetization parallel to itself in a time of the order of the spin-lattice relaxation time for the protons, that is, in two or three seconds. Once the sample is polarized the current to the coil is shut off and the magnetization begins to precess about the earth's field at the characteristic Larmor frequency which is about 2000 Hz for a field strength of about 0.5 gauss. The precessing magnetization induces an alternating emf of about a microvolt in the coil. This minute voltage is amplified by conventional electronic methods and coupled to a frequency measuring circuit. Since the nuclear signal decays within a few seconds, the frequency of the protons must be measured within this period of time. In practice, it is possible to measure the frequency to a precision of 1 part in 50,000. The magnetic field strength is, of course, proportional to the frequency and the instrument can be calibrated to indicate the field strength to an accuracy of five figures.

In the proton magnetometer an absolute measurement of field strength is obtained by comparing the proton precession frequency with that of a stable electronic oscillator. When hunting for archaeological remains, one searches for significant differences in magnetic field over distances comparable with the dimensions of the object concerned. One therefore need only compare the proton precession frequencies from two samples carried at either end of a rod about 2 m in length. The two proton signals

---

[11] S. H. Hall, "Measuring Magnetic Fields in Space," *Contemporary Physics*, **8** (September 1967), 447.

[12] M. J. Aitken, *Physics and Archaeology* (New York: Interscience Publishers Ltd., 1961).

can be added together and the beat frequency[13] observed. If the two samples are in identical fields and if the proton signals are arranged to be in phase, the resultant signal has twice the amplitude of the individual signals and decays away in the usual way in about 3 seconds. If the field strengths at the two samples are sufficiently different, beats will occur (see Problem 40). For archaeological surveying it is sufficient to amplify the combined signal and use an earphone or loudspeaker to detect the beats. Such a device, known as a **bleeper**, has been used, for example, in connection with the search for Dead Sea Scrolls.

QUESTIONS AND PROBLEMS

1. How could you deduce the sign of the nuclear magnetogyric ratio from a resonance experiment?

2. An accurate measurement of the magnetogyric ratio of the proton is of fundamental significance. Discuss an experiment to obtain this quantity.[14]

3. A system of noninteracting magnetic moments is in equilibrium in a static magnetic field $B_0\mathbf{k}$. An oscillating radio-frequency magnetic field of amplitude B_1 at the resonance frequency is applied along the x axis for a sufficient time to rotate the magnetization $\mathbf{M}_0$ to the $-\mathbf{k}$ direction. What is the minimum value t_w of the magnetic field pulse? Such a pulse is called a π pulse. Deduce a numerical value for t_w, assuming protons in a static field of 1.0 Wb·m^{-2}, for an oscillating field of amplitude 0.0050 Wb·m^{-2}.

*4. A system of noninteracting magnetic moments is in an inhomogeneous magnetic field. The fraction Δf of magnetic moments experiencing a magnetic field of magnitude between B and $B + \Delta B$ is

$$\Delta f = p(B)\, \Delta B$$

where

$$p(B) = \frac{1}{2a} \quad \text{for} \quad B_0 - a < B < B_0 + a$$
$$= 0 \quad \text{otherwise.}$$

Assuming that the inhomogeneity gives a spread in field with no change in direction, determine the magnetization in the x direction perpendicular to the static field as a function of time if at $t = 0$ the magnetization M_0 is in the x direction.

5. Show that

$$\frac{W_\uparrow}{W_\downarrow} = \exp\left(-\frac{\gamma \hbar B_0}{kT}\right)$$

---

[13] *MWTP*, Section 16.5.
[14] P. Vigoureux, "The Gyromagnetic Ratio of the Proton," *Contemporary Physics*, **2** (June 1961), 360.

where $W_\uparrow$ is the probability per second per spin of a transition to the state $m = -\frac{1}{2}$ due to the spin-lattice interaction and $W_\downarrow$ the probability of a transition to the state $m = +\frac{1}{2}$ (see Section 16.3).

6. Show that the equation

$$\frac{d(n_+ - n_-)}{dt} = 2(n_- W_\downarrow - n_+ W_\uparrow)$$

may be written as

$$\frac{d(\Delta n)}{dt} = \frac{\Delta n_0 - \Delta n}{T_1}$$

where

$$\Delta n = n_+ - n_-, \quad n_0 = n\left(\frac{W_\downarrow - W_\uparrow}{W_\downarrow + W_\uparrow}\right) \quad \text{and} \quad \frac{1}{T_1} = (W_\downarrow + W_\uparrow).$$

7. Starting from the model for the lattice indicated in Fig. 16.8, compute $W_\uparrow$ and $W_\downarrow$ and thereby show that

$$\frac{W_\uparrow}{W_\downarrow} = \frac{n_a}{n_b}$$

where n_a and n_b are the population densities of the lattice states labeled a and b.

8. The absorption of power by a spin system may be calculated from the expression

$$P = -\overline{\mathbf{M} \cdot \left(\frac{d\mathbf{B}}{dt}\right)}.$$

Take the radio-frequency magnetic field to be linearly polarized in the x direction, assume that the response of the system is linearly proportional to the stimulation by the radio-frequency field, and calculate an expression for the power absorbed.

*9. Consider a tube filled with water with a coil wrapped around it and situated in a static magnetic field of 1.0 Wb·m^{-2}. Assume that the coil diameter is 1.0 cm, that it consists of 10 turns, and that it has a quality factor Q of 100. (The voltage V induced in the coil by a precessing nuclear magnetization is $V = Q\mathcal{E}$ where $\mathcal{E}$ is the induced emf.) The resonant frequency of the coil is 42.6 MHz. A radio-frequency $\pi/2$ pulse is applied to the protons in the sample. Compute the voltage available at the terminals of the coil given that the equilibrium magnetization of the sample is 3.0×10^{-3} A·m^{-1}.

10. Show that the steady-state solution of the Bloch equation yields for the components of the magnetization in the rotating reference frame

$$\tilde{M}_x = \frac{\Delta \omega \gamma B_1 T_2^2}{1 + (T_2 \Delta \omega)^2 + \gamma^2 B_1^2 T_1 T_2} M_0$$

$$\tilde{M}_y = \frac{\gamma B_1 T_2}{1 + (T_2 \Delta \omega)^2 + \gamma^2 B_1^2 T_1 T_2} M_0$$

$$M_z = \frac{1 + (\Delta \omega T_2)^2}{1 + (T_2 \Delta \omega)^2 + \gamma^2 B_1^2 T_1 T_2} M_0.$$

11. Show from the steady-state solution of the Bloch equation that in the limit $\gamma^2 B_1^2 T_1 T_2 \ll 1$ a radio-frequency susceptibility $\chi(\omega) = \chi'(\omega) - i\chi''(\omega)$ can be defined with components

$$\chi'(\omega) = \frac{\chi_0}{2}\omega_0 T_2 \frac{\Delta\omega T_2}{1 + (\Delta\omega T_2)^2}$$

$$\chi''(\omega) = \frac{\chi_0}{2}\omega_0 T_2 \frac{1}{1 + (\Delta\omega T_2)^2}$$

where $M_0 = \chi_0 H_0$ defines the static susceptibility χ_0.

12. A pulsed nuclear magnetic resonance experiment is designed to study the relaxation times in metallic sodium in a homogeneous static field of 1.0×10^4 gauss (γ(sodium-23) $= 7.08 \times 10^3$ sec$^{-1} \cdot$ gauss^{-1}.)
    (a) Calculate the approximate resonance frequency of the sodium spins in MHz.
    (b) Calculate the strength of a linearly polarized radio-frequency field which, if applied perpendicular to the static field, would produce a $\pi/2$ pulse in 5 μsec.
    (c) The spin-lattice relaxation time is measured by applying a π pulse followed at a time τ later by a $\pi/2$ pulse. Write equations for the time variation of the components of magnetization parallel and perpendicular to the static field following each pulse. Assume that the pulse times are short compared to the relaxation times.
    (d) The spin-spin relaxation time is measured to be 220 μsec using a single $\pi/2$ pulse. Sketch the free induction decay signal following the pulse. Calculate the half-width of the line shape function in gauss. Note: Assume that the Bloch equation is valid for this system.

*13. If the line shape function is a Lorentzian curve of the form

$$f_{T_2}(\Delta\omega) = \frac{T_2}{\pi} \frac{1}{1 + (\Delta\omega T_2)^2},$$

show that the free induction decay is an exponential with time constant T_2.

14. The spin-echo technique enables one to measure the spin-spin relaxation time in the presence of inhomogeneities in the external magnetic field. Write a note on this novel technique.[15]

15. What are **wiggles**?

16. Show that the equation of motion for the complex magnetization $\mathfrak{M}$ is

$$\frac{d\mathfrak{M}}{dt} + \alpha\mathfrak{M} = -i\gamma B_1 M_0$$

as stated in Section 16.5.

17. In Section 16.5 the steady-state solution of the Bloch equation modified for chemical exchange was stated as

$$M = -i\gamma B_1 M_0 \frac{\tau_A + \tau_B + \tau_A\tau_B(\alpha_A p_B + \alpha_B p_A)}{(1 + \alpha_A\tau_A)(1 + \alpha_B\tau_B) - 1}.$$

Deduce this solution.

[15] E. L. Hahn, "Free Nuclear Induction," *Physics Today*, November, 1953.

*18. Show that in the slow exchange limit of chemical exchange the imaginary part of the complex magnetization is

$$\tilde{M}_{Ay} = -\gamma B_1 M_0 \frac{p_A T'_{2A}}{1 + (T'_{2A})^2(\omega_A - \omega)^2}$$

with

$$\frac{1}{T'_{2A}} = \frac{1}{T_{2A}} + \frac{1}{\tau_A}$$

as stated in Section 16.5.

*19. Show that in the rapid exchange limit of chemical exchange the imaginary part of the complex magnetization is

$$\tilde{M}_y = -\gamma B_1 M_0 \frac{T'_2}{1 + (T'_2)^2(p_A\omega_A + p_B\omega_B - \omega)^2}$$

with

$$\frac{1}{T'_2} = \frac{p_A}{T_{2A}} + \frac{p_B}{T_{2B}}$$

as stated in Section 16.5.

20. Show that in the intermediate exchange regime, assuming equal populations, equal lifetimes $\tau_A = \tau_B = 2\tau$ and large spin-spin relaxation times T_{2A} and T_{2B} the imaginary part of M is given by

$$\tilde{M}_y = -\gamma B_1 M_0 \frac{\tau(\omega_A - \omega_B)^2}{[(\omega_A + \omega_B) - 2\omega]^2 + 4\tau^2(\omega_A - \omega)^2(\omega_B - \omega)^2}.$$

21. Consider the following three possible structures for the diketene molecule (see Fig. 16.32). Comment on the nature of the proton magnetic resonance spectra to be expected in each case. Consider only chemical shifts.

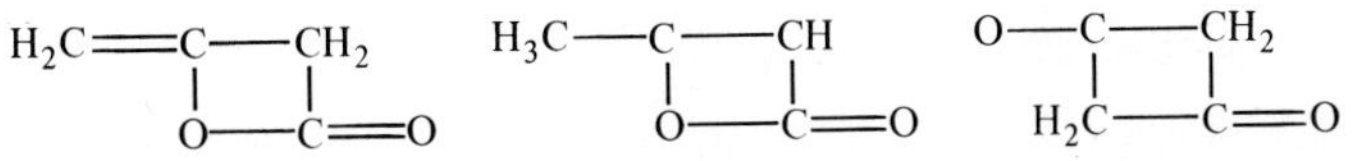

$$\begin{array}{ccc}
H_2C=C-CH_2 & H_3C-C-CH & O-C-CH_2 \\
| \ \ \ \ | & | \ \ \ \ | & | \ \ \ \ | \\
O-C=O & O-C=O & H_2C-C=O
\end{array}$$

Fig. 16.32. Three possible structures for the diketene molecule.

*22. (a) Deduce the nature of the phosphorous magnetic resonance spectrum of the PF_3 molecule.
(b) Deduce the nature of the fluorine magnetic resonance spectrum of the PF_3 molecule. Hint: The three fluorine atoms are chemically equivalent.
(c) Generalize your result to deduce the nature of the spectra characteristic of a set of n_A equivalent nuclei of type A and spin $\frac{1}{2}$ interacting with n_x equivalent nuclei of type x and spin $\frac{1}{2}$.

23. Deduce the proton magnetic resonance spectra of $CHCl_2CH_2Cl$ and CH_3CH_2Cl. Assume that the chemical shift differences are large compared to the electron coupled spin splittings and neglect couplings between protons and chlorine nuclei.

24. Fig. 16.33(a) shows the proton spectrum of ordinary liquid ammonia; Fig. 16.33(b) shows the spectrum of superdry ammonia. Comment on these spectra.

25. The spin-coupling constant J_{HH} of the hydrogen molecule (H_2) has been deduced from the experimentally measured value of the spin-coupling constant $J_{HD} = 43$ Hz of the hydrogen deuteride molecule (HD) by means of the relation

$$J_{HH} = \frac{\gamma_P}{\gamma_D} J_{HD}$$

where γ_D and γ_P are the deuteron and proton magnetogyric ratios, respectively. Deduce the value for J_{HH}. Why could this constant not be measured directly using the hydrogen molecule?

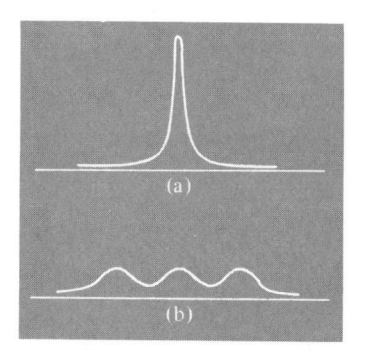

Fig. 16.33. (a) The proton NMR spectrum of ordinary liquid ammonia. (b) The proton spectrum of superdry ammonia.

26. Discuss the **Knight shift**.

27. Calculate the second and fourth moments for a Gaussian line shape function

$$f(\omega - \omega_0) = \frac{1}{\Delta(2\pi)^{1/2}} \exp\left[-\frac{(\omega - \omega_0)^2}{2\Delta^2}\right].$$

What is the ratio $M_4/(M_2)^2$? Express the half-width of the line at half intensity in terms of M_2.

28. Calculate the second moment for a Lorenztian line shape function. Generalize your result to all higher even moments. Comment on the physical significance of the results.

*29. Consider the two forms of benzene shown in Fig. 16.34. The replacement of a proton by a deuteron will reduce the contribution of that site to the proton second moment by the factor

$$\frac{4}{9} \frac{\gamma_D^2 I_D(I_D + 1)}{\gamma_P^2 I_P(I_P + 1)}.$$

This is known as the $\frac{4}{9}$ **effect**. Derive an expression for the second moment M_{2D} of the deuterated form of benzene in terms of the intramolecular M_2' and intermolecular M_2'' contributions to the second moment M_2 of the normal form of benzene. Assume that the sample is composed of randomly oriented crystallites.

*30. Predict the density dependence of the nuclear spin-lattice relaxation time in a gas of monatomic molecules such as helium-3.

*31. Assuming that the intramolecular dipolar interaction dominates the relaxation, deduce a power law relation for the temperature variation of T_1 for a gas of diatomic molecules. State any assumptions that you use in your derivation.

Fig. 16.34. The molecular structure of the molecules C_6H_6 and $C_6H_3D_3$.

32. In Section 16.9 we considered the electrostatic energy of interaction of an ion surrounded by an array of negative charges. Show that the result of the tetragonal distortion is to split the $d\epsilon$ orbital into a singlet and a doublet. In order that the singlet be lowest should the negative charges on the z axis be displaced toward or away from the central ion?

33. Although the direct process for quadrupolar spin-lattice relaxation has been experimentally observed in EPR experiments, it has not been observed in NMR experiments. Suggest a reason for this.

34. Discuss the importance of the hyperfine interaction in EPR studies.

35. EPR is a useful tool for the study of free radicals.[16] Discuss this application of magnetic resonance.

*36. A **proton spin refrigerator** is a device for achieving an enhanced proton polarization. The essential element of the device is a single crystal of yttrium ethyl sulfate $Y(C_2H_5SO_4)_3 9H_2O$ containing a small percentage of paramagnetic ytterbium Yb^{3+} ions. These impurity ions constitute a two-level electron spin system having a very anisotropic g-factor given by

$$g(\theta) = [g_\parallel^2 \cos^2 \theta + g_\perp^2 \sin^2 \theta]^{1/2}$$

where $g_\parallel = 3.35$, $g_\perp \simeq g_n = 0.00304$, and θ is the angle between the external magnetic field and the symmetry axis of the crystal. The electron spin-lattice relaxation time for these ions is also anisotropic and proportional to $(\cos^2 \theta \sin^2 \theta)^{-1}$ with a minimum value $\simeq 10^{-3}$ sec for $\theta = 45°$. The proton relaxation time is of the order of minutes. The refrigerator is operated by placing the crystal in a field $\simeq 1$ Wb·m^{-2} at 1.4°K and rotating the crystal so that θ varies from 0 to 2π every $\frac{1}{60}$ sec. As a result the populations of the proton spin levels are dramatically altered from their equilibrium values. In practice, values of the ratio $N_\downarrow/N_\uparrow \simeq 4/5$ have been achieved. Give a qualitative explanation of how this device might produce these enhanced proton polarizations. Justify the name "proton spin refrigerator."

37. The isotope indium-113 has nuclear spin of $\frac{9}{2}$. Draw an energy-level diagram for such a nucleus located in an axially symmetric electric field gradient.

38. Six identical point charges q are located at the coordinates $(0, 0, \pm a)$, $(0, \pm a, 0)$ and $(\pm a, 0, 0)$. A point nucleus with a quadrupole moment eQ is located at the origin. Calculate the energy of interaction of the nuclear quadrupole moment and the electric field gradient.

39. The temperature variation of NQR frequencies can yield lattice dynamical information. Discuss this statement with reference to cuprous oxide (Cu_2O).[17]

---

[16] J. Turkevich, "Electron Spin Resonance of Stable Free Radicals," *Physics Today*, July, 1965.

[17] G. L. Baker and R. L. Armstrong, "Temperature Variation of Nuclear Quadrupole Resonance Frequencies," *American Journal of Physics*, **36** (January 1968), 33.

40. Estimate the smallest difference in magnetic field strength that can be detected using a proton magnetometer consisting of two samples separated by 2 meters.

41. Write an essay on the application of proton magnetometers to the problem of mineral exploration.

42. Discuss the importance of nuclear magnetic double resonance to chemistry.

43. Write an account of ferromagnetic and antiferromagnetic resonance.

44. Write an account of the application of magnetic resonance to biology.

45. **Molecular beam magnetic resonance** constitutes another type of magnetic resonance technique that was not discussed in this chapter. Discuss the principles and applications of this technique.[18]

46. The **Overhauser effect** provides one example of **dynamic nuclear orientation**. Enlarge upon this statement.

47. The article, "Dynamic Polarization and Biology,"[19] provides an interesting example of the way that a physicist might approach a biological problem. Comment on this article.

---

[18] O. R. Frisch, "Molecular Beams," *Scientific American*, May, 1965.
[19] S. P. Heins, "*Dynamic Polarization and Biology*," *American Journal of Physics*, **38** (September 1970), 1128.

THOMAS YOUNG

17 Double beam interference

17.1 INTRODUCTION

The mathematical description of the interference of electromagnetic waves is based upon the **principle of linear superposition**[1] proposed by Thomas Young in 1802. The principle states that the resultant electric field **E** and the resultant magnetic field **B** produced at a point in space by n sources of electromagnetic waves are given by the vector sums

$$\mathbf{E} = \sum_{i=1}^{n} \mathbf{E}_i, \quad \mathbf{B} = \sum_{i=1}^{n} \mathbf{B}_i$$

where $\mathbf{E}_i$ is the electric field and $\mathbf{B}_i$ the magnetic field produced at that point by source i. We showed in Section 15.3 that when an electromagnetic wave is present in a medium, the ratio of the magnetic to the electric force on a charged particle is V/v where V is the speed of the charged particle and v is the speed of the wave in the medium. In almost all cases $V \ll v$ and the force due to the magnetic field can be neglected.

An experimental test of the validity of the principle of superposition involves allowing two waves to interact and comparing the separated components before and after the interaction. If the separate waves are unaffected by the interaction, then the principle is valid; otherwise, it is not. Although so far always found to be true in a vacuum, the superposition principle is only approximately true in matter. Observable deviations will always occur for sufficiently high wave intensities (see Chapter 20).

The generalized double beam interference experiment is depicted in Fig. 17.1. Radiation from a source S travels via paths SAP and SBP to the point of observation P. Points A and B can be considered as secondary sources. The problem is to determine the dependence of the intensity at P on the difference in path lengths and the coherence properties of the source. In this chapter we consider various aspects of this problem, assuming the validity of the principle of linear superposition and dealing exclusively with the electric field vector **E**.

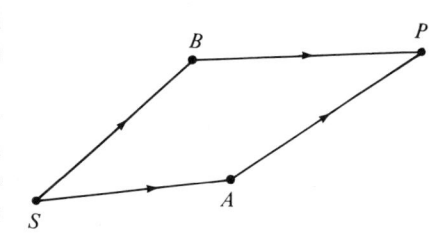

Fig. 17.1. The generalized double beam interference experiment.

17.2 INTERFERENCE OF COHERENT RADIATION

In the simplest case, S is taken as a point source of coherent, linearly polarized, and monochromatic plane wave radiation of constant amplitude. A source is said to be **coherent** if all of the electric charges in it oscillate in phase. Electronic oscillators utilizing transistors, klystrons, etc.

[1] *MWTP*, Section 16.1.

provide coherent sources for radio waves and microwaves. Lasers provide coherent sources for optical and infrared radiation. A source may be realistically considered as a point source if its dimensions are small compared to the wavelength of the radiation produced. If the distance from any source to the point of observation is sufficiently large, the waves arriving at the point of observation can be taken as plane waves.

The component $\mathbf{E}_A$ of the electric field arriving at P and appearing to come from the secondary source A may be expressed as

$$\mathbf{E}_A(P) = \mathbf{E}_{A0} \exp[i(\omega t - \mathbf{k}_A \cdot \mathbf{r} + \phi_A)].$$

The magnitude of $\mathbf{E}_{A0}$ is the amplitude of the radiation leaving A and the direction of $\mathbf{E}_{A0}$ is the direction of polarization of the radiation. Also, $\mathbf{k}_A$ is the wave vector and ω the angular frequency of the radiation traveling between A and P, ϕ_A is the phase angle introduced by the traversal of the path SA from the source to point A, and $\mathbf{r}$ is the position vector of the point P. Similarly, the component $\mathbf{E}_B$ of the electric field arriving at P from the secondary source B may be expressed as

$$\mathbf{E}_B(P) = \mathbf{E}_{B0} \exp[i(\omega t - \mathbf{k}_B \cdot \mathbf{r} + \phi_B)]$$

where the various symbols have meanings corresponding to those just described for the previous equation.

The intensity of radiation at P is given, to within a proportionality constant, by (see Section 9.7)

$$I(P) = |E(P)|^2$$

with

$$\mathbf{E}(P) = \mathbf{E}_A(P) + \mathbf{E}_B(P).$$

Therefore,

$$\begin{aligned}I(P) &= |\mathbf{E}_A(P)|^2 + |\mathbf{E}_B(P)|^2 + \mathbf{E}_A^*(P) \cdot \mathbf{E}_B(P) + \mathbf{E}_A(P) \cdot \mathbf{E}_B^*(P) \\ &= |\mathbf{E}_A(P)|^2 + |\mathbf{E}_B(P)|^2 + 2Re[\mathbf{E}_A^*(P) \cdot \mathbf{E}_B(P)].\end{aligned}$$

It is the **interference term** $2Re[\mathbf{E}_A^*(P) \cdot \mathbf{E}_B(P)]$ which gives rise to interference phenomena. Let us examine this term in more detail. Since

$$\begin{aligned}\mathbf{E}_A^*(P) \cdot \mathbf{E}_B(P) &= \mathbf{E}_{A0} \cdot \mathbf{E}_{B0} \exp[-i(\omega t - \mathbf{k}_A \cdot \mathbf{r} + \phi_A)] \\ &\quad \times \exp[i(\omega t - \mathbf{k}_B \cdot \mathbf{r} + \phi_B)] \\ &= \mathbf{E}_{A0} \cdot \mathbf{E}_{B0} \exp\{i[(\mathbf{k}_A - \mathbf{k}_B) \cdot \mathbf{r} - (\phi_A - \phi_B)]\},\end{aligned}$$

$$2Re[\mathbf{E}_A^*(P) \cdot \mathbf{E}_B(P)] = 2\mathbf{E}_{A0} \cdot \mathbf{E}_{B0} \cos[(\mathbf{k}_A - \mathbf{k}_B) \cdot \mathbf{r} - (\phi_A - \phi_B)].$$

To evaluate $(\mathbf{k}_A - \mathbf{k}_B) \cdot \mathbf{r}$ we refer to Fig. 17.2 and note that the magnitude of the vector $\Delta \mathbf{k} = \mathbf{k}_A - \mathbf{k}_B$ is approximately $k\theta$ where $k = |\mathbf{k}_A| \simeq |\mathbf{k}_B|$, and the angle θ (measured in radians) is approximately s/x if $|\mathbf{r}| = |x\mathbf{i} + y\mathbf{j}| \approx x$. Since $\Delta \mathbf{k}$ is approximately in the direction $\mathbf{j}$, it follows that

$$(\mathbf{k}_A - \mathbf{k}_B) \cdot \mathbf{r} \simeq \left(\frac{ks}{x}\mathbf{j}\right) \cdot (x\mathbf{i} + y\mathbf{j}) = \frac{ksy}{x}.$$

SEC. 17.2 INTERFERENCE OF COHERENT RADIATION

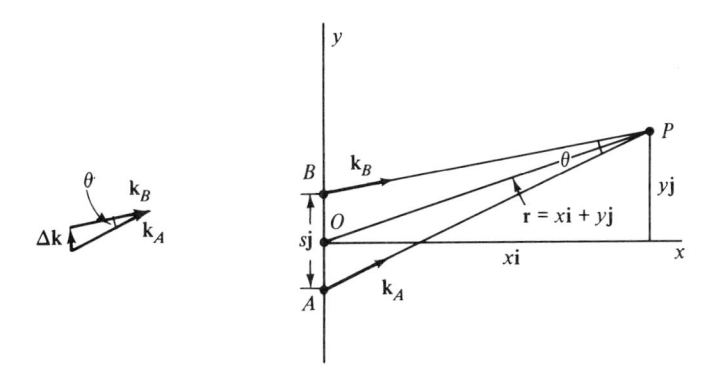

Fig. 17.2. Geometry for analyzing a double beam interference experiment.

Therefore,

$$I(P) = |\mathbf{E}_{A0}|^2 + |\mathbf{E}_{B0}|^2 + 2\mathbf{E}_{A0} \cdot \mathbf{E}_{B0} \cos\left(\frac{ksy}{x} - \Delta\phi\right)$$

with

$$\Delta\phi = \phi_A - \phi_B \quad \text{and} \quad y \ll x.$$

If the secondary sources A and B produce radiation of the same amplitude and polarization, then

$$\mathbf{E}_{A0} = \mathbf{E}_{B0}$$

and

$$|\mathbf{E}_{A0}|^2 = |\mathbf{E}_{B0}|^2 = \mathbf{E}_{A0} \cdot \mathbf{E}_{B0} = I_0$$

where I_0 is the intensity of each of the separate waves. If $\Delta\varphi = 0$, the intensity at P is given by

$$I(P) = 2I_0\left[1 + \cos\left(\frac{ksy}{x}\right)\right].$$

The intensity takes on values in the range 0 to $4I_0$ depending on the value of $\cos(ksy/x)$. At positions of maximum intensity $I(P) = 4I_0$; these occur for

$$\frac{ksy}{x} = 2n\pi$$

with n an integer. At positions of minimum intensity $I(P) = 0$; these occur for

$$\frac{ksy}{x} = (2n+1)\pi$$

with n an integer.

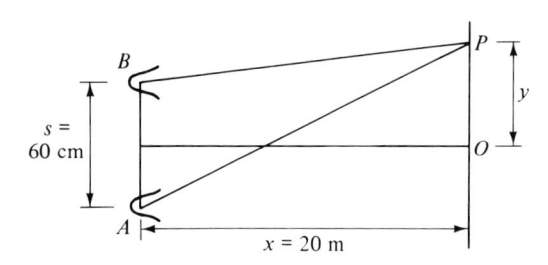

Fig. 17.3. Waves from A and B interfere to give a point of maximum intensity at P.

Example. Two klystron oscillators A and B, both emitting 3.0 cm microwaves, are placed 60 cm apart along a line parallel to a screen that is 20 m distant as shown in Fig. 17.3. (Note that the distance AB has been exaggerated for clarity.) If the radiation from the two sources is in phase, a maximum in the interference pattern will be observed at point O. Point P represents the position of the interference maximum adjacent to the one occurring at O. Deduce the distance y between O and P.

Solution. The distance y is given by the relation

$$\frac{ksy}{x} = 2n\pi$$

with $k = 2\pi/\lambda = 2\pi/3.0$ cm^{-1}, $s = 60$ cm, $x = 20$ m, and $n = 1$. That is,

$$y = \frac{2n\pi x}{ks} = \frac{(2\pi)(1)(20)}{\left(\frac{2\pi}{3.0}\right)(60)} = 1.0 \text{ m.}$$

Note that since klystron oscillators are coherent sources of radiation, two independent sources operating at the same frequency produce radiation that differs by a constant phase factor. In this case it is not necessary to start with a single source of radiation in order to obtain observable interference effects.

17.3 TEMPORAL COHERENCE

If the source for the interference experiment is an ordinary source of visible light, such as a sodium lamp, the amplitude and phase of the emitted radiation will vary rapidly and randomly with time. In this case the intensity may be defined as a time average[2]

$$I(P) = \lim_{T \to \infty} \frac{\int_0^T |\mathbf{E}(P)|^2 \, dt}{\int_0^T dt}$$

$$= \lim_{T \to \infty} \frac{1}{T} \int_0^T \{|\mathbf{E}_A^2(t)| + |\mathbf{E}_B^2(t+\tau)| + 2Re[\mathbf{E}_A^*(t) \cdot \mathbf{E}_B(t+\tau)]\} \, dt$$

$$\equiv \overline{|\mathbf{E}_A^2(t)|} + \overline{|\mathbf{E}_B^2(t+\tau)|} + 2Re\overline{[\mathbf{E}_A^*(t) \cdot \mathbf{E}_B(t+\tau)]}.$$

In this expression τ is the difference in the time required for the light to reach P via the two different paths. All time averages are taken to be independent of the choice of the origin of time; that is, the fluctuating quantities are assumed to belong to the class of functions known as **stationary random functions**. For simplicity we will take the radiation to be

---

[2] *MWTP*, Section 7.4.

polarized perpendicular to the plane of the page (see Fig. 17.1) and assume that

$$\overline{|E_A^2(t)|} = \overline{|E_B^2(t+\tau)|} = I_0.$$

The function

$$\Gamma_{AB}(\tau) \equiv \overline{[E_A^*(t)E_B(t+\tau)]}$$

is called the **mutual coherence function** or the **mutual correlation function** of the fields $\mathbf{E}_A$ and $\mathbf{E}_B$. The intensity at P may be written in terms of a **normalized correlation function** $C_{AB}(\tau)$ as

$$I(P) = 2I_0 \left\{ 1 + \frac{Re[\Gamma_{AB}(\tau)]}{I_0} \right\}$$
$$= 2I_0[1 + C_{AB}(\tau)].$$

As the point P moves away from O the time difference increases from zero, the correlation function $C_{AB}(\tau)$ varies in a periodic manner, and a pattern of interference fringes is observed. We assume that $C_{AB}(\tau)$ can be written in the form

$$C_{AB}(\tau) = C_{AB}^0 \cos\left(\frac{\tau}{\tau_0}\right)$$

where τ_0 is a constant characteristic of the source called the **coherence time** and C_{AB}^0 a constant called the **degree of coherence**. The intensity in the interference pattern varies between the limits

$$I_{\max} = 2I_0(1 + C_{AB}^0) \quad \text{and} \quad I_{\min} = 2I_0(1 - C_{AB}^0).$$

If $C_{AB}^0 = 1$, we have the case of **complete coherence** considered in the previous section. If $C_{AB}^0 = 0$, we have **complete incoherence** and no interference pattern is observable. In general, $0 < C_{AB}^0 < 1$ and a state of **partial coherence** results.

A useful parameter to describe the contrast between the regions of maximum intensity and the regions of minimum intensity is the **fringe visibility**

$$\mathcal{V} = \frac{I_{\max} - I_{\min}}{I_{\max} + I_{\min}}$$
$$= C_{AB}^0.$$

That is, the fringe visibility is equal to the degree of coherence for the case of equal amplitudes in a two beam interference experiment.

To proceed further, we consider a simple and idealized model of the field produced by a radiating atom that is undergoing collisions. The field is assumed to be of constant amplitude and to vary sinusoidally with frequency ω_0 for an indefinite series of equal time intervals τ_0. Each interval is terminated by a **strong collision** that results in the establishment of a random phase relation between the oscillations in adjacent intervals. That is, the electric field of a radiating atom is taken as

$$E = E_0 \exp\{i[\omega_0 t + \phi(t)]\}$$

where $\phi(t)$ suffers abrupt and random changes for $t = n\tau_0$ with n an integer (see Fig. 17.4). If this source is used for an interference experiment, the correlation function $C_{AB}(\tau)$ is given by (see Problem 13)

$$C_{AB}(\tau) = \cos(\omega_0 \tau) Re \left\{ \lim_{T \to \infty} \frac{1}{T} \int \exp\{-i[\phi_A(\tau) - \phi_B(t + \tau)]\} dt \right\}.$$

To evaluate this expression we refer to Fig. 17.4 and note that $\phi_A(t) - \phi_B(t + \tau) = \Delta(t)$ is zero for values of t given by

$$n\tau_0 + \tau < t < (n+1)\tau_0; \quad n = 0, 1, 2, 3, \ldots$$

and assumes some random value between 0 and 2π for values of t given by

$$n\tau_0 < t < n\tau_0 + \tau; \quad n = 0, 1, 2, 3, \ldots.$$

The integral can therefore be written as the sum of two series of integrals — one series for the nonzero values of $\Delta_n(t)$ and one for the zero values of $\Delta_n(t)$ where the subscript has been added to $\Delta(t)$ to indicate that we are picking out one of the time intervals described above, the first few of which are shown in Fig. 17.4. The integral now becomes

$$\lim_{T \to \infty} \frac{1}{T} \int \exp[-i\Delta(t)] dt$$

$$= \lim_{T \to \infty} \sum_{n=0}^{n=T/\tau_0} \frac{1}{T} \int_{n\tau_0}^{n\tau_0 + \tau} \exp[-i\Delta_n(t)] dt + \lim_{T \to \infty} \sum_{n=0}^{n=T/\tau_0} \frac{1}{T} \int_{n\tau_0 + \tau}^{(n+1)\tau_0} dt.$$

Since $\Delta_n(t)$ is a random number, the first term averages to zero. The second term gives $(\tau_0 - \tau)/\tau_0$ for $\tau < \tau_0$; otherwise, it does not exist. Therefore,

$$C_{AB}(\tau) = \left(1 - \frac{\tau}{\tau_0}\right) \cos \omega_0 t \quad \tau < \tau_0$$
$$= 0 \quad \tau > \tau_0.$$

The intensity in the interference pattern varies between the limits

$$I_{max} = 2I_0\left(2 - \frac{\tau}{\tau_0}\right) \quad \text{and} \quad I_{min} = 2I_0 \frac{\tau}{\tau_0}$$

and the fringe visibility is

$$\mathcal{V} = 1 - \frac{\tau}{\tau_0}.$$

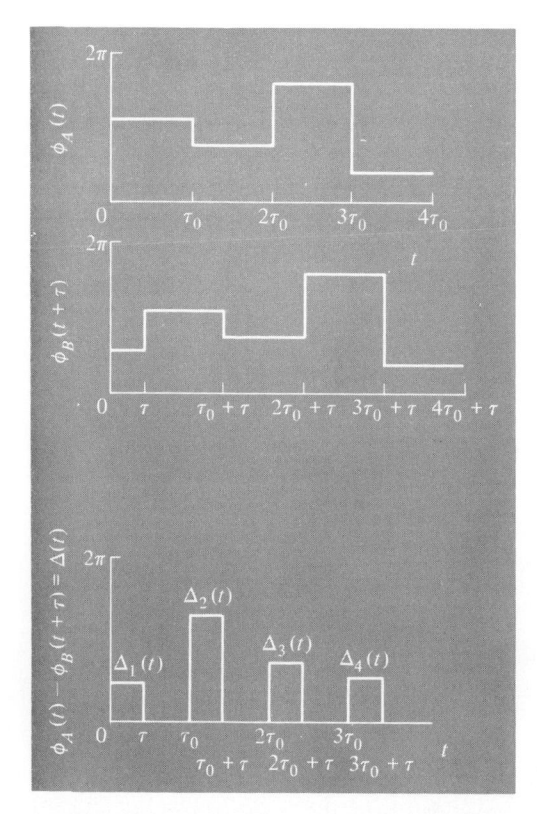

Fig. 17.4. Relevant variations of phase with t for a radiating atom used as a source for an interference experiment.

SEC. 17.3 TEMPORAL COHERENCE

In order to observe an interference pattern it is necessary that the path difference l traveled by the two beams satisfies the condition

$$l < v\tau < v\tau_0 = l_c$$

where v is the wave speed in the medium. The quantity l_c is called the **coherence length**; it represents the length of one uninterrupted wave train.

The coherence time of a light source is related to the line width of the radiation emitted by the source. In fact, for a single wave train of finite duration τ_0, we can easily show that the width $\Delta\omega$ of the associated frequency distribution is related to τ_0 by

$$\Delta\omega = \frac{2\pi}{\tau_0}.$$

This may also be stated as

$$\Delta\nu = \frac{1}{\tau_0}.$$

We begin the development of these relations by noting that the wave train shown in Fig. 17.5 may be described by

$$f(t) = \exp(i\omega_0 t) \qquad -\frac{\tau_0}{2} < t < \frac{\tau_0}{2}$$

$$= 0 \qquad \text{otherwise.}$$

Using **Fourier's integral theorem**[3] we can write

$$g(\omega) = \frac{1}{2\pi} \int_{-\infty}^{\infty} f(t) \exp(-i\omega t)\, dt$$

where $g(\omega)$ describes the wave train as a function of frequency rather than time. Upon substitution

$$g(\omega) = \frac{1}{2\pi} \int_{-\tau_0/2}^{\tau_0/2} \exp[-i(\omega-\omega_0)t]\, dt$$

$$= \frac{1}{2\pi} \frac{\exp[-i(\omega-\omega_0)t]}{-i(\omega-\omega_0)}\bigg|_{-\tau_0/2}^{\tau_0/2}$$

$$= \frac{1}{\pi(\omega-\omega_0)} \frac{\exp\left[i(\omega-\omega_0)\frac{\tau_0}{2}\right] - \exp\left[-i(\omega-\omega_0)\frac{\tau_0}{2}\right]}{2i}$$

$$= \frac{\sin\left[(\omega-\omega_0)\frac{\tau_0}{2}\right]}{\pi(\omega-\omega_0)}.$$

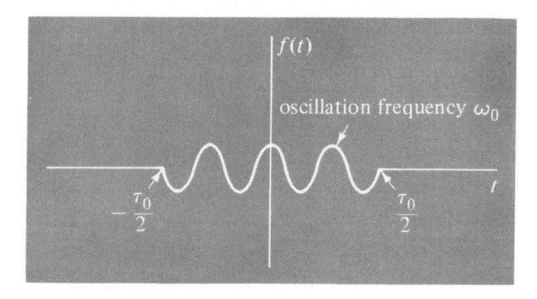

Fig. 17.5. A wave train of finite duration.

[3] H. Margenau and G. M. Murphy, *The Mathematics of Physics and Chemistry* (Princeton, New Jersey: D. Van Nostrand Co., Inc., 1956), p. 253.

The **power spectrum** $G(\omega)$ of the source is defined as

$$G(\omega) = |g(\omega)|^2 = \frac{\sin^2\left[(\omega - \omega_0)\frac{\tau_0}{2}\right]}{\pi^2(\omega - \omega_0)^2}.$$

This function, plotted in Fig. 17.6, gives the frequency distribution of the energy radiated by the source. $G(\omega)$ has a maximum for $\omega = \omega_0$ and falls to zero for $(\omega - \omega_0)\tau_0/2 = \pi$. Since

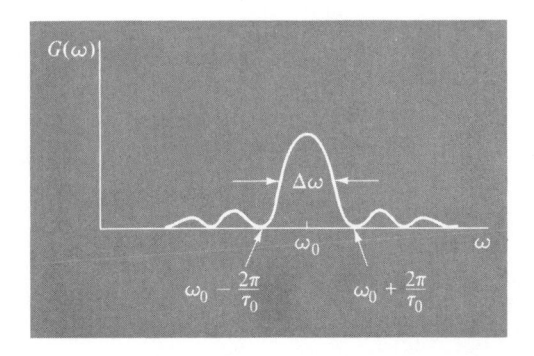

$$\omega = 2\pi\nu,$$
$$\Delta\omega = 2\pi\Delta\nu = \omega - \omega_0$$

Fig. 17.6. The power spectrum of a finite wave train.

and

$$\frac{(\omega - \omega_0)\tau_0}{2} = \frac{2\pi\Delta\nu\tau_0}{2} = \pi$$

so that

$$\Delta\omega = \frac{2\pi}{\tau_0}$$

or

$$\Delta\nu = \frac{1}{\tau_0}$$

as stated above.

The frequency and wavelength of the wave are related through the equation

$$v = \nu\lambda$$

where v is the wave speed in the medium. Rearranging and differentiating gives

$$\nu = \frac{v}{\lambda}$$

$$d\nu = -\frac{v}{\lambda^2}d\lambda$$

or

$$\Delta\nu = \frac{v}{\lambda^2}\Delta\lambda$$

for finite differences. (We omit the minus sign in the last equation since we are concerned only with the magnitude of the difference. The minus sign merely indicates that an increase in wavelength is accompanied by a decrease in frequency, or vice versa.) The coherence length may now be written as

$$l_c = v\tau_0 = \frac{v}{\Delta v} = \frac{\lambda^2}{\Delta \lambda}.$$

Example. A discharge tube emitting green light at a wavelength of 540 nm has a line width of 0.10 nm.
(a) What path differences are allowable in an interference experiment using this discharge tube before fringe visibility would become vanishingly small?
(b) Determine the coherence time of the source.

Solution.
(a) The discharge tube is characterized by the wavelength $\lambda = 540 \times 10^{-9}$ m and the line width $\Delta \lambda = 1.0 \times 10^{-10}$ m. The coherence length is

$$l_c = \frac{\lambda^2}{\Delta \lambda} = \frac{(540)^2 \times 10^{-18}}{1.0 \times 10^{-10}} = 2.9 \times 10^{-3} \text{ m} = 2.9 \text{ mm}.$$

For path differences much larger than the coherence length, the fringe visibility would become vanishingly small. Therefore, path differences up to $\simeq 10$ mm should produce visible fringes.
(b) The coherence time is

$$\tau_0 = \frac{1}{\Delta v} = \frac{\lambda^2}{v \Delta \lambda} = \frac{l_c}{v}.$$

The coherence time depends on the medium in which the radiation from the source is traveling. The coherence time is minimum in a vacuum for which

$$\tau_0 = \frac{l_c}{c} = \frac{2.9 \times 10^{-3}}{3.0 \times 10^8} = 9.7 \times 10^{-12} \text{ sec}.$$

Therefore, the minimum value of τ_0 is 9.7×10^{-12} sec, and would be $2\tau_0 = 1.9 \times 10^{-11}$ sec for a medium with index of refraction $n = 2$.

17.4 SPATIAL COHERENCE

We now turn to the problem of the finite extent of a real source. Initially, we will concern ourselves with the problem of two point sources S_1 and S_2 within an extended source (see Fig. 17.7). The phases of these sources will be assumed to vary randomly and independently. That is, S_1 and S_2 are **mutually incoherent** sources. We wish to deduce the coherence of the radiation arriving at A with respect to that arriving at B. For our discussion we will take the line joining A and B to be parallel to the line joining S_1 and S_2. In this case the coherence between the radiation arriving at A and that arriving at B is called the **lateral spatial coherence** of the

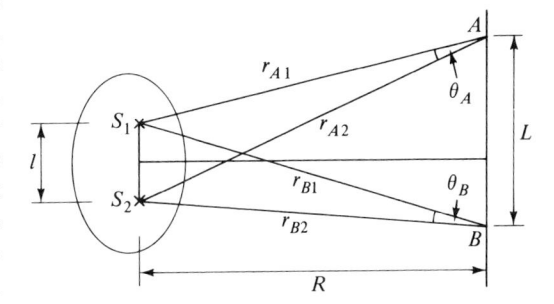

Fig. 17.7. Diagram for the calculation of the lateral coherence produced by two point sources.

field. The fields $\mathbf{E}_A$ and $\mathbf{E}_B$ at A and B are given by

$$\mathbf{E}_A = \mathbf{E}_{A1} + \mathbf{E}_{A2}$$
$$\mathbf{E}_B = \mathbf{E}_{B1} + \mathbf{E}_{B2}$$

where $\mathbf{E}_{A1}$ is the contribution to $\mathbf{E}_A$ from the source S_1, $\mathbf{E}_{A2}$ is the contribution to $\mathbf{E}_A$ from the source S_2, etc. If we assume that the radiation is polarized perpendicular to the plane of the paper (Fig. 17.7), we may drop the vector designation. The relevant normalized correlation function is

$$\begin{aligned} C_{AB}(\tau) &= \frac{Re\overline{[E_A^*(t)E_B(t+\tau)]}}{(I_A I_B)^{1/2}} \\ &= \frac{Re\overline{[E_{A1}^*(t) + E_{A2}^*(t)][E_{B1}(t+\tau) + E_{B2}(t+\tau)]}}{(I_A I_B)^{1/2}} \\ &= \frac{Re\overline{[E_{A1}^*(t)E_{B1}(t+\tau)]}}{(I_A I_B)^{1/2}} + \frac{Re\overline{[E_{A2}^*(t)E_{B2}(t+\tau)]}}{(I_A I_B)^{1/2}} \end{aligned}$$

where the cross terms are zero since sources S_1 and S_2 are mutually incoherent. Once again we will use the simple model invoked in the previous section. That is, we will take

$$E_A = E_{A0}\{\exp i[\omega_0 t + \phi_A(t)]\}$$
$$E_B = E_{B0}\{\exp i[\omega_0 t + \phi_B(t)]\}.$$

It then follows that (see Problem 15)

$$Re\overline{[E_{A1}^*(t)E_{B1}(t+\tau_1)]} = (I_A I_B)^{1/2}\left(1 - \frac{\tau_1}{\tau_0}\right)\cos(\omega_0 \tau_1)$$

with

$$\tau_1 = \frac{r_{A1} - r_{B1}}{v}$$

and that

$$Re\overline{[E_{A2}^*(t)E_{B2}(t+\tau_2)]} = (I_A I_B)^{1/2}\left(1 - \frac{\tau_2}{\tau_0}\right)\cos(\omega_0 \tau_2)$$

with

$$\tau_2 = \frac{r_{A2} - r_{B2}}{v}$$

However, we cannot simply insert these two average quantities back into the expression for $C_{AB}(\tau)$. Since radiation from both S_1 and S_2 is received at both A and B, correlations between the fields at A and B are actually counted twice in the calculation of $C_{AB}(\tau)$. Therefore, we must divide by 2 to obtain the true correlation function which is

$$\begin{aligned} C_{AB}(\tau) &= \frac{1}{2}\left[\left(1 - \frac{\tau_1}{\tau_0}\right)\cos(\omega_0\tau_1) + \left(1 - \frac{\tau_2}{\tau_0}\right)\cos(\omega_0\tau_2)\right] \\ &\simeq \frac{1}{2}\left(1 - \frac{\tau_1}{\tau_0}\right)(\cos\omega_0\tau_1 + \cos\omega_0\tau_2) \end{aligned}$$

SEC. 17.4   SPATIAL COHERENCE

if $\tau_1 - \tau_2 \ll \tau_1, \tau_2$

$$= \left(1 - \frac{\tau_1}{\tau_0}\right) \cos\left[\frac{\omega_0(\tau_1 + \tau_2)}{2}\right] \cos\left[\frac{\omega_0(\tau_1 - \tau_2)}{2}\right].$$

The term $\cos[\omega_0(\tau_1 + \tau_2)/2]$ represents the average oscillation frequency of the correlation function. The slow variation of the amplitude superposed on this rapid fluctuation is given by

$$\left(1 - \frac{\tau_1}{\tau_0}\right) \cos\left[\frac{\omega_0(\tau_1 - \tau_2)}{2}\right]$$

The behavior of this quantity depends critically on the value of $(\tau_1 - \tau_2)$. The amplitude goes to zero when

$$\frac{\omega_0(\tau_1 - \tau_2)}{2} = \frac{\pi}{2}.$$

Now

$$\tau_1 - \tau_2 = \frac{r_{A1} - r_{B1}}{v} - \frac{r_{A2} - r_{B2}}{v}$$

$$= \frac{r_{A1} - r_{A2}}{v} - \frac{r_{B1} - r_{B2}}{v}.$$

For $l, L \ll R$, we have

$$r_{A1} - r_{A2} \simeq l\theta_A, \qquad r_{B1} - r_{B2} \simeq l\theta_B$$

so that

$$\tau_1 - \tau_2 \simeq \frac{l}{v}(\theta_A - \theta_B).$$

But

$$\theta_A - \theta_B \simeq \frac{L}{R}$$

and

$$\tau_1 - \tau_2 \simeq \frac{lL}{vR}.$$

Therefore,

$$\omega_0(\tau_1 - \tau_2) = \frac{\omega_0 lL}{vR} = \pi.$$

The **lateral coherence width** L_w is defined as

$$L_w = \frac{2vR\pi}{\omega_0 l}$$

$$= \frac{R\lambda}{l}$$

and gives, approximately, the extent of the region of good lateral coherence. The lateral coherence is illustrated in Fig. 17.8.

For an extended source the above result applies, with the simple change that l is interpreted as a dimension of the source and the formula is multiplied by a numerical factor that depends on the shape of the source. For example, for a circular source the numerical factor is 1.22.

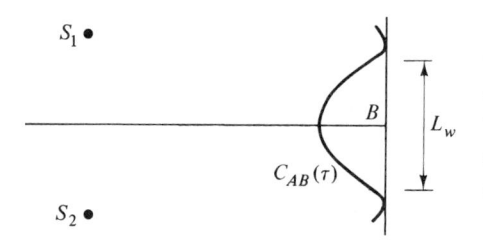

Fig. 17.8. Lateral coherence produced by two point sources.

Example. A circular aperture of diameter 0.10 mm is to be used to define a source for an optical interference experiment. The secondary sources are defined by two slits placed 1.0 m from the source. If the radiation used is incoherent and has a wavelength of 500 nm, what is the maximum allowable separation between the slits if interference is to be observed?

Solution. The experimental arrangement is shown in Fig. 17.9. The slit separation should be less than the lateral coherence width L_w which for a circular aperture of diameter D is given by

$$L_w = \frac{1.22 R \lambda}{D}$$

$$= \frac{1.22 \times 1.0 \times 500 \times 10^{-9}}{0.10 \times 10^{-3}}$$

$$= 6.1 \times 10^{-3} \text{ m}$$

$$= 6.1 \text{ mm}.$$

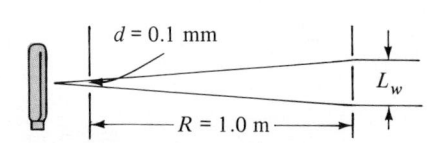

Fig. 17.9. Experimental arrangement for an optical interference experiment.

The slit separation should be less than 6 mm for interference to be observed.

17.5 INTERFEROMETERS IN RADIO ASTRONOMY

The simplest interferometer for use in radio astronomy (see Fig. 17.10) consists of two antennas separated by a distance L that is large compared with the operating wavelength. Coaxial cables carry the signals from the antennas A and B to a junction point P where they are superposed. For large antenna separations the coaxial cables are replaced by radio relay links. Very long baseline interferometry[4,5,6] ($L =$ several thousand kilometers) has been achieved by using high-quality tape recorders at the antenna sites with time markers controlled by atomic clocks. The tape recorders are physically brought together and the time markers used to synchronize them.

[4] N. Broten et al., "Long Baseline Interferometry Using Atomic Clocks and Tape Recorders," *Science,* **156** (June 1967), 1592.
[5] B. F. Burke, "Long Baseline Interferometry," *Physics Today,* July, 1969.
[6] K. I. Kellermann, "Intercontinental Radio Astronomy," *Scientific American,* (February 1972).

SEC. 17.5   INTERFEROMETERS IN RADIO ASTRONOMY

An astronomical source S sweeps across the sky causing the path lengths SA and SB from source to antennas to change continuously. Since the sources are very distant, the paths SA and SB are essentially parallel and the path difference to the two antennas is $L \sin \theta$. If the lengths of cable are equal, the condition for a maximum response at the receiver is

$$n\lambda = L \sin \theta$$

with n an integer. The response of an antenna is directional. As an antenna is rotated through 360° about an axis perpendicular to the direction of a source, the output power due to a distant source varies in intensity. A polar plot of output power as a function of the angle from the direction of the source usually shows several regions of maximum output power separated by regions of minimum output. The regions of maximum output are called **lobes**. An example of a polar plot of signal strength as a function of polar angle is given in Fig. 19.11. If each of the two antennas is oriented to have a maximum response when $\theta = 0$, then the lobe of the interference pattern corresponding to $n = 0$ will be the largest. The other lobes will decrease progressively with increasing n. A typical interferometer response for a point source is shown in Fig. 17.11.

Actual radio sources are of course of finite extent. Only if the angular width of the source is small compared with the width of one of the lobes of the interference pattern will the received pattern be substantially that of a point source. A typical fringe pattern for a source with an angular width approximately equal to the width of a lobe is shown in Fig. 17.12. The quantity P is proportional to the total power incident on the antennas; the fringe amplitude F is determined by the angular width of the source relative to the width of a lobe. By taking a series of recordings on an interferometer for different baseline separations L, one can obtain a graph of F/P vs L and from such a graph deduce the angular width of the source.

The simple interferometer is limited by the presence of **cosmic radio noise**. For a weak source, the interference pattern appears only as a small oscillation superposed on the galactic background. The

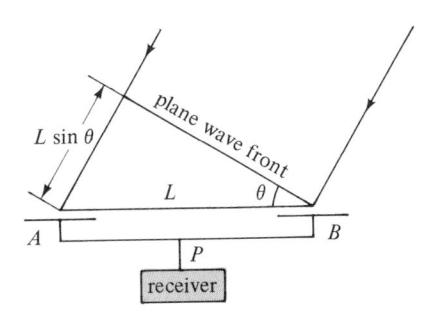

Fig. 17.10. A radio-frequency interferometer.

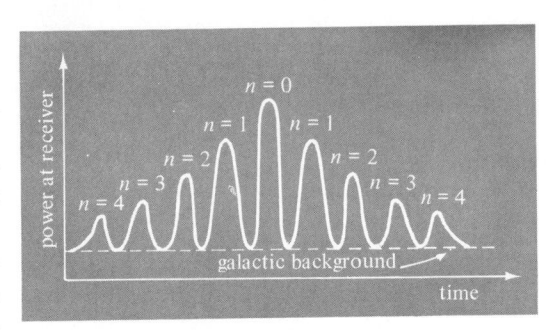

Fig. 17.11. A typical interferometer response from a point radio source.

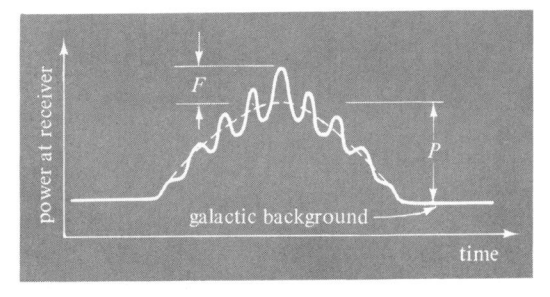

Fig. 17.12. Fringe pattern for a radio source with an angular width approximately equal to the width of a lobe of the interference pattern.

contribution from such noise may be greatly reduced by the use of a **phase-switched interferometer**. If the signal from one arm of a simple interferometer is shifted in phase by π rad, the entire interference pattern is inverted. That is, the maxima occur in the positions taken by the minima when no phase shift is introduced and vice versa. In the operation of the phase-switched interferometer a phase shift of π rad is periodically introduced and removed by electronic means. The difference signal between the receiver output with and without the phase shift present is recorded. If the radiation received by the antennas comes from a very broad source, then that radiation will enter the antennas through all of the lobes simultaneously. The shift in lobe direction causes essentially no change in the noise power

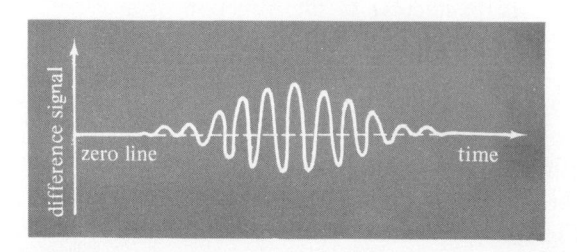

Fig. 17.13. A typical interference pattern obtained from a phase-switched interferometer.

received. Therefore, when the difference signal is recorded, cosmic noise is canceled out.

For a small source, the radiation enters the antennas through a single lobe. If the source is in the direction of a lobe maximum with no phase shift, then following a phase shift of π rad the source is in the direction of a lobe minimum. The difference signal is a maximum. If with no phase shift the small source is in the direction of a lobe minimum, it is easily seen that the difference signal is again a maximum but of opposite sign. Between the two directions the difference signal goes to zero. A typical interference pattern for a phase-switching interferometer is shown in Fig. 17.13. The fringes are symmetric with respect to a zero signal line. Since the total power is not recorded with this interferometer, it cannot be used to determine source dimensions.

17.6 INTENSITY INTERFEROMETRY

The method outlined in the previous section for the determination of the angular diameters of radio sources may also be applied to the determination of the diameters of optical sources. Even for stars that are relatively nearby, angular diameters are exceedingly small—of the order of hundredths of a second of arc. The corresponding lateral coherence widths at optical wavelengths are of the order of meters. This poses a significant technical problem with respect to the optical system necessary for carrying out such measurements. An experimental arrangement that has been used successfully for the determination of stellar diameters is shown in Fig. 17.14. The converging lens is required to bring the two light beams together so that they will interfere along the plane of observation. The practical limitation of the method is determined by the ability of the experimenter to align his optical system.

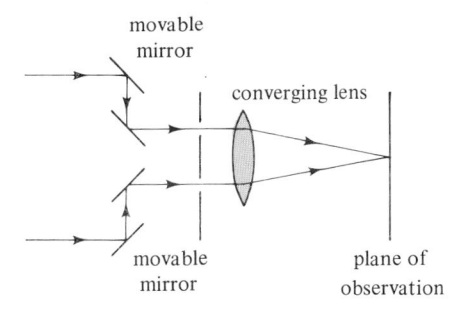

An alternate approach called **intensity interferometry** has been proposed by Hanbury-Brown and Twiss.[7] Radiation falls on two photomultiplier tubes separated by distance L. The photomultiplier currents are amplified, multiplied in a mixer, and integrated in a recorder. The experiment relies on the fact that, because of the finite coherence lengths of the wave trains, the intensity and hence the photomultiplier currents fluctuate in time. The cross correlation function observed is

$$\overline{[\Delta I_1(t)\, \Delta I_2(t+\tau)]}$$

where $\Delta I_1(t)$ and $\Delta I_2(t+\tau)$ are the instantaneous fluctuations in current in photomultipliers 1 and 2 at times t and $t+\tau$, respectively. The relation between intensity correlations and wave coherence is by no means obvious. The nontrivial and, in fact, initially strongly contested result of the Hanbury-Brown and Twiss experiment was that the correlation for intensity fluctuations is commensurate with wave coherence.

The obvious advantage of intensity interferometry is that the two branches of the interferometer are connected by coaxial cables rather than by optical paths. Angular diameters as small as 0.0005 sec of arc have been measured with such instruments.

17.7 THE MICHELSON INTERFEROMETER

One of the most versatile interferometric devices is the **Michelson interferometer** shown schematically in Fig. 17.15. The radiation travels from the source to the detector via the two paths indicated. The half-silvered mirror is designed to reflect 50% of the light incident upon it and to transmit the remaining 50%. The two contributions arriving at the

---

[7] R. Hanbury-Brown and R. Q. Twiss, "Correlation Between Photons in Two Coherent Beams of Light," *Nature* **177** (January 1956), 27.

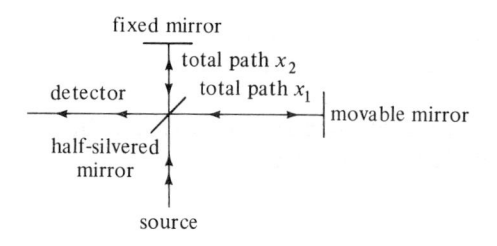

Fig. 17.15. A Michelson interferometer (schematic).

detector have a constant phase relation to one another for any fixed position of the movable mirror. For a monochromatic source, the radiation from the source can be written as

$$E_s = E_0 \exp[i(\omega t - kx)].$$

The radiation arriving at the detector can be expressed as

$$E_D = A'\{\exp[i(\omega t - kx_1)] + \exp[i(\omega t - kx_2)]\}$$

where A' is a constant of the apparatus. Note that the amplitudes of the two contributions are equal. The power P_D received by the detector is proportional to $|E_D|^2$. That is,

$$\begin{aligned}P_D &\propto [\exp(-ikx_1) + \exp(-ikx_2)][\exp(ikx_1) + \exp(ikx_2)]\\&= 2 + \exp[ik(x_1 - x_2)] + \exp[-ik(x_1 - x_2)]\\&= 2[1 + \cos k(x_1 - x_2)].\end{aligned}$$

The **interferogram** $I(\delta)$ is defined as

$$I(\delta) = C \cos k\delta$$

where $\delta = x_1 - x_2$ is the path difference for the two arms of the interferometer and C is a constant.

For a nonmonochromatic source, the interferogram is a superposition of such contributions each characterized by its own weighting factor $B(k)$. That is,

$$I(\delta) = \int_0^\infty B(k) \cos k\delta \, dk.$$

$B(k)$, the power spectrum of the source, may be obtained from $I(\delta)$ by means of Fourier's integral theorem and is

$$B(k) = \frac{1}{2\pi} \int_0^\infty I(\delta) \cos k\delta \, d\delta.$$

To determine $B(k)$ requires a knowledge of $I(\delta)$ for δ ranging from 0 to ∞. In practice, the mirror is movable only over some finite range corresponding to path differences from 0 to δ_0. We therefore define the function

$$B'(k) = \frac{1}{2\pi} \int_0^\infty I(\delta) R(\delta) \cos k\delta \, d\delta$$

where

$$R(\delta) = 1, \quad \delta \leq \delta_0$$
$$ 0, \quad \delta > \delta_0.$$

The relation between $B'(k)$ and $B(k)$ is indicated in Fig. 17.16 for an essentially monochromatic source.

SEC. 17.7 THE MICHELSON INTERFEROMETER

In a modern, automated system the data from the interferometer is directly coupled into an electronic computer that continually computes the spectrum and displays it on an oscilloscope. As the path difference increases, the spectral features appearing on the oscilloscope become narrower and the detailed structural features appear. When the oscilloscope display no longer changes with further increase in path difference, the scan is stopped. The interferogram resulting from the absorption of infrared radiation by the molecular rotational motions of water molecules in water vapor is shown in Fig. 17.17. The pure rotational

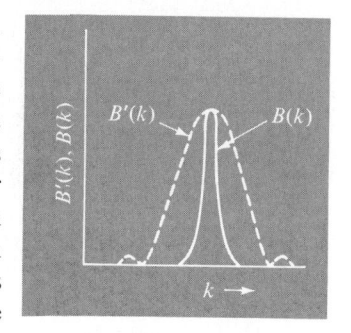

Fig. 17.16. The relation between $B'(k)$ and $B(k)$ for an essentially monochromatic source.

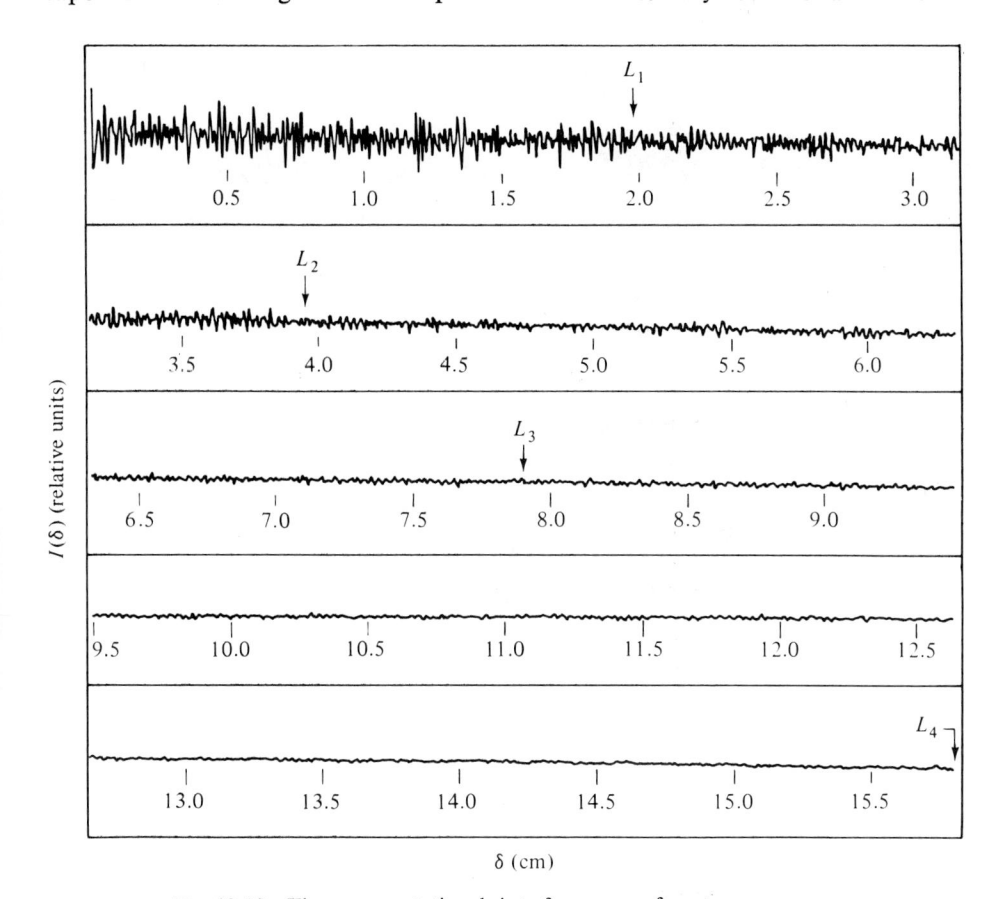

Fig. 17.17. The pure rotational interferogram of water vapor. [From the Proceedings of the International Conference on Fourier Spectroscopy (Aspen, Colorado: Aspen Institute Conference Center, 1970), p. 143.]

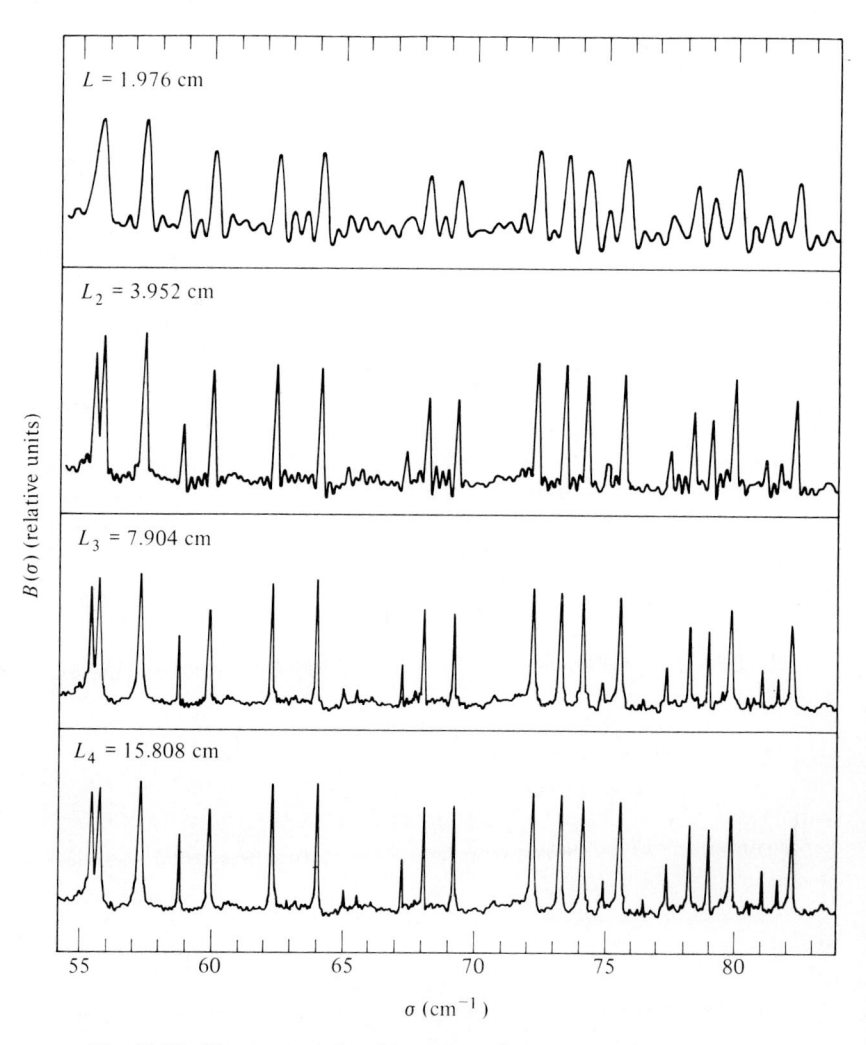

Fig. 17.18. The pure rotational spectrum of water vapor between 54 and 84 cm^{-1}. The spectrum is computed from the interferogram shown in Fig. 17.17 truncated at $x = 1.976$ cm, 3.952 cm, 7.904 cm, and 15.808 cm. [From the Proceedings of the International Conference on Fourier Spectroscopy (Aspen, Colorado: Aspen Institute Conference Center, 1970), p. 152.]

spectrum of water vapor computed from the interferogram of Fig. 17.17 truncated for four path differences is shown in Fig. 17.18. Very little change in the spectrum occurs once δ becomes greater than 7.9 cm.

17.8 THE MICHELSON-MORLEY EXPERIMENT

In the nineteenth century physicists assumed that the universe was pervaded by a medium that they referred to as the **luminiferous ether** and that had the sole property of supporting the propagation of electromagnetic waves. A reference frame attached to the ether was thought by Isaac Newton to constitute a preferred inertial frame. It was predicted that observers moving with respect to this frame would measure the speed of light to be dependent on their state of motion. A critical experiment[8] for testing this thesis was proposed by Clerk Maxwell and performed first by Albert Michelson alone and later with the assistance of Edward Morley.

The Michelson–Morley experiment was designed to measure the earth's motion relative to the ether. If the speed of light is determined by the properties of the ether, the apparent speed as measured by an earth-based observer, according to the Galilean transformation,[9] should be dependent on the velocity of the earth through the ether. The experiment employs a Michelson interferometer similar in principle to that shown in Fig. 17.15 but with both end mirrors fixed a distance L from the half-silvered mirror. In practice, the end mirrors are not exactly perpendicular to the light beams incident upon them. Therefore, a phase difference occurs between rays reflected from different regions of each mirror giving rise to an interference pattern at the detector. It may be shown (see Problem 28) that, on the basis of the Galilean transformation, a time difference exists for light to travel equal distances along the mutually perpendicular paths for a specific orientation of the interferometer. The phase difference producing the interference pattern seen at the detector is then due partially to the difference in travel time in the two arms of the apparatus. If the interferometer is rotated through 90°, the transit times in the two arms are reversed and the effect of travel time on the phase difference is also reversed and should lead to a change in the interference pattern. To carry out the experiment an observer views the interference pattern as the apparatus is rotated through 90° and watches for a movement of the interference fringes across his field of view. Taking 30 km·sec^{-1}, the earth's orbital speed, as the relative speed of the interferometer and the ether, Michelson and Morley calculated that the expected shift in the fringe pattern would be about 200 times the minimum observable shift. At some time during the day or night of some day during the year, the speed of the earth relative to the ether must be at least 30 km·sec^{-1} even if the ether is not stationary. No shift has ever been observed, however, although the experiment has been repeated many times with increasingly more sensitive apparatus since the initial attempt in 1880. The **null result** is no longer disputed; it is now firmly believed that the motion of the earth through space cannot be detected as a motion of the earth relative to the ether.

---

[8] R. S. Shankland, "The Michelson–Morley Experiment," *Scientific American*, November, 1964. Available as *Scientific American Offprint 321* (San Francisco: W. H. Freeman and Co., Publishers).

[9] *MWTP*, Section 4.4.

One possible explanation for the null result of the Michelson–Morley experiment is that the earth drags an envelope of the ether along with it, this envelope being at rest relative to the earth. Other experimental evidence, however, rules out this possibility.

The presently accepted explanation is that the Lorentz transformation,[10] and not the Galilean transformation, correctly relates measurements in frames of reference in uniform relative motion. According to the Lorentz transformation, the speed of light is a constant independent of the relative motion of the source and the observer. Therefore, the experimental detection of the ether is precluded and the possibility of identifying a preferred inertial reference frame is lost. In his special theory of relativity Albert Einstein demonstrated that the seemingly innocuous hypothesis of the ether had led to a completely incorrect interpretation of time and space.

QUESTIONS AND PROBLEMS

1. Show that the principle of linear superposition is a direct consequence of Maxwell's equations.

2. Two oscillators separated by a distance d and producing radiation of wavelength λ are out of phase by an angle ϕ. Show that

$$\cos\theta = \left(m - \frac{\phi}{2\pi}\right)\frac{\lambda}{d}$$

where θ is the angle between one direction of maximum intensity and the line joining the oscillators and m is an integer.

3. In a double-slit experiment light of 500 nm is used to illuminate two slits separated by 0.10 mm. Calculate the separation between adjacent fringes if the interference pattern is observed on a screen 1.0 m from the slits.

4. Calculate the interference pattern that would be obtained using three identical and coherent sources as shown in Fig. 17.19. Plot your results.

5. The arrangement shown in Fig. 17.20, known as the Lloyd's **single-mirror experiment**, may be used to demonstrate interference using visible radiation. For a source of wavelength 550 nm situated 1.0 mm above the mirror and a plane of observation 5.0 m from the source, calculate the fringe separation.

6. A microwave transmitter of wavelength λ is located on the x axis a distance $d = 3\lambda$ from a conducting plane (yz plane). Deduce the **nodal lines** (loci of points of minimum intensity) for the case that the electric field is in the z direction. Will any change in the pattern occur if the magnetic field rather than the electric field is in the z direction?

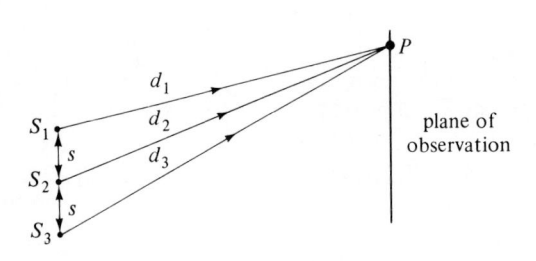

Fig. 17.19. The radiation from three identical and coherent sources interferes at the point P.

[10]*MWTP*, Section 4.7.

7. A circular lens rests on a flat plate. The bottom of the lens is spherical of radius R and the top is flat. Show that concentric dark circles of radii $(mR\lambda)^{1/2}$ where $m = 0, 1, 2 \ldots$ will appear when the lens is illuminated with light of wavelength λ and viewed from above. Assume that R is much larger than the width of the lens. The interference pattern produced in this way is known as **Newton's rings**.

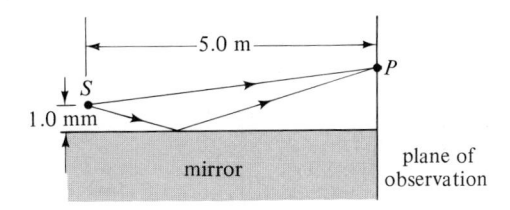

Fig. 17.20. Lloyd's single-mirror experiment. The distance of the source above the mirror is exaggerated for clarity.

*8. The diffraction pattern resulting from the illumination of a slit of width D can be determined by replacing the slit with an array of N point sources spaced a distance d from each other (see Fig. 17.21). If each point source contributes a wave $\text{Re}\left\{\dfrac{E_0}{r} \exp i(\omega t - kr)\right\}$ and the number of sources is increased without limit in such a way that the intensity remains fixed, that is,

$$\lim_{N \to \infty} \left(\frac{E_0 N}{d}\right) = C,$$

show that

$$E_{\text{total}} = \text{Re}\left[C \exp(i\omega t) \int_{-D/2}^{+D/2} \frac{(\exp ikr)}{r} dy\right].$$

Make suitable approximations for r in the integral and show that

$$E_{\text{total}} = \frac{DC}{R} \frac{\sin\left(\frac{kD}{2} \sin\theta\right)}{\frac{kD}{2} \sin\theta} \cos(\omega t - kR).$$

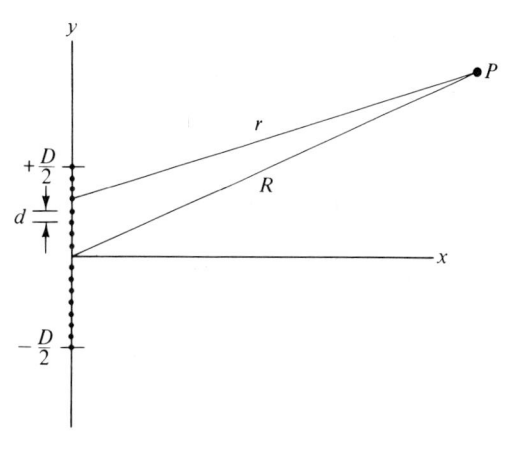

Fig. 17.21. A single slit diffraction experiment.

9. The radiation from a laser operating at 693.4 nm has a coherence length of 1.0×10^4 m. Deduce the line width of the radiation. State your answer in both nm and Hz.

10. The monochromator illustrated schematically in Fig. 17.22 is used to obtain approximately monochromatic light from a white-light source. Given that the linear dispersion of the device is 2.5 nm·mm^{-1} and that the exit slit is 0.15 mm, deduce the coherence time and coherence length of light leaving the monochromator with mean wavelength 400 nm.

11. The classic experiment that demonstrates interference with visible light was first carried out by Thomas Young in 1802. An interference experiment of the Young type is illustrated in Fig. 17.23. The pinhole of diameter 1.0 mm acts as the source of coherent radiation of 589 nm. Deduce the maximum distance between the slits so that interference fringes will just be observable.

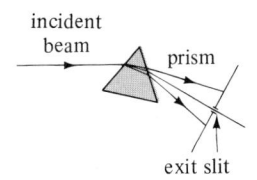

Fig. 17.22. Schematic view of a monochromator.

12. In Section 17.3 we showed that the fringe visibility is equal to the degree of coherence for the case of equal amplitudes in a two beam interference experiment. Generalize this result for arbitrary amplitudes of the two radiation fields.

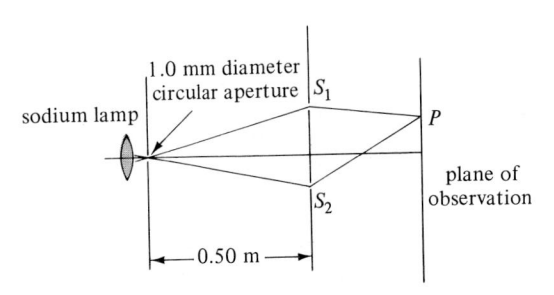

Fig. 17.23. The Young double slit interference experiment.

13. Show that the correlation function for the idealized source introduced in Section 17.3 is

$$C_{AB}(\tau) = \cos \omega_0 \tau \, Re \left\{ \lim_{T \to \infty} \frac{1}{T} \right.$$
$$\left. \times \int_0^T \exp\{-i[\phi_A(t) - \phi_B(t+\tau)]\} \, dt \right\}$$

as stated.

14. Estimate the lateral coherence width of sunlight.

15. In Section 17.4 it is stated that

$$\overline{Re[E_{A1}^*(t)E_{B1}(t+\tau_1)]}$$
$$= (I_A I_B)^{1/2} \left(1 - \frac{\tau_1}{\tau_0}\right) \cos(\omega_0 \tau_1)$$

with

$$\tau_1 = \frac{r_{A1} - r_{B1}}{v}$$

Prove this result.

16. A damped electromagnetic wave has the form

$$E(t) = E_0 \exp[-(\alpha - i\omega_0)t] \quad \text{for } t \geq 0$$
$$= 0 \quad \text{for } t < 0.$$

Evaluate the power spectrum of the wave. Plot both the wave and its power spectrum.

17. Repeat Problem 16 for a Gaussian pulse of electromagnetic radiation. Hint: For a Gaussian pulse

$$E(t) = E_0 \exp[-(\alpha t - i\omega_0)t].$$

18. Write a detailed account of long baseline interferometry.

19. Discuss the use of interference methods for the measurement of the angular diameters of stars.[11]

20. It has been proposed[12,13] that radiofrequency astronomical interferometers may someday be used to give very precise measurements of the fluctuations in the rotational period of the earth. Discuss this statement.

21. Write a short essay on the radio sources that have been observed in the solar system.[14]

---

[11]R. Hanbury Brown, "Measuring the Angular Diameters of Stars," *Contemporary Physics*, **12** (July 1971), 357.

[12]T. Gold, "Radio Method for the Precise Measurement of the Rotation Period of the Earth," *Science*, **157** (July 1967), 302.

[13]G. J. F. MacDonald, "Implications for Geophysics of the Precise Measurement of the Earth's Rotation," *Science*, **157** (July 1967), 304.

[14]M. R. Kundu, *Solar Radio Astronomy* (New York: Interscience Publishers, 1965).

22. The telescopic image of a star is in reality a diffraction pattern. The angular width of the central maximum is approximately $1.22\lambda/D$ with λ the wavelength of the radiation and D the diameter of the aperture of the telescope. This angle also represents the best theoretically possible resolution of the telescope. Use this formula to estimate the resolution of (a) a 5.0 m diameter optical telescope at 500 nm, (b) a parabolic antenna 70 m in diameter used for 21 cm radiation, and (c) two individual antennas 1.0 km apart used for 3 cm radiation.

23. If an interferometer were constructed using the moon and the earth as end points, what resolution would be obtained at 21.0 cm?

24. The lateral coherence width associated with the radiation from a star is found to be 15.0 m for 550 nm wavelength light using a Michelson stellar interferometer. Calculate the diameter of the star if it is known to be 4.0×10^{14} m distant from the earth. How large a baseline would be required to measure this diameter if the observations were carried out at 21 cm?

25. The insertion of a thin film of transparent material with refractive index $n = 1.60$ into one of the beams of a Michelson interferometer causes a displacement of 15 fringes when light of 530 nm wavelength is used. Calculate the thickness of the film.

26. Describe how you could measure the separation of the two **sodium "D" lines** using a Michelson interferometer. The "D" lines are a closely spaced doublet in the visible portion of the spectrum.

27. An evacuated cell of length 10 cm is placed in one arm of a Michelson interferometer using a light source of 589 nm. When methane gas is allowed to enter the cell, the interference pattern is observed to shift by 75 fringes as the pressure increases to atmospheric. Calculate the index of refraction of methane at this frequency.

28. Suppose that a Michelson interferometer is aligned so that light incident upon the half-silvered mirror from the source is in the direction of the positive x, x' axes of coordinate systems attached to the ether and to the earth, respectively. Assuming the Galilean transformation, deduce expressions for the time required for light to travel from the half-silvered mirror to each end mirror and back again. Therefore, calculate the fringe shift expected on the basis of the ether theory. Obtain a numerical value for the fringe shift expected given that the distance between the half-silvered mirror and each of the other mirrors is 10 m and that the earth's orbital speed is 3.0×10^4 m·sec^{-1}.

AUGUSTIN FRESNEL

18 Multiple beam interference

18.1 INTRODUCTION

The interference that results from the multiple reflections of a beam of light between the two surfaces of a thin film of semitransparent material is responsible for the familiar color effects shown by films of oil on water and by soap bubbles. In 1816 the French physicist Augustin Jean Fresnel derived an equation to relate the reflection coefficient to the refractive indexes of the media concerned. The phenomenon of multiple interference provides the basis for an interferometer that is capable of high-precision wavelength measurements—the **Fabry–Perot interferometer**. Further, the principles of multiple interference are utilized in the design of antireflecting films and interference filters.

18.2 MULTIPLE REFLECTIONS FROM A THIN PLATE

We begin our discussion of multiple beam interferometry by considering a beam of light incident upon a thin nonabsorbing and semitransparent plate of thickness d. The electric vector is taken to be perpendicular to the plane of the page (see Fig. 18.1). The incident radiation (electric vector of magnitude E_0) makes an angle θ with the normal to the surface [Fig. 18.1(a)]. At the upper air-glass interface [Fig. 18.1(b)] the beam is partly reflected and partly transmitted. If the **coefficient of reflection** is r and the **coefficient of transmission** is t, the amplitude of the reflected portion is $E_0 r$ and the amplitude of the transmitted portion is $E_0 t$. The transmitted beam proceeds to the lower interface [Fig. 18.1(c)] where it is partially reflected to produce a component of amplitude $(E_0 t)r$ and partially transmitted to produce a component of amplitude $(E_0 t)t$ (assuming r and t to be the same for a glass-air as for an air-glass interface). The reflected component travels back to the upper interface where it in turn is partially reflected to produce a component of amplitude $(E_0 tr)r$ and partially transmitted to produce a component of amplitude $(E_0 tr)t$ [Fig. 18.1(d)]. The process continues indefinitely.

Each successive transmitted component travels an additional distance $2d \cos \phi$ where ϕ is the angle that the ray traveling inside the plate makes with the normal to the surface. The associated phase

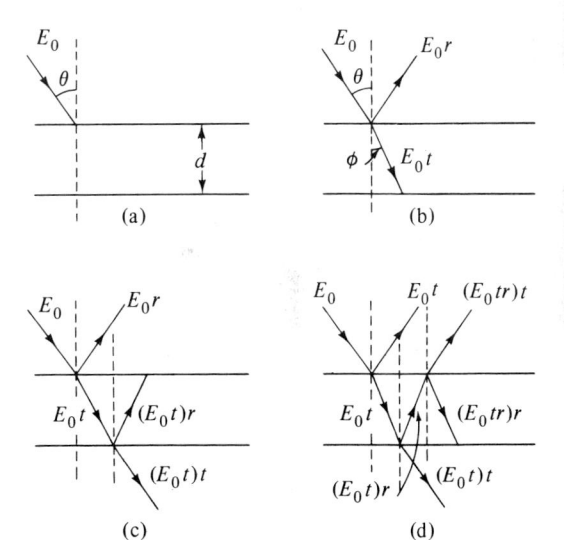

Fig. 18.1. (a) A beam of light incident upon a semitransparent plate. (b) The beam is partly transmitted and partly reflected at the upper surface. (c) The first transmitted beam is itself partly transmitted and partly reflected at the lower surface. (d) The beam once reflected from the lower surface is in turn partly transmitted and partly reflected at the upper surface.

difference δ between successive rays is

$$\delta = 2\pi \left(\frac{2d \cos \phi}{\lambda} \right)$$

(see Problem 1). To obtain the total transmitted beam E_T we must sum the contributions of all of the transmitted components taking into account both the variations in amplitude and in phase between successive components. That is,

$$\begin{aligned} E_T &= E_0 t^2 + E_0 t^2 r^2 \exp(i\delta) + E_0 t^2 r^4 \exp(2i\delta) + \ldots \\ &= E_0 t^2 [1 + r^2 \exp(i\delta) + r^4 \exp(2i\delta) + \ldots] \\ &= \frac{E_0 t^2}{1 - r^2 \exp(i\delta)}. \end{aligned}$$

The intensity of the transmitted light I_T is therefore

$$\begin{aligned} I_T &= |E_T|^2 \\ &= \frac{I_0 t^4}{|1 - r^2 \exp(i\delta)|^2}. \end{aligned}$$

In general, a phase change is induced by a reflection (Section 14.4), so that r is a complex number. It is convenient to denote r by

$$r = |r| \exp\left(\frac{i\delta_r}{2} \right)$$

where $(\delta_r/2)$ is the phase change induced by a single reflection. Therefore

$$\begin{aligned} r^2 &= |r^2| \exp(i\delta_r) \\ &\equiv R \exp(i\delta_r) \end{aligned}$$

where $R = |r^2|$ is by definition the **reflectance** of the surface. We also define the **transmittance** T as

$$T \equiv |t^2|$$

Putting $\delta_r + \delta = \Delta$, we have

$$\begin{aligned} I_T &= \frac{I_0 T^2}{|1 - R \exp i\Delta|^2} \\ &= \frac{I_0 T^2}{(1 - R \cos \Delta)^2 + R^2 \sin^2 \Delta} \\ &= \frac{I_0 T^2}{1 + R^2 - 2R \cos \Delta} \\ &= \frac{I_0 T^2}{1 - 2R + R^2 + 4R \sin^2 \frac{\Delta}{2}} \\ &= \frac{I_0 T^2}{(1 - R)^2} \frac{1}{1 + \left[\frac{4R}{(1 - R)^2} \right] \sin^2 \frac{\Delta}{2}}. \end{aligned}$$

THE FABRY-PEROT INTERFEROMETER

This expression shows that the intensity of the transmitted radiation varies with the path difference between successive transmitted components. The maxima and minima of the interference pattern occur for $\Delta = 2n\pi$ and $\Delta = (2n + 1)\pi$, respectively, with n an integer. The maximum and minimum transmitted intensities are

$$I_T^{max} = I_0 \frac{T^2}{(1-R)^2}$$

$$I_T^{min} = I_0 \frac{T^2}{(1-R)^2} \frac{1}{1 + \left[\frac{4R}{(1-R)^2}\right]} = I_0 \frac{T^2}{(1+R)^2}.$$

Since absorption within the plate is neglected,

$$T = 1 - R$$

and

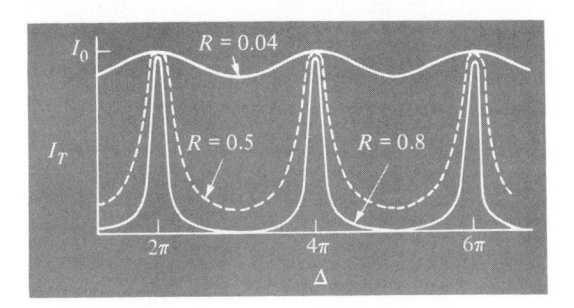

$$I_T^{max} = I_0$$

$$I_T^{min} = I_0 \left(\frac{1-R}{1+R}\right)^2.$$

Fig. 18.2. Intensity contours in a transmitted beam due to multiple reflections within a semitransparent plate, for three values of the reflectance R.

A plot of the intensity distribution as a function of Δ for monochromatic radiation and for several values of R is shown in Fig. 18.2. We see that the fringes become sharper **as the value of R approaches unity.**

18.3 THE FABRY-PEROT INTERFEROMETER

The principle of multiple beam interference finds application in the **Fabry–Perot interferometer.** In essence, the Fabry–Perot interferometer consists of two optically flat, partially reflecting plates of glass. (An optically flat piece of glass has irregularities in its surface that are much smaller than the wavelength of light. The glass used in a Fabry–Perot interferometer must have a flatness of 0.05λ or less where λ is the wavelength of the light being transmitted.) The plates are mounted so as to be accurately parallel with the highly reflecting surfaces inward. The space between the plates may be fixed in which case the device is called an **etalon** or the space between the plates may be varied mechanically in which case the device is called an **interferometer**. A Fabry–Perot interferometer set up to produce circular interference fringes is shown in Fig. 18.3. The lens is an essential part of the arrangement and is required to bring the set of parallel rays resulting from the multiple reflections together for interference. If the condition for reinforcement, namely,

$$2d \cos \theta = n\lambda, \qquad n = \text{an integer,}$$

is fulfilled at point P, it is also fulfilled for all points on the circle through P with center O defined by the line through the center of the lens and perpendicular to the plates of the interferometer.

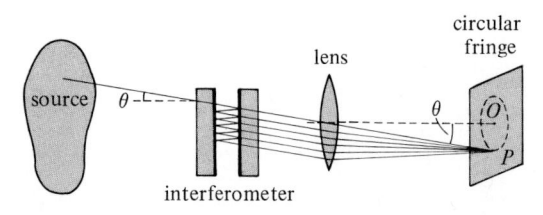

Fig. 18.3. The formation of circular interference fringes by multiple reflections within a Fabry–Perot interferometer.

A fringe pattern that would result from light of a single frequency ν incident upon the interferometer is illustrated in Fig. 18.4. The intensity distribution along any diameter of the set of concentric circular fringes is as shown in Fig. 18.2. The positions of the maxima along a diameter are not evenly spaced, however, due to the action of the lens. Positions corresponding to increases of 2π in the value of Δ become progressively closer together as the distance from the center of the fringe pattern increases.

A parameter of interest in assessing the quality of a Fabry–Perot interferometer is the **free spectral range** $\Delta \nu_f$, defined by

$$\Delta \nu_f = \frac{c}{2d \cos \phi}.$$

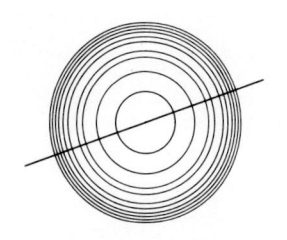

Fig. 18.4. Typical fringe pattern for monochromatic light incident upon a Fabry–Perot interferometer.

We leave it as an exercise for the reader (see Problem 4) to show that $\Delta \nu_f$ is the difference in frequency corresponding to two adjacent maxima in the interference pattern. For small angles of incidence, $\phi \simeq 0$ and

$$\Delta \nu_f \simeq \frac{c}{2d}.$$

The free spectral range is often quoted in units of wavelength and denoted by $\Delta \lambda_f$. Since

$$\Delta \lambda = \frac{\lambda^2}{c} \Delta \nu,$$

we have

$$\Delta \lambda_f = \frac{\lambda^2}{c} \frac{c}{2d \cos \phi} = \frac{\lambda^2}{2d}$$

for $\phi \simeq 0$.

SEC. 18.3   THE FABRY-PEROT INTERFEROMETER

For $\lambda \simeq 500$ nm and $d = 0.1$ cm,

$$\Delta\lambda_f = \frac{500 \times 500 \times 10^{-18}}{2 \times 0.1 \times 10^{-2}}$$

$$\simeq 0.1 \text{ nm.}$$

In such an interferometer the free spectral range at 500 nm is 0.1 nm.

If the source consists of two components that differ slightly in frequency, the intensity distribution along a diameter of the resulting fringe system will be as shown in Fig. 18.5. As the difference in frequency between the components ν and ν' decreases, the separation $\Delta'_n - \Delta_n$ decreases. Another parameter of importance for an interferometer is its **resolving power**, that is, its ability to separate two nearly coincident spectral components. The resolving power clearly depends on the width of the interference fringes and therefore on the reflectance of the plates of the interferometer. By convention, the **half-width** Δ_h of a fringe is defined as the value of Δ for which the intensity is one-half of its maximum value. This is illustrated in Fig. 18.6.

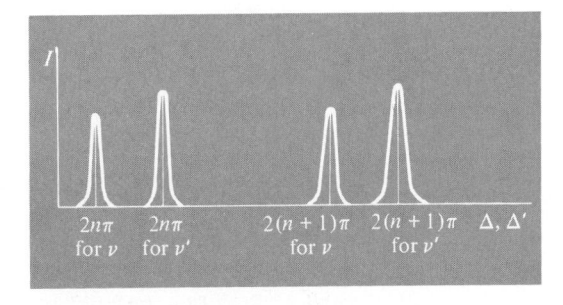

Fig. 18.5. Intensity contours corresponding to two adjacent orders of interference for a source with frequency components ν and ν'.

The intensity I_T as a function of Δ is given by

$$I_T = I_T^0 \frac{1}{1 + \left[\dfrac{4R}{(1-R)^2}\right] \sin^2 \dfrac{\Delta}{2}}$$

where I_T^0 is the maximum intensity. As we see from Fig. 18.6, the intensity falls to $I_T^0/2$ when $\Delta = 2n\pi \pm \Delta_h$. Therefore,

$$\frac{I_T}{I_T^0} = \frac{1}{2}$$

$$= \frac{1}{1 + \left[\dfrac{4R}{(1-R)^2}\right] \sin^2 \left[\dfrac{(2n\pi \pm \Delta_h)}{2}\right]}$$

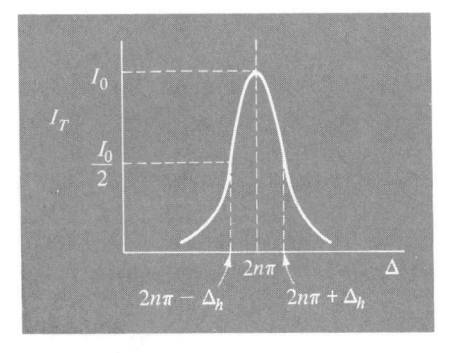

Fig. 18.6. The intensity in an interference maximum falls to one-half of the peak value for $\Delta = 2n\pi \pm \Delta_h$.

or

$$1 + \left[\frac{4R}{(1-R)^2}\right] \sin^2\left(n\pi \pm \frac{\Delta_h}{2}\right) = 2$$

$$\sin^2\left(n\pi \pm \frac{\Delta_h}{2}\right) = \frac{(1-R)^2}{4R}.$$

$$\sin\left(n\pi \pm \frac{\Delta_h}{2}\right) = \pm\frac{(1-R)}{2R^{1/2}}.$$

For large R, the maxima are narrow and Δ_h is small so that to a good approximation,

$$\Delta_h = \frac{1-R}{R^{1/2}}.$$

The half-width Δ_h provides a convenient criterion for the resolving power of an interferometer. As can be seen from Fig. 18.7, two spectral components of equal intensity that are separated by $2\Delta_h$ can be distinguished.

Example. A Fabry–Perot interferometer has a free spectral range of 0.1 nm at 500 nm and a reflectance $R = 0.9$. Can interference maxima from radiation at wavelengths of 500.006 and 500.004 nm be resolved by this instrument?

Solution. The half-width of an interference maximum is

$$\Delta_h = \frac{1-R}{R^{1/2}} = \frac{0.1}{(0.9)^{1/2}} = 0.11 \text{ rad}.$$

The free spectral range $\Delta\lambda_f$ corresponds to two adjacent maximum in the interference pattern for a single wavelength. These maxima are separated in phase by 2π rad. Two different wavelengths that have adjacent maxima at least $2\Delta_h$ different in phase can be resolved. Therefore,

$$\frac{\Delta\lambda_f}{2\pi} = \frac{\Delta\lambda}{2\Delta_h}$$

where $\Delta\lambda$ is the wavelength separation of two component waves which can just be resolved. For this interferometer,

$$\Delta\lambda = \Delta\lambda_f \frac{\Delta_h}{\pi} = 0.1 \times \frac{0.11}{\pi} = 3.5 \times 10^{-3} \text{ nm}.$$

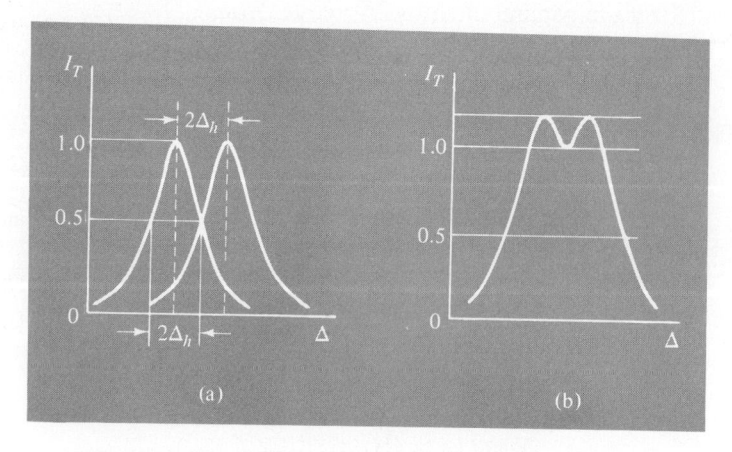

Fig. 18.7. (a) Intensity contours of two frequency components of equal strength and separated by $2\Delta_h$. (b) The summed contour showing the central dip of 17%.

The incident radiation at 500 nm has components differing in wavelength by $0.002 = 2 \times 10^{-3}$ nm. Therefore, these components cannot be resolved by this interferometer.

18.4 MULTILAYER FILMS

Let us consider a slab of dielectric material of index of refraction n' and thickness L. We shall assume the media on either side of the slab to be of infinite extent and characterized by indices of refraction n_i and n_t. For simplicity we shall take the light beam to be normally incident upon the slab (see Fig. 18.8) and the electric field vector to lie in the plane of the page. Subscripts i, r, and t are used to indicate incident, reflected, and transmitted quantities, respectively. Propagation vectors to the right in Fig. 18.8 are positive and propagation vectors to the left are negative. The boundary condition that the tangential component of the electric field must be continuous at an interface requires that

$$E_i + E_r = E'_i + E'_r$$

at interface A and

$$E'_i \exp(ik'L) + E'_r \exp(-ik'L) = E_t,$$

at interface B. The phase factors $\exp(ik'L)$ and $\exp(-ik'L)$ occur since the wave travels a distance L from one interface to the other. Similarly, the boundary condition that the tangential component of the magnetic field must be continuous at an interface requires that

$$B_i - B_r = B'_i - B'_r$$
$$B'_i \exp(ik'L) - B'_r \exp(-ik'L) = B_t.$$

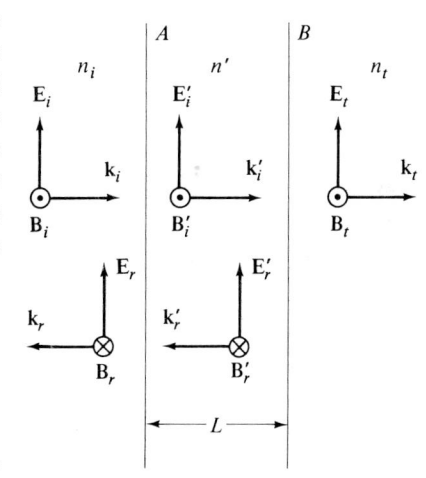

Fig. 18.8. Electromagnetic field vectors and wave vectors for the case of light normally incident upon a single dielectric layer.

In a nonmagnetic medium of index of refraction n'

$$B = n'(\mu_0 \varepsilon_0)^{1/2} E = n'cE$$

(see Sections 14.3 and 14.4). The above equations may therefore be written

$$n_i E_i - n_i E_r = n' E'_i - n' E'_r$$
$$n' E'_i \exp(ik'L) - n' E'_r \exp(-ik'L) = n_t E_t.$$

The amplitudes E'_i and E'_r can be eliminated from the four equations involving the electric field and the remaining two equations rearranged to give the equations

$$1 + r = \left[\cos k'L - i\left(\frac{n_t}{n'}\right) \sin k'L\right] t$$

$$1 - r = \left[-i\left(\frac{n'}{n_i}\right) \sin k'L + \frac{n_t}{n_i} \cos k'L\right] t$$

where $r = E_r/E_i$ and $t = E_t/E_i$ are the reflection and transmission coefficients, respectively. Eliminating t from these two equations, gives

$$n_i(1 - r)\left[\cos k'L - i\left(\frac{n_t}{n'}\right) \sin k'L\right] = (-in' \sin k'L + n_t \cos k'L)(1 + r)$$

or

$$r = \frac{(n_i - n_t) \cos k'L - i\left(\frac{n_i n_t}{n'} - n'\right) \sin k'L}{(n_i + n_t) \cos k'L - i\left(\frac{n_i n_t}{n'} + n'\right) \sin k'L}$$

(see Problem 15). If we take $kL = \pi/2$, that is, if we make the optical thickness of the film equal to $\lambda/4$, then

$$r = \frac{\left(\frac{n_i n_t}{n'} - n'\right)}{\left(\frac{n_i n_t}{n'} + n'\right)}$$

$$= \frac{n_i n_t - n'^2}{n_i n_t + n'^2}.$$

Let us consider the problem of coating a glass plate with an antireflecting film. That is, we take $n_i = 1$ and set $R = |r|^2 = 0$. This requires

$$n' = n_t^{1/2}.$$

That is, the reflectance of the plate for light of wavelength λ may be reduced to zero by coating it with a thin film of index of refraction $n' = n_t^{1/2}$. For a typical glass, $n_t = 1.5$ so that $n' = 1.2(3)$. In practice, magnesium fluoride for which $n' = 1.35$ is often used as a coating material. This reduces the reflectance by a factor of 4.

We now wish to generalize this result to multilayer films. This is done most simply by introducing the **transfer matrix** of a film. The two equations involving r and t can be combined in matrix form (see Appendix D) as follows:

$$\begin{bmatrix} 1 \\ n_i \end{bmatrix} + \begin{bmatrix} 1 \\ -n_i \end{bmatrix} r = \begin{bmatrix} \cos k'L & -\frac{i}{n'} \sin k'L \\ -in' \sin k'L & \cos k'L \end{bmatrix} \begin{bmatrix} 1 \\ n_t \end{bmatrix} t$$

$$= M \begin{bmatrix} 1 \\ n_t \end{bmatrix} t.$$

SEC. 18.4　　MULTILAYER FILMS

For an N-layer film, the ith layer may be represented by a transfer matrix M_i of the above form with appropriate values of n', k', and L. The transfer matrix of the film is the product of the individual matrices.

Let us consider a multilayer film with alternate layers of large index n_L and small index n_S (see Fig. 18.9). We assume each layer to be of thickness $\lambda_0/4$. All M_i are of the same form. The product $M_L M_S$ of two adjacent layers for radiation of wavelength λ_0 is

$$\begin{bmatrix} 0 & -\dfrac{i}{n_L} \\ -in_L & 0 \end{bmatrix} \begin{bmatrix} 0 & -\dfrac{i}{n_S} \\ -in_S & 0 \end{bmatrix} = \begin{bmatrix} -\dfrac{n_S}{n_L} & 0 \\ 0 & -\dfrac{n_L}{n_S} \end{bmatrix}.$$

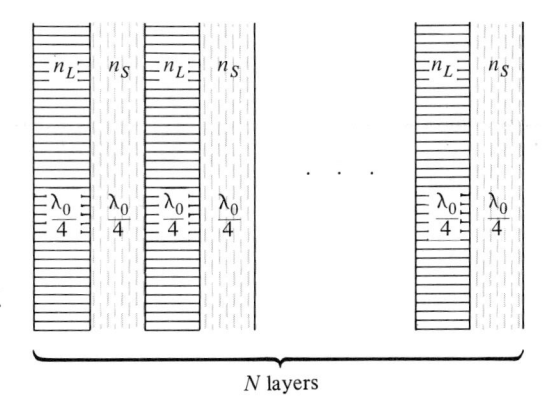

Fig. 18.9. A multilayer film consisting of alternate $\lambda_0/4$ layers of large and small index materials.

Therefore, for an N-layer film,

$$M = \begin{bmatrix} -\dfrac{n_S}{n_L} & 0 \\ 0 & -\dfrac{n_L}{n_S} \end{bmatrix}^{N/2}$$

$$= \begin{bmatrix} \left(-\dfrac{n_S}{n_L}\right)^{N/2} & 0 \\ 0 & \left(-\dfrac{n_L}{n_S}\right)^{N/2} \end{bmatrix}.$$

If we now take $n_i = n_t = 1$, we obtain

$$1 + r = \left(-\dfrac{n_S}{n_L}\right)^{N/2} t$$

$$1 - r = \left(-\dfrac{n_L}{n_S}\right)^{N/2} t$$

and hence

$$r = \dfrac{\left(\dfrac{n_S}{n_L}\right)^{N/2} - \left(\dfrac{n_L}{n_S}\right)^{N/2}}{\left(\dfrac{n_S}{n_L}\right)^{N/2} + \left(\dfrac{n_L}{n_S}\right)^{N/2}}.$$

Note that $R = |r|^2$ approaches unity for large N. Such a multilayer film has arbitrarily high reflectance for one wavelength of incident radiation. As a specific example let us consider the case $n_L = 2.32$ (zinc sulphide) and $n_S = 1.35$ (magnesium fluoride). If we wish to have $R > 0.999$ at a particular wavelength, then from the above equation it follows that

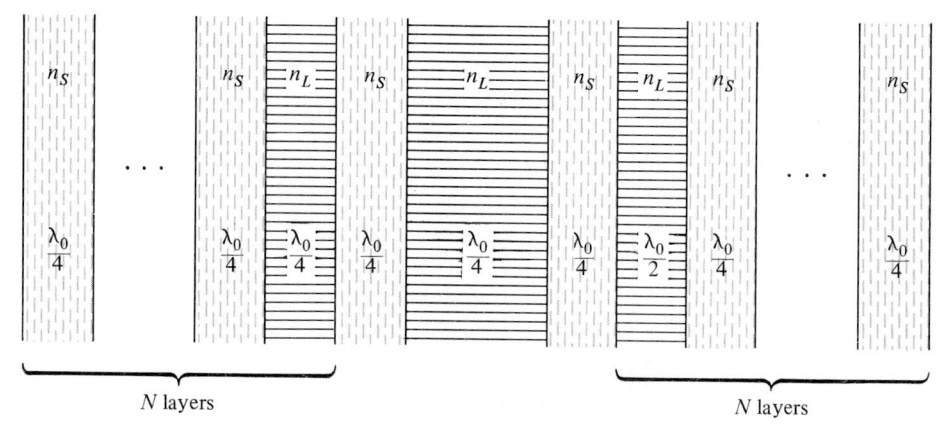

Fig. 18.10. A multilayer interference filter.

$N \geq 30$. In practice, it is possible to manufacture films with high reflectance over a relatively broad range of wavelengths by using layers of varying thickness.

As a second example let us consider a multilayer film that consists of a layer of dielectric of thickness $\lambda_0/2$ (for some wavelength) and is bounded on both sides by partially reflecting surfaces (see Fig. 18.10). The product $M_S(\lambda_0/4)M_L(\lambda_0/2)M_S(\lambda_0/4)$ for radiation of wavelength λ_0 is

$$\begin{bmatrix} 0 & -\dfrac{i}{n_S} \\ -in_S & 0 \end{bmatrix} \begin{bmatrix} -1 & 0 \\ 0 & -1 \end{bmatrix} \begin{bmatrix} 0 & -\dfrac{i}{n_S} \\ -in_S & 0 \end{bmatrix} = \begin{bmatrix} 1 & 0 \\ 0 & 1 \end{bmatrix}.$$

Therefore, for the multilayer film of Fig. 18.10,

$$M = \begin{bmatrix} \left(-\dfrac{n_L}{n_S}\right)^{N/2} & 0 \\ 0 & \left(-\dfrac{n_S}{n_L}\right)^{N/2} \end{bmatrix} \begin{bmatrix} 1 & 0 \\ 0 & 1 \end{bmatrix} \begin{bmatrix} \left(-\dfrac{n_S}{n_L}\right)^{N/2} & 0 \\ 0 & \left(-\dfrac{n_L}{n_S}\right)^{N/2} \end{bmatrix}$$

$$= \begin{bmatrix} 1 & 0 \\ 0 & 1 \end{bmatrix}.$$

If we now take $n_i = n_t = 1$, we obtain

$$1 + r = t = 1 - r$$

and hence $r = 0$ and $t = 1$. That is, for this multilayer film all incident radiation of wavelength λ_0 is transmitted. The variation of the transmission characteristics of this film as a function of λ in the vicinity of λ_0 may be shown to be that of a Fabry–Perot etalon; as a result such a multilayer film is known as a **Fabry–Perot interference filter**. The spectral width of the transmission band decreases as the number of $\lambda_0/4$ layers increases.

Today multilayer films are widely used both in industry and pure research to produce optical surfaces having almost any desired reflectance and transmittance characteristics.[1,2] Some familiar applications of these films are to antireflecting coatings for lenses, to heat-reflecting and heat-transmitting mirrors, and to one-way mirrors.

QUESTIONS AND PROBLEMS

1. Show that the phase difference between two successive rays is in fact

$$\delta = 2\pi\left(\frac{2d \cos \theta}{\lambda}\right)$$

as stated in Section 18.2.

2. Use Fresnel's equations for normal incidence of an electromagnetic wave on a nonmagnetic dielectric medium to deduce the coefficient of transmission t at the interface between the dielectric medium and air. State the result in terms of the index of refraction n_d of the dielectric and the index of refraction n_0 of air.

3. Show that the transmittance T for normal incidence of a system consisting of a plate of thickness d and of index of refraction n_1 between media of indices of refraction n_0 and n_2 (see Fig. 18.11) is given by

$$T = \frac{(1 - r_1^2)(1 - r_2^2)}{1 + r_1^2 r_2^2 - 2r_1 r_2 \cos 2\phi}.$$

In this expression r_1 is the coefficient of reflection at the $n_0 \to n_1$ interface, r_2 is that at the $n_1 \to n_2$ interface, and $\phi = 2\pi n_1 d/\lambda$.

4. Show that the free spectral range of a Fabry–Perot interferometer is the difference in frequency corresponding to two adjacent maxima in the interference pattern.

5. A Fabry–Perot interferometer has plates with reflectance 0.90. What plate separation must be used to resolve two spectral lines separated by 0.014 nm and centered at 656 nm? Deduce the resulting free spectral range in nm.

6. A Fabry–Perot interferometer has been used to resolve spectral components separated by 150 MHz in the output of a helium-neon gas laser. Deduce the maximum allowable plate spacing if $R = 0.999$.

7. A particular Fabry–Perot interferometer can just resolve interference maxima from radiation at wavelengths 530.242 nm and 530.238 nm. If the reflectivity is 0.90, calculate the free spectral range at 530 nm.

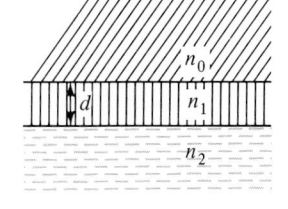

Fig. 18.11. A semitransparent plate bounding two media.

[1] H. Henderson, "Interference Filters," *Contemporary Physics*, **1** (August 1960), 467.
[2] P. Baumeister and G. Pincus, "Optical Interference Coatings," *Scientific American*, December, 1970.

8. Calculate the ratio of the intensity of a maximum to the intensity midway between maxima for a Fabry–Perot interferometer with plates of reflectance 0.90. This quantity is called the **fringe contrast**.

9. The **finesse** of an interferometer is defined as the ratio of the phase shift between two maxima to the phase shift across the half-width. Derive an expression for the finesse of a Fabry–Perot interferometer. Calculate the finesse of an interferometer with plates of reflectance 0.90.

10. What is a **Lummer-Gehrcke interferometer**? Write a short description of the mechanism by which it works.

11. A slab of high-index material constitutes a solid Fabry–Perot etalon. Calculate the resolving power of a cm slab of stibnite (Sb_2S_3) of index of refraction 4.46. Calculate the ratio I_{max}/I_{min}.

12. What is a **cold mirror**?

13. Discuss several applications of optical band-pass filters.

14. What is meant by the term **blooming** when it is used in the discussion of lenses?

*15. Solve the four equations involving the electric fields inside and on either side of a dielectric layer as given in Section 18.4 to yield the result

$$r = \frac{(n_i - n_t)\cos k'L - i\left[\left(\frac{n_i n_t}{n'}\right) - n'\right]\sin k'L}{(n_i + n_t)\cos k'L - i\left[\left(\frac{n_i n_t}{n'}\right) + n'\right]\sin k'L}.$$

16. A glass coverplate of index of refraction 1.50 for a scuba mask is to be designed so that the reflectance is zero for 530 nm wavelength light. Calculate the thickness and index of refraction of a thin-film coating that will meet the design requirements. The refractive index of water is 1.33.

17. A multilayer film is made from alternate layers of zinc sulfide ($n_L = 2.32$) and magnesium fluoride ($n_S = 1.38$). Deduce the maximum reflectance for an 8-layer stack (that is, 4 layers of high and 4 layers of low index material). By how much is the maximum reflectance increased by adding 4 more layers? By how much is the maximum reflectance increased by using titanium dioxide of index 2.40 rather than zinc sulfide?

18. A glass plate coated with a layer of magnesium fluoride ($n = 1.38$) is viewed in reflected light. What color will the plate appear if the thickness of the film is 8.6×10^{-6} cm? What color will the plate appear if the thickness of the film is 12.4×10^{-6} cm?

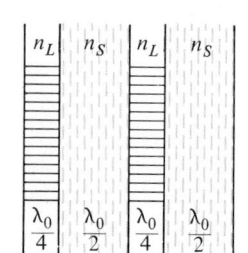

Fig. 18.12. A multilayer film.

19. Calculate the transfer matrix for the multilayer system shown in Fig. 18.12.

*20. A three-layer antireflection coating is shown in Fig. 18.13. Calculate the reflectance of the coating at wavelength λ_0. Such coatings have a lower reflectance than single-layer coatings over most of the visible spectrum.

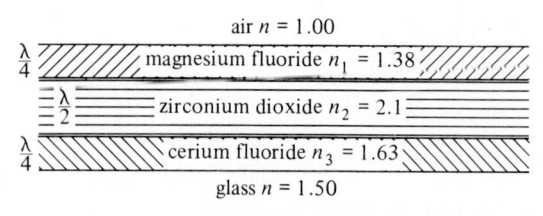

Fig. 18.13. A three-layer antireflection coating.

JOSEPH VON FRAUNHOFER

Diffraction 19

19.1 INTRODUCTION

Diffraction and interference are both terms that have been introduced to describe departures of electromagnetic waves from straight-line propagation. They were first used in the description of optical phemonema and their use was later extended to other regions of the electromagnetic spectrum where similar phenomena were observed. The use of the term diffraction or the term interference to describe a particular phenomenon is primarily a matter of custom. There is no intrinsic physical distinction between them. The shadow cast by an opaque object does not have a sharp edge; this departure from straight-line propagation is an example of diffraction. Radiation passing through a double slit produces a pattern of bright and dark interference bands (or fringes) on a screen placed beyond the slits; this departure from straight-line propagation is an example of interference.

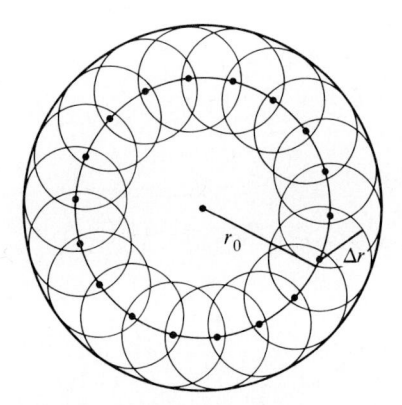

Fig. 19.1. The construction of a spherical wave front according to Huygen's principle.

Huygen's principle provides a qualitative explanation for the essential features of diffraction phenomena and a prescription for the determination of the manner in which an electromagnetic wave propagates. Each point on a wave front is assumed to act as the source of a secondary wave that spreads out in all directions. The new wave front is the envelope of all the secondary waves. This is illustrated in Fig. 19.1 for spherical waves. A burst of radiation from a point source spreads out uniformly in all directions so that after time t_0 the leading edge of the burst defines a spherical wave front of radius r_0. According to Huygen's principle, each point on this wave front serves as a source of spherical waves which, after a further time Δt, have radius Δr. Several of these secondary waves are shown in Fig. 19.1. The envelope of all secondary waves is a spherical surface of radius $r + \Delta r$ and this envelope defines the wave front at a time $t + \Delta t$.

We note that both an ingoing and an outgoing wave are predicted by Huygen's principle. It is possible to formulate a theory in which the ingoing or backward wave is eliminated by the introduction of an **obliquity factor** that cancels out the backward wave.

Huygen's principle, when cast into a precise mathematical form, gives rise to the **Fresnel–Kirchhoff formula** which provides a satisfactory quantitative description of all diffraction phenomena.

19.2 FRAUNHOFER AND FRESNEL DIFFRACTION

There are two general situations that are usually considered in the discussion of diffraction. In the first situation both the incident and diffracted waves can be treated as plane waves; this is termed **Fraunhofer diffraction**. In the second situation either the source or observation point, or both, is close enough to the object causing the diffraction that the curvature of the corresponding wave front must be taken into account; this is termed **Fresnel diffraction**. Clearly, there is no sharp distinction between the two.

We can establish a quantitative criterion for classifying a given diffraction phenomenon as either Fraunhofer or Fresnel. The geometrical arrangement for the general diffraction problem is shown in Fig. 19.2. Radiation traveling from the source S to the reception point R via the lower edge of the aperture travels a shorter distance than that which travels via the upper edge of the aperture. The difference in the distance traveled in these two extreme cases is

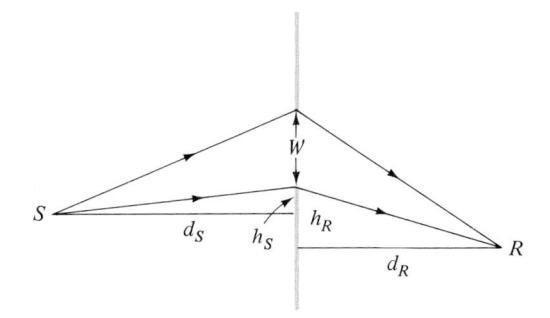

Fig. 19.2. The geometrical arrangement for the general diffraction problem. The source is at S and the receiver is at R. Diffraction occurs due to the presence of an aperture of dimension W in a barrier.

$$[d_S^2 + (h_S + W)^2]^{1/2} + [d_R^2 + (h_R + W)^2]^{1/2}$$
$$- [d_S^2 + h_S^2]^{1/2} - [d_R^2 + h_R^2]^{1/2}$$
$$= \left(\frac{h_S}{d_S} + \frac{h_R}{d_R}\right)W + \frac{1}{2}\left(\frac{1}{d_S} + \frac{1}{d_R}\right)W^2 + \cdots$$

where we have used the binomial theorem to expand the quantities under the square root sign (see Problem 5). The quantities $1/d_S$ and $1/d_R$ provide a measure of the curvature of the wave fronts of the incident and diffracted waves, respectively. Therefore, if the second term in the expansion is small compared to the wavelength λ of the incident radiation, both incident and diffracted waves may be taken to be plane waves. That is, the condition for Fraunhofer diffraction is

$$\frac{1}{2}\left(\frac{1}{d_S} + \frac{1}{d_R}\right)W^2 \ll \lambda.$$

If this condition is not satisfied, the curvature of at least one of the incident or diffracted waves is important and the diffraction is classified as Fresnel diffraction.

Example. In Fig. 19.3 we see the geometric arrangement for a diffraction experiment using a point source of visible light of wavelength 500 nm. Determine whether a Fraunhofer or Fresnel diffraction pattern will be observed.

Solution. For $d_S = d_R = 4.0$ m and $W = 4 \times 10^{-3}$ m,

$$\frac{1}{2}\left(\frac{1}{d_S} + \frac{1}{d_R}\right)W^2 = \frac{1}{2}\left(\frac{1}{4} + \frac{1}{4}\right)(16 \times 10^{-6})$$
$$= 4 \times 10^{-6} \text{ m}.$$

Therefore, for $\lambda = 500$ nm $= 5 \times 10^{-7}$ m,

$$\frac{1}{2}\left(\frac{1}{d_S} + \frac{1}{d_R}\right)W^2 \gg \lambda$$

and Fresnel diffraction applies.

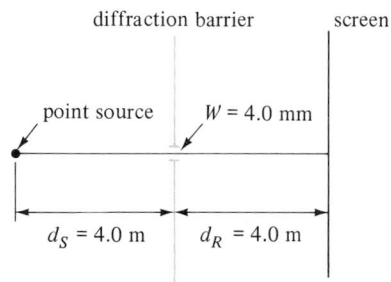

Fig. 19.3. The geometric arrangement for a diffraction experiment.

19.3 FRAUNHOFER DIFFRACTION BY A SINGLE SLIT (QUALITATIVE)

A slit is defined as a rectangular aperture whose length is large compared to its width. An experimental arrangement for Fraunhofer diffraction is shown schematically in Fig. 19.4. The slit is placed with its

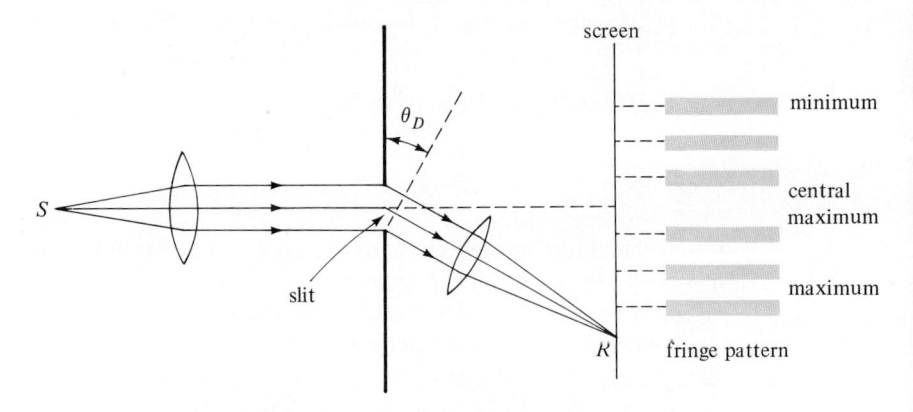

Fig. 19.4. Experimental arrangement to observe Fraunhofer diffraction by a single slit. The diffraction pattern has a broad central maximum. The subsidiary maxima become weaker in intensity as the diffraction angle θ_D increases.

long dimension perpendicular to the page. Note that converging lenses have the property that incident radiation from a point source situated in their focal plane becomes plane wave radiation, and that incident plane wave radiation is focused in their focal plane.[1] The nature of the diffraction pattern produced by the single slit on a screen is also shown in Fig. 19.4. In this section we will present a qualitative discussion of the diffraction pattern produced by a single slit; the correct mathematical description will be given in Section 19.4.

In Fig. 19.5 we show diffracted rays from a single slit for several different diffraction angles. In Fig. 19.5(a) we single out the radiation that passes straight through the slit. The diffraction angle is zero and all secondary waves are in phase and interfere constructively to give a central maximum (intensity I_0, say) in the diffraction pattern. In Fig. 19.5(b) we show the radiation which, after passing through the slit, is traveling in the direction $\theta_D = \sin^{-1}(\lambda/W)$ where λ is the wavelength of the radiation (assumed to be monochromatic). In this direction radiation traveling via the upper edge of the slit travels one wavelength farther than that traveling via the lower edge of the slit. Note that for any arbitrarily chosen ray passing through the lower half of the slit, there exists a ray passing through the upper half of the slit which travels an additional one-half wavelength before reaching the receiving point R. Therefore, we consider the beam to consist of two halves that interfere destructively with each other to produce a minimum of zero intensity in the diffraction pattern.

In Fig. 19.5(c) we show the radiation diffracted through the angle $\theta_D = \sin^{-1}(3\lambda/2W)$. In this case we consider the beam to be divided into thirds; the radiation from two of these thirds interferes destructively leaving the radiation from the remaining third to produce a subsidiary maximum of intensity $I_0/(3)^2 = I_0/9$ in the diffraction pattern.

In Fig. 19.5(d) we show the radiation diffracted through the angle $\theta_D = \sin^{-1}(2\lambda/W)$. We consider the radiation to be divided into quarters, with the radiation from adjacent quarters interfering destructively to produce a second minimum of zero intensity in the diffraction pattern.

In Fig. 19.5(e) we show the radiation diffracted through the angle $\theta_D = \sin^{-1}(5\lambda/2W)$. We consider the radiation divided into fifths; the radiation from four of the fifths interferes destructively leaving the remaining fifth to produce a second subsidiary maximum of intensity $I_0/(5)^2 = I_0/25$ in the diffraction pattern.

Finally, in Fig. 19.5(f) we show the radiation diffracted through the angle $\theta_D = \sin^{-1}(3\lambda/W)$ which produces a third minimum of zero intensity in the diffraction pattern. This procedure can be continued indefinitely.

This analysis predicts the diffraction pattern to have a central maxi-

---

[1] *Physical Science Study Committee, Physics*, 2nd ed. (Boston: D. C. Heath & Company, 1965), Sections 13.10 and 13.11.

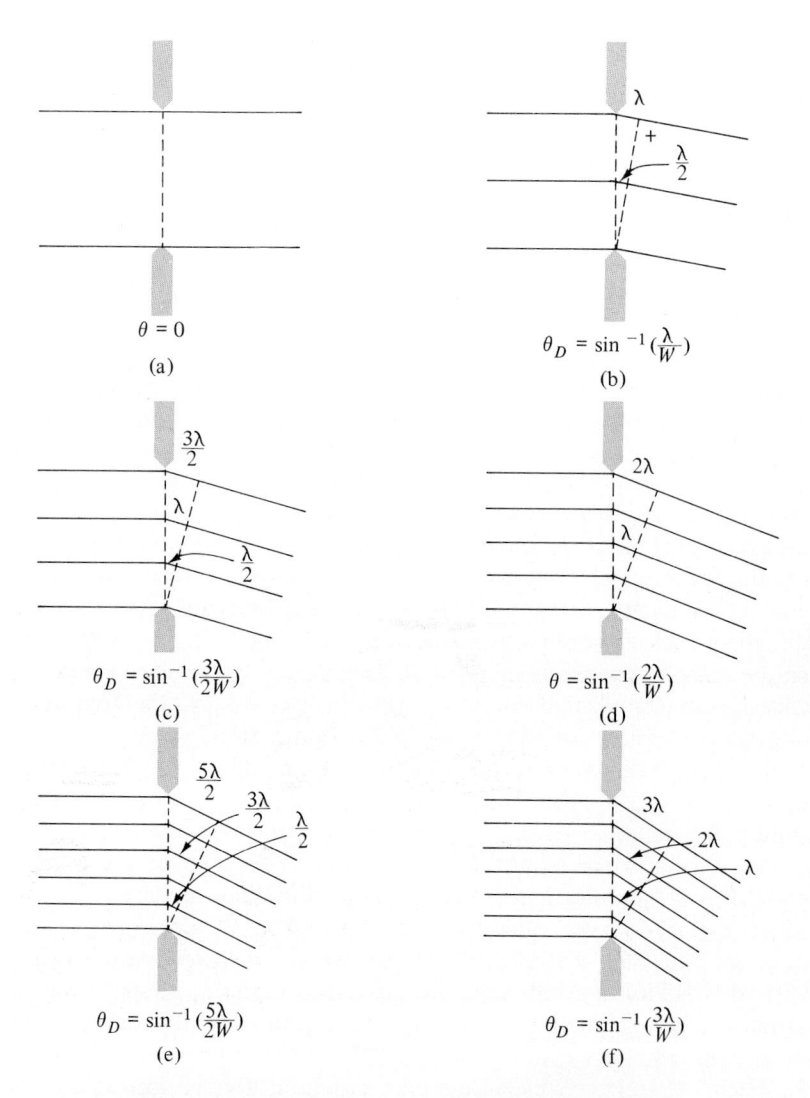

Fig. 19.5. Diffraction by a single slit, showing directions for maxima [(a), (c), and (e)] and for minima [(b), (d), and (f)] in the diffraction pattern.

mum in the straight-through direction and secondary maxima of rapidly diminishing intensities, separated by minima of zero intensity, on either side of the central maximum. For a given wavelength, the angular width of the diffraction pattern increases when the slit width decreases (and vice versa). The intensity of the central maximum increases with the slit width. These predictions are in qualitative agreement with the diffraction pattern produced by a single slit but they are not in quantitative agreement. In particular, this simple picture does not predict correctly the positions of the subsidiary maxima, although it does predict correctly the positions of zero intensity. In addition, the relative intensities of the subsidiary maxima are not given correctly.

19.4 THE FRESNEL-KIRCHHOFF FORMULA FOR FRAUNHOFER DIFFRACTION

The correct mathematical description of diffraction is given by the **Fresnel–Kirchhoff formula**. This formula follows from the application of a very general mathematical theorem (called Green's theorem) to the problem of the diffraction of electromagnetic waves. For the experimental arrangement for the observation of Fraunhofer diffraction shown in Fig. 19.4, the Fresnel–Kirchhoff formula assumes the following simple form for the amplitude E_R of the diffracted wave at the point R

$$E_R = C' \int_A \frac{\exp(ikr)}{r} dA \simeq C \int_A \exp(ikr) \, dA$$

where C and C' are constants, r is the distance from the aperture (which need not be a slit but may assume any general shape) to the point of observation, and the integration is carried out over the area of the aperture. The second integral is a good approximation since the variation in the value of r as the point R moves over the field of observation is much less than the variation in the exponential factor $\exp(ikr)$.

The Fresnel–Kirchhoff formula states that the amplitude of the diffracted wave at a given point is obtained by summing over the area of the aperture the secondary spherical waves, $\exp(ikr)/r$, arriving at that point from the aperture. In fact, in most cases we can ignore the factor r^{-1} and merely sum the phase factors $\exp(ikr)$ over the source. The Fresnel–Kirchhoff formula is a mathematical statement of Huygen's principle.

The single slit

We shall consider a slit of length L and width W. The geometry for the calculation of the diffracted amplitude for a given direction θ is shown in Fig. 19.6. The element of area dA is

$$dA = L \, dz$$

and the distance r is

$$r = r_0 + z \sin \theta$$

where r_0 is the distance from the source to the plane of observation via the

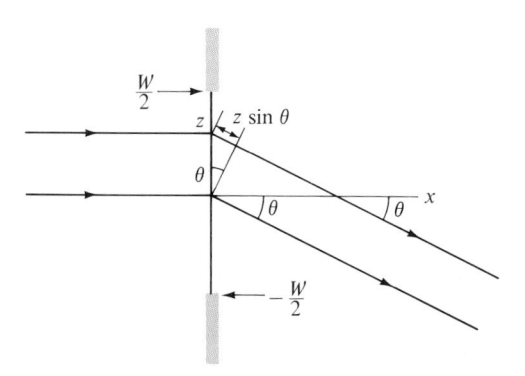

Fig. 19.6. Geometric factors for the calculation of Fraunhofer diffraction by a single slit.

center of the aperture ($z = 0$). The amplitude E_R for the radiation diffracted through angle θ is then

$$E_R = CL \exp(ikr_0) \int_{-W/2}^{W/2} \exp(ikz \sin\theta)\, dz$$

$$= CL \exp(ikr_0) \cdot \left. \frac{\exp(ikz \sin\theta)}{ik \sin\theta} \right|_{-W/2}^{W/2}$$

$$= \frac{CL \exp(ikr_0)}{k \sin\theta} \left[\frac{\exp\left(\frac{ikW \sin\theta}{2}\right) - \exp\left(\frac{-ikW \sin\theta}{2}\right)}{i} \right]$$

$$= \frac{CLW \exp(ikr_0)}{\frac{kW \sin\theta}{2}} \left[\frac{\exp\left(\frac{ikW \sin\theta}{2}\right) - \exp\left(\frac{-ikW \sin\theta}{2}\right)}{2i} \right]$$

$$= E_0 \left[\frac{\sin\left(\frac{kW \sin\theta}{2}\right)}{\frac{kW \sin\theta}{2}} \right] = E_0 \left(\frac{\sin\alpha}{\alpha} \right)$$

where $E_0 = CLW \exp(ikr_0)$ and $\alpha = (kW \sin\theta)/2$.

The intensity I_R of the diffraction pattern over the field of observation (the focal plane of the lens) is

$$I_R = |E_R|^2 = I_0 \left(\frac{\sin\alpha}{\alpha} \right)^2$$

where $I_0 = (CLW)^2$ independent of the diffraction angle θ. This function is plotted in Fig. 19.7. The maximum value is I_0 and occurs for $\theta = \alpha = 0$, the forward or straight-through direction. Zeroes in the intensity pattern occur for

$$\alpha = \pm n\pi$$

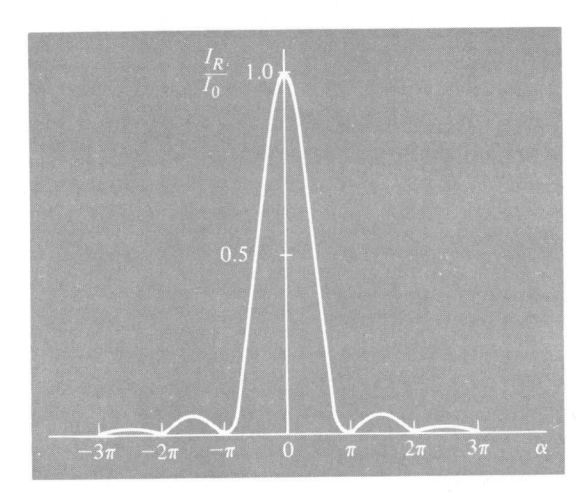

Fig. 19.7. Intensity distribution in the Fraunhofer diffraction pattern of a single slit.

where n is an integer. Subsidiary maxima occur at positions where

$$\frac{dI_R}{d\alpha} = \frac{2I_0 \sin \alpha}{\alpha} \left[\frac{\cos \alpha}{\alpha} - \frac{\sin \alpha}{\alpha^2} \right] = 0,$$

that is, where

$$\tan \alpha = \alpha.$$

This transcendental equation can be solved graphically as indicated in Fig. 19.8. Maxima occur for values of α equal to zero (central maximum), to slightly less than $3\pi/2$ (first subsidiary maximum), to slightly less than $5\pi/2$ (second subsidiary maximum), etc. More exact values, which can be obtained from graphs such as that in Fig. 19.8 but with greatly expanded scales, place the first three subsidiary maxima at values of α equal to 4.494 rad, 7.725 rad, and 10.904 rad, respectively. The corresponding values of I_R/I_0 are 0.0472, 0.0165, and 0.0088, respectively. These results are in good agreement with experiment.

Fig. 19.8. Graphical solution of the equation $\tan \alpha = \alpha$.

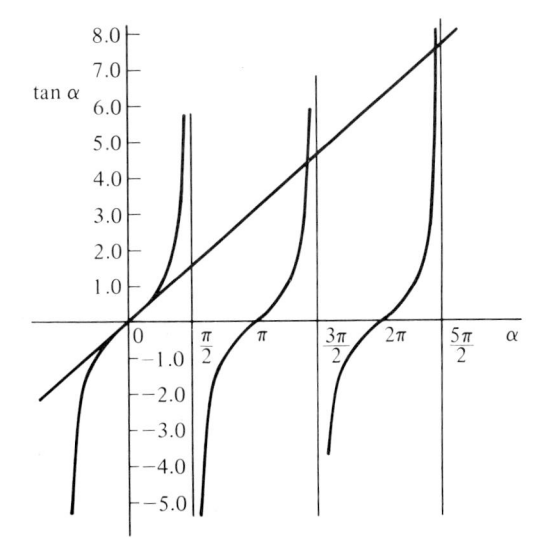

Example. A plane electromagnetic wave of a microwave frequency is diffracted from a slit in a conducting plane. Deduce the nature of the diffraction pattern assuming that the electric field is parallel to the slit and that the slit width $W = (2)^{1/2}\lambda$ where λ is the wavelength of the radiation.

Solution. The geometry of the experiment is illustrated in Fig. 19.9. The slit may be considered to be an array of line sources each emitting radiation of the form

$$E_z = E_0 \left(\frac{\lambda}{r}\right)^{1/2} \cos\theta \cos(\omega t - \mathbf{k}\cdot\mathbf{r})$$

$$= E_0 \left(\frac{\lambda}{r}\right)^{1/2} \cos\theta \cos\left[\omega\left(t - \frac{r}{c}\right)\right]$$

(see Problem 7). We wish to calculate the electric field at a point P in the xy plane. Let this point have coordinates (r_0, θ_0) with respect to the center of the slit (see Fig. 19.10). The electric field at P is given by

$$E(P) = E_0 \int_{-W/2}^{W/2} \left(\frac{\lambda}{r}\right)^{1/2} \cos\theta \cos\left[\omega\left(t - \frac{r}{c}\right)\right] dy$$

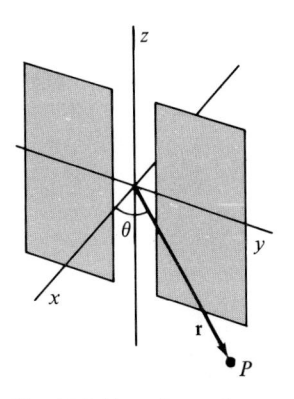

Fig. 19.9. Experimental geometry for the diffraction of microwaves by a single slit.

where

$$r = (r_0^2 - 2r_0 y \sin\theta_0 + y^2)^{1/2}$$

$$\cos\theta = \frac{r_0 \cos\theta_0}{(r_0^2 - 2r_0 y \sin\theta_0 + y^2)^{1/2}}.$$

If $r_0 \gg W$, then θ is very nearly constant over the slit and $(\lambda/r)^{1/2}$ may be approximated by $(\lambda/r_0)^{1/2}$. Therefore,

$$E(P) \simeq E_0 \left(\frac{\lambda}{r_0}\right)^{1/2} \cos\theta_0 \int_{-W/2}^{W/2} \cos\omega\left(t - \frac{r_0}{c} + \frac{y}{c}\sin\theta_0\right) dy$$

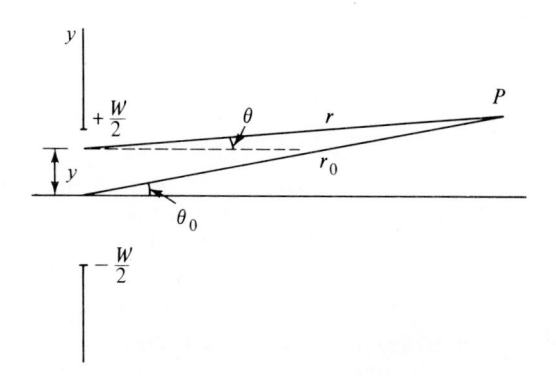

Fig. 19.10. Definition of the polar coordinates of point P.

where the binomial theorem has been used to obtain the approximate value of r in the phase factor. The integration is now straightforward and leads to the expression

$$E(P) = \frac{E_0}{\pi} \frac{\lambda}{W} \left(\frac{\lambda}{r_0}\right)^{1/2} \cot\theta_0 \sin\left(\frac{\pi W}{\lambda} \sin\theta_0\right) \cos\left[\omega\left(t - \frac{r_0}{c}\right)\right].$$

The signal detected at P will be proportional to $E^2(P)$. It is customary to illus-

THE FRESNEL–KIRCHHOFF FORMULA

trate the dependence of the signal upon θ_0 by means of a **polar plot** with $E^2(P)$ as the radius. Such a plot is shown in Fig. 19.11 for $W = (2)^{1/2}\lambda$.

The double slit

As a second example of the application of the Fresnel–Kirchhoff formula to Fraunhofer diffraction, we shall consider two identical slits, each of length L and width W, separated by a distance S. The geometry for the calculation is shown in Fig. 19.12. The amplitude of the wave diffracted in the direction θ is

$$E_R = CL \exp(ikr_0)\left[\int_0^W \exp(ikz\sin\theta)\,dz + \int_S^{S+W} \exp(ikz\sin\theta)\,dz\right]$$

$$= \frac{CL\exp(ikr_0)}{ik\sin\theta}\left[\exp(ikz\sin\theta)\Big|_0^W + \exp(ikz\sin\theta)\Big|_S^{S+W}\right]$$

$$= \frac{CL\exp(ikr_0)}{ik\sin\theta}\{\exp(ikW\sin\theta) - 1 + \exp[ik(S+W)\sin\theta]$$

$$\quad - \exp(ikS\sin\theta)\}$$

$$= CL\exp(ikr_0)\left[\frac{\exp(ikW\sin\theta) - 1}{ik\sin\theta}\right][1 + \exp(ikS\sin\theta)].$$

Fig. 19.11. A polar plot of the dependence of the signal at P on the angle θ_0 for $W = (2)^{1/2}\lambda$.

This expression can be simplified further by factoring $\exp[(ikW\sin\theta)/2]$ out of the terms in the first bracket, and by factoring $\exp[(ikS\sin\theta)/2]$ out of the terms in the second bracket. After a little rearranging, the resulting expression for E_R becomes

$$E_R = 2CLW\exp(ikr_0)$$
$$\times \exp(i\alpha)\exp(i\beta)\frac{\sin\alpha}{\alpha}\cos\beta$$

where

$$\alpha = \frac{1}{2}kW\sin\theta \quad \text{and} \quad \beta = \frac{1}{2}kS\sin\theta$$

(see Problem 9).

The corresponding intensity distribution function is

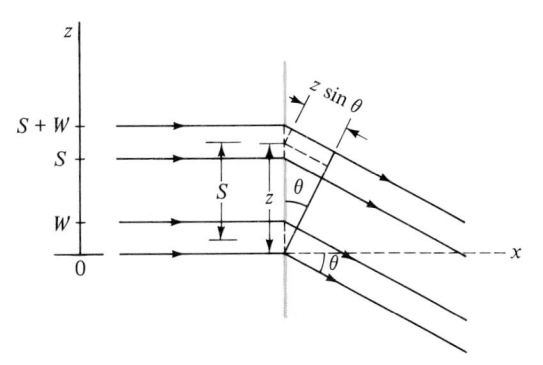

Fig. 19.12. Geometry for the calculation of the Fraunhofer diffraction pattern produced by a double slit.

$$I_R = |E_R|^2 = 4C^2L^2W^2\left(\frac{\sin\alpha}{\alpha}\right)^2\cos^2\beta$$

or

$$I_R = I_0\left(\frac{\sin\alpha}{\alpha}\right)^2\cos^2\beta$$

where, again, I_0 is the intensity in the straight-through direction.
The factor $(\sin\alpha/\alpha)^2$ is the distribution function found previously for

the single slit. The factor $\cos^2 \beta$ oscillates more rapidly than does $\sin^2 \alpha$ since $S > W$. The result is an interference pattern where the positions of the maxima are determined by the $\cos^2 \beta$ term while the intensities of these maxima are modulated by the $(\sin \alpha/\alpha)^2$ term. This is shown in Fig. 19.13. Maxima occur for $\beta = \pm n\pi$ with n an integer. The angular separation between fringes is π rad so that

$$\Delta\beta = \pi \simeq \frac{kS}{2} \cos\theta \, \Delta\theta \simeq \frac{kS \, \Delta\theta}{2}$$

in magnitude, or

$$\Delta\theta \simeq \frac{2\pi}{kS} = \frac{\lambda}{S}$$

which is equivalent to the result obtained in the analysis of the double beam interference experiment as discussed in Section 17.2.

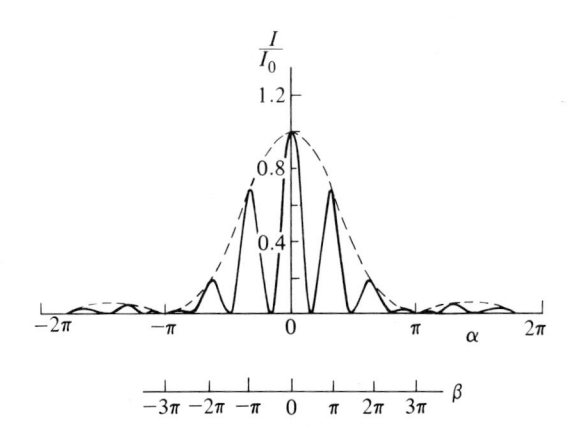

Fig. 19.13. Fraunhofer diffraction pattern for a double slit. The intensity of the pattern is modulated by the factor $(\sin \alpha/\alpha)^2$ (the dotted line) which is the distribution function for a single slit.

19.5 DIFFRACTION GRATINGS

Any device that can be represented by an equivalent set of identical, parallel, equally spaced slits in its interaction with electromagnetic waves is called a **diffraction grating**.[2] Gratings are powerful devices for spectroscopic studies. We can treat the diffraction grating used in the manner shown in Fig. 19.14 as an extension of the double slit problem of Section 19.4; this configuration is known as the **Fraunhofer arrangement**.

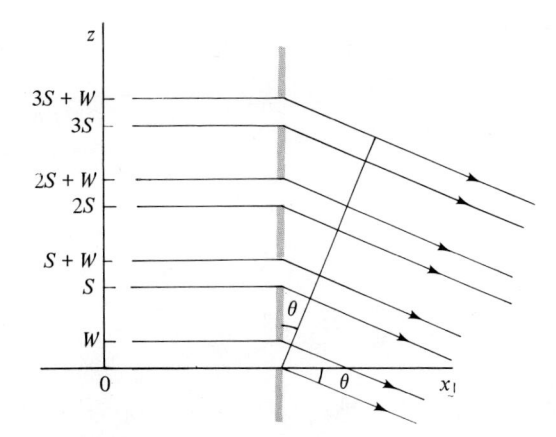

Fig. 19.14. A diffraction grating.

The amplitude of the waves diffracted in the direction θ is (refer to Section 19.4)

$$E_R = CL \exp(ikr_0)\left[\int_0^W + \int_S^{S+W} + \int_{2S}^{2S+W} + \ldots \right.$$

$$\left. + \int_{(N-1)}^{(N-1)S+W} \right] \exp(ikz \sin\theta) \, dz$$

$$= CL \exp(ikr_0) \frac{\exp(ikW \sin\theta) - 1}{ik \sin\theta} \{1 + \exp(ikS \sin\theta) + \ldots$$

$$+ \exp[ik(N-1)S \sin\theta]\}$$

[2] E. W. Palmer and J. F. Verrill, "Diffraction Gratings," *Contemporary Physics,* **9** (May 1968), 257.

SEC. 19.5   DIFFRACTION GRATINGS

$$= CL \exp(ikr_0) \left[\frac{\exp(ikW\sin\theta) - 1}{ik\sin\theta} \right] \left[\frac{1 - \exp(ikNS\sin\theta)}{1 - \exp(ikS\sin\theta)} \right]$$

$$= \frac{CLW}{2} \exp(ikr_0) \exp(i\alpha) \exp(iN\beta) \left(\frac{\sin\alpha}{\alpha} \right) \left(\frac{\sin N\beta}{\beta} \right)$$

where

$$\alpha = \frac{1}{2} kW\sin\theta, \qquad \beta = \frac{1}{2} kS\sin\theta.$$

The intensity distribution is given by

$$I_R = I_0 \left(\frac{\sin\alpha}{\alpha} \right)^2 \left(\frac{\sin N\beta}{N\beta} \right)^2.$$

The factor N^2 has been added in the denominator as a normalization factor so that $I_R = I_0$ when $\theta = 0$.

Maxima in the diffraction pattern are determined by the term containing β while the term containing α modulates the intensity of the principal maxima as shown in Fig. 19.15. Principal maxima occur for

$$\beta = n\pi, \qquad n = 0, 1, 2, \ldots.$$

That is,

$$n\pi = \frac{kS}{2}\sin\theta$$

or

$$n\lambda = S\sin\theta, \qquad n = 0, 1, 2, \ldots.$$

This is the **grating equation** which relates the wavelength λ to the angle of diffraction θ; the integer n is called the **order of diffraction**. Subsidiary maxima (of very low intensity) occur in the diffraction pattern for values

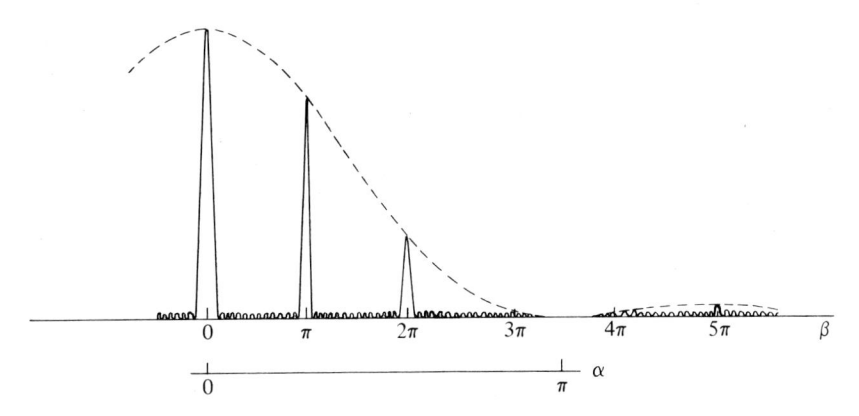

Fig. 19.15. Typical Fraunhofer diffraction pattern for a 15-slit grating. The positions of the maxima are determined by β and the intensities of the principal maxima by α.

of β for which $\sin N\beta = 1$ and minima occur for values of β for which $\sin N\beta = 0$.

The angular width of a principal maximum, defined as the separation between the peak and the adjacent minimum, is found by putting $\Delta(N\beta) = \pi$. That is,

$$\Delta\beta = \frac{\pi}{N} = \frac{1}{2}kS \cos\theta\, \Delta\theta$$

and

$$\Delta\theta = \frac{\lambda}{NS \cos\theta}.$$

As N increases, $\Delta\theta$ decreases, and the maxima become very narrow. By differentiating the grating equation and taking finite differences, we obtain

$$n\Delta\lambda = S \cos\theta\, \Delta\theta$$

or

$$\Delta\theta = \frac{n\, \Delta\lambda}{S \cos\theta}.$$

The two equations for $\Delta\theta$ can be combined to give the **resolving power** $\lambda/\Delta\lambda$ of a grating as

$$\frac{\lambda}{\Delta\lambda} = nN.$$

Example. A typical grating used for optical spectroscopy has of the order of 60,000 lines or slits. By a suitable shaping of the grooves most of the diffracted light can be directed into one order, thereby increasing the efficiency of the grating. Calculate the resolving power of such a grating in the 10th order. What is the minimum separation of two lines centered at 600 nm that can just be resolved with this instrument?

Solution. The resolving power of the grating is

$$\frac{\lambda}{\Delta\lambda} = nN = 10 \times 60{,}000 = 600{,}000.$$

For $\lambda = 600$ nm

$$\Delta\lambda = \frac{\lambda}{nN} = \frac{600}{600{,}000} = 0.001 \text{ nm}.$$

Two lines centered at 600 nm and separated by 0.001 nm can just be resolved by the grating.

19.6 FRESNEL DIFFRACTION

Fresnel diffraction is mathematically more difficult to treat than Fraunhofer diffraction since we are no longer dealing with plane waves. Rather than consider the more general form of the Fresnel–Kirchhoff formula, we shall look at a few simple cases that can be explained by graphical means.

Circular aperture

The experimental arrangement for viewing Fresnel diffraction by a circular aperture is shown in Fig. 19.16. The source is located at S and the point of observation at P so that the line SOP is perpendicular to the plane of the aperture. The point O is in the plane of the aperture; the distance from O to any arbitrary point A in the circular aperture is r. The distance SAP can be written as

$$(a + \Delta a) + (b + \Delta b)$$
$$= (a^2 + r^2)^{1/2} + (b^2 + r^2)^{1/2}$$
$$= (a + b) + \frac{r^2}{2}\left(\frac{1}{a} + \frac{1}{b}\right) + \cdots$$

or

$$\Delta = \Delta a + \Delta b \simeq \frac{r^2}{2}\left(\frac{1}{a} + \frac{1}{b}\right),$$
$$a, b \gg r.$$

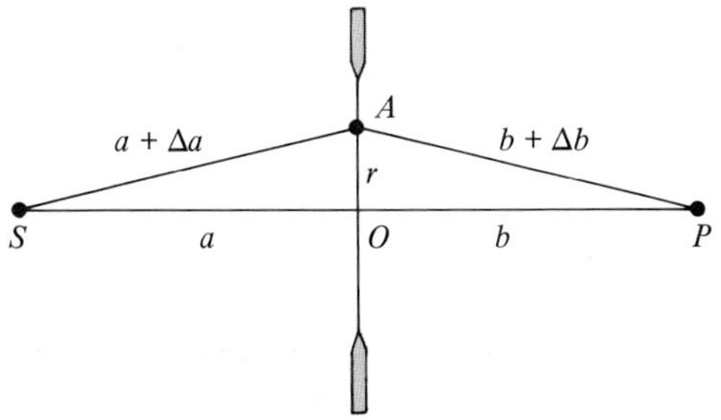

Fig. 19.16. Geometric arrangement for the study of Fresnel diffraction by a circular aperture.

As r increases in magnitude, Δ increases and the path difference SAP–SOP increases.

It is convenient at this point to introduce **Fresnel zones**. These zones are defined by concentric circles drawn about the point O in such a manner that the value of Δ increases by $\lambda/2$ when we move from one circle to the next larger circle; λ is the wavelength of the radiation under study. The nth Fresnel zone is defined by the two concentric circles of radii r_n and r_{n+1} where

$$\Delta_n = \frac{r_n^2}{2}\left(\frac{1}{a} + \frac{1}{b}\right) = \frac{n\lambda}{2}.$$

The radius r_n of the nth zone is then

$$r_n = (n\lambda)^{1/2}\left(\frac{1}{a} + \frac{1}{b}\right)^{-1/2}$$

and the area of the nth zone is

$$\pi r_{n+1}^2 - \pi r_n^2 = \pi\lambda\left(\frac{1}{a} + \frac{1}{b}\right)^{-1} = \pi r_1^2$$

so that the area of a zone is a constant independent of n.

The amplitude of the diffracted wave at point P is the algebraic sum of the contributions $E_1, E_2, \ldots$ from the various Fresnel zones. This sum has magnitude $|E_P|$ where

$$|E_P| = |E_1| - |E_2| + |E_3| - \cdots \pm |E_n| \mp \cdots.$$

Alternate terms are of opposite sign since the mean phase change between adjacent Fresnel zones is π rad. Since the areas of all zones are identical, the contribution to the amplitude at P must be nearly the same for all zones. If the aperture includes exactly N zones, the value of $|E_P|$ would be approximately zero for N even and approximately $|E_1|$ for N odd. However, the Fresnel–Kirchhoff formula contains a radial distance factor that causes the value of $|E_n|$ to decrease slowly as n increases. If the decrease is slow enough, we should expect the contribution $|E_n|$ from the nth zone to be approximately equal to the mean value of the contributions from the two adjacent zones. In this case it is useful to group the terms in the expression for $|E_P|$ as follows:

$$|E_P| = \frac{|E_1|}{2} + \left[\frac{|E_1|}{2} - |E_2| + \frac{|E_3|}{2}\right] + \left[\frac{|E_3|}{2} - |E_4| + \frac{|E_5|}{2}\right] + \cdots$$

The terms in brackets are each very nearly zero.

For the case of no aperture at all ($n = \infty$), we have

$$|E_P| = \frac{|E_1|}{2}$$

while for a small circular aperture with $n = 1$

$$|E_P| = |E_1|.$$

Therefore, the intensity at P for an aperture containing one Fresnel zone is four times greater than the intensity at P for no aperture at all. A complete mathematical treatment using the Fresnel–Kirchhoff formula predicts a circular diffraction pattern at P; this is indeed observed.

Circular obstacle

Diffraction produced by a circular obstacle placed in the path of an electromagnetic wave can be treated in a manner analogous to the circular aperture. In this case the construction of unobstructed Fresnel zones is started at the edge of the obstacle. Following the same reasoning as for the circular aperture, we conclude that the value of $|E_P|$ is one-half the contribution from the first unobstructed zone. That is, at the center of the shadow cast by a circular, opaque obstacle, there is a bright spot that has very nearly the same intensity as would occur if the obstacle were not present at all.

A complete investigation of the diffraction produced by a circular obstacle shows that a faint circular diffraction pattern surrounds the bright spot in the center of the shadow, and that another circular diffraction pattern borders the outside of the shadow.

19.7 DIFFRACTION OF X-RAYS AND PARTICLES

Electromagnetic waves of wavelength $\simeq 0.1$ nm lie in the x-ray region of the electromagnetic spectrum. Suppose we were to attempt to view the diffraction of x-rays of wavelength 0.1 nm with a multiple-slit grating of the type shown in Fig. 19.14. The grating equation

$$n\lambda = S \sin \theta, \quad n = 0, 1, 2, \ldots$$

applies and, in particular, for first order diffraction

$$\lambda = S \sin \theta.$$

For $\lambda = 0.1$ nm $= 10^{-10}$ m,

$$S \sin \theta = 10^{-10}.$$

Since θ must obviously be a small angle, we can replace $\sin \theta$ by θ and write

$$S\theta = 10^{-10}.$$

In order to observe first-order diffraction at, say, $\theta = 1$ mrad, a slit width $S = 10^{-7}$ m is required. It is no surprise that the earliest attempts to detect x-ray diffraction using a conventional grating were not successful but only suggested that x-rays must have wavelengths in the region of 0.1 nm.

In 1912 Max von Laue conceived the idea of using the regularly spaced atoms in a crystalline solid as a three-dimensional diffraction grating. The internuclear separations in crystals are of the order of a few tenths of a nanometer. By assuming that each atom in the crystal would serve as a scattering center for x-rays and applying the principle of interference, von Laue was able to show that the diffracted radiation would interfere constructively in certain specified directions so that diffracted maxima should be observed. This idea formed the basis for subsequent developments in both x-ray spectroscopy and the analysis of crystal structures.

The positions of the principal maxima in the diffraction pattern formed by a conventional multiple-slit grating depend only on the slit width S and not on the details of the shape of the slits (see Section 19.5). Similarly, the details of size and shape of the lattice particles are not important in determining the directions of the principal maxima in a diffraction pattern produced by a space-lattice grating. However, the relative intensities of the maxima do depend on these details through an **atomic form factor** or **structure factor**. In this treatment we shall concern ourselves only with the directions of the principal maxima and not with their relative intensities.

An elementary description of diffraction from a crystal lattice was first given by William L. Bragg who considered the problem from the viewpoint of reflection. We suppose that a plane wave of x-ray radiation is incident upon a simple cubic lattice at an angle θ as shown in Fig. 19.17. This wave interacts with the successive layers of atoms in the lattice, thereby giving

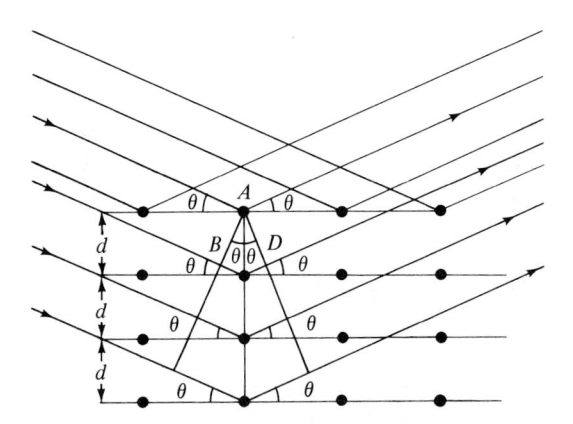

Fig. 19.17. Illustration of the reflection of x-rays from successive layers of a crystal lattice.

rise to a series of scattered waves, each of which leaves the crystal at an angle θ as shown (for clarity, only one scattered wave from each layer below the surface is shown). In order that the waves scattered from successive layers of atoms interfere constructively to give a diffraction maximum, it is necessary that the path difference $2d \sin \theta$ between successive waves be an integral mutiple of the x-ray wavelength λ. That is, **Bragg's law** for a diffraction maximum is

$$n\lambda = 2d \sin \theta, \qquad n = 1, 2, 3, \ldots$$

where n is the order of the diffraction. The low orders are those of practical importance as the intensity decreases rapidly with increasing n due principally to the finite spatial extent of the lattice particles and to their thermal vibrational motion about their equilibrium positions.

The formation of a reflected beam is dependent only on the existence of equally spaced layers of particles in the crystal and not on the regularity of arrangement of the particles within these layers. Many differently oriented sets of planes may be found in a crystal lattice. This is illustrated for a two-dimensional lattice in Fig. 19.18. Each of these sets of planes can be equally well regarded as a set of reflecting planes. Since a fixed number of lattice particles must be accounted for, those sets of planes that have a large number of atoms per unit area have a much larger separation d than those sets of planes that have only a few atoms per unit area. Densely populated planes serve very well as reflecting planes. The low intensity of reflection from sparsely populated planes is attributable to their small separation. The spatial extent of the electron "clouds" associated with the lattice atoms is comparable to the plane separation. This spatial extent, combined with the thermal vibrations of the lattice atoms, reduces the intensity of radiation reflected from these sets of planes.

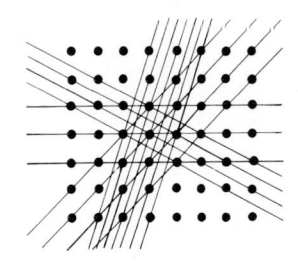

Fig. 19.18. Some of the various families of atomic planes of a two-dimensional lattice.

There is a fundamental difference between diffraction by a crystal lattice and diffraction by a conventional grating. For a space lattice and a specific angle of incidence for the x-rays, there is a single direction of diffraction for any set of reflecting planes, and a single wavelength is diffracted in a given order. For a conventional grating and a specific direction of incidence, there is a continuous range of directions in which a continuous range of wavelength is diffracted in a given order. That is, a **spectrum** is formed by a conventional grating but not by a crystal lattice. A conventional grating is essentially a two-dimensional device and a crystal lattice is three-dimensional. The constraint of the third dimension of a lattice prevents a spectrum being formed when the angle of incidence is fixed. A continuous range of angles of incidence on a crystal lattice can be achieved by keeping the x-ray beam fixed and rotating the crystal; this produces a spectrum of diffracted radiation.

From Bragg's law we see that, for a given order, for small θ, short wavelength x-rays are diffracted and that the wavelength of the diffracted radiation increases with θ to a maximum value at $\theta = \pi/2$ where, for $n = 1$, the wavelength $\lambda = 2d$. The maximum possible value for d in a simple cubic crystal is a, the lattice constant, so that the maximum wavelength that can be diffracted by such a crystal is $\lambda = 2a$. The lower limit to the diffracted wavelength is set by the smallest value of θ that can be obtained in practice and by the smallest value of d for a set of reflecting planes that will produce a diffracted beam of sufficient intensity to be detected. The range of x-ray wavelengths that can be measured by diffraction from crystals extends from $\simeq 0.01$ nm to $\simeq 2$ nm.

The Laue equations

Let us turn to a more general treatment in which we consider the elastic scattering of x-rays by the individual atoms of a crystal. In Fig. 19.19 we show two atoms whose separation is represented by the vector **r**. The incident and reflected waves are represented by unit vectors $\mathbf{s}_i$ and $\mathbf{s}_r$, respectively. The path difference between the waves scattered from the two atoms is

$$r \cos \theta_r - r \cos \theta_i = \mathbf{r} \cdot \mathbf{s}_r - \mathbf{r} \cdot \mathbf{s}_i$$
$$= \mathbf{r} \cdot (\mathbf{s}_r - \mathbf{s}_i) = \mathbf{r} \cdot \mathbf{S}.$$

The phase difference ϕ between the two scattered waves is

$$\phi = \frac{2\pi}{\lambda} \mathbf{r} \cdot \mathbf{S}.$$

If $\phi = 2n\pi$ or $n\lambda = \mathbf{r} \cdot \mathbf{S}$, the two waves will be in phase. From Fig. 19.20 we see that **S** is perpendicular to the plane that reflects $\mathbf{s}_i$ into $\mathbf{s}_r$; it is the plane of the reflecting atoms for the Bragg derivation. Since $\mathbf{s}_i$ and $\mathbf{s}_r$ are

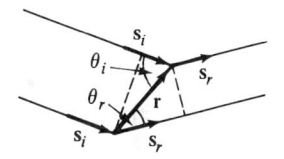

Fig. 19.19. Diagram for the determination of the phase difference between waves scattered from two atoms.

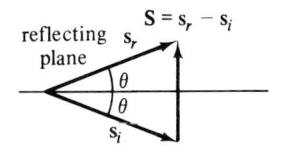

Fig. 19.20. The vector **S** is perpendicular to the plane that reflects $\mathbf{S}_i$ into $\mathbf{S}_r$.

unit vectors

$$S = 2 \sin \theta$$

where 2θ is the angle between $\mathbf{s}_i$ and $\mathbf{s}_r$.

In a crystal the primitive translation vectors[3] $\mathbf{a}$, $\mathbf{b}$, $\mathbf{c}$ are the nearest neighbor distances. Therefore, it is reasonable to express the phase difference in terms of adjacent atoms that are separated by these vectors. We write

$$\phi_a = \frac{2\pi}{\lambda} \mathbf{a} \cdot \mathbf{S} = 2\pi h, \qquad \phi_b = \frac{2\pi}{\lambda} \mathbf{b} \cdot \mathbf{S} = 2\pi k,$$

$$\phi_c = \frac{2\pi}{\lambda} \mathbf{c} \cdot \mathbf{S} = 2\pi l$$

where h, k, l are integers. These are the **Laue equations**; they must all be satisfied for constructive interference. They may also be written in the form

$$\mathbf{a} \cdot \mathbf{S} = 2a\alpha \sin \theta = \lambda h, \qquad \mathbf{b} \cdot \mathbf{S} = 2b\beta \sin \theta = \lambda k,$$

$$\mathbf{c} \cdot \mathbf{S} = 2c\gamma \sin \theta = \lambda l$$

where α, β, γ are the direction cosines[4] $\cos(\mathbf{a}, \mathbf{S})$, $\cos(\mathbf{b}, \mathbf{S})$ and $\cos(\mathbf{c}, \mathbf{S})$, respectively. Therefore, in a direction of constructive interference the direction cosines are proportional to h/a, k/b, and l/c.

Question. From the results of the above derivation it would appear that any plane described by an arbitrary set of Miller indices[5] will reflect x-rays. What then is the value of x-ray diffraction for the determination of crystal structures?

Solution. We began the derivation of the Laue equations by considering the scattering of x-rays by atoms. If there are no atoms, there is no scattering. From the observed directions of reflection and the relative intensities of the scattered radiation, we can deduce the crystal structure.

Particle diffraction

Material particles possessing momentum have an associated de Broglie wavelength[6]

$$\lambda = \frac{h}{p}$$

where p is the particle momentum and h is Planck's constant. Particles of atomic or nuclear dimensions often have de Broglie wavelengths in the range from $\simeq 0.01$ nm to $\simeq 2$ nm and should, therefore, be expected to be

[3] *MWTP*, Section 21.2.
[4] The direction cosine $\cos(\mathbf{r}, \mathbf{R})$ is the cosine of the angle between the vectors $\mathbf{r}$ and $\mathbf{R}$.
[5] *MWTP*, Section 21.4.
[6] *MWTP*, Section 18.5.

diffracted by a crystal lattice in a manner identical to that of x-rays. Indeed, this does occur and particle diffraction, particularly involving electrons and neutrons, has proven invaluable in the determination of the structure of a large variety of crystals including those composed of giant molecules.

19.8 SPATIAL FILTERING[7]

In this section we present two examples to illustrate the general result that any Fraunhofer diffraction pattern can be interpreted as the Fourier series coefficients or the Fourier transform of the object which gives rise to the diffraction pattern.

First we consider a diffraction grating with periodicity S and slit width W. The grating, located in the yz plane with its rulings parallel to the z axis, is illuminated from behind by a coherent plane wave of wavelength λ traveling in the x direction. The effect of the grating is to multiply the incident wave by the amplitude transmission function $g(y)$ of the grating. We take $g(y)$ to be as shown in Fig. 19.21, namely, unity at the positions of the slits and zero elsewhere. Whereas the incident plane wave propagates in the x direction, the diffracted wave propagates in the xy plane. To obtain the amplitude distribution in the xy plane we carry out a Fourier analysis[8] of the amplitude transmission function. It follows that (Problem 31)

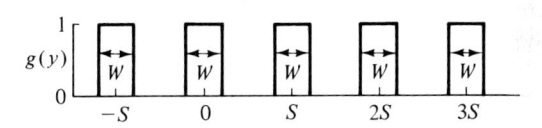

Fig. 19.21. The amplitude transmission function of the grating.

$$g(y) = \sum_{n=-\infty}^{\infty} \left(\frac{1}{n\pi}\right) \sin\left(\frac{n\pi W}{S}\right) \exp\left(\frac{i2\pi n y}{S}\right).$$

The transmitted wave E_t is obtained by multiplying the incident wave $E_i = E_0 \exp[i(\omega t - k_x x)]$ by the function $g(y)$. It follows that

$$E_t = E_0 \left[\sum_{n=-\infty}^{\infty} \left(\frac{1}{n\pi}\right) \sin\left(\frac{n\pi W}{S}\right) \exp\left(\frac{i2\pi n y}{S}\right) \right] \exp[i(\omega t - k_x x)]$$

$$= \frac{2W}{S} E_0 \left[\sum_{n=0}^{\infty} \frac{\sin\left(\frac{n\pi W}{S}\right)}{\frac{n\pi W}{S}} \cos\left(\frac{2\pi n}{S} y\right) \right] \exp[i(\omega t - k_x x)]$$

(see Appendix C). The quantities $k_y = 2\pi n/S$ are referred to as **spatial frequencies**. This expression represents an infinite series of plane waves propagating in the xy plane. The nth wave makes an angle with the x axis

---
[7] J. C. Brown, "Fourier Analysis and Spatial Filtering," *American Journal of Physics*, **39** (July 1971), 797.

[8] *MWTP*, Section 16.8.

given by
$$\theta_n = \tan^{-1}\left(\frac{2\pi n/S}{kx}\right) = \tan^{-1}\left(\frac{n\lambda}{S}\right).$$

Note that for small angles
$$\sin\theta_n \simeq \tan\theta_n$$
so that
$$S\sin\theta_n = n\lambda$$
which is the grating equation obtained in Section 19.5.

The interpretation of this result is that the object forces the incident plane wave to take on a specific amplitude distribution proportional to the amplitude transmission function for the object, with the result that spatial frequencies $2\pi n/S$ are introduced having amplitudes proportional to the Fourier coefficients $\sin(n\pi W/S)/(n\pi W/S)$. Since each Fourier component represents a plane wave traveling in a specific direction, each component will be imaged in the focal plane of the lens; the intensity of each image will be proportional to the square of the appropriate Fourier coefficient.

Spatial filtering is achieved by inserting stops in the focal plane of the lens to remove certain components from the spectrum of spatial frequencies of the object. In Fig. 19.22 we see a converging lens of focal length f located a distance l from the object plane. The Fraunhofer diffraction pattern is found in the focal plane of the lens. A stop located in this plane permits only the spatial frequencies $-4\pi/S$, $-2\pi/S$, 0, $2\pi/S$, $4\pi/S$ to be transmitted to the image plane; **low-pass optical filtering** is thereby achieved. The image function is observed at a distance l_i from the lens. The object and image functions are illustrated in Fig. 19.23.

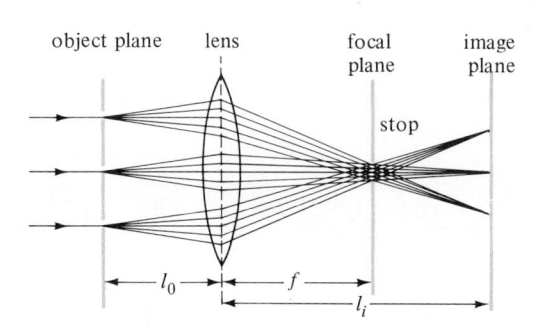

Fig. 19.22. The stop in the focal plane of the converging lens permits only the spatial frequencies $-4\pi/S$, $-2\pi/S$, 0, $2\pi/S$, $4\pi/S$ to be transmitted to the image plane.

If in Fig. 19.22 a stop is placed in the focal plane of the lens that permits only the high-frequency spatial components to be transmitted, **high-pass optical filtering** is achieved. The object and image functions are shown in Fig. 19.24 for a stop that eliminates the frequencies $-4\pi/S$, $-2/\pi S$, 0, $2\pi/S$, $4\pi/S$.

As an example of an object that does not possess a periodic structure, we consider a single slit of width W illuminated by a coherent plane wave of wavelength λ. The amplitude transmission function of the slit is

$$g(y) = 1 \qquad -\frac{W}{2} \leq y \leq \frac{W}{2}$$
$$= 0 \qquad \text{elsewhere.}$$

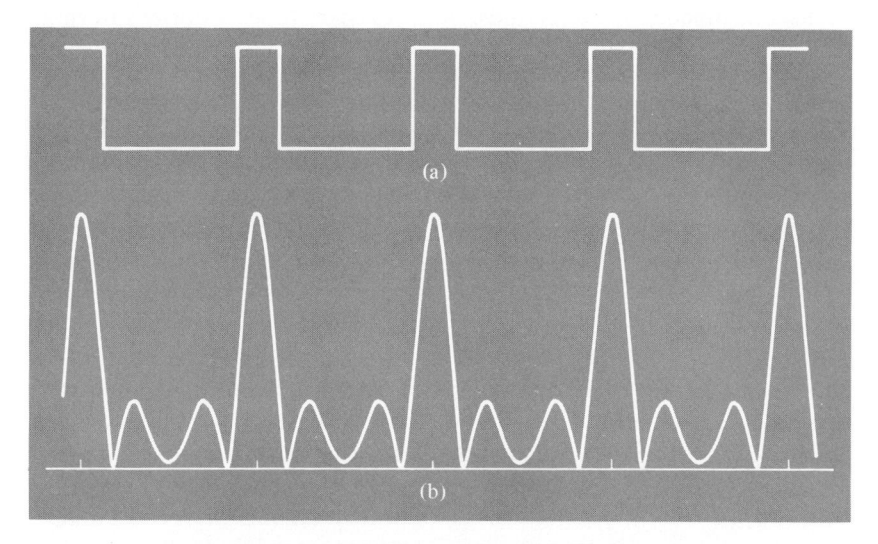

Fig. 19.23. The result of low-pass optical filtering: (a) the object function, (b) the image function.

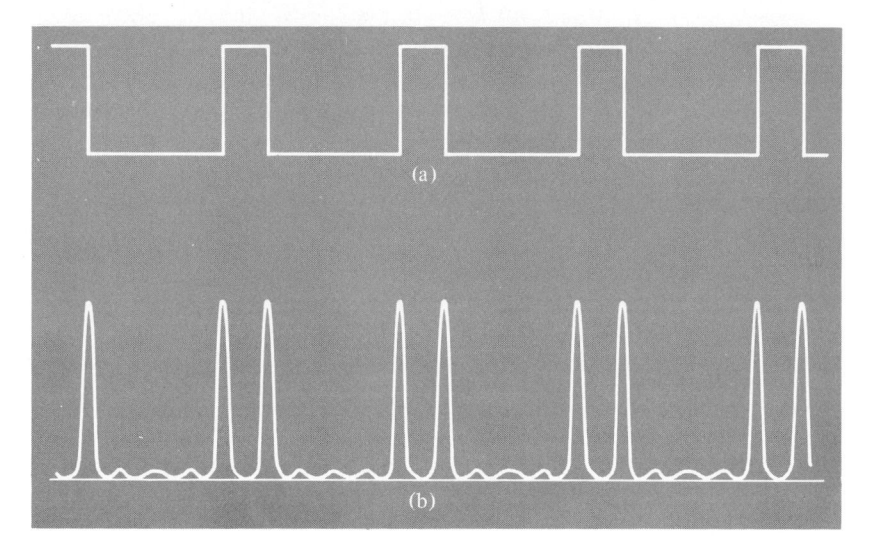

Fig. 19.24. The result of high-pass optical filtering: (a) the object function, (b) the image function.

Since this function is not periodic, it is necessary to use Fourier's integral theorem to take the Fourier transform (see Section 17.3) of $g(y)$ to determine the spatial frequencies k_y introduced by the slit. In particular,

$$f(k_y) = \int_{-\infty}^{\infty} g(y) \exp(-ik_y y)\, dy$$

$$= \int_{-W/2}^{W/2} \exp(-ik_y y)\, dy$$

$$= W \frac{\sin\left(\dfrac{k_y W}{2}\right)}{\dfrac{k_y W}{2}}.$$

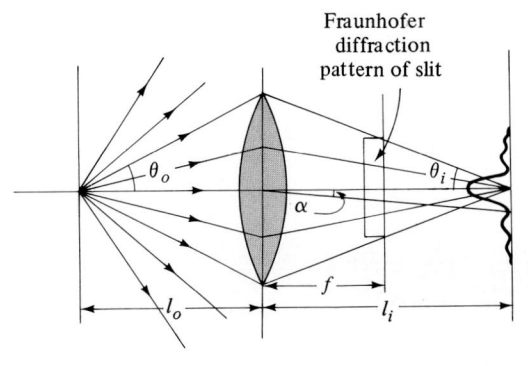

Fig. 19.25. The image of a very narrow slit formed by a lens of diameter D.

For a specified k_y, the value of $f(k_y)$ is analogous to the Fourier series coefficient for a given k_y; that is, it gives the amplitude of the component wave with that spatial frequency.

Let us suppose that the radiation from a very narrow slit is intercepted by a lens of diameter D a distance l_0 from the slit (Fig. 19.25) and that the image is viewed a distance l_i from the lens. In the limit of $W \to 0$, $f(k_y) \to C$ and the wave after passing through the slit is made up of all spatial frequencies with equal amplitudes. However, the lens intercepts only those components that satisfy the condition

$$|\tan\theta_0| \leq \frac{D}{2l_0}.$$

After passing through the lens, these component waves are traveling at angles such that

$$|\tan\theta_i| \leq \frac{D}{2l_i} = \frac{k_y^{\max}}{k_x}.$$

Therefore, the image of the slit will be composed of all spatial frequencies from $k_y = -k_x D/2l_i$ to $k_y = k_x D/2l_i$ with equal amplitudes. In this example the lens acts as a low-pass optical filter. That portion of the Fraunhofer diffraction pattern of the slit formed in the focal plane of the lens is indicated schematically in Fig. 19.25. The image of the slit is the Fourier transform of this function, namely,

$$E_i \propto \int_{-\pi D/l_i \lambda}^{\pi D/l_i \lambda} \exp(ik_y y)\, dk_y$$

$$\propto \frac{\sin\left(\dfrac{\pi D y}{l_i \lambda}\right)}{\dfrac{\pi D y}{l_i \lambda}}.$$

This is equivalent to the diffraction pattern of a single slit of width D. The first minimum in the pattern occurs for $y = \lambda_i l/D$; this displacement subtends an angle $\alpha = y/l_i = \lambda/D$ at the lens. The **Rayleigh criterion** for the resolution of two closely spaced line sources is that their angular separation $\delta \geq \alpha = \lambda/D$. The above treatment shows clearly that the lack of definition in the image arises from the elimination of the high spatial frequency components of the object due to the finite dimensions of the lens.

19.9 HOLOGRAPHY

The technique of **holography**[9,10,11,12,13] provides the means to store *all* of the information about a system that comes to us in the form of electromagnetic waves that have interacted with that system. Then, by making use of the **hologram** that is produced, we can at any later time reconstruct the original wave front. For example, if we use holography to record a visual object, the hologram allows us to perceive that object in three dimensions exactly as it was when the hologram was recorded. This may be contrasted with conventional photography in which only a two-dimensional intensity distribution of an image of the object is recorded. Although the theoretical basis for the holographic technique was presented by Dennis Gabor in 1948, it is only in the past five years or so that the method has become widely-known and used. One reason for this delay was the requirement of a light source that provides radiation with significant temporal and spatial coherence; the laser (see Section 20.3) provides just such a source.

In order to produce a hologram one starts with a coherent beam of radiation and splits it into two components. One component falls on the object and is diffracted as a result of its interaction with the object. The **object wave** so produced is directed toward a suitable detector, such as a photographic plate. The other component, termed the **reference wave**, travels directly to the detector. Since the object and reference waves are mutually coherent, they produce a stable interference pattern at the detec-

---

[9] E. N. Leith and J. Upatnieks, "Wavefront Reconstruction Photography," *Physics Today*, August, 1965.

[10] A. E. Ennos, "Holography and its Applications," *Contemporary Physics*, **8** (March 1967), 153.

[11] E. N. Leith and J. Upatnieks, "Photography by Laser," *Scientific American*, June, 1965. Available as *Scientific American Offprint 300* (San Francisco: W. H. Freeman and Co., Publishers).

[12] K. S. Pennington, "Advances in Holography," *Scientific American*, February, 1968.

[13] E. N. Leith and J. Upatnieks, "Progress in Holography," *Physics Today*, (March 1972).

tor. The complex microscopic details of this interference pattern are unique to the particular object wave front. The permanent record of the interference pattern is called a **hologram**. The phase information carried by the diffracted wave is preserved on the hologram since the reference wave provides a standard of comparison. When at some later time the hologram is illuminated with a **reconstruction wave**, similar to the radiation used to produce the hologram, the diffracted radiation consists of three components, one of which exactly duplicates the original object wave front. By viewing this reconstructed wave front one sees an exact replica of the original object, even though the object is not present during the process of reconstruction. Unfortunately, a second component wave is of the same amplitude as the object wave but is of opposite phase in relation to the reference wave. This **twin wave** gives rise to a pseudoscopic view of the original object. The third component is simply a portion of the reconstruction wave itself. It was the presence of the twin wave almost as much as the lack of sufficiently coherent sources of radiation in either the visible or x-ray regions that made holography a scientific curiosity rather than a viable scientific technique until the mid 1960's.

In 1962 Emmett Leith and Juris Upatnieks at the University of Michigan suggested a simple method that allowed a complete separation of the reconstructed object wave and the twin wave. They showed that if the object wave and the reference wave strike the detector at a nonzero relative angle, the reconstruction process produces two waves which are separated in space and which correspond to the reconstructed object wave and the twin wave. The hologram is essentially a diffraction grating and the two waves produced in the reconstruction process are the two first-order diffraction patterns. In order to illustrate this method of split-beam holography we assume that the object wave and the reference wave are both plane waves at the photographic plate as shown in Fig. 19.26. The object wave may be described mathematically as

$$A_0 \exp[ik(-x\cos\theta + y\sin\theta)]\exp(i\omega t)$$

and the reference wave as

$$A_R \exp[ik(-x\cos\theta - y\sin\theta)]\exp(i\omega t).$$

Assuming perfect coherence, the amplitudes of the two waves add at the photographic plate to produce a total wave amplitude

$$\begin{aligned} A_T &= A_0 \exp[ik(-x\cos\theta + y\sin\theta)]\exp(i\omega t) \\ &+ A_R \exp[ik(-x\cos\theta - y\sin\theta)]\exp(i\omega t) \\ &= A_0\left[1 + \frac{A_R}{A_0}\exp(-i2k_y\sin\theta)\right]\exp[ik(-x\cos\theta + y\sin\theta)]\exp(i\omega t). \end{aligned}$$

The intensity distribution at the photographic plate is

$$I_T = |A_T|^2 = A_0^2\left[1 + \left(\frac{A_R}{A_0}\right)^2 + 2\left(\frac{A_R}{A_0}\right)\cos(2ky\sin\theta)\right]$$

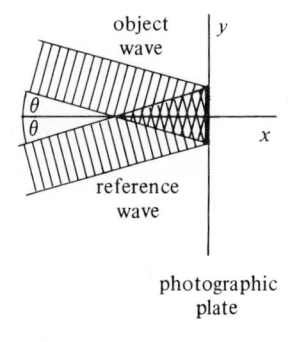

Fig. 19.26. The recording of a plane wave hologram.

SEC. 19.9   HOLOGRAPHY

$$= \left[\frac{A_0^2}{1+\left(\frac{A_R}{A_0}\right)^2}\right]\left[1 + \frac{2\left(\frac{A_R}{A_0}\right)}{1+\left(\frac{A_R}{A_0}\right)^2} \cos(2ky \sin\theta)\right].$$

The intensity varies periodically in the y direction in the plane of the photographic plate with interference maxima separated by a distance $\lambda/(2\sin\theta)$. In general, the photographic plate is inclined at an angle ϕ to the bisector of the object and reference beams and the resulting hologram is a diffraction grating with characteristic spacing $\lambda/(2\sin\theta\sin\phi)$. In order to reconstruct the object wave a duplicate of the reference wave is allowed to interact with the hologram. We assume that the transmission T through the hologram is a linear function of I_T; that is,

$$T = a + bI_T$$

where a and b are constants. Therefore, the radiation transmitted by the hologram is given by

$$A_R \exp[ik(-x\cos\theta - y\sin\theta)]\exp(i\omega t)(a + bI_T).$$

Without loss of physical generality we may set $a = 0$ and $b = 1$. Substitution for I_T into this expression then yields (see Problem 38)

$$A_R A_0^2\left[1 + \left(\frac{A_R}{A_0}\right)^2\right]\exp[ik(-x\cos\theta - y\sin\theta)]\exp(i\omega t)$$

$$+ A_R A_0^2\left(\frac{A_R}{A_0}\right)\exp[ik(-x\cos\theta + y\sin\theta)]\exp(i\omega t)$$

$$+ A_R A_0^2\left(\frac{A_R}{A_0}\right)\exp[ik(-x\cos\theta - 3y\sin\theta)]\exp(i\omega t).$$

The situation is illustrated in Fig. 19.27. The first term represents a plane wave in the direction of the reference wave. The second term, often called the **primary diffracted wave**, is a plane wave in the direction of the original object wave and corresponds to the reconstruction of the object wave. The third term, often referred to as the **secondary diffracted wave**, constitutes the twin wave.

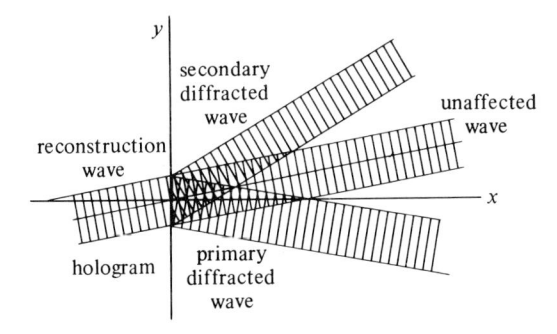

Fig. 19.27. The diffraction of a plane wave from a plane wave hologram.

Practical applications of the holographic technique are under active investigation at the present time. Many proposed uses have been suggested in such areas as motion pictures and television, microscopy, interferometry, information storage, and character recognition. Many of the suggested applications represent new methods of performing certain measurements; some of the suggested applications represent true innovations. In the latter case the fundamental difference between holography and photography is exploited—a hologram is a record of an object wave front whereas a photograph is a record of the intensity distribution in an image.

QUESTIONS AND PROBLEMS

1. Classical optics was a field endowed with controversy since its origin. Write a short historical summary of its development making particular reference to the concept of the **ether** and the particulate nature of light.[14]

2. Discuss several examples of meteorological phenomena that involve diffraction.

3. Use a geometrical construction based on Huygen's principle to deduce the law of reflection from a plane surface.

4. Use a geometrical construction based on Huygen's principle to deduce the law of refraction at an interface bounding media with different indices of refraction.

5. Show that, for the geometrical arrangement of Fig. 19.2, the path difference for radiation traveling via the upper edge of the aperture as compared to radiation traveling via the lower edge is given by

$$\left(\frac{h_S}{d_S} + \frac{h_R}{d_R}\right)W + \frac{1}{2}\left(\frac{1}{d_S} + \frac{1}{d_R}\right)W^2 + \cdots.$$

6. Consider the example discussed in Section 19.2. How could you change the experimental arrangement so that Fraunhofer diffraction would be observed?

7. In the example in Section 19.4 the radiation from a narrow slit in a conducting plane is quoted as

$$E_z = E_0\left(\frac{\lambda}{r}\right)^{1/2} \cos\theta \cos\left[\omega\left(t - \frac{r}{c}\right)\right]$$

for the case that the electric field is parallel to the slit. Justify this result.

8. Consider the case of microwave diffraction by a single slit assuming that the magnetic field of the incident wave is parallel to the slit. Compare the result

[14]M. Born and E. Wolf, *Principles of Optics*, 2nd ed. (Oxford: Pergamon Press, 1964).

with that obtained in the example discussed in Section 19.4 assuming that $W \ll \lambda$.

9. Show that the amplitude of the wave diffracted in the direction θ by the double slit in Fig. 19.9 is as stated in Section 19.4, namely,

$$E_R = 2CLW \exp(ikr_0) \exp(i\alpha) \exp(i\beta) \frac{\sin \alpha}{\alpha} \cos \beta.$$

10. Consider a double slit for which $S = 3W$. Plot the intensity functions $\cos^2 \beta$, $(\sin \alpha/\alpha)^2$ and $2(\sin \alpha/\alpha)^2 \cos^2 \beta$.

11. Show that the double-slit diffraction pattern will reduce to that of a single slit of width $2W$ when $W = S$.

12. Determine the number of peaks occurring under the central maximum of a double-slit pattern when the separation is S and the width of each slit is W.

13. The 5th secondary maximum in the Fraunhofer diffraction pattern of a double slit is observed to be missing. Deduce the possible ratios of the slit width to the slit separation.

*14. Use the Fresnel–Kirchhoff formula to calculate the Fraunhofer diffraction pattern of a rectangular aperture of dimensions H and L. Sketch the resultant pattern for $H = 2L$.

*15. Use the Fresnel–Kirchhoff formula to calculate the Fraunhofer diffraction pattern for a circular aperture. Hint: You will need some knowledge of **Bessel functions** and in particular the formula

$$\int_{-1}^{+1} \exp[i\rho(1-u^2)^{1/2}] \, du = \frac{\pi J_1(\rho)}{\rho}$$

where $J_1(\rho)$ is a **Bessel function of the first kind** of order one.

16. A ruby laser produces a beam of diameter 2.0 mm at 694.3 nm wavelength. How wide will the beam be after having traveled 100 m?

17. Discuss the relative merits of a diffraction grating and a Fabry–Perot interferometer as devices for spectroscopic studies.

18. Calculate the minimum resolvable wavelength separation attainable at 550 nm with a grating 2.0 cm wide having 1500 lines per mm and used in the 8th order.

19. Determine the minimum number of lines required on a diffraction grating being used to resolve in the first order the doublet ($n = 2 \rightarrow n = 1$) produced by a mixture of hydrogen and deuterium.

20. What is a **zone plate**?

21. Suppose that $|E_p|$ is a sum of contributions from an odd number N of Fresnel zones. Making the approximation that amplitudes from adjacent zones are approximately equal, show that

$$|E_p| = \frac{|E_1|}{2} - \frac{|E_N|}{2}.$$

22. A point source A ($\lambda = 500$ nm) is 1.0 m from an aperture as shown in

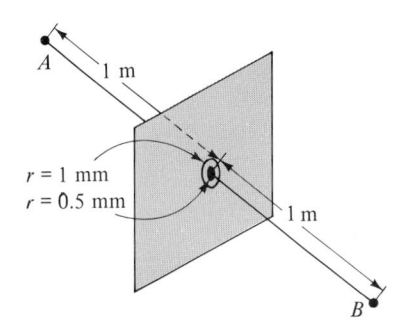

Fig. 19.28. Diffraction by a circular aperture with an opaque circular disc in the center.

Fig. 19.28. The aperture is a hole 1.00 mm in radius with an opaque circular disc 0.50 mm in radius in the center. Calculate the ratio of the intensity at a point B directly opposite the aperture and 1.0 m distant from it to that if the aperture were removed.

23. Since a crystal is fully packed with atoms, it might seem that all planes would be full of atoms and thereby reflect x-rays. Why is this not so?

24. Show how a crystal lattice can be used to produce a nearly monochromatic beam of x-rays from a source producing a broad continuum of wavelengths.

25. Starting from the Laue equations obtain a relation between x-ray wavelengths that are effectively scattered and the lattice constant for a simple cubic lattice.

26. When x-rays of wavelength 0.15 nm are incident upon a simple cubic crystal, a first-order reflection is observed at 16°. Calculate the nearest neighbor distance for atoms in this crystal.

*27. Show the equivalence of the Bragg and Laue treatments of x-ray diffraction.

28. Discuss the process of **apodization**.

29. In recent years the technique of optical data processing has been receiving increased attention. Many of the optical data processing techniques are based on the concept of optical spatial filtering. Discuss this statement.

*30. State **Babinet's principle** as applied to Fraunhofer diffraction and interference. Using the technique introduced in Section 19.8 and assuming a periodic amplitude transmission function prove Babinet's principle. Show that Babinet's principle is equivalent to the statement that an object and its complement have identical Fourier components with a change in sign but referred to different base lines.

31. Perform a Fourier analysis of the amplitude transmission function

$$g(y) = 1, \quad -\frac{W}{2} \leq y \leq \frac{W}{2}$$

$$0, \quad -\frac{S}{2} < y < -\frac{W}{2}, \quad \frac{W}{2} < y < \frac{S}{2}.$$

*32. Calculate the distribution of spatial frequencies resulting from the coherent illumination of an opaque narrow strip of width W.

*33. Calculate the amplitude of the wave transmitted by a grating whose amplitude transmission function is shown in Fig. 19.29.

*34. The amplitude transmission function of a narrow strip can be described by

$$U(x) = (2\pi)^{-1/2} \exp(-x^2)$$

where x is the distance from the center. Calculate the resultant spatial frequencies.

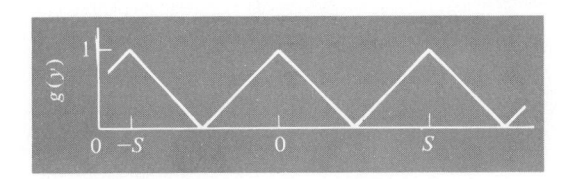

Fig. 19.29. The amplitude transmission function for a particular grating.

35. Discuss the application of holography to microscopy. Include in your discussion the distinction between **holographically augmented microscopy** and **holographic magnification**.

36. By irradiating an object with acoustical radiation instead of with electromagnetic radiation, acoustical holograms can be produced that then provide three-dimensional pictures when viewed by visible radiation.[15] Discuss.

37. **Diagnostic cytology** and **radiology** are included among the possible biomedical applications of holography.[16] Discuss.

38. Show that the diffraction of a plane wave from a plane wave hologram may be written as the sum of an undiffracted wave, a primary diffracted wave, and a secondary diffracted wave (see Section 19.9).

[15] A. F. Metherell, "Acoustical Holography," *Scientific American*, October, 1969.
[16] E. J. Feleppa, "Biomedical Applications of Holography," *Physics Today*, July, 1969.

CHARLES H. TOWNES

20 Nonlinear optics and lasers

20.1 INTRODUCTION

The passage of light through a material can be affected by applying an electric or magnetic field to the material. Michael Faraday discovered that the plane of polarization of a beam of light is rotated when the light travels along lines of magnetic force. This **Faraday effect** can be explained by considering the beam of plane-polarized light to consist of two circularly polarized beams whose electric vectors rotate in opposite directions (see Section 14.8). The effect of the magnetic field is to alter the index of refraction of the material for one of the beams so that the phase difference of the two beams increases uniformly with time (or distance through the material), resulting in a rotation of the plane of polarization.

John Kerr produced double refraction of light (see Section 14.10) in glass by applying a strong electric field. In the presence of the electric field the molecules within the sample are aligned so that it behaves optically as if it were a uniaxial crystal with the direction of the optic axis defined by the electric field. This phenomenon is known as the **Kerr electro-optic effect**; it is observed in both solids and liquids and has a magnitude that is proportional to the square of the electric field strength.

The magnetic analogue to the Kerr electro-optic effect occurs in liquids; it is known as the **Cotton-Mouton magneto-optic effect**. In this case the double refraction is attributed to the lining up of magnetically anisotropic molecules in the direction of an applied magnetic field. The magnitude of the effect is proportional to the square of the applied magnetic field.

Since light is an electromagnetic wave, we should expect the oscillating electric and magnetic fields in the light wave itself to be capable of altering the index of refraction of the material through which it is passing. Before 1960 there were no light sources of sufficient intensity (that is, with high enough electric field amplitude) to produce such alterations in the materials that supported their passage. However, in 1960 a light source known as a **laser** was developed. A laser produces an essentially monochromatic beam of coherent radiation that can be of great intensity. Lasers were immediately applied to the study of the nonlinear responses of various materials to the passage of light waves.

20.2 NONLINEAR OPTICAL EFFECTS

An oscillating electric field applied to a crystal will displace the positive and negative charges in opposite directions and thereby induce an oscillating electric dipole. Since the nuclei are very massive relative to the electrons, the nuclear motions are negligible compared to the electronic motions. Therefore, to a first approximation, we may think of the electrons oscillating about fixed nuclear positions. For small amplitude oscillations resulting from the application of small electric fields, the restoring forces experienced by the electrons are proportional to their displacements from equilibrium and the electrons execute simple harmonic motion[1] at the frequency of the imposed electric field. However, for sufficiently large applied fields the electrons execute large amplitude oscillations and their motion becomes significantly anharmonic.[2] Their energy of oscillation is no longer proportional to the square of their amplitude of oscillation. The explanation for this departure from harmonic behavior is that each electron is subject to an internal electric field due to neighboring electrons and nuclei as well as to the external field, and when the strength of the applied field begins to become comparable to the strength of the internal field, the opposition to motion increases.

The term **nonlinear optics**[3,4,5] refers to all those optical phenomena that result when a light beam has sufficient intensity that the material through which it passes does not respond harmonically to its presence.

The above microscopic picture can be translated into a macroscopic one by consideration of the vector sum of all the electric dipoles in the crystal as the polarization **P**. A simple nonlinear relation between the magnitude P of the polarization and the magnitude E of the applied electric field is the power series expansion

$$P = \alpha E + \beta E^2 + \gamma E^3 + \delta E^4 \ldots$$

where $\alpha, \beta, \gamma, \delta \ldots$ are constant coefficients. The relative orders of magnitudes of these coefficients are given by

$$\frac{\alpha}{\beta} \simeq \frac{\beta}{\gamma} \simeq \frac{\gamma}{\delta} \simeq \frac{E_{\text{int}}}{E}.$$

Internal electric fields have magnitudes E_{int} of the order of 10^{10} V·m^{-1} or higher so that very large magnitudes E of the external field are required before the term in E^2 becomes appreciable.

An oscillating field of magnitude

$$E = E_0 \sin \omega t$$

applied to such a nonlinear material will produce a polarization

$$P = \alpha E_0 \sin \omega t + \beta E_0^2 \sin^2 \omega t + \gamma E_0^3 \sin^3 \omega t + \ldots$$
$$= \alpha E_0 \sin \omega t + \beta E_0^2 \frac{1 - \cos 2\omega t}{2} + \gamma E_0^3 (3 \sin \omega t - \sin 3\omega t) + \ldots$$

---

[1] *MWTP*, Section 11.3.
[2] *MWTP*, Section 22.7.
[3] J. A. Giordmaine, "The Interaction of Light with Light," *Scientific American*, April, 1964.
[4] J. K. Wright, "Non-Linear Optics," *Contemporary Physics*, **6**, (October 1964), 1.
[5] J. A. Giordmaine, "Nonlinear Optics," *Physics Today*, January, 1969.

SEC. 20.2 NONLINEAR OPTICAL EFFECTS

$$= \left(\frac{\beta E_0^2}{2} + \ldots\right) + (\alpha E_0 + 3\gamma E_0^3 + \ldots)\sin \omega t$$
$$+ \left(-\frac{\beta E_0^2}{2} + \ldots\right)\cos 2\omega t + (-\gamma E_0^3 + \ldots)\sin 3\omega t + \ldots.$$

The nonlinear terms in the relation between P and E generate **harmonic polarization waves** in the crystal at frequencies $2\omega, 3\omega, \ldots$ as well as a polarization wave at the fundamental frequency ω (see Fig. 20.1). In addition, a constant polarization term (or **bias polarization**) is generated by the even power terms. In Fig. 20.2 we show a graphical analysis of a nonlinear polarization wave in a dielectric medium where only the coefficient β of the second-order term in the polarization is nonzero. Some crystals have a center of symmetry or a center of inversion.[6] For such crystals, the electric polarization must change sign when the applied electric field changes sign. Therefore, there can be no even powers of E in the expansion for P and the coefficients $\beta, \delta, \ldots$ are zero. These crystals should show no bias polarization and no even harmonic polarization

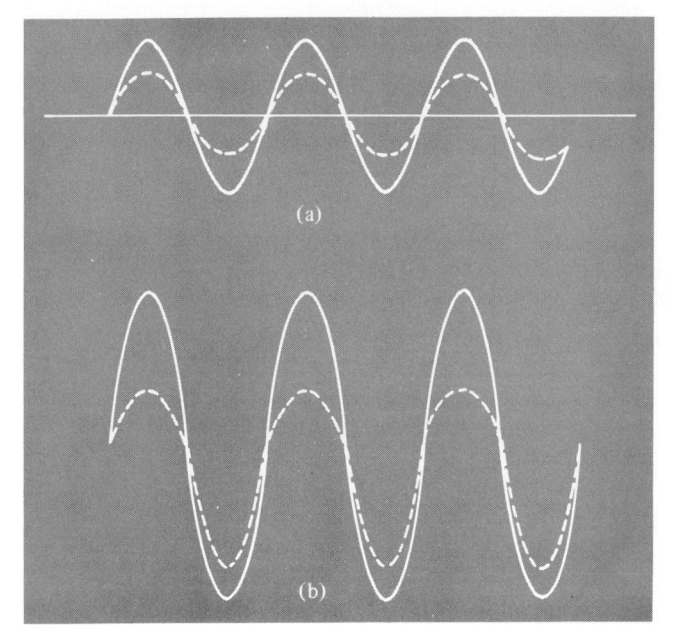

Fig. 20.1. (a) A linear polarization wave (dashed line) is created by a light wave of moderate intensity (full line) in a dielectric medium. (b) A nonlinear polarization wave is created by a light wave of sufficient intensity in a dielectric medium.

[6] *MWTP*, Section 21.2.

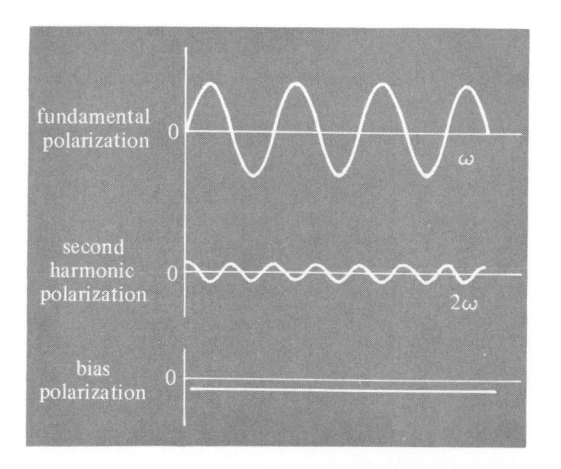

Fig. 20.2. Analysis of a nonlinear polarization wave in a dielectric medium where only α and β in the expression for the polarization are nonzero.

waves. Noncrystalline isotropic media such as liquids and glasses are expected to behave like crystals with a center of inversion.

We now consider the effect of passing two light waves of different frequency simultaneously through a nonlinear medium. For simplicity we assume that only α and β in the expansion for P are significant and that the waves are traveling in the $+x$ direction. The oscillating field is written as

$$E = E_1 \sin(\omega_1 t - k_1 x) + E_2 \sin(\omega_2 t - k_2 x)$$

so that

$$\begin{aligned}P &= \alpha E_1 \sin(\omega_1 t - k_1 x) + \alpha E_2 \sin(\omega_2 t - k_2 x) \\ &+ \beta E_1^2 \sin^2(\omega_1 t - k_1 x) \\ &+ \beta E_2^2 \sin^2(\omega_2 t - k_2 x) \\ &+ 2\beta E_1 E_2 \sin(\omega_1 t - k_1 x) \sin(\omega_2 t - k_2 x) \\ &= \alpha E_1 \sin(\omega_1 t - k_1 x) + \alpha E_2 \sin(\omega_2 t - k_2 x) \\ &+ \frac{\beta E_1^2}{2}[1 - \cos(2\omega_1 t - 2k_1 x)] + \frac{\beta E_2^2}{2}[1 - \cos(2\omega_2 t - 2k_2 x)] \\ &+ \beta E_1 E_2 \{\cos[(\omega_1 - \omega_2)t - (k_1 - k_2)x] \\ &- \cos[(\omega_1 + \omega_2)t - (k_1 - k_2)x]\}.\end{aligned}$$

The term in βE^2 gives rise to **polarization waves** inside the crystal at the original frequencies of ω_1 and ω_2 and at the new frequencies zero (bias polarization), $2\omega_1$, $2\omega_2$, $(\omega_1 + \omega_2)$ and $(\omega_1 - \omega_2)$. This is shown schematically in Fig. 20.3.

Each polarization wave, with the exception of the one at zero frequency, radiates an electromagnetic wave at its particular frequency. The zero frequency term results in the establishment of a static potential difference across the crystal. We suppose that the crystal is in the form of a slab of thickness L in the x direction and consider as an example the amplitude of the wave of frequency $2\omega_1$ at the exit face of the crystal. Each infinitesimal length dx will radiate at frequency $2\omega_1$; the total amplitude of the emergent wave will be the vector sum of the contributions from the

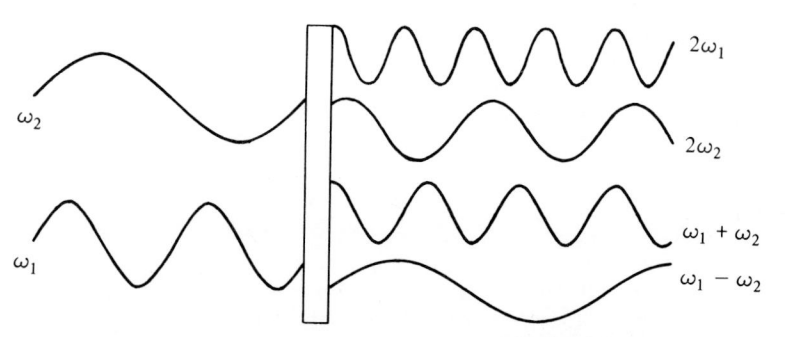

Fig. 20.3. Production of second harmonic, sum, and difference frequencies in a second-order nonlinear crystal.

infinitesimal elements. Assuming that the incident radiation is not significantly attenuated on passing through the crystal either by the non-linearities or by absorption, the displacement $E(2\omega_1, L)$ of the wave leaving the crystal at time t at frequency $2\omega_1$ will be proportional to

$$\int_0^L \cos[2\omega_1(t-t') - 2k_1 x]\,dx.$$

The time t' is the time required for the wave of frequency $2\omega_1$ and propagation constant k_3 and originating at position x to travel to the exit face. Therefore,

$$t' = \frac{(L-x)k_3}{2\omega_1}.$$

The substitution of this expression into the integral which is then evaluated shows that the intensity $I(2\omega_1)$ of the emergent wave at frequency $2\omega_1$ is given by

$$I(2\omega_1) = C \frac{\sin^2\left[\frac{(2k_1 - k_3)L}{2}\right]}{(2k_1 - k_3)^2}$$

where C is a constant (Problem 4). From this result we see that if $2k_1 = k_3$, the intensity at frequency $2\omega_1$ is proportional to the square of the thickness of the slab. Usually, however, because of dispersion within the crystal $2k_1 \neq k_3$ and the intensity will vary periodically with L with identical intensity maxima at intervals of length $2\Delta L$ where

$$2\Delta L = \frac{2\pi}{(2k_1 - k_3)}.$$

The length ΔL is the maximum useful length for building up oscillations at frequency $2\omega_1$. This length is the **coherence length for second harmonic generation**. Since ΔL is typically of the order of 10^{-5} m, it might seem unlikely that large amplitude oscillations can be built up at frequency $2\omega_1$ by traversing macroscopic lengths of crystal. However, several schemes have been devised to overcome the effect of dispersion and permit large amplitude oscillations of one of the polarization waves to build up. These will be discussed in Section 20.6.

20.3 THE PRINCIPLE OF THE LASER

The device that has made possible the observation of nonlinear optical effects is the laser.[7] The word laser is an acronym for "light amplification by stimulated emission of radiation." In an ordinary light

---

[7] *Lasers and Light, Readings from Scientific American* (San Francisco: W. H. Freeman and Co., Publishers). See also the film, *Laser Light*, available from *Scientific American*, 415 Madison Ave., New York, N.Y.

source, such as an incandescent lamp, fluorescent lamp, or neon tube, atoms are continuously "pumped" into excited electronic states from which they very quickly decay to the ground state, and in the process they emit radiation in the visible (and often ultraviolet and infrared) region of the electromagnetic spectrum. The individual atoms radiate their excess energy randomly in time giving rise to spatially incoherent radiation (see Section 17.4). Also, the energy of an ordinary thermal light source (such as an incandescent lamp) is spread over a large range of frequencies so that very little power appears in any given narrow range of frequencies. Even gas discharge tubes which emit light at a few well-defined frequencies are very limited as sources of power at any individual frequency in comparison to lower-frequency electronic oscillators which can produce powerful coherent radiofrequency and microwave radiation.

In order to generate a coherent light wave of high power, a method of synchronizing a large number of excited atoms to radiate together is required. In 1958 Arthur L. Schawlow and Charles H. Townes proposed that this could be accomplished by stimulated emission of radiation from excited atoms in a special kind of **resonant cavity**. The first successful operation of a laser based upon the principles proposed by Schawlow and Townes was announced in 1960. Within a few years large numbers of lasers of various kinds were in operation and producing coherent radiation at over 100 different wavelengths.

We have already discussed the processes of emission and absorption of radiation in Section 15.6. Emission of radiation from an excited state may be either spontaneous or stimulated. Spontaneous emission is characterized by a coefficient A and stimulated emission by a coefficient B; the relation between the coefficients is

$$\frac{A}{B} = \frac{8\pi h v^3}{c^3}$$

where v is the frequency of the emitted radiation. Stimulated emission occurs only under the influence of external radiation of precisely the frequency v. A photon of frequency v emitted spontaneously by one atom (or molecule) could stimulate a second atom to emit its photon prematurely. The probability of stimulated emission depends on the energy density ρ_v of photons of frequency v and is $\rho_v B$. The ratio of probabilities for stimulated emission to spontaneous emission is

$$\frac{\rho_v B}{A} = \frac{\rho_v c^3}{8\pi h v^3} = \text{const}\left(\frac{\rho_v}{v^3}\right).$$

In ordinary light sources the density ρ_v is so low that stimulated emission is negligible.

When a wave of frequency v interacts with an atom in an excited state which can deexcite via the emission of radiation of frequency v, oscillations at frequency v build up which are in phase with oscillations in the incident wave. As a result a second wave of frequency v is emitted, and the two waves are in phase or coherent. Therefore, the possibility arises of using stimulated emission to produce high-power coherent radiation at optical frequencies. A first requirement is to devise some means of raising the radiation density to the point where stimulated emission will predominate over spontaneous emission. It is for this reason that a form of resonant cavity is needed to confine the radiation. The second requirement is to achieve a **population inversion**; that is, the excited state or states of the atoms or molecules must be more densely populated than the terminal state (which may be another excited state or the ground state). This is necessary so that the absorption of photons of frequency v by atoms in the lower state, which is characterized by the same coefficient B, does not predominate over the stimulated emission.

Since at ordinary temperatures virtually all atoms are in their ground states, a large amount of energy is required to pump atoms into the appropriate excited states. There are at least three methods commonly used to accomplish this: (a) intense pulses of light from flash tubes (called **optical pumping**); (b) electrical current; (c) chemical reactions. These methods will be discussed more fully in Section 20.5 with respect to their uses in various types of lasers.

Once a population inversion has been produced, radiation emitted spontaneously by one atom is capable of stimulating other atoms to emit photons with the same frequency and phase. The pumping and subsequent stimulated emission is pictured schematically in Fig. 20.4. For successful laser action to occur, the population must be maintained so that photons are not lost due to absorption by atoms in the ground level. In some media, particularly gases, a continuous population inversion can be maintained so that coherent radiation is emitted continuously. In other media the population inversion cannot be maintained continuously, and the coherent radiation is emitted only in short bursts.

In a typical laser the material (solid, liquid, or gas) in which the coherent beam is produced is confined in a cylinder whose diameter is normally much less than its length [see Fig. 20.5(a)]. Mirrors are attached to the ends of the cylinder and are aligned very accurately parallel to each other and perpendicular to the axis of the cylinder, thereby producing the resonant cavity. Energy is added to

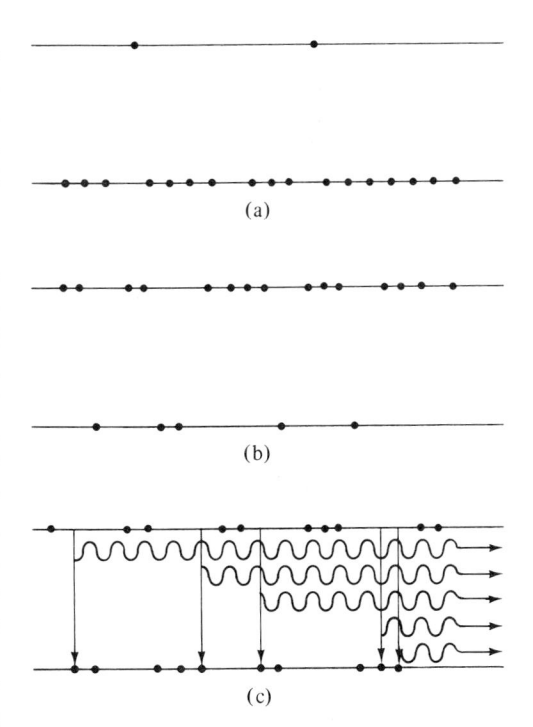

Fig. 20.4. (a) Normal population of a two-level atomic system. (b) Pumping has produced a population inversion in which there are more atoms in the excited level than in the ground level. (c) Spontaneous emission by one atom causes stimulated emission of photons of the same frequency and phase by other atoms.

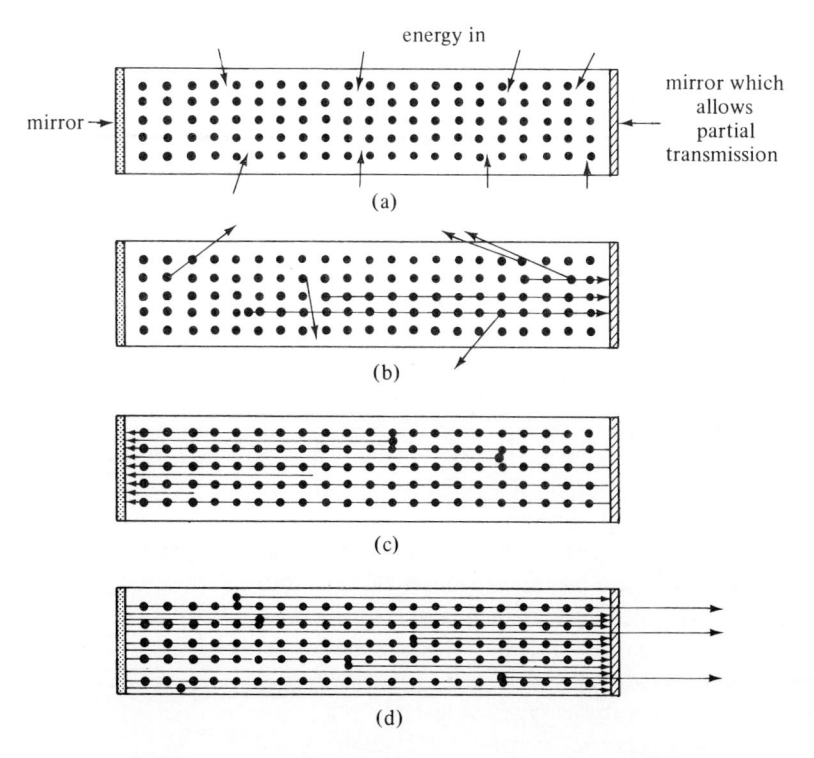

Fig. 20.5. (a) Energy is poured into the laser material to produce a population inversion. (b) Spontaneous emission occurs in a few atoms. A photon traveling parallel to the axis of the cylinder causes a coherent beam to build up. (c) The coherent beam is mostly reflected by the right-hand mirror and produces further stimulated emission before being totally reflected by the left-hand mirror. (d) The process of reflections and further stimulated emissions repeats. After many reflections, an intense coherent beam of radiation is passing out through the right-hand mirror.

the laser medium to produce the population inversion. Spontaneous emission of a photon parallel to the cylinder axis causes a coherent beam to build up through stimulated emission [Fig. 20.5(b)]. The beam is reflected back and forth between the mirrors [Fig. 20.5(c) and (d)] and a very much amplified beam is built up. One of the end mirrors is not totally reflecting but allows a small percentage of the beam to pass through. The emerging beam is coherent, has small divergence (that is, is very nearly parallel so that it spreads out very little as it travels away from the source), and can have very high intensity dependent on the type of laser.

20.4 AMPLIFICATION IN A MEDIUM AND THE THRESHOLD CONDITION FOR LASER ACTION

We consider a medium characterized by two energy states $E_2 > E_1$. We assume that a population inversion exists, that is, that the population per unit volume n_2 of state E_2 exceeds the population per unit volume n_1 of state E_1. A parallel beam of light of frequency $v = (E_2 - E_1)/h$ propagates through the medium in the x direction. Because of the finite widths of the energy levels, only Δn_1 of the n_1 atoms in state E_1 are available for absorption and Δn_2 of the n_2 atoms in state E_2 are available for stimulated emission in a specified frequency interval Δv. Therefore, the rate of upward transitions is $B\rho_v \Delta n_1$, the rate of induced downward transitions is $B\rho_v \Delta n_2$, and the net time rate of change of the radiation energy density $\rho_v \Delta v$ in the frequency interval Δv is given by

$$\frac{d}{dt}(\rho_v \Delta v) = (\Delta n_2 - \Delta n_1) B \rho_v h v.$$

Since the energy density is related to the intensity I_v through the relation $\rho_v = I_v/c$, and since the wave travels a distance $dx = c\, dt$ in time dt, the equation can be rewritten as

$$\frac{dI_v}{dx} = \frac{h}{c}\left(\frac{v}{\Delta v}\right)(\Delta n_2 - \Delta n_1) B I_v$$
$$= \alpha_v I_v$$

where α_v is the **gain constant** at frequency v. This differential equation has as its solution

$$I_v = I_{0v} \exp(\alpha_v x)$$

where I_{0v} is the intensity of the incident beam. Amplification in the medium occurs provided that $\Delta n_2 > \Delta n_1$.

A more useful expression for α_v may be deduced (see Problem 6) by assuming the Maxwell–Boltzmann distribution of molecular speeds[8] from elementary kinetic theory and taking account of the Doppler effect.[9] In particular,

$$\alpha_v = \left(\frac{m}{2\pi k T}\right)^{1/2} (n_2 - n_1) h B \exp[-\beta(v - v_0)^2]$$

where

$$\beta = \frac{mc^2}{2kT v_0^2},$$

m is the atomic mass, and v_0 is the central frequency of the transition. According to this result, the gain of the amplifying medium is proportional to a Gaussian function $\exp[-\beta(v - v_0)^2]$ centered at frequency v_0.

[8] *MWTP*, Section 17.3.
[9] *MWTP*, Section 15.8.

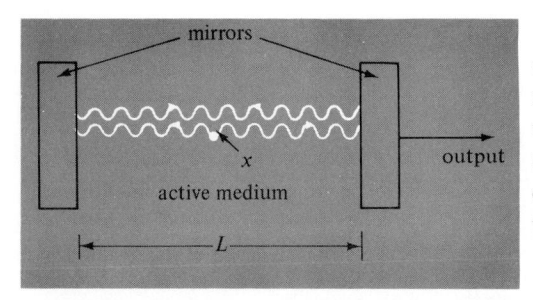

Fig. 20.6. A wave in a laser cavity starts at x, travels back and forth between the mirrors, and returns to x. (The return wave has been displaced for clarity.)

We now consider a wave in the resonant cavity of a laser; we take the length of the amplifying medium to be L (see Fig. 20.6). For plane mirrors, the optical cavity constitutes a Fabry–Perot etalon (Section 18.3). The nth resonance frequency of the cavity for longitudinal modes of oscillation satisfies the condition

$$\nu_n = \frac{nc}{2L}.$$

Let us consider a wave that starts at some point x in the cavity and travels back and forth between the cavity mirrors returning to point x. While traveling the distance $2L$ the wave loses some fraction Δ of its energy by scattering, reflection loss, useful output, etc. In order for the laser to operate, the gain $(I_\nu - I_{0\nu})$ must be equal to or greater than the loss $(\Delta I_{0\nu})$; that is,

$$I_\nu - I_{0\nu} = I_{0\nu}[\exp{(2\alpha_\nu L)} - 1] \geq \Delta I_{0\nu}$$

or

$$\exp{(2\alpha_\nu L)} - 1 \geq \Delta.$$

If $2\alpha_\nu L \ll 1$, this threshold condition becomes

$$2\alpha_\nu L \geq \Delta.$$

Let us suppose that at a certain frequency the gain is greater than the loss. Then, of course, the wave will grow in amplitude and continue to grow so long as $2\alpha_\nu L > \Delta$. The fractional loss is practically constant for all wave amplitudes. The gain, as reflected by the population difference $n_2 - n_1$, decreases until $2\alpha_\nu L = \Delta$ at which time an equilibrium state obtains. This depletion occurs at the frequency or frequencies of oscillation of the laser cavity which are above threshold. It is called **hole burning** and results in the modification of the gain curve indicated in Fig. 20.7. In this particular case the width of the gain curve relative to the separation between the longitudinal modes of the resonant cavity is such that the laser oscillates simultaneously at two frequencies.

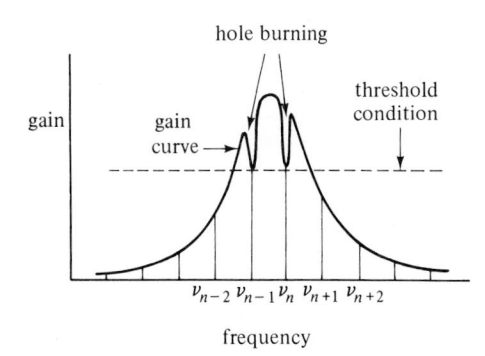

Fig. 20.7. The effect of hole burning on the gain curve of a particular laser.

20.5 SOME BASIC KINDS OF LASERS

Lasers have been built that produce radiation at wavelengths as short as 240 nm (1.25×10^{15} Hz) and as long as 3.4×10^5 nm (8.6×10^{11}

Hz). The possibility exists of spanning the range from 100 nm to 10^7 nm provided suitable laser materials can be found. In addition, coherent radiation at even higher frequency can be produced via harmonic generation as suggested in Section 20.2. Experimental results on harmonic generation will be discussed in Section 20.6. We shall devote this section to a discussion of some of the basic kinds of lasers as given in Table 20.1 and the pumping methods employed in their operation.

Table 20.1 The basic sources of laser radiation

Physical state	Active material
solid	ions
	semiconductors
gas	ions
	molecules
	neutral atoms
liquid	ions
	molecules

Solid lasers

The first operating laser used ruby as the active medium. Ruby is an aluminum oxide (Al_2O_3) crystal in which some of the aluminum atoms have been replaced by chromium ions (Cr^{3+}). A partial energy-level scheme for chromium ions in a ruby crystal is shown in Fig. 20.8. The chromium ions in ruby absorb ultraviolet and green-yellow radiation and thereby populate the rather broad levels E_2 and E_3 at energies of about 2.2 and 3.1 eV, respectively. Although these levels are very short lived, deexcitation directly to the ground state does not occur. Rather, the atoms lose some energy to the crystal lattice and fall into a metastable (long-lived) level

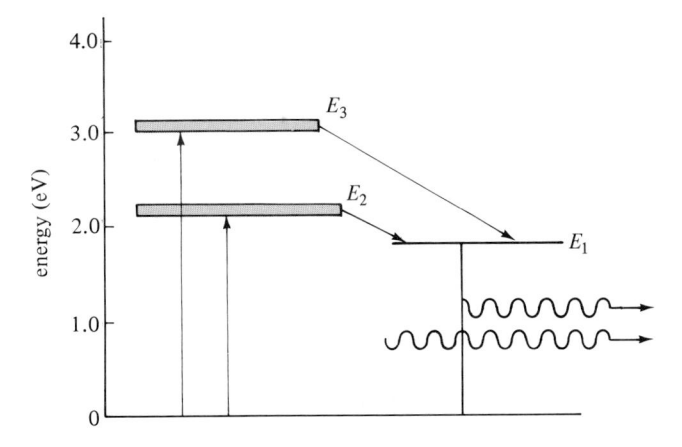

Fig. 20.8. Partial energy-level diagram for chromium ions in a ruby crystal. Absorption excites the chromium ions to states E_2 or E_3. These states are very short-lived, the atoms very quickly giving up some energy to the crystal lattice and falling into state E_1 which is long-lived. Stimulated emission of radiation from this level is responsible for the laser action in ruby.

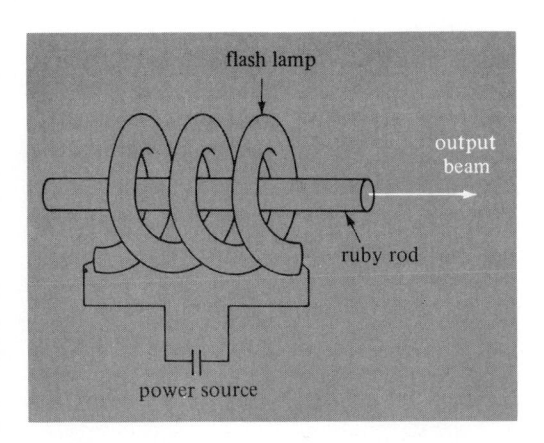

Fig. 20.9. Schematic diagram of a ruby laser.

E_1 at 1.79 eV. The mean life of atoms in this metastable level is a few milliseconds unless they are subjected to stimulation. The long lifetime of this metastable level makes ruby a very good medium for laser action since it is relatively easy to create a population inversion in this level. The wavelength of radiation emitted when the atom returns to its ground state is (at room temperature) 694.3 nm.

The ruby crystal used in a laser is in the form of a rod a few centimeters long and about 0.5 cm in diameter. The ends of the rod are normally polished optically flat and parallel and are silvered to form the mirrors required to produce optimum laser action. Pumping of atoms into the states at 2.2 and 3.1 eV is carried out optically through the use of a flash lamp as shown schematically in Fig. 20.9. The ruby rod is sometimes cooled to liquid nitrogen temperature. The flash lamp is normally flashed in a short pulse of high intensity which pumps most of the chromium atoms into the metastable state. Stimulated emission then takes place and an intense, red laser beam lasting about 0.5 millisec is produced. The efficiency for conversion of optical flash-tube energy into coherent laser energy is less than 1%. The power in the beam during the pulse is several kilowatts (kW). All crystal lasers operate in essentially the same manner as the ruby laser.

The power in the beam during a pulse can be raised significantly by shortening the length of the pulse. This is accomplished by using some form of shutter to delay the release of the energy stored in the laser medium until the population inversion has become maximum. At this time the shutter is suddenly opened and the laser emission proceeds in one giant coherent pulse. This process is known as the **Q-switching technique**. One method of Q-switching employs a light absorbing, bleachable dye between the laser material and the end mirror through which the beam emerges. During pumping, energy is stored in the laser medium until the spontaneous, incoherent emissions from the active medium bleach the shutter material, which results in the release of the stored energy through laser emission. A second method makes use of a rotating mirror instead of a fixed mirror at one end of the active medium. Energy is stored in the medium during pumping. The relative timing of pumping and rotation of the mirror in a pulsed laser is such that the rotating mirror becomes parallel to the partially transparent mirror at the other end of the active material at the instant the stored energy is a maximum. Again, a very short, intense burst of coherent radiation is emitted.

When Q-switching is applied to a solid laser such as ruby, pulses lasting about 10 nsec are produced. The power level in the beam during the pulse can be several hundred megawatts (MW). Q-switching can also be used with continuously pumped lasers. In this case a series of pulses is produced, one for each rotation of the mirror. Q-switching is of great practical importance since it raises the electric field in coherent laser beams to the magnitude required to produce the nonlinear effect discussed in Section 20.2.

Semiconductor lasers

A second form of solid laser makes use of a *p-n* junction (see Section 12.8) in a suitable semiconductor.[10,11] The junction in the semiconductor is produced in such a manner that the Fermi level is below the top of the valence band in the *p*-type region and above the bottom of the conduction band in the *n*-type region when there is no voltage applied across the junction (see Fig. 20.10a). This leaves the levels in the valence band in the *p*-type region above the Fermi level empty. The application of a forward

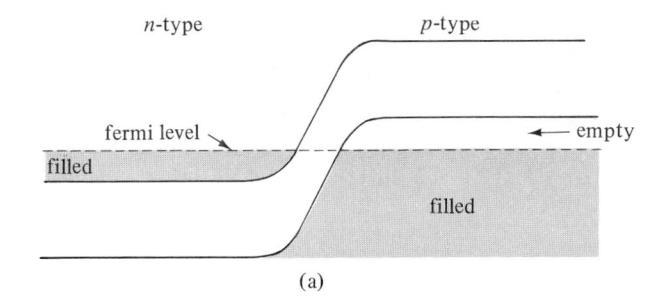

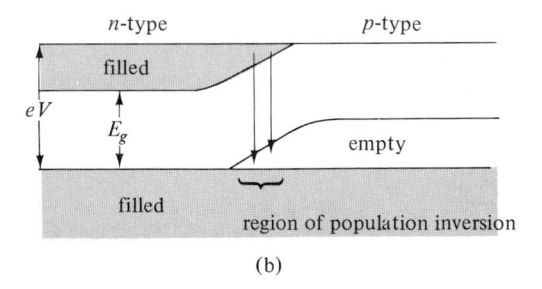

Fig. 20.10. (a) *p-n* junction in a semiconductor diode. (b) Application of a forward bias across the junction, where eV is greater than the gap energy E_g, produces a population inversion in the region of the junction.

[10] R. H. Rediker, "Semiconductor Lasers," *Physics Today*, February, 1965.
[11] M. B. Parrish and I. Hayashi, "A New Class of Diode Lasers," *Scientific American*, July, 1971.

bias V across the junction, where eV is somewhat greater than the gap energy E_g, produces a population inversion in the junction region as illustrated in Fig. 20.10(b). In fact, it is quite easy to show that a necessary condition for population inversion is (see Problem 12)

$$eV > h\nu$$

where $h\nu$ is the energy of the photon emitted or absorbed in a transition between the valence and conduction bands.

The factors determining the successful initiation of laser action in a semiconductor are rather complicated for they involve such effects as recombination rates of holes and electrons and the mode of propagation of electromagnetic waves in the conducting laser medium. We shall have to content ourselves with the statement that laser action does indeed occur in many semiconductors.

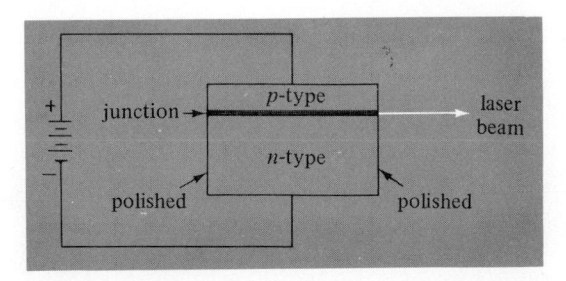

Fig. 20.11. A diode laser. Two ends are polished and the sides left unpolished in order to produce laser emission preferentially in one direction.

A typical semiconductor (or diode) laser is shown schematically in Fig. 20.11. The pumping is electrical in this laser. A battery produces the required forward bias and pumps electrons into the conduction band. A diode laser can operate continuously and the efficiency for conversion is 10% or greater and, theoretically, could even approach 100%. However, the stimulated radiation is neither as monochromatic nor as directional as that from other kinds of lasers.

Gas lasers

The first gas laser consisted of a mixture of helium and neon gas. The stimulated emission occurs between two excited states in the neon atom (see Fig. 20.12). The pumping of neon atoms into the proper excited state is a two-stage process in this laser. An electrical discharge is produced in the gas and the helium atoms are raised to an excited metastable state at about 20 eV by collisions with electrons. Since neon atoms have energy

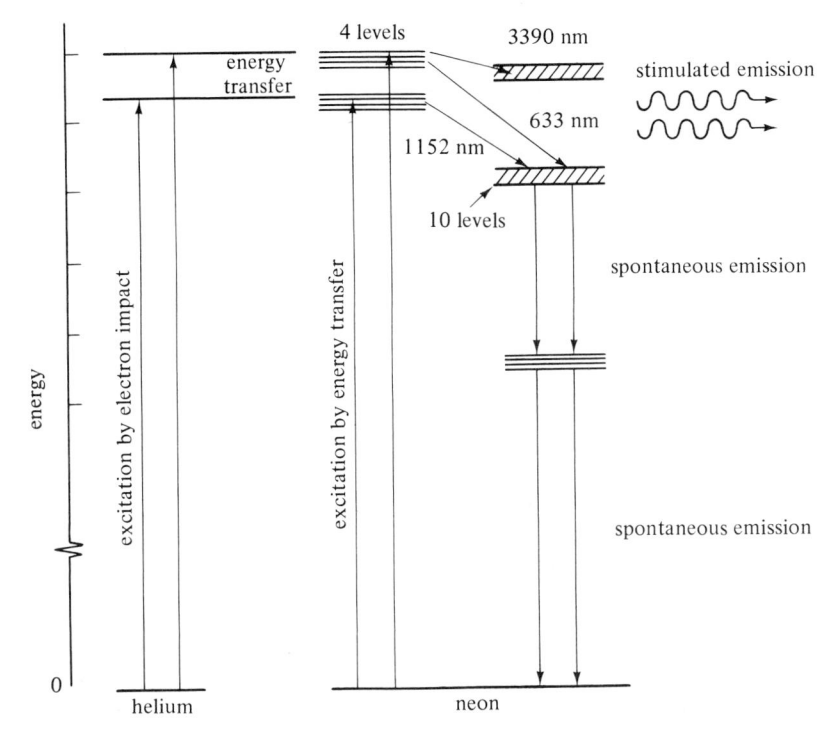

Fig. 20.12. Partial energy-level diagrams for helium and neon atoms. Stimulated emission takes place between two excited states in the neon atom.

levels close to the metastable states of helium, there is a high probability that a collision between a helium atom in a metastable state and an unexcited neon atom will result in a radiationless energy transfer between the two atoms. Under suitable conditions a population inversion of the relevant neon levels will result. Stimulated emission then takes place between two excited states in the neon atom. The transitions at 632.8, 1152.3, and 3390 nm are the most important ones for laser action in this system. Subsequent transitions of the neon atom back to the ground state are spontaneous and do not contribute to the laser action.

The operation of a He-Ne laser is continuous and can be at quite low-power levels (as low as 50 W) even though the efficiency for conversion of input energy into laser beam energy is less than 1%. The optimum conditions for laser action in this system correspond to a total gas pressure of about 1 Torr[12] and a helium to neon ratio of about 7 to 1.

Much more powerful gas lasers can be built when molecules are used

[12] 1 Torr $\equiv 10^{-3}$ m of Hg.

for the laser medium. For example, the carbon dioxide laser can produce a continuous coherent beam of infrared radiation at a power level of many kilowatts.[13] Molecules can rotate and vibrate and so possess a complicated set of energy levels.[14] For each electronic energy level, there exists a set of vibrational levels that are closely spaced in energy compared to the separation between electronic levels; for each vibrational level, there exists a set of rotational levels that are closely spaced in energy compared to the separation between vibrational levels. Transitions are allowed between energy levels whose vibrational and rotational quantum numbers differ by one. These vibrational-rotational transitions are usually in the infrared region of the spectrum and are the basis of high-power molecular gas lasers.

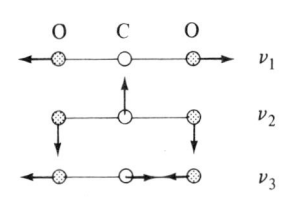

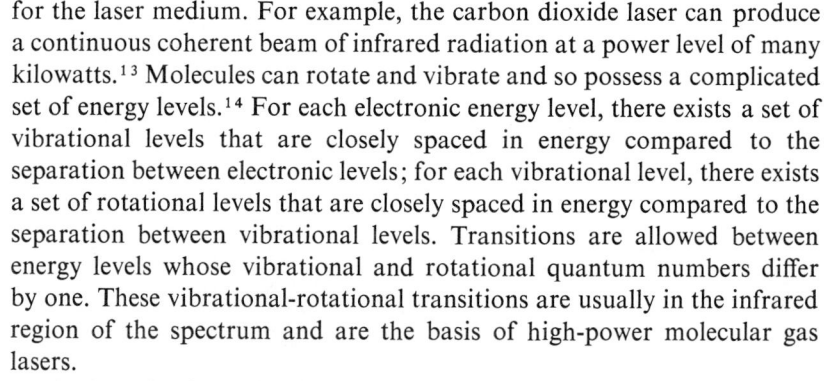

Carbon dioxide (CO_2) is a linear molecule and possesses four normal modes of vibration.[15] However, two of these modes are degenerate so that only three distinguishable vibrational motions occur (see Fig. 20.13). The vibrational state of a CO_2 molecule is characterized by three quantum numbers (v_1, v_2, v_3) which refer to the modes of frequency v_1, v_2, and v_3, respectively. Laser action in a CO_2 molecule is based upon transitions between the (001) vibrational level and the (100) or (020) vibrational levels.

The pumping mechanism in a CO_2 laser is of the two-stage variety described above for the He-Ne laser (see Fig. 20.14). Nitrogen gas (N_2) is mixed with the CO_2 gas. Since N_2 is a diatomic molecule, it has only one

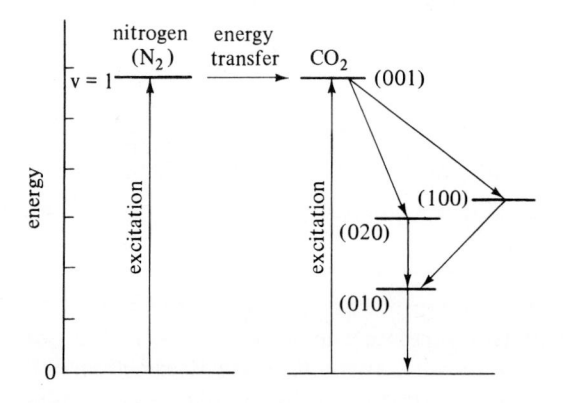

Fig. 20.13. The normal modes of vibration of a carbon dioxide molecule. The mode of frequency v_2 is two-fold degenerate.

Fig. 20.14. Partial energy-level diagram for N_2 and CO_2 molecules. Stimulated emission takes place between the (001) and (020) or (100) vibrational levels in CO_2.

[13]C. K. N. Patel, "High-Power Carbon Dioxide Lasers," *Scientific American*, August, 1968.
[14]*MWTP*, Section 18.8.
[15]*MWTP*, Section 14.7.

mode of vibration. The first vibrational level (quantum number $v = 1$) in N_2 is at almost the same energy as the (001) level in CO_2. The N_2 molecules are very efficiently excited into the first vibrational level via collisions with electrons in a discharge. The N_2 molecules preferentially lose this excess energy through collisions with the CO_2 molecules in which the (001) vibrational level in CO_2 is excited. Stimulated emission to the (100) or (020) vibrational levels in CO_2 takes place with the emission of coherent beams of infrared radiation. Since each vibrational level has a set of rotational levels associated with it, there are actually very many separate transitions closely spaced in energy involved in these two stimulated emission processes.

The (001) vibrational level in CO_2 can be selectively populated in a properly constructed CO_2-N_2 laser. As a result efficiencies of 30% or more for the conversion of input energy to output energy can be attained. This high efficiency, coupled with the construction of very lengthy lasers (up to several hundred feet), produces the continuous output beam of several kilowatts mentioned earlier. The use of Q-switching via a rotating mirror with a CO_2 laser produces very high peak powers (up to several MW) in bursts of about 150 nsec duration at rates of 400 bursts per second.

An alternate method of pumping for gas lasers has been discovered.[16] When some organic molecules are excited by photons from a flash lamp, a chemical reaction takes place and the molecule is dissociated into two or more fragments. An example of such a molecule is trifluoromethiodide (CF_3I). The carbon—iodine bond in this molecule breaks when a photon is absorbed. As a result of the reaction, the iodine atom is left preferentially in a state of electronic excitation. The population inversion occurs as the result of a redistribution of energy which accompanies the rupture of a chemical bond.

Laser emission in hydrogen chlorine (HCl) mixtures occurs when hydrogen and chlorine interact to form HCl. The process is initiated by a brief flash of light but then proceeds chemically as follows:

(a) $Cl_2 + hv \rightarrow Cl + Cl$
(b) $Cl + H_2 \rightarrow HCl + H$
(c) $H + Cl_2 \rightarrow HCl^* + Cl +$ energy
(d) $HCl^* + hv \rightarrow HCl + 2hv$.

In (a) the initial flash produces free Cl atoms. The Cl atoms react with H_2 molecules to produce H atoms (b) which in turn react with Cl_2 molecules to produce excited HCl molecules and more Cl atoms (c). The HCl molecules are left in a vibrational energy level from which stimulated emission proceeds (d). A supply of Cl atoms is built up during steps

[16] G. C. Pimental, "Chemical Lasers," *Scientific American*, April, 1966. Available as *Scientific American Offprint 303* (San Francisco: W. H. Freeman and Co., Publishers).

(a) and (c) and the pumping proceeds via the cycle (b) → (c) which continues to produce Cl atoms and excited HCl molecules. The efficiency of the HCl laser is about 15%.

Liquid lasers

Laser action has also been produced in liquids.[17,18] Rare-earth ions are especially suitable for use in liquid lasers. The rare-earth ions in some liquid lasers are incorporated into organic molecules. Pumping occurs via excitation of energy levels belonging to certain subsections or groups within the organic molecule and the subsequent transfer of energy to the rare-earth ion in which the population inversion occurs. In other liquid lasers the rare-earth ion is surrounded by an array of inorganic solvent molecules that effectively isolate it from interactions with the solvent. This isolation factor permits the population inversion to be attained in the rare-earth ions and stimulated emission results.

20.6 NONLINEAR OPTICAL EFFECTS OBSERVED WITH LASERS

The first laser operated in 1960 and the first nonlinear optical effect was observed in 1961 with the use of the beam from a ruby laser. In this experiment a 3 kW pulse of light of wavelength 694.3 nm passed through a quartz crystal (see Fig. 20.15). The radiation emerging from the crystal was found to consist of one part in 10^8 of radiation of wavelength 347.15 nm. This latter radiation was associated with the second harmonic polarization wave produced by the nonlinear susceptibility of the quartz

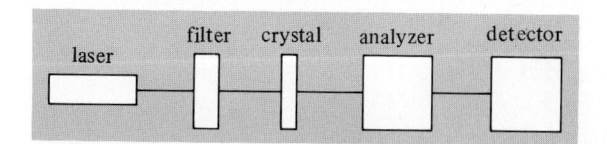

Fig. 20.15. Schematic diagram of equipment required to produce and detect second harmonic generation in a nonlinear crystal. The analyzer could be a simple filter, prism, monochromator, or polarization analyzer while the detector could be a photographic plate, photocell, or photomultiplier tube.

[17] A. Lempicki and H. Samelson, "Liquid Lasers," *Scientific American*, June, 1967.
[18] A. Heller, "Laser Action in Liquids," *Physics Today*, November, 1967.

crystal as we discussed in Section 20.2. With better understanding of nonlinear effects and the production of better nonlinear materials, it is now possible to convert almost all of the energy of a laser beam into second harmonic radiation in an appropriate material.[19]

Third harmonic production in crystals, such as calcite, which possess a center of inversion (Section 20.2) has been observed. It is possible to produce second harmonic output from such crystals by applying a constant electric field across the crystal. This applied field removes the center of inversion and the crystal can produce even harmonic polarization waves.

A difficulty in the observation of the second harmonic radiation occurs due to dispersion in the crystal. The fundamental and higher-order polarization waves travel through the crystal in phase with the input laser beam of frequency ω. The second harmonic polarization wave radiates light waves at frequency 2ω in the direction in which it is traveling. Due to dispersion in the crystal, the light wave at frequency 2ω travels at a different speed from the light wave at frequency ω. Since the polarization wave at frequency 2ω is in phase with the wave at frequency ω, there is a continuous slippage in phase between the second harmonic polarization wave and the second harmonic radiated light wave. The two waves are 180° out of phase after one coherence length, as we have already noted in Section 20.2 in our discussion of the generation of sum and difference frequencies. The very short coherence length limits the widths of crystals used to generate harmonic waves to about 10^{-5} m, unless some method is used to avoid the difficulties of the coherence length.

The most common technique used to overcome this problem is the use of doubly refracting crystals (see Section 14.10). In such crystals there are certain directions of travel for light waves in which the speed of both the ordinary and extraordinary waves is the same. For example, in a crystal of potassium dihydrogen phosphate (KDP) ordinary fundamental light of wavelength 694.3 nm travels at the same speed as extraordinary second harmonic light of wavelength 347.15 nm at an angle of 50° to the optic axis (see Fig. 20.16). This technique permits the use of crystals of one centimeter or more in thickness with an increase in conversion efficiency of 10^6, or more.

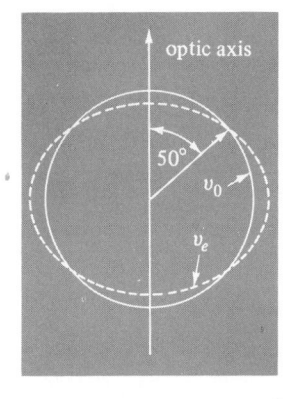

Fig. 20.16. In a crystal of KDP the speed v_0 of ordinary waves of wavelength 694.3 nm equals the speed v_e of extraordinary waves of wavelength 347.15 nm at an angle of 50° to the optic axis.

Stimulated Raman emission

When light from an ordinary source is incident upon a molecule, occasionally a molecule in the ground state will absorb a photon of arbitrary energy, be excited to a higher energy level, and then reemit the energy remaining from the incident photon as a photon of lower frequency. This is known as the **Raman effect** and is illustrated in Fig. 20.17(a).

[19]See footnote 3 in this Chapter.

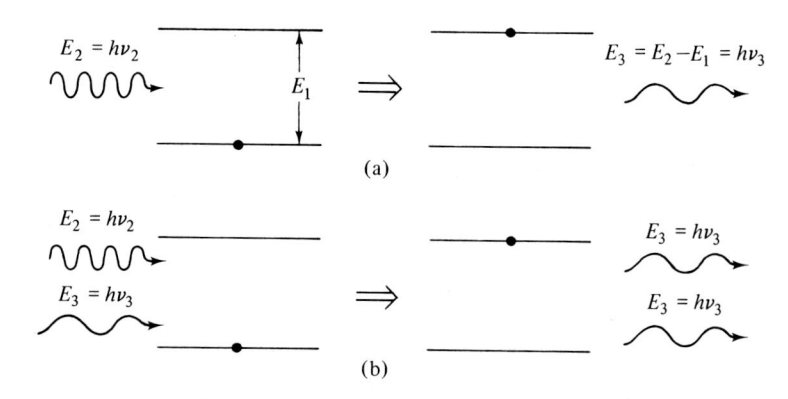

Fig. 20.17. (a) Schematic illustration of the ordinary Raman effect. (b) In stimulated Raman emission the production of the reemitted (or scattered) photon is stimulated by the presence of another photon of exactly the same energy.

Stimulated Raman emission was discovered experimentally during measurements on the light output of a pulsed ruby laser. Approximately 10% of the radiation expected at 694.3 nm was missing, but it was soon found to be in a coherent beam at 766.0 nm. The difference in wavelength corresponds to a difference in frequency of 4.04×10^{13} Hz which is a characteristic frequency of vibration of the nitrobenzene molecules through which the laser beam was passing.

In the ordinary Raman effect photons of frequency v_3 are produced from photons of frequency v_2 after absorption of energy E_1 [see Fig. 20.17(a)]. A laser beam supplies enormous numbers of photons all at frequency v_2. Very large numbers of photons of frequency v_3 are produced via the ordinary Raman effect. If these photons do not escape from the sample quickly (that is, if they are reflected back and forth between mirrors, for example), they stimulate further Raman emission at frequency v_3 as indicated in Fig. 20.17(b). A large fraction of the incident beam can be converted into lower-frequency radiation through stimulated Raman emission.

Generation of sum and difference frequencies

We showed in Section 20.2 that two light waves of different frequencies passing simultaneously through a nonlinear medium should produce waves at frequencies equal to the sum and the difference between the frequencies of the incident waves. Generation of sum waves was observed in 1962 using two ruby lasers of slightly different frequencies (the difference being produced by a difference in operating temperatures for the two lasers). The frequencies needed to be different so that the sum wave could be distinguished from second harmonic wave. Soon, sum frequencies produced by light from two different types of lasers were observed. Beause of the short coherence length (see Section 20.4) for the generated radiation, some compensating device, such as the doubly refracting crystal mentioned earlier in this section, must be employed to permit measurable amounts of power at the sum frequency.

The observation of waves at the difference frequency is rather more difficult. The efficiency is proportional to $(\omega_1 - \omega_2)/\omega_1$ and is small for ω_1 not too different from ω_2. However, if $\omega_1 - \omega_2$ is small, the difference frequency is in the microwave region of the spectrum where high-sensitivity detectors are available. If $\omega_1 - \omega_2$ is increased to increase the efficiency of conversion, detection becomes more difficult. For example, a doubly refracting crystal that is transparent to both visible and infrared radiation may be required. The first difference frequency detected was in the microwave region at 2.964 GHz using two ruby lasers operating at slightly different frequencies.

Self-focusing

In early experiments on stimulated Raman emission it was noted that the conversion of energy into the lower-frequency radiation occurred in a much shorter length of sample than had been expected. A close examination of what was happening showed that the incident laser beam collapsed into self-trapped filaments of dimensions from 5×10^{-6} to 5×10^{-5} m and of extremely high power density. This effect is called **self-focusing** and is due to a molecular Kerr effect in which the intense electric field in the coherent laser beam aligns the molecular dipoles in the medium. The passage of an electromagnetic wave through a dielectric material in which the molecular dipoles are so affected by the wave becomes unstable and the wave collapses into the very thin filaments just described. The details of self-focusing are not yet well understood.

Optical modulation

Modulation of the amplitude[20] or frequency of radiofrequency and microwave electromagnetic waves is used at present almost exclusively for the transmission of information between distant points. Since light is an electromagnetic wave, we might expect that amplitude and frequency modulation could be applied to laser light in a similar way to provide new systems for communication. This is indeed so[21,22] and much effort is being expended on the development of practical systems. The range of frequencies for carrier waves available with lasers is much greater than the range

[20]*MWTP*, Section 16.6.
[21]S. E. Miller, "Communication by Laser," *Scientific American*, January, 1966. Available as *Scientific American Offprint 302* (San Francisco: W. H. Freeman and Co., Publishers).
[22]D. F. Nelson, "The Modulation of Laser Light," *Scientific American*, June 1968.

available at radio- or microwave frequencies. There is already a scarcity of available channels for communication in the lower-frequency range so that the possibility of using optical and infrared waves for communication is very attractive. Details of the problems to be solved and the approaches being taken to develop reliable communications systems at optical frequencies are given in the references referred to in footnotes 21 and 22.

QUESTIONS AND PROBLEMS

1. When viewed in the light of a laser an (apparently) smooth surface appears granular. That is, there are numerous bright and dark spots on the surface. What is the explanation of this **granular effect**?

2. What are **Brewster windows**?

3. What is a **Kerr cell**?

*4. In Section 20.2 a formula was stated for the intensity of an emergent wave from a nonlinear crystal at twice the frequency of the incident wave. Show that the result quoted is correct.

5. Deduce the effect of passing two light waves of frequencies ω_1 and ω_2 simultaneously through a nonlinear medium possessing a center of symmetry. For simplicity, assume that only α and γ in the expansion for P are non-zero.

6. Derive the expression for the gain constant quoted in Section 20.4.

7. The laser cavity has, in addition to the longitudinal modes discussed in Section 20.4, transverse modes. Describe these modes and the physical reasons for their existence.[23]

8. The cross section of some lasers is elliptical with the lasing material at one focus and the flash lamp at the other. Why is this arrangement used?

9. List at least ten solid-state laser materials.

10. Discuss several kinds of gas lasers other than those mentioned in Section 20.5. Tabulate the principal laser wavelengths for the systems that you discuss.

11. A **ring laser** is an example of a technological application of lasers. Describe the operation of this device.

*12. Show that a necessary condition for population inversion in a semiconductor laser is

$$eV > h\nu$$

where $h\nu$ is the energy of the photon emitted or absorbed in a transition between the valence and conduction bands.

[23] A. Yariv, *Quantum Electronics* (New York: John Wiley & Sons Inc., 1967).

QUESTIONS AND PROBLEMS

13. In a certain ruby (Al_2O_3) laser there is 0.05% by weight Cr_2O_3. The laser wavelength is 693.4 nm and the line width 0.10 nm. Estimate the gain constant per cm at the center of the spectral line for the laser. Assume that 48% of the Cr^{3+} ions are in the first excited state and 43% in the ground state. The lifetime of the upper state for spontaneous emission to the ground state is 3.0×10^{-3} sec. The density of Al_2O_3 is 3.95×10^3 kg·m^{-3}.

14. A helium-neon gas laser operates at 633 nm at a temperature of 350°K. The gain constant of the laser at the line center is 2% per meter. Deduce the population difference per m³. Assume that the lifetime of the upper state for spontaneous emission to the lower state is 1.0×10^{-7} sec.

15. Consider a medium with an index of refraction n which is dependent upon the square of the electric field intensity

$$n = n_0 + \alpha E^2$$

where n_0 and α are constants. If a beam of light of diameter d and wavelength λ is passing through such a medium, show that the condition for a stable **self-trapped filament** is

$$\frac{E^2}{\lambda^2} \geq \frac{0.744}{n_0 \alpha d^2}.$$

Hint: Balance the diffraction effect ($\theta = 1.22\lambda/d$) against the critical angle for total internal reflection.

16. Calculate the maximum photon flux of a Q-switched ruby laser operating at 693.4 nm. Take the power to be 500 MW and the beam diameter to be 5×10^{-6} m in a self-trapped filament. Assume that the power goes into 100 filaments. Calculate the electric field magnitude within such a filament given that $K = 10.5$ and $n = 1.76$ for ruby.

17. Very intense laser beams can produce higher-order **Raman effects**, particularly at infrared frequencies. Discuss the major qualitative features of these emissions.

18. Write a short essay on **stimulated Brillouin scattering**.

19. Discuss communications systems at optical frequencies.

20. The future of lasers was the subject of a panel discussion held in Esfahan, Iran late in 1971 and reported in Physics Today.[24] Comment on this discussion.

---

[24]"The Future of Lasers", *Physics Today*, (March 1972).

APPENDICES

APPENDICES

Some useful A vector relations

A.1 GRADIENT AND DIVERGENCE IN CURVILINEAR COORDINATES

We consider a point P with coordinates (x, y, z) in a rectangular coordinate system and coordinates (u_1, u_2, u_3) in a curvilinear coordinate system. The position vector of the point P is

$$\mathbf{r} = x\mathbf{i} + y\mathbf{j} + z\mathbf{k} = \mathbf{r}(u_1, u_2, u_3).$$

Tangent vectors to the u_1, u_2, and u_3 curves at P are

$$\frac{\partial \mathbf{r}}{\partial u_1}, \quad \frac{\partial \mathbf{r}}{\partial u_2}, \quad \text{and} \quad \frac{\partial \mathbf{r}}{\partial u_3},$$

respectively. Unit vectors in these directions are

$$\mathbf{e}_1 = \frac{\left(\frac{\partial \mathbf{r}}{\partial u_1}\right)}{\left|\frac{\partial \mathbf{r}}{\partial u_1}\right|} = \frac{1}{h_1}\left(\frac{\partial \mathbf{r}}{\partial u_1}\right),$$

$$\mathbf{e}_2 = \frac{\left(\frac{\partial \mathbf{r}}{\partial u_2}\right)}{\left|\frac{\partial \mathbf{r}}{\partial u_2}\right|} = \frac{1}{h_2}\left(\frac{\partial \mathbf{r}}{\partial u_2}\right),$$

and

$$\mathbf{e}_3 = \frac{\left(\frac{\partial \mathbf{r}}{\partial u_3}\right)}{\left|\frac{\partial \mathbf{r}}{\partial u_3}\right|} = \frac{1}{h_3}\left(\frac{\partial \mathbf{r}}{\partial u_3}\right),$$

respectively, where the quantities h_1, h_2, and h_3 are called **scale factors**. If $\mathbf{e}_1$, $\mathbf{e}_2$, and $\mathbf{e}_3$ are mutually perpendicular, the curvilinear coordinate system is said to be **orthogonal**.

If V is a scalar function and $\mathbf{E} = E_1\mathbf{e}_1 + E_2\mathbf{e}_2 + E_3\mathbf{e}_3$ is a vector function of orthogonal curvilinear coordinates, then

$$\nabla V = \operatorname{grad} V = \frac{1}{h_1}\left(\frac{\partial V}{\partial u_1}\right)\mathbf{e}_1 + \frac{1}{h_2}\left(\frac{\partial V}{\partial u_2}\right)\mathbf{e}_2 + \frac{1}{h_3}\left(\frac{\partial V}{\partial u_3}\right)\mathbf{e}_3$$

$$\nabla \cdot \mathbf{E} = \operatorname{div} \mathbf{E} = \frac{1}{h_1 h_2 h_3}\left[\frac{\partial}{\partial u_1}(h_2 h_3 E_1) + \frac{\partial}{\partial u_2}(h_3 h_1 E_2) + \frac{\partial}{\partial u_3}(h_1 h_2 E_3)\right].$$

Spherical coordinates (r, θ, ϕ)

For spherical coordinates

$$h_r = 1, \quad h_\theta = r, \quad h_\phi = r \sin \theta.$$

Therefore,

$$\nabla V = \left(\frac{\partial V}{\partial r}\right)\mathbf{e}_r + \frac{1}{r}\left(\frac{\partial V}{\partial \theta}\right)\mathbf{e}_\theta + \frac{1}{r \sin \theta}\left(\frac{\partial V}{\partial \phi}\right)\mathbf{e}_\phi$$

$$\nabla \cdot \mathbf{E} = \frac{1}{r^2}\frac{\partial}{\partial r}(r^2 E_r) + \frac{1}{r \sin \theta}\frac{\partial}{\partial \theta}(\sin \theta E_\theta) + \frac{1}{r \sin \theta}\left(\frac{\partial E_\phi}{\partial \phi}\right)$$

where $\mathbf{e}_r$, $\mathbf{e}_\theta$, $\mathbf{e}_\phi$ are unit vectors in the direction of increasing r, θ, and ϕ respectively.

A.2 THE DIVERGENCE THEOREM

The divergence theorem is expressed by the equation

$$\int_S \mathbf{E} \cdot d\mathbf{S} = \int_v \mathbf{E} \cdot d\mathbf{v}$$

where the left-hand side is the surface integral of $\mathbf{E}$ over a closed surface and the right-hand side is the volume integral of the divergence of $\mathbf{E}$ throughout the volume enclosed by the surface. The right-hand integral can be written as

$$\int_v \nabla \cdot \mathbf{E}\, dv = \iiint \left(\frac{\partial E_x}{\partial x} + \frac{\partial E_y}{\partial y} + \frac{\partial E_z}{\partial z}\right) dx\, dy\, dz$$

$$= \iiint \left(\frac{\partial E_x}{\partial x}\right) dx\, dy\, dz + \iiint \left(\frac{\partial E_y}{\partial y}\right) dx\, dy\, dz$$

$$+ \iiint \left(\frac{\partial E_z}{\partial z}\right) dx\, dy\, dz$$

$$= \iint (E_{2x} - E_{1x})\, dy\, dz + \iint (E_{2y} - E_{1y})\, dx\, dz$$

$$+ \iint (E_{2z} - E_{1z})\, dx\, dy$$

where the subscripts 2 and 1 refer to values of E_x, E_y, or E_z at the upper or lower limits of the variable of integration, respectively.

The quantity $(E_{2x} - E_{1x})\, dy\, dz$ is a rectangular prism of cross-sectional area $dy\, dz$ which is parallel to the x axis and reaches from one surface of the volume to the opposite surface. The surface elements cut out by the ends of this prism are dS_2 and dS_1. Therefore,

$$E_{2x}\mathbf{i} \cdot d\mathbf{S}_2 = E_{2x}\, dy\, dz$$

and

$$E_{1x}\mathbf{i} \cdot d\mathbf{S}_1 = -E_{1x}\, dy\, dz$$

so that

$$\iint (E_{2x} - E_{1x})\, dy\, dz = \int_S E_x\mathbf{i} \cdot d\mathbf{S}.$$

Similarly,

$$\iint (E_{2y} - E_{1y})\, dx\, dz = \int_S E_y \mathbf{j} \cdot d\mathbf{S}$$

$$\iint (E_{2z} - E_{1z})\, dx\, dy = \int_S E_z \mathbf{k} \cdot d\mathbf{S}$$

and

$$\int_v \mathbf{\nabla} \cdot \mathbf{E}\, dv = \int_S (E_x \mathbf{i} + E_y \mathbf{j} + E_z \mathbf{k}) \cdot d\mathbf{S} = \int_S \mathbf{E} \cdot d\mathbf{S}$$

which proves the divergence theorem.

A.3 STOKES' THEOREM

Stokes' theorem is expressed by the equation

$$\int_S (\mathbf{\nabla} \times \mathbf{B}) \cdot d\mathbf{S} = \oint \mathbf{B} \cdot d\mathbf{l}$$

where the left-hand side is the surface integral of the curl of **B** over the surface and the right-hand side is the line integral of **B** along a closed curve bounding the surface. We begin by considering a surface element $d\mathbf{S} = \mathbf{k}\, dx\, dy$ and locate the origin of the coordinate system at the center of the element (see Fig. A.1). We assume that **B** has the value $\mathbf{B} = \mathbf{i}B_x + \mathbf{j}B_y + \mathbf{k}B_z$ at the origin. The values of the x component of **B** at points D and F are

$$B_x - \frac{1}{2}\left(\frac{\partial B_x}{\partial y}\right) dy$$

and

$$B_x + \frac{1}{2}\left(\frac{\partial B_x}{\partial y}\right) dy,$$

respectively, while the values of the y component of **B** at points A and C are

$$B_y + \frac{1}{2}\left(\frac{\partial B_y}{\partial x}\right) dx$$

and

$$B_y - \frac{1}{2}\left(\frac{\partial B_y}{\partial x}\right) dx,$$

respectively. The line integral $\mathbf{B} \cdot d\mathbf{l}$ around this element is then

$$\mathbf{B} \cdot d\mathbf{l} = \left(B_x - \frac{1}{2}\frac{\partial B_x}{\partial y} dy\right) dx + \left(B_y + \frac{1}{2}\frac{\partial B_y}{\partial x} dx\right) dy$$
$$- \left(B_x + \frac{1}{2}\frac{\partial B_x}{\partial y} dy\right) dx - \left(B_y - \frac{1}{2}\frac{\partial B_y}{\partial x} dx\right) dy$$

SEC. A.3 — STOKES' THEOREM

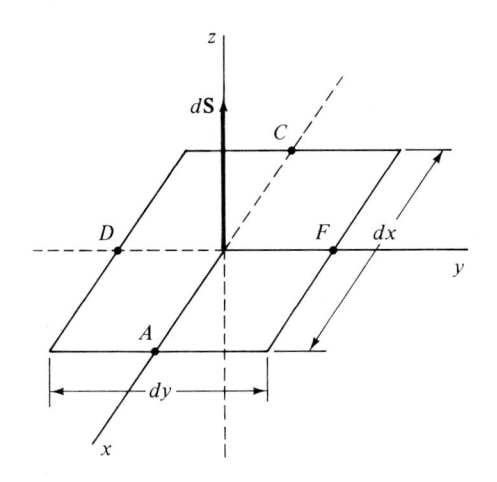

Fig. A.1. A surface element $d\mathbf{S} = \mathbf{k}\, dx\, dy$ at the origin of a coordinate system.

$$= \left(\frac{\partial B_y}{\partial x} - \frac{\partial B_x}{\partial y}\right) dx\, dy.$$

Now, for this surface element,

$$(\mathbf{\nabla} \times \mathbf{B}) \cdot d\mathbf{S} = (\mathbf{\nabla} \times \mathbf{B}) \cdot \mathbf{k}\, dx\, dy$$
$$= \left(\frac{\partial B_y}{\partial x} - \frac{\partial B_x}{\partial y}\right) dx\, dy.$$

so that

$$(\mathbf{\nabla} \times \mathbf{B}) \cdot d\mathbf{S} = \mathbf{B} \cdot d\mathbf{l}$$

for the surface element under consideration. But any surface can be considered as the sum of infinitesimal rectangular surface elements. The line integral around this surface will be the sum of the line integrals around each of the infinitesimal rectangular surface elements. However, all interior boundaries give contributions to the line integral that cancel each other so that only those contributions from exterior boundaries give nonzero contributions to the line integral. Therefore, we conclude that

$$\int_S (\mathbf{\nabla} \times \mathbf{B}) \cdot d\mathbf{S} = \oint \mathbf{B} \cdot d\mathbf{l}$$

which proves Stokes' theorem.

B Rotating coordinate axes

Let us consider a vector **r** such as the position vector of a particle. As the particle moves, the vector **r** varies with time. The manner of its variation will often depend on the coordinate system to which the observations are referred. Of special interest is the case in which two coordinate systems rotate relative to each other. The change $d\mathbf{r}$ in vector **r** as observed in one system (considered to rotate) is given by

$$(d\mathbf{r})_{\text{rot}} = (d\mathbf{r})_f + (d\mathbf{r})_r$$

where $(d\mathbf{r})_f$ is the change in **r** relative to the fixed axes and $(d\mathbf{r})_r$ is the change in **r** due to the rotation of the first system.

An expression for the vector $(d\mathbf{r})_r$ can be obtained by reference to Fig. B.1. The magnitude of $(d\mathbf{r})_r$ is (to first approximation)

$$|(d\mathbf{r})_r| = r \sin \theta \, d\phi.$$

A clockwise rotation through angle $d\phi$ pictured in Fig. B.1 can be specified by the vector $d\boldsymbol{\phi}$ whose direction is specified by the right-hand rule. Note, however, that the clockwise rotation of the vector **r** corresponds to a counterclockwise rotation of the coordinate ϕ. Therefore, the vector $(d\mathbf{r})_r$ has magnitude $r \sin \theta \, d\phi$ and is seen to be perpendicular to both $d\boldsymbol{\phi}$ and **r**. Therefore, we write

$$(d\mathbf{r})_r = \mathbf{r} \times d\boldsymbol{\phi}$$

so that

$$(d\mathbf{r})_{\text{rot}} = (d\mathbf{r})_f + \mathbf{r} \times d\boldsymbol{\phi}.$$

or

$$(d\mathbf{r})_f = (d\mathbf{r})_{\text{rot}} - \mathbf{r} \times d\boldsymbol{\phi} = (d\mathbf{r})_{\text{rot}} + d\boldsymbol{\phi} \times \mathbf{r}.$$

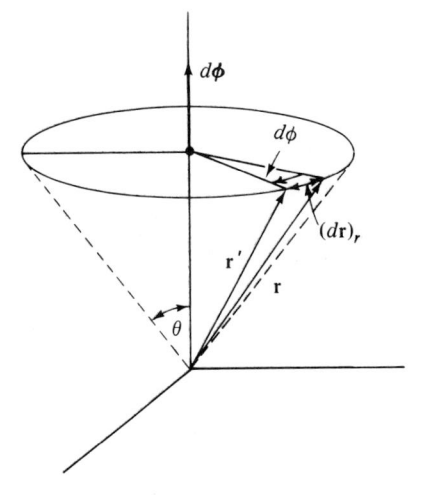

Fig. B.1. The change $d\mathbf{r}_r$ in a vector **r** due to rotation.

If we divide this equation by dt, the time required for the differential changes $(d\mathbf{r})_{\text{rot}}$, $(d\mathbf{r})_f$, and $d\boldsymbol{\phi}$, we obtain

$$\left(\frac{d\mathbf{r}}{dt}\right)_f = \left(\frac{d\mathbf{r}}{dt}\right)_{\text{rot}} + \frac{d\boldsymbol{\phi}}{dt} \times \mathbf{r}$$

$$= \left(\frac{d\mathbf{r}}{dt}\right)_{\text{rot}} + \boldsymbol{\omega} \times \mathbf{r}$$

where $\boldsymbol{\omega} = d\boldsymbol{\phi}/dt$ is the **angular frequency** of the rotation. This equation may also be written as

$$\mathbf{v}_f = \mathbf{v}_{\text{rot}} + \boldsymbol{\omega} \times \mathbf{r}$$

where $\mathbf{v}_f$ is the velocity of the particle in the system considered fixed and $\mathbf{v}_{\text{rot}}$ is the velocity of the particle in the system rotating with respect to the fixed system.

The equation involving the time rate of change of **r** can be viewed as an **operator equation**

$$\left(\frac{d}{dt}\right)_f = \left(\frac{d}{dt}\right)_{rot} + \boldsymbol{\omega} \times$$

which operates on the vector **r**. It is essentially a statement of how the time derivative transforms between two coordinate systems. We can use this operator equation to operate on any vector in the fixed system, for example, the velocity $\mathbf{v}_f$. We then obtain

$$\left(\frac{d\mathbf{v}_f}{dt}\right)_f = \left(\frac{d\mathbf{v}_f}{dt}\right)_{rot} + \boldsymbol{\omega} \times \mathbf{v}_f$$

so that

$$\mathbf{a}_f = \left[\frac{d}{dt}(\mathbf{v}_{rot} + \boldsymbol{\omega} \times \mathbf{r})\right]_{rot} + \boldsymbol{\omega} \times (\mathbf{v}_{rot} + \boldsymbol{\omega} \times \mathbf{r})$$

$$= \mathbf{a}_{rot} + \boldsymbol{\omega} \times \left(\frac{d\mathbf{r}}{dt}\right)_{rot} + \boldsymbol{\omega} \times \mathbf{v}_{rot} + \boldsymbol{\omega} \times (\boldsymbol{\omega} \times \mathbf{r})$$

$$= \mathbf{a}_{rot} + 2\boldsymbol{\omega} \times \mathbf{v}_{rot} + \boldsymbol{\omega} \times (\boldsymbol{\omega} \times \mathbf{r})$$

where $\mathbf{a}_f$ is the acceleration of the particle in the fixed system and $\mathbf{a}_{rot}$ is the acceleration of the particle in the rotating system. The quantity $2\boldsymbol{\omega} \times \mathbf{v}_{rot}$ is called the **Coriolis acceleration**[1] and the quantity $\boldsymbol{\omega} \times (\boldsymbol{\omega} \times \mathbf{r})$ is the **centrifugal acceleration**.[1]

The equation of motion of the particle in the fixed system can be written as

$$\mathbf{F} = m\mathbf{a}_f = m\left(\frac{d\mathbf{v}_f}{dt}\right).$$

The equation of motion in the rotating system can be written as

$$\mathbf{F}_{eff} = \mathbf{F} - 2m(\boldsymbol{\omega} \times \mathbf{v}_{rot}) - m\boldsymbol{\omega} \times (\boldsymbol{\omega} \times \mathbf{r})$$
$$= m\mathbf{a}_{rot}$$

where the effective force $\mathbf{F}_{eff}$ is different from the force **F** due to the presence of the Coriolis and centrifugal forces in the rotating coordinate system.

[1] *MWTP*, Section 10.3.

Complex numbers C

The quantity
$$Z = a + ib$$
where a and b are real and $i = (-1)^{1/2}$ is said to be a **complex quantity** or **complex number**. The real numbers a and b are called the **real** and **imaginary components or parts**, respectively. Any complex quantity can be put in the standard form $a + ib$. For example, if
$$Z = \frac{m + in}{p + iq},$$
we can put Z into the standard form by multiplying both numerator and denominator by $p - iq$ which is called the **complex conjugate** of $p + iq$. Carrying through this operation, we obtain
$$\begin{aligned} Z &= \frac{m + in}{p + iq} \times \frac{p - iq}{p - iq} \\ &= \frac{mp + nq}{p^2 + q^2} + i\frac{np - mq}{p^2 + q^2} \\ &= c + id \end{aligned}$$
which is the standard form.

It is often convenient to express a complex number in an exponential form. The complex number
$$Z = a + ib$$
can also be written as
$$Z = p \cos \theta + ip \sin \theta$$
where
$$p = (a^2 + b^2)^{1/2}$$
$$\tan \theta = \frac{b}{a}.$$

The exponential quantity $\exp(i\theta)$ is, by definition,
$$\begin{aligned} \exp(i\theta) &= 1 + i\theta + \frac{(i\theta)^2}{2!} + \frac{(i\theta)^3}{3!} + \cdots \\ &= \left(1 - \frac{\theta^2}{2!} + \frac{\theta^4}{4!} - \cdots\right) + i\left(\theta - \frac{\theta^3}{3!} + \cdots\right) \\ &= \cos \theta + i \sin \theta. \end{aligned}$$
Therefore, we can write
$$\begin{aligned} Z &= p \cos \theta + ip \sin \theta \\ &= p \exp(i\theta). \end{aligned}$$

The number p is called the **absolute value** or **modulus** of Z, and the angle θ is called the **argument** or **phase** of Z.

APPENDIX C — COMPLEX NUMBERS

The rules for addition, subtraction, multiplication, and division are:

Addition

$$Z = Z_1 + Z_2$$
$$= (a_1 + ib_1) + (a_2 + ib_2)$$
$$= (a_1 + a_2) + i(b_1 + b_2)$$
$$= a + ib.$$

Subtraction

$$Z = Z_1 - Z_2$$
$$= (a_1 + ib_1) - (a_2 + ib_2)$$
$$= (a_1 - a_2) + i(b_1 - b_2)$$
$$= a + ib.$$

Division

$$Z = \frac{Z_1}{Z_2} = \frac{a_1 + ib_1}{a_2 + ib_2} = \frac{a_1 + ib_1}{a_2 + ib_2} \times \frac{a_2 - ib_2}{a_2 - ib_2}$$
$$= \frac{a_1 a_2 + b_1 b_2}{a_2^2 + b_2^2} + i\frac{b_1 a_2 - a_1 b_2}{a_2^2 + b_2^2}$$
$$= a + ib$$

or

$$Z = \frac{Z_1}{Z_2} = \frac{p_1 \exp(i\theta_1)}{p_2 \exp(i\theta_2)} = \frac{p_1}{p_2} \exp i(\theta_1 - \theta_2) = p \exp(i\theta)$$

Multiplication

$$Z = Z_1 Z_2 = (a_1 + ib_1)(a_2 + ib_2)$$
$$= (a_1 a_2 + b_1 b_2) + i(a_1 b_2 + b_1 a_2)$$
$$= a + ib$$

or

$$Z = Z_1 Z_2 = p_1 \exp(i\theta_1) p_2 \exp(i\theta_2)$$
$$= p_1 p_2 \exp i(\theta_1 + \theta_2)$$
$$= p \exp(i\theta)$$

Note that the exponential form is particularly well-suited for division and multiplication.

The square of a complex number Z is found by multiplying the complex number by its complex conjugate Z^*. Therefore, the square of $Z = a + ib = p \exp(i\theta)$ is

$$Z^2 = ZZ^* = (a + ib)(a - ib) = a^2 + b^2.$$

The square of a complex number is equal to the sum of the squares of the real and imaginary parts of the complex number.

The real and imaginary parts of a complex number $Z = a + ib$ are

often written as $Re(Z)$ and $Im(Z)$, respectively. Therefore,
$$Re(Z) = a, \quad Im(Z) = b.$$
Let us consider two complex numbers $Z_1 = a_1 + ib_1$ and $Z_2 = a_2 + ib_2$. It then follows that
$$Z_1 Z_2^* + Z_1^* Z_2 = 2Re(Z_1 Z_2^*) = 2Re(Z_1^* Z_2).$$

Finally, we shall give useful expressions for $\sin \theta$ and $\cos \theta$ in terms of complex numbers. We showed above that $\exp(i\theta) = \cos \theta + i \sin \theta$. The complex conjugate of this expression is $\exp(-i\theta) = \cos \theta - i \sin \theta$. By adding and subtracting these two expressions in turn, we obtain

$$\cos \theta = \frac{\exp(i\theta) + \exp(-i\theta)}{2}$$

$$\sin \theta = \frac{\exp(i\theta) - \exp(-i\theta)}{2i}.$$

D Matrix algebra

A **matrix** is a rectangular array of numbers. The general form of a matrix is

$$\begin{pmatrix} u_{11} & u_{12} & \cdots & u_{1m} \\ u_{21} & u_{22} & \cdots & u_{2m} \\ \vdots & \vdots & & \vdots \\ u_{n1} & u_{n2} & \cdots & u_{nm} \end{pmatrix}$$

where n and m can take on any integral values beginning with unity, and indicate the number of rows or columns, respectively, in the matrix. Matrices can be added together, one subtracted from a second, or multiplied together. For simplicity these algebraic operations will be illustrated using 2×2 matrices (that is, matrices with two rows and two columns).

Addition

The matrices

$$A = \begin{pmatrix} u_{11} & u_{12} \\ u_{21} & u_{22} \end{pmatrix}, \qquad B = \begin{pmatrix} v_{11} & v_{12} \\ v_{21} & v_{22} \end{pmatrix}$$

are added together to give a third matrix C as follows:

$$C = A + B = \begin{pmatrix} u_{11} + v_{11} & u_{12} + v_{12} \\ u_{21} + v_{21} & u_{22} + v_{22} \end{pmatrix}.$$

The result is a third 2×2 matrix C whose elements are the sums of the corresponding elements of matrices A and B.

Subtraction

The matrix D formed by subtracting matrix B from matrix A is

$$D = A - B = \begin{pmatrix} u_{11} - v_{11} & u_{12} - v_{12} \\ u_{21} - v_{21} & u_{22} - v_{22} \end{pmatrix}.$$

The result is a 2×2 matrix D whose elements are the differences of the corresponding elements of matrices A and B.

Multiplication

The matrix E formed by multiplying matrices A and B together is

$$E = AB = \begin{pmatrix} u_{11} & u_{12} \\ u_{21} & u_{22} \end{pmatrix} \begin{pmatrix} v_{11} & v_{12} \\ v_{21} & v_{22} \end{pmatrix}$$
$$= \begin{pmatrix} u_{11}v_{11} + u_{12}v_{21} & u_{11}v_{12} + u_{12}v_{22} \\ u_{21}v_{11} + u_{22}v_{21} & u_{21}v_{12} + u_{22}v_{22} \end{pmatrix}.$$

The result is a 2×2 matrix whose elements are formed by multiplying the elements of one column in matrix B by the elements of one row in matrix A and adding the resultant products. In general, two matrices can be multiplied together **only if the number of columns in the first matrix is equal to the number of rows in the second matrix.** For example, if

$$A = \begin{pmatrix} u_{11} & u_{12} & u_{13} \\ u_{21} & u_{22} & u_{23} \\ u_{31} & u_{32} & u_{33} \end{pmatrix} \quad \text{and} \quad B = \begin{pmatrix} v_{11} & v_{12} \\ v_{21} & v_{22} \\ v_{31} & v_{32} \end{pmatrix},$$

the product AB can be formed but the product BA cannot. In general, the product has the number of rows of the first matrix and the number of columns of the second matrix.

Note also that matrix multiplication is not **commutative**. That is, if two matrices A and B are multiplied together, the product AB is not, in general, equal to the product BA, even if both products can be formed.

When more than two matrices are being multiplied together, multiplication starts from the left and proceeds to the right. For example,

$$\begin{pmatrix} u_{11} & u_{12} \\ u_{21} & u_{22} \end{pmatrix} \begin{pmatrix} v_{11} & v_{12} \\ v_{21} & v_{22} \end{pmatrix} \begin{pmatrix} w_{11} & w_{12} \\ w_{21} & w_{22} \end{pmatrix}$$

$$= \begin{pmatrix} u_{11}v_{11} + u_{12}v_{21} & u_{11}v_{12} + u_{12}v_{22} \\ u_{21}v_{11} + u_{22}v_{21} & u_{21}v_{12} + u_{22}v_{22} \end{pmatrix} \begin{pmatrix} w_{11} & w_{12} \\ w_{21} & w_{22} \end{pmatrix}$$

$$= \begin{pmatrix} [(u_{11}v_{11} + u_{12}v_{21})w_{11} + (u_{11}v_{12} + u_{12}v_{22})w_{21}] & [(u_{11}v_{11} + u_{12}v_{21})w_{12} \\ & \quad + (u_{11}v_{12} + u_{12}v_{22})w_{22}] \\ [(u_{21}v_{11} + u_{22}v_{21})w_{11} + (u_{21}v_{12} + u_{22}v_{22})w_{21}] & [(u_{21}v_{11} + u_{22}v_{21})w_{12} \\ & \quad + (u_{21}v_{12} + u_{22}v_{22})w_{22}] \end{pmatrix}.$$

This procedure can be repeated indefinitely so long as the matrix on the left at the end of each multiplication has the same number of columns as there are rows in the next matrix to be multiplied.

Division

There is no division in matrix algebra.

Periodic table E
of the elements

Masses given in the table are average nuclear masses on the ^{12}C scale. For elements for which no stable isotope exists, the mass of the most stable isotope is given in brackets. The recently produced elements of atomic numbers 103, 104, 105, and 106 are omitted because their position in the table is still uncertain. In each entry in the table the symbol for the element is listed first, with the atomic number immediately below the symbol, and the nuclear mass below the atomic number.

The elements

Element	Symbol	Element	Symbol	Element	Symbol
Actinium	Ac	Gold	Au	Praseodymium	Pr
Aluminum	Al	Hafnium	Hf	Promethium	Pm
Americium	Am	Helium	He	Protoactinium	Pa
Antimony	Sb	Holmium	Ho	Radium	Ra
Argon	Ar	Hydrogen	H	Radon	Rn
Arsenic	As	Indium	In	Rhenium	Re
Astatine	At	Iodine	I	Rhodium	Rh
Barium	Ba	Iridium	Ir	Rubidium	Rb
Berkelium	Bk	Iron	Fe	Ruthenium	Ru
Beryllium	Be	Krypton	Kr	Samarium	Sm
Bismuth	Bi	Lanthanum	La	Scandium	Sc
Boron	B	Lead	Pb	Selenium	Se
Bromine	Br	Lithium	Li	Silicon	Si
Cadmium	Cd	Lutetium	Lu	Silver	Ag
Calcium	Ca	Magnesium	Mg	Sodium	Na
Californium	Cf	Manganese	Mn	Strontium	Sr
Carbon	C	Mendeleevium	Md	Sulfur	S
Cerium	Ce	Mercury	Hg	Tantalum	Ta
Cesium	Cs	Molybdenum	Mo	Technetium	Tc
Chlorine	Cl	Neodymium	Nd	Tellurium	Te
Chromium	Cr	Neon	Ne	Terbium	Tb
Cobalt	Co	Neptunium	Np	Thallium	Tl
Copper	Cu	Nickel	Ni	Thorium	Th
Curium	Cm	Niobium	Nb	Thulium	Tm
Dysprosium	Dy	Nitrogen	N	Tin	Sn
Einsteinium	Es	Nobelium	No	Titanium	Ti
Erbium	Er	Osmium	Os	Tungsten	W
Europium	Eu	Oxygen	O	Uranium	U
Fermium	Fm	Palladium	Pd	Vanadium	V
Fluorine	F	Phosphorus	P	Xenon	Xe
Francium	Fr	Platinum	Pt	Ytterbium	Yb
Gadolinium	Gd	Plutonium	Pu	Yttrium	Y
Gallium	Ga	Polonium	Po	Zinc	Zn
Germanium	Ge	Potassium	K	Zirconium	Zr

PERIODIC TABLE OF THE ELEMENTS

H 1 1.00797																	He 2 4.0026
Li 3 6.939	Be 4 9.0122											B 5 10.811	C 6 12.01115	N 7 14.0067	O 8 15.9994	F 9 18.9984	Ne 10 20.183
Na 11 22.9898	Mg 12 24.312				Transition elements							Al 13 26.9815	Si 14 28.086	P 15 30.9738	S 16 32.064	Cl 17 35.453	Ar 18 39.948
K 19 39.102	Ca 20 40.08	Sc 21 44.956	Ti 22 47.90	V 23 50.942	Cr 24 51.996	Mn 25 54.938	Fe 26 55.847	Co 27 58.933	Ni 28 58.71	Cu 29 63.54	Zn 30 65.37	Ga 31 69.72	Ge 32 72.59	As 33 74.9216	Se 34 78.96	Br 35 79.909	Kr 36 83.80
Rb 37 85.47	Sr 38 87.62	Y 39 88.905	Zr 40 91.22	Nb 41 92.906	Mo 42 95.94	Tc 43 99	Ru 44 101.07	Rh 45 102.905	Pd 46 106.4	Ag 47 107.870	Cd 48 112.40	In 49 114.82	Sn 50 118.69	Sb 51 121.75	Te 52 127.60	I 53 126.9044	Xe 54 131.30
Cs 55 132.905	Ba 56 137.34	Lu 71 174.97	Hf 72 178.49	Ta 73 180.948	W 74 183.85	Re 75 186.2	Os 76 190.2	Ir 77 192.2	Pt 78 195.09			Tl 81 204.37	Pb 82 207.19	Bi 83 208.98	Po 84 [210]	At 85 [210]	Rn 86 [222]
Fr 87 [223]	Ra 88 [226]																

Rare Earths

La 57 138.91	Ce 58 140.12	Pr 59 140.907	Nd 60 144.24	Pm 61 [145]	Sm 62 150.35	Eu 63 151.96	Gd 64 157.25	Tb 65 158.924	Dy 66 162.50	Ho 67 164.930	Er 68 167.26	Tm 69 168.934	Yb 70 173.04	
Ac 89 [227]	Th 90 232.038	Pa 91 [231]	U 92 238.03	Np 93 [237]	Pu 94 [242]	Am 95 [243]	Cm 96 [247]	Bk 97 [249]	Cf 98 [251]	Es 99 [254]	Fm 100 [253]	Md 101 [258]	No 102 [254]	

Physical constants and conversion factors

F.1 PHYSICAL CONSTANTS

Constant	Symbol	Value
Speed of light	c	2.9979×10^8 m·sec^{-1}
Electron rest mass	m_e	9.1091×10^{-31} kg
Proton rest mass	m_p	1.6725×10^{-27} kg
Neutron rest mass	m_n	1.6748×10^{-27} kg
Planck's constant	h	6.6257×10^{-34} J·sec
	$\hbar = h/2\pi$	1.0545×10^{-34} J·sec
Boltzmann's constant	k	1.3805×10^{-23} J·(°K)$^{-1}$
Avogadro's number	N_0	6.0225×10^{26} (kg-mole)$^{-1}$
Gravitational constant	G	6.670×10^{-11} N·m^2·kg^{-2}
Vacuum permittivity	ϵ_0	8.8538×10^{-12} C^2·N^{-1}·m^{-2}
	$\dfrac{1}{4\pi\epsilon_0}$	8.9880×10^9 N·m^2·C^{-2}
Vacuum permeability	μ_0	1.2566×10^{-6} kg·m·C^{-2}
	$\mu_0/4\pi$	1.0000×10^{-7} kg·m·C^{-2}
Electron charge	e	-1.6021×10^{-19} C

F.2 CONVERSION FACTORS

Force

$$1 \text{ newton (N)} = 10^5 \text{ dynes (dyn)}$$

Charge

$$1 \text{ coulomb (C)} = 2.9979 \times 10^9 \text{ statcoulombs}$$
$$= 0.1 \text{ abcoulomb}$$

Magnetic field

$$1 \text{ weber} \cdot \text{m}^{-2} \text{ (Wb} \cdot \text{m}^{-2}) \equiv 1 \text{ tesla (T)} = 10^4 \text{ gauss (G)}$$

Mass and Energy

$$
\begin{aligned}
1 \text{ u} &= 1.660 \times 10^{-27} \text{ kg} \\
1 \text{ u} &= 1.492 \times 10^{-10} \text{ J} \\
1 \text{ u} &= 931.0 \quad\quad \text{MeV} \\
1 \text{ kg} &= 6.025 \times 10^{26} \text{ u} \\
1 \text{ kg} &= 8.989 \times 10^{16} \text{ J} \\
1 \text{ kg} &= 5.611 \times 10^{35} \text{ eV} \\
1 \text{ eV} &= 1.602 \times 10^{-19} \text{ J} \\
1 \text{ eV} &= 1.074 \times 10^{-9} \text{ u} \\
1 \text{ eV} &= 1.782 \times 10^{-36} \text{ kg} \\
1 \text{ J} &= 6.704 \times 10^9 \text{ u} \\
1 \text{ J} &= 6.242 \times 10^{18} \text{ eV}
\end{aligned}
$$

$$1 \text{ J} = 1.112 \times 10^{17} \text{ kg}$$
$$1 \text{ keV} = 10^3 \text{ eV}$$
$$1 \text{ MeV} = 10^6 \text{ eV}$$
$$1 \text{ GeV} = 10^9 \text{ eV}$$

Answers to odd-numbered problems

CHAPTER 2

5. 5.83
7. 2.30×10^{-8} N; 1.12×10^6 m·sec^{-1}
11. 1.6°, 16°, 110°
13. 2.17×10^{-13} m; 1.75×10^{-7}
15. 2.0×10^3
17. 1.05 mA
19. $\mathbf{F}_{12} = -\mathbf{F}_{21} = k_m \mathbf{I} \cdot \mathbf{I}_2 \hat{\mathbf{r}}/r$

CHAPTER 3

3. 1.25×10^{15} N·C^{-1}, south
5. $+\infty$, $-\infty$, 2.1 cm from the smaller charge toward the larger
7. a parabola
9. $\left(\dfrac{2ek_e\lambda}{m_e}\right)^{1/2}$
13. $(v_u + v_d)\left[\dfrac{72\pi^2\eta^2 v_d}{E^2(\rho_0 - \rho_a)g}\right]^{1/2}$
15. $\dfrac{1}{2\pi}\left(\dfrac{pE}{I}\right)^{1/2}$
17. $F = \mathbf{p} \cdot \nabla E$
21. 0.44D, 0.61D, 0.68D, 0.78D
25. $2k_e\lambda(\ln r - \ln \infty)$
27. $k_e q \left\{ \dfrac{3(x-1)}{[(x-1)^2 + y^2]^{3/2}} - \dfrac{2x}{[x^2 + (y+1)^2]^{3/2}} + \dfrac{(x-1)}{[(x-1)^2 + (y-1)^2]^{3/2}} \right\}$
29. $E_r = \dfrac{9qd^2}{16\pi\epsilon_0 r^4}(2\cos^2\theta - 1)$, $E_\theta = \dfrac{3qd^2}{8\pi\epsilon_0 r^4}\sin 2\theta$, $E_\varphi = 0$
31. $\tfrac{4}{3}\pi R^3 \sigma_0$

CHAPTER 4

3. zero
15. $g = \dfrac{mG}{r^2 + a^2}$ within star, $g = \dfrac{GM}{r^2}$ outside star
21. 3.67×10^3 kg·m^{-3}

CHAPTER 5

3. $\left(\dfrac{\epsilon_0 A}{d}\right)(n-1)$
9. 4.0×10^3 V·m^{-1}; 80 V·m^{-1}; 3.6×10^{-8} C·m^{-2}
11. $\dfrac{4\pi\epsilon_0 ab}{b-a}$
13.

Capacitor	Q (coulombs $\times 10^4$)	E (joules $\times 10^2$)
1	0.6	0.08
2	1.7	0.74
3	1.2	0.17
4	1.7	0.74
5	4.0	2.0
6	4.0	2.0

15. 1.2 μF
17. 3.28×10^{-6} J·m^{-1}
19. (a) $\frac{\epsilon_0 A}{d}[(1-K)x + Kb]$; (b) $\frac{\epsilon_0 a V_0^2(K-1)}{2d}$

CHAPTER 6

5. The displacement of the particle is given by
$$x(t) = x_0 - \frac{mv_\perp}{qB}\cos\omega t$$
$$y(t) = y_0 + \frac{mv_\perp}{qB}\sin\omega t$$
$$z(t) = z_0 + v_\| t$$
where $\omega = qB/m$ and the uniform magnetic field is in the z direction
7. 2.4×10^{-15} N, down and to the northeast
9. $0.99995c$, yes
11. 1.2×10^{-2} m

CHAPTER 7

3. $\frac{\mu I d}{2\pi r(d-r)}$ where r is the distance from one of the wires
5. $\frac{\mu_0 I}{\pi}\left\{\frac{1}{a} + \frac{1}{b} + \frac{1.414}{a}\sin\left[\tan^{-1}\left(\frac{a}{b}\right)\right] + \frac{1.414}{b}\sin\left[\tan^{-1}\left(\frac{b}{a}\right)\right]\right\}$
where $a = 1.414L - 2x$
$b = 1.414L + 2x$
$x =$ distance along a diagonal from the center of the coil to the field point
7. $\frac{\mu_0 q\Omega}{2\pi R}$; $\frac{q\Omega R^2}{4}$
9. $\frac{\mu_0 i}{2}$ where i is the linear current density in A·m^{-1}
11. $\frac{\mu_0 IL}{2\pi}\ln\left(1+\frac{L}{d}\right)$ **13.** $\frac{N\mu_0 I}{2\pi l}\sin\left(\frac{\pi}{N}\right)$
15. 8.0 A; 5.9×10^{-2} Wb·m^{-2}; 5.3×10^{-2} Wb·m^{-2}
23. 1.3×10^{29} J   **27.** 5.0×10^{-5} J
29. $\mu \tanh\left(\frac{\mu B}{kT}\right)$
31. 13.9 Wb·m^{-2}; 1.92×10^{-23} A·m^2
33. 2.5β; 7.9β **37.** 5.81×10^{-2} mm

CHAPTER 8

7. 6.75×10^{-4} Wb·m^{-2} **9.** $\frac{-\mu_0 l}{2\pi}\left(\frac{dI}{dt}\right)\ln\left[\frac{(1+w/a)}{(1+w/b)}\right]$
11. $\frac{\mu_0 \mu}{2\pi R^3} 2\pi v A \sin(2\pi v t + \varphi)$ **15.** 6.67 Wb·m^{-2}

17. 5.0×10^{-4} V·m^{-1}; 6.25×10^{-4} V·m^{-1}; 3.12×10^{-4} V·m^{-1}
18. varies as r^{-1} **23.** 2.3×10^9 A
25. $\mu_0 n_1 n_2 AL$ **27.** $\frac{1}{2}LI^2$, $L = \frac{\mu_0}{2\pi} \ln\left(\frac{r_2}{r_1}\right)$
29. 7.9×10^{-2} J·m^{-1} **31.** $L = \dfrac{L_1 L_2}{L_1 + L_2}$

CHAPTER 9

7. (a) $\dfrac{a^3}{2}$; (b) zero; (c) zero; (d) zero

13. $\nabla \times \mathbf{B} = \mu_0 \left(\mathbf{j}_e + \epsilon_0 \dfrac{\partial \mathbf{E}}{\partial t}\right)$ $\nabla \cdot \mathbf{E} = \dfrac{\rho_e}{\epsilon_0}$

$\nabla \times \mathbf{E} = \mathbf{j}_m - \dfrac{\partial \mathbf{B}}{\partial t}$ $\nabla \cdot \mathbf{B} = \dfrac{\rho_m}{\mu_0}$

15. $\oint \mathbf{B} \cdot d\mathbf{l} = \mu_0 \int_S \left(\mathbf{j} + \epsilon_0 \dfrac{\partial \mathbf{E}}{\partial t}\right) \cdot d\mathbf{S}$ $\int_S \mathbf{E} \cdot d\mathbf{S} = \dfrac{1}{\epsilon_0} \int_v \rho \, dv$

$\oint \mathbf{E} \cdot d\mathbf{l} = -\int_S \dfrac{\partial \mathbf{B}}{\partial t} \cdot d\mathbf{S}$ $\int_S \mathbf{B} \cdot d\mathbf{S} = 0$

25. 1.23×10^4 V·m^{-1}; 4.09×10^{-5} Wb·m^{-2}; 5.03×10^{23} photons·sec^{-1}

29. Phase difference for E_θ is zero, phase difference for B_φ is $-\dfrac{\pi}{2}$

CHAPTER 10

7. $\epsilon_0 A[K_1 K_2/(d_1 K_2 + d_2 K_1)]$; $(2\epsilon_0 A/d) K_1 K_2/(K_1 + K_2)$
9. $(\epsilon_0 A/d)(K_2 - K_1)/\ln(K_2/K_1)$
11. 8.8×10^{-8} J; 1.5×10^{-8} J
17. 0.0992 eV; 6.19×10^{-10} m
19. $4\pi\epsilon_0 R^3 [(K-1)/(K+2)] E_0$; $[3/(K+2)] E_0$
23. Inside sphere:
$\mathbf{D} = (Q/4\pi r^2)\hat{\mathbf{r}}$, $\mathbf{E} = (Q/4\pi\epsilon_0 K r^2)\hat{\mathbf{r}}$, $\mathbf{P} = [(K-1)/K](Q/4\pi r^2)\hat{\mathbf{r}}$
Outside sphere:
$\mathbf{D} = (Q/4\pi r^2)\hat{\mathbf{r}}$, $\mathbf{E} = (Q/4\pi\epsilon_0 r^2)\hat{\mathbf{r}}$
$[(K-1)/K]Q$; $-[(K-1)/K]Q$; Q/K
25. 2.3×10^{-41} C^2·sec^2·kg^{-1}; 8.9×10^{-41} C^2·sec^2·kg^{-1}
27. 2.5×10^{-10} C·m^{-2}

CHAPTER 11

11. 5.75×10^9 N·m^{-2}
15. 4.69×10^{-14} sec, 7.38×10^{-8} m, 3.15×10^{-3} m, 4.18×10^{-13} ohm·cm
17. $m\mathcal{E}$, mr, $\dfrac{m\mathcal{E}}{R + mr}$ **19.** $\left(\dfrac{R}{R+r}\right)\mathcal{E}$, 0.2%
21. 0.1 A
25. 2.2 W, 3.6 W, -2.4 W; 1.21 W, 1.21 W, 0.02 W, 0.64 W, 0.32 W

29. 1) $q = q_0\left(1 + \frac{R}{2L}t\right)\exp\left(-\frac{R}{2L}t\right)$

2) $q = q_0 + \frac{q_0 b}{2a}\left[\exp\left(\frac{-b+a}{2}\right)t - \exp\left(\frac{-b-a}{2}\right)t\right]$

where $a = \left(\frac{R^2}{L^2} - \frac{4}{LC}\right)^{1/2}$, $b = \frac{R}{L}$

3) $q = \frac{q_0}{\cos\delta}\exp\left(-\frac{bt}{2}\right)\cos\left[\frac{at}{2} - \delta\right]$

where $a = \left(\frac{4}{LC} - \frac{R^2}{L^2}\right)^{1/2}$, $b = \frac{R}{L}$, $\delta = \tan^{-1}\left(-\frac{b}{a}\right)$

31. 0.693

33. (a) $10^{-5}\exp(-t)$ C·sec^{-1} (b) $10^{-4}\exp(-2t)$ W (c) 1.35×10^{-3} W

CHAPTER 12

9. $E_F = \left(\frac{E_d + E_g}{2}\right) + \frac{kT}{2}\log\left[\frac{n_d}{2(2\pi m_e kT/h^2)^{3/2}}\right]$

13. 3.2×10^{-4} V **21.** $\lambda_L = \left(\frac{m_e}{n_s\mu_0 e^2}\right)^{1/2}$; 17 nm

CHAPTER 13

1. 4.0×10^2 **7.** 3.4×10^{-21} m^2; $1.2a_0^2$

CHAPTER 14

3. 377 ohms **7.** 38.7°
9. (a) $n = \sin[(\phi + \delta_m)/2]/\sin(\phi/2)$; (b) $\delta_b - \delta_r = (n_b - n_r)\phi$; (c) 0.3°
11. (b) $R = C/\sin\theta_i$ **13.** 34.3 cm
15. $E_3/E_1 = \tan(\theta_1 - \theta_2)/\tan(\theta_1 + \theta_2)$
19. 1.4×10^8 Hz; 1.4×10^6 Hz; 1.4×10^2 Hz
21. $\mu\rho\sigma\omega/2\pi[1.41a(\mu\sigma\omega)^{1/2} - 1]$ **25.** 2.6×10^8 m·sec^{-1}; 0.53 MeV
29. fast axis horizontal $\exp(i\pi/4)\begin{bmatrix}1 & 0\\0 & -i\end{bmatrix}$

fast axis vertical $\exp(i\pi/4)\begin{bmatrix}1 & 0\\0 & i\end{bmatrix}$

35. twice the angle resulting from a single traversal.

CHAPTER 15

1. $0.693/\mu$ **3.** 1.4×10^7 Hz; 7.1×10^{-8} sec
5. 7.2×10^{-3} eV; 29.7 cm^{-1} **11.** 1.68 eV; 505 nm
17. 0.4 MeV/c; 0.4 eV **19.** 10 : 1
25. $2.9 \times 10^4 : 1$

CHAPTER 16

3. $t_w = 2\pi/\gamma B_1$; 4.7 μsec   **9.** 0.08 V

21. (a) two resonance lines of equal intensity
(b) two resonance lines with intensities in the ratio 3 : 1
(c) one resonance line

23.

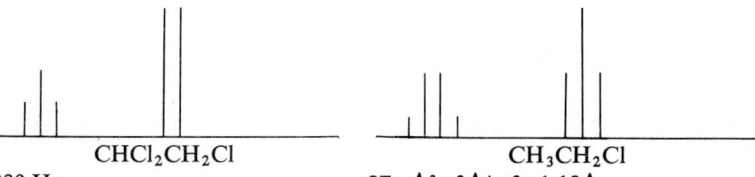

    CHCl$_2$CH$_2$Cl   CH$_3$CH$_2$Cl

25. 280 Hz   **27.** Δ^2; $3\Delta^4$; 3; 1.18Δ

29. $0.5(1 + \beta)M'_2 + (0.97\beta + 0.035)M''_2$

31. $T_1 \propto T^{-1/2}$

37. $\pm\frac{9}{2}$ ─────────

$4\Delta E$

$\pm\frac{7}{2}$ ─────────

$3\Delta E$

$\pm\frac{5}{2}$ ─────────

$2\Delta E$

$\pm\frac{3}{2}$ ─────────

ΔE

$\pm\frac{1}{2}$ ─────────

M

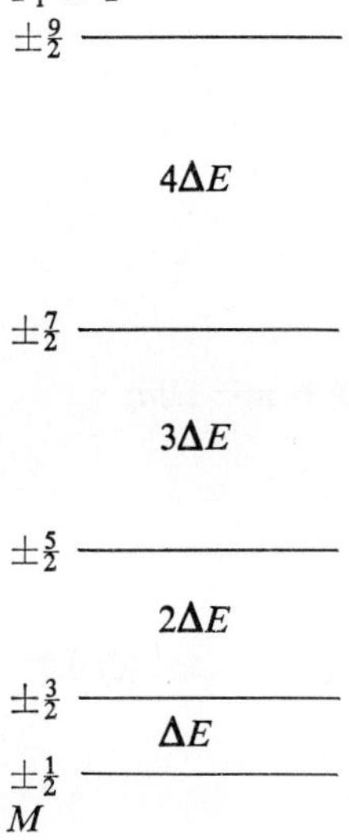

CHAPTER 17

3. 2.5 mm   **5.** 0.69 mm

9. 4.8×10^{-8} nm; 3×10^4 Hz   **11.** 3.6 mm

17. $(E_0^2/4\alpha\pi) \exp\{-2\alpha[(\omega - \omega_0)/2\alpha]^2\}$

23. 5.5×10^{-10} rad   **25.** 0.013 mm

27. 1.00044

CHAPTER 18

5. 0.54 mm; 0.80 nm   **7.** 1.2×10^{11} Hz

9. $\pi R^{1/2}/(1 - R)$; 29   **11.** 0.944; 5.50

17. 0.939; 0.053; 0.014   **19.** $\begin{bmatrix} -1 & 0 \\ 0 & -1 \end{bmatrix}$

CHAPTER 19

13. 5, 5/2, 5/3, 5/4
15. $I = I_0[2J_1(p)/p]^2$ with $I_0 = (C\pi R^2)^2$
19. 3670 lines
25. $\lambda = 2a \sin\theta/(h^2 + k^2 + l^2)^{1/2}$
31. $g(y) = W/S + 2 \sum_{n=1}^{\infty} (1/n\pi) \sin(n\pi W/S) \cos(2n\pi y/S)$
33. $E_0[\frac{1}{2} + \sum_{n=1,3,5\cdots}^{\infty} (2/n\pi)^2 \cos(2n\pi y/S)]$

CHAPTER 20

5. Output frequencies: ω_1, ω_2, $(2\omega_1 - \omega_2)$, $(2\omega_2 - \omega_1)$, $(2\omega_1 + \omega_2)$ and $(2\omega_2 + \omega_1)$
13. 0.80 cm^{-1}

Index

Index

INDEX

Abampere, 15
Abcoulomb, 15
Absorption
    by atoms and molecules, 296f
    optical, 295f
    by solids, 297f
Absorption coefficient, 295
Absorption edges, 302
Absorption process, 310
Acceptor, 227
Action-at-a-distance, 3, 20
Adenine, 29
Adenosine diphosphate (ADP), 307
Adenosine triphosphate (ATP), 307
Ampere, 14
Ampère, André Marie, 2, 13
Amino acid, 308
Ampère's law, 95
    Maxwell's second equation, 139f
Amplification in a medium, 433f
Angular spectrum, 277
Anomalous dispersion, 273
Antibody, 198
Antiferromagnetism, 255f
Antigen, 198
Asymmetry parameter, 344
Atomic form factor, 409
Atomic magnetic moment, 110
Atomic number, 8
Atomic photoelectric effect, 301
Attenuation length, 305

Back emf, 128
Bacon, Francis, 256
Bardeen, John, 232, 234
Basov, Nikolai, 5
BCS theory (*see* Superconductivity)
Beta-ray spectrograph, 134
Betatron, 123f
Bias polarization, 427
Biaxial crystal, 286
Biomagnetism, 87
Biophysics, 29
Biot, Jean, 92
Biot-Savart law, 92
Bipolar junction transistor, 232f
Birefringence, 286
Blackbody radiation, 311
Bleaney, Brebis, 317
Bleeper, 349
Bloch, Felix, 5, 317, 326
Bloch's equation, 326f
    chemical exchange, 330f
    response to a pulse, 330
    steady state solution, 327f
Bond moment additivity, 29
Bohr magneton, 109
    effective number, 249

Bohr, Niels, 4
Born, Max, 5
Bound charge, 174
Boundary conditions
    at dielectric surfaces, 181
    at magnetic surfaces, 244f
Bragg, William L., 409
Bragg's law, 410
Brattain, Walter H., 232
Brewster angle, 285
Brewster, Sir David, 285
Broglie, Louis de, 5

Capacitance, 60
    of capacitors connected in parallel, 65
    of capacitors connected in series, 66
    of cylindrical capacitor, 64
    of parallel plate capacitor, 61
    in the presence of a dielectric, 176f
Capacitor, 60
    cylindrical, 64
    energy stored, 67f
    parallel plate, 60
Cathode-ray tube, 62
    writing rate, 62
Cavendish, Robert, 53
Cerenkov, Paul, 276
Cerenkov radiation, 276f
CGS electromagnetic system of units, 15
CGS electrostatic system of units, 15
Characteristic impedance, 266
Charge, 8
    bound, 174
    elementary, 23f
    magnetic, 74
    polarization (bound), 174
Charge density, 34
Chemical exchange, 330f
    rapid exchange limit, 332
    slow exchange limit, 332
Chemical shift, 333f
    table of proton shifts in some simple
        gases, 334
Chlorophyll, 307
Chromosome, 309
Circuit element
    nonohmic, 196
    ohmic, 195
Circular polarization, 280
Coercive force, 255
Coherence
    complete, 361
    degree of, 361
    lateral width, 367
    partial, 361
    spatial, 365f
        lateral, 365
    temporal, 360f

INDEX

Coherence function
  mutual, 361
Coherence length, 363
  for second harmonic generation, 429
Coherence time, 361
  and line width, 363
Coherent source, 357
Color center, 298
Complete incoherence, 361
Complex conjugate, 460
Complex number, 460f
  absolute value, 460
  argument, 460
  imaginary part, 460
  modulus, 460
  phase, 460
  real part, 460
  rules for arithmetic operations, 461
Compton, Arthur H., 302
Compton effect, 302f
Compton wavelength, 304
Conduction band, 223
Conductivity
  electrical, 195
    temperature variation, 203f
  intrinsic semiconductor, 226
Conductor, 173, 222f
  index of refraction, 272
  propagation of plane waves, 271f
Continental drift, 256f
Cooper, Leon, 234
Cooper pairs, 236
Corpuscular theory of light, 3
Correlation function
  mutual, 361
  normalized, 361
Cosmic rays, 77f
  and the magnetic field of the earth
    east-west effect, 79
    latitude effect, 79
  primary, 77
  secondary, 77
  and supernovae, 78
Cosmology
  big-bang, 165
  steady-state, 165
Cotton-Mouton magneto-optic effect, 425
Coulomb, 14
Coulomb, Charles Augustin, 2, 9
Coulomb scattering, 10f
  impact parameter, 11
  scattering angle, 11
Coulomb's law, 9f
  in a dielectric, 180
Crick, Francis, 29
Crystal field theory, 341f
  medium field case, 341
  weak field case, 341
Curie law, 248

Curie, Pierre, 248
Curie temperature, 187, 248, 254
Curie-Weiss law, 248
Curl, 137
Current, 8
  displacement, 141, 179f
  forces between, 13
Current density, 205
Cyclotron, 18, 89
Cyclotron frequency, 81, 230
Cyclotron resonance, 82, 230
Cytoplasm, 309
Cytosine, 29

Debye unit, 28
Degree of coherence, 361
Dehmelt, H., 317
Del, 35, 137
Density
  charge, 34
  current, 205
  electric energy, 69
  magnetic energy, 130, 243
  magnetic flux, 74
Deoxyribonucleic acid (DNA), 29, 308
Diamagnetic susceptibility, 247
Diamagnetism, 245f
Dielectric, 173
  nonpolar, 183f
  polar, 184f
  propagation of plane waves, 264f
  speed of electromagnetic waves, 265
Dielectric constant, 173
  for a gas, 275
  and molecular polarizability, 184, 186
  for some nonpolar gases, 184
  for some polar gases, 186
  principal values, 286
Difference frequencies
  generation of, 445
Diffraction, 4
  circular aperture, 407
  circular obstacle, 408
  double slit, 403f
  Fraunhofer, (see Fraunhofer diffraction)
  Fresnel, (see Fresnel diffraction)
  particle, 412
  single slit, 396f, 399f
  x-rays, 409f
Diffraction grating, 404f
  resolving power, 406
  three-dimensional, 409
Dimer
  thymine-thymine, 31, 309
Diode laser, 438
Dipole-dipole interaction, 336f

INDEX

Dipole moment
   electric, (*see* Electric dipole moment)
   magnetic (*see* Magnetic dipole moment)
   molecular, 28
Dirac, Paul, 4, 5
Direct process, 343
Disc dynamo, 126
Dispersion, 265, 272*f*
   anomalous, 273
   mass, 83
   normal, 273
   rotary, 283
Displacement
   electric, 176, 181
Displacement current, 141
   in presence of dielectric, 179*f*
Dissymmetric molecular structure, 283
Divergence, 137
   curvilinear co-ordinates, 452
   spherical co-ordinates, 453
Divergence theorem, 138, 453*f*
Domain
   ferroelectric, 187
   ferromagnetic, 252, 254*f*
Donor, 227
Doping, 228
Double refraction, 286*f*
Drift speed, 194

Earth's magnetic field, 78, 125f, 127
East-west effect, 79
Effective Bohr magneton number, 249
Effective mass, 230
Einstein, Albert, 4, 310, 376
Einstein coefficients, 310*f*
Electric dipole moment, 25, 36*f*
   induced, 183
   lines of force, 27
   of nuclei and elementary particles, 40*f*
   oscillating, 154
   for some polar molecules, 186
   potential due to, 27
   torque on, 27
Electric dipole radiation, 154*f*
Electric displacement, 176
   boundary condition at dielectric surfaces, 181
Electric energy, 68
Electric energy density, 69
Electric field, 3, 4, 20*f*
   about an accelerated charge, 161
   boundary condition at dielectric surfaces, 181
   about a charge in uniform motion, 161
   due to a charged conducting spherical shell, 52
   due to a large plane sheet of uniform

Electric field *(cont.)*
   charge density, 53
   due to a long linear uniform charge distribution, 54
   around an oscillating dipole, 155*f*
   relation to magnetic field, 97*f*
   due to a spherical charge distribution, 51
   stored energy, 68, 178*f*
Electric field gradient parameter, 344
Electric flux, (*see* Flux)
Electric force, 21
Electric intensity, 21
Electric lines of force, 21, 74
   refraction at dielectric surfaces, 182
Electric potential, 32*f*
   isolated point charge, 33
   system of point charges, 33
Electric potential energy, 32
Electric quadrupole, 37*f*
   linear, 37
   potential of, 40
   torque experienced, 39
Electric quadrupole moment
   of nuclei, 41*f*
Electric susceptibility, 175
Electrical conductivity (*see* Conductivity)
Electromagnetic induction, 3
   Faraday's law, 118, 120*f*
Electromagnetic radiation, 154*f*
Electromagnetic spectrum, 150
Electromagnetic waves, 4, 137, 146*f*
   energy flow, 149*f*
   relation between electric and magnetic fields, 147*f*
   from space, 162*f*
   spectrum, 150
   speed in a medium, 190, 265
   speed in vacuum, 147
Electromotive force, 118, 204*f*
   back, 128
   Hall, 229
   induced, 119*f*
Electron, 8
   solvated, 277
Electron-coupled spin interaction, 334*f*
Electron paramagnetic resonance, 317
   crystal field theory, 341*f*
   exchange narrowing, 340
   quadrupolar spin-lattice relaxation, 343*f*
Electron spin *g*-factor, 109
Electron spin resonance, 342
Electrostatic force, 9
Electrostatics, 72
Elementary charge, 23*f*
Elliptical polarization, 281
Energy
   electric, 32, 68, 178*f*

486 INDEX

Energy *(cont.)*
  of electric dipole moment, 27
  exchange, 252
  Fermi, 199
  magnetic, 130
  of a magnetic dipole moment, 108
  stored in a capacitor, 67f
  stored in an electric field, 68, 178f
  stored in a magnetic field, 130
Energy absorption
  by spin system, 326
Energy band, 219
Energy density
  in magnetic field, 243
Energy flow
  via electromagnetic waves, 149f
Energy gap
  semiconductor, 224
  superconductor, 235
Enzyme, 309
Ether, 4, 374
Evanescent wave, 269, 277
Exchange
  chemical (*see* Chemical exchange)
Exchange energy, 252
Exchange narrowing, 340
Exciton, 297
Extrinsic semiconductor, 228

F-center, 298
Fabry-Perot etalon, 384
Fabry-Perot interference filter, 390
Fabry-Perot interferometer, 383f
  finesse, 392
  free spectral range, 384
  fringe contrast, 392
  resolving power, 385
Farad, 60
Faraday, 16
Faraday, Michael, 3, 21, 118, 126, 425
Faraday's law, 118, 120f
  Maxwell's third equation, 143f
Fermat, P., 288
Fermat's principle, 288
Fermi-Dirac distribution, 199
Fermi electron gas, 199f
  electron energy distribution, 202
Fermi energy, 199
Fermi speed, 203
Ferrimagnetism, 256
Ferrite, 256
Ferroelectric crystal, 187
  hysteresis effect, 188
  polarization catastrophe, 187
Ferroelectric domain, 187
Ferromagnetic domain, 252, 254f
  anisotropy energy, 254
Ferromagnetism, 252f

Ferromagnetism *(cont.)*
  coercive force, 255
  Curie temperature, 254
  hysteresis loop, 255
  remanence, 255
Feynman, Richard, 4
Field, 20
  electric (*see* Electric field)
  magnetic (*see* Magnetic field)
Field theory, 3
Finesse, 392
Fizeau experiment, 270
Flash photolysis, 291
Fluorescence, 299
Flux
  electric, 49
  magnetic, 74, 120f
    trapped, 237
Forbidden region, 221
Force
  coercive, 255
  between currents, 13
  electric, 21
  electromotive (*see* Electromotive force)
  electrostatic, 9
  Lorentz, 77
  magnetic, (*see* Magnetic force)
Four-ninth effect, 353
Frank, Ilya, 277
Fraunhofer diffraction, 395f
  double slit, 403f
  Fresnel-Kirchhoff formula, 399f
  by single slit, 396f, 399f
Free induction decay, 330
Free spectral range, 384
Frequency
  cyclotron, 81, 230
  difference
    generation of, 445
  Larmor, 319
  of LC circuit, 132
  of LCR circuit, 212
  spatial, 413
  sum, 444
Fresnel, Augustin Jean, 4, 381
Fresnel diffraction, 395f, 406f
  circular aperture, 407
  circular obstacle, 408
Fresnel-Kirchhoff formula, 395
  for Fraunhofer diffraction, 399f
Fresnel zone, 407
Fresnel's equations, 269f
Fringe contrast, 392
Fringe visibility, 361
Fusion power reactors, 104f

Gabor, Dennis, 417

INDEX

Gain constant, 433
Gas laser, 438f
Gauss, 76
Gauss, Karl Friedrich, 3, 49
Gauss' law, 49f
   applications, 51f
   for a dielectric, 175
   at dielectric surfaces, 181
   for electric displacement, 180f
   and the electric field, 138f
   and the gravitational interaction, 55
   interface between two magnetic media, 244
   and the magnetic field, 139
Gene, 308
Gerlach, Walter, 112
Gradient, 35, 137
   curvilinear coordinates, 452
   spherical coordinates, 453
Grating equation, 405
Gravitational radiation, 164
Guanine, 29

Hall coefficient, 229
   table of values, 229
Hall, Edwin, 228
Hall effect, 90, 228f
Hall emf, 229
Hanbury-Brown, R., 371
Hansen, William, 317
Harmonic polarization waves, 427
Heisenberg, Werner, 5
Helmholtz coils, 114
Henry, Joseph, 118, 129
Hertz, Heinrich, 4, 137, 147
Hess, Victor, 77
Hexadecapole moment, 44
High-frequency choke, 135
Hole, 225
Hole burning, 434
Hologram, 417, 418
Holography, 417f
   object wave, 417
   primary diffracted wave, 419
   reconstruction wave, 418
   reference wave, 417
   secondary diffracted wave, 419
   twin wave, 418
Hubble, Edwin P., 165
Hubble's law, 163
Hund's rules, 259
Huygens, Christian, 3
Huygens' principle, 394
Hydrogen
   hyperfine levels, 163
   spin-flip transition, 163
Hysteresis effect, 188
Hysteresis loop, 255

Immune response, 198
Impact parameter, 11
Impedance
   characteristic, 266
Impurity semiconductor (*see* Semiconductor, impurity)
Incoherence
   complete, 361
Index of refraction, 265
   absorptive component, 276
   of conductor, 272
   dispersive component, 276
   extraordinary, 286
   ordinary, 286
   principal values, 286
   table for some common crystals, 287
Induced electric dipole moment, 183
Induced electromotive force, 119f
Inductance, 127f
   mutual, 128
   self, 127
Induction
   electromagnetic (*see* Electromagnetic induction)
   magnetic, 74
Inductor, 131
Insulator, 173, 222f
Intensity interferometry, 370f
Interference, 4
Interference filter, 390
Interference term, 358
Interferogram, 372
   of water vapor, 373
Interferometer
   Fabry-Perot, (*see* Fabry-Perot interferometer)
   Michelson, 371f
   phase-switched, 370
   in radio astronomy, 368f
   simple radio-frequency, 369
Interferometry
   intensity, 370f
   long baseline, 368
Internal resistance, 207
Intrinsic semiconductor (*see* Semiconductor, intrinsic)
Isotope effect, 235

Josephson, Brian, 237
Josephson junctions, 237

Kerr electro-optic effect, 425, 445
Kerr, John, 425
Kerst, Donald W., 123
Kirchhoff, Gustav Robert, 217
Kruger, H., 317

488   INDEX

Lande g-factor, 111
Larmor frequency, 319
Larmor precession, 245f
Larmor radius, 90
Laser, 5, 429
    diode, 438
    gas, 438f
    liquid, 442
    nonlinear optical effects, 442f
    principle of, 429f
    semiconductor, 437f
    solid, 435f
    threshold condition, 433f
Lateral coherence width, 367
Latitude effect, 79
Laue equations, 411f
Laue, Max von, 409
LC circuit, 131
    frequency of oscillation, 132
LCR circuit, 211f
    critically damped charging, 211
    overdamped charging, 211
    underdamped charging, 211
        frequency of oscillation, 212
Leith, Emmett, 418
Lenz, Heinrich, 121
Lenz's principle, 121
Light pipe, 288
Line shape function, 325
Line width
    and coherence time, 363
    natural, 312
Lines of force, 3
    electric, 21, 74
        refraction at dielectric surfaces, 182
    about an electric dipole, 27
    magnetic, 74
    about a point charge, 22
Liquid laser, 442
Lloyd's single-mirror experiment, 376
Lobe, 369
London, F., 240
London, H., 240
Long-baseline interferometry, 368
Lord Rayleigh's blue sky law, 297
Lorentz force, 77
Lorentzian form, 328
LR circuit, 210f
    time constant, 211
Luminescence, 299

Magnetic bubbles, 255
Magnetic charge, 74
Magnetic dipole moment, 90, 107
    of a free atom, 110
    in an inhomogeneous magnetic field, 111

Magnetic dipole moment *(cont.)*
    of an orbiting electron, 108
    of a spinning electron, 109
    of a nucleus, 109
    potential energy in a magnetic field, 108
    spin, 109
    torque experienced in a magnetic field, 107
Magnetic dipole radiation, 159f
    from pulsars, 163f
Magnetic energy, 130
Magnetic energy density, 130, 243
Magnetic field, 3, 4, 74
    on the axis of a circular current loop, 106
    boundary condition at magnetic surfaces, 244f
    about charge in uniform motion, 161
    of the earth, 78, 125f
        reversals, 127
    energy stored in, 130
    motion of a charge in, 79f
    near a long, straight current, 92
    about a neutron star, 165
    about oscillating electric dipole, 155
    relation to electric field, 97f
Magnetic field intensity, 242
    boundary condition at magnetic surfaces, 245
Magnetic field strength, 74
Magnetic flux *(see* Flux)
Magnetic flux density, 74
Magnetic force, 13
    a current in a magnetic field, 86
    on a moving charge, 74f
    between two long, parallel currents, 94
Magnetic induction, 74
Magnetic lines of force, 74
Magnetic mirror, 85
Magnetic moment *(see* Magnetic dipole moment)
Magnetic permeability, 243
Magnetic polarization, 242
Magnetic resonance, 5, 317f
Magnetic resonance spectrum, 331
Magnetic susceptibility, 242
Magnetization, 242
Magnetogyric ratio, 252
Magnetron
    parallel-plate, 88
Maser
    optical, 5
Mass
    effective, 230
Mass dispersion, 83
Mass spectrometer, 82f
    age determination of rocks, 83
    in biology, 84

Mass spectrometer *(cont.)*
  variable field, 89
Matched circuits, 208
Matrix, 464
  rules for arithmetic operations, 464*f*
Maxwell, James Clerk, 3, 53, 137, 375
Maxwell's equations, 4
  summary of, 144
  for free space, 144
  general form, 263
  invariance, 150*f*
Maxwell's first equation, 139
Maxwell's fourth equation, 139
Maxwell's second equation, 142
  and Ampere's law, 139*f*
Maxwell's third equation, 144
  and Faraday's law, 143*f*
Mean free path, 195
Meissner effect, 235
Metallic crystals, 219*f*
Michelson, Albert, 375
Michelson interferometer, 371*f*
Michelson-Morley experiment, 374*f*
Microfarad, 60
Miller, Stanley, 307
Millikan oil-drop experiment, 23*f*
Millikan, Robert A., 23
Mirror point, 85
Modulation
  optical, 445*f*
Molecular dipole moment, 28
Molecular dissymmetry, 283
Molecular field coefficient, 260
Molecular polarizability, 183
  and dielectric constant, 184, 186
Monopole, 74
Monopole moment, 44
Morley, Edward, 375
Motional narrowing, 337
Multilayer film, 387*f*
  transfer matrix, 388
Multiple reflections
  from a thin plate, 381*f*
Mutual coherence function, 361
Mutual correlation function, 361
Mutual inductance, 128

Natural line width, 312
Negative crystal, 286
Neutron, 8
Neutron star, 164
Newton, Isaac, 3, 375
Newton's rings, 377
Nodal line, 376
Nonlinear optics, 426, 442*f*
Nonohmic circuit element, 196
Nonpolar dielectric media, 183*f*
Nonpolar molecule, 28

Normal dispersion, 273
Normalized correlation function, 361
n-type semiconductor, 227
Nuclear g-factor, 110
Nuclear magnetic resonance, 317
Nuclear magneton, 110
Nuclear paramagnetism, 252
Nuclear quadrupole resonance, 317, 344*f*
  and molecular structure, 346
  and the nature of chemical bonds, 346
  quadrupole coupling constants and
    asymmetry parameters, 345
  ratio of nuclear quadrupole moments,
    345
  structural phase transitions, 347
Nucleon, 8

Obliquity factor, 395
Octupole moment, 44
Oersted, Hans Christian, 2, 13, 92
Ohm, 195
Ohm, Georg Simon, 194
Ohmic circuit element, 195
Ohm's law, 193*f*
Onnes, Kamerlingh, 234
Optic axis, 286
Optical absorption, 295*f*
Optical activity, 282*f*
Optical filtering
  high-pass, 414
  low-pass, 414
Optical modulation, 445*f*
Optical pumping, 431
Optics, 3
Orbach process, 344
Order of diffraction, 405
Oscillating electric dipole, 154
  radiation pattern, 159
Oscillations
  LC circuits, 131*f*
  LCR circuit, 212
Oscilloscope, 61*f*
  deflection sensitivity, 62
  dual-beam, 62
  linear time base, 62
  rise time, 70
  sampling, 63
Ozone layer, 309

Packard, Martin, 317
Pair production, 304*f*
Pair production cross section, 305
Paramagnetic susceptibility, 248*f*
Paramagnetism, 247*f*
  some iron-group ions, 250
  nuclear, 252
  Pauli, 251
  some rare-earth ions, 250

Partial coherence, 361
Particle diffraction, 412
Pauli paramagnetism, 251
Pauli, Wolfgang, 5, 251
Penetration depth, 240
Periodic table, 468
Permeability
   free space, 91
   magnetic, 243
Permittivity constant, 14
Permittivity, 177
   vacuum, 14
Phase change upon reflection, 270
Phase-switched interferometer, 370
Phonon, 343
Phosphor, 299
Phosphorescence, 299
Photoconductivity, 299f
Photoelectric cross section, 301
Photoelectric effect, 300f
   atomic, 301
   properties of some metals, 301
Photon, 294
Photoreactivation, 309
Photosynthesis, 306
Pi/2 pulse, 320
Picofarad, 60
Piezoelectricity, 191
Pinch effect, 105
Planck, Max, 4
Plane wave, 148
   using complex notation, 263f
   in conducting media, 271f
   in isotropic dielectrics, 264f
p-n junction, 231
   current vs voltage, 232
   forward bias, 231
   reverse bias, 231
p-n junction diode, 232
Point charge
   line of force about, 22
Poisson's equation, 168
Polar dielectric media, 184f
Polar molecule, 28, 183
   electric dipole moment, 186
Polarizability
   molecular (*see* Molecular polarizability)
Polarization, 174, 280f
   bias, 427
   circular, 280
   elliptical, 281
Polarization
   magnetic, 242
   by reflection, 285
Polarization catastrophe, 187
Polarization charge, 174
Polarizing angle, 285
Population inversion, 431

Positive crystal, 286
Positron, 17
Positronium, 17
Potential
   electric, 32f
   electric quadrupole, 40
   scalar, 169
   vector, 169
Potential drop, 206
Potential energy
   electric, 32
   of electric dipole moment, 27
   of magnetic dipole moment, 108
Pound, Robert, 317
Power
   condition for maximum transfer, 208
   electrical, 204f
   fusion reactors, 104f
Power spectrum, 339, 364
Poynting vector, 267
Primeval fireball, 166
Principle of least time, 288
Principle of linear superposition, 357
Prokhorov, Aleksandr, 5
Propagation constant, 264
Propagation vector, 264
Proton, 8
Proton magnetometer, 348f
Proton spin refrigerator, 354
p-type semiconductor, 227
Pulsar, 164
Pulse radiolysis, 278f
Purcell, Edward, 5, 317

Q-switching technique, 436
Quadrupolar spin-lattice relaxation, 343f
   direct process, 343
   Orbach process, 344
   Raman process, 343
Quadrupole
   electric (*see* Electric quadrupole)
Quadrupole coupling constant, 345
Quadrupole moment
   electric, 37f
   scalar, 38
Quadrupole moment tensor, 44
Quantization
   spatial, 112
Quantum number space, 200
Quantumelectrodynamics, 4
Quark, 24
Quasar, 166
Quasi-stellar radio source, 166
Quenching of orbital angular momentum, 250

Rabi, Isidor, 5

INDEX

Radiation
    from accelerated charge, 160f
    blackbody, 311
    Cerenkov, 276f
    electric dipole, 154f
    electromagnetic, 154f
        from primeval fireball, 165f
    gravitational, 164
    magnetic dipole, 159f
        from pulsars, 163f
    synchrotron, 162
    21.1 cm, 162f
Radiation damage, 31
Radiation pattern
    oscillating dipole, 159
Radio-frequency susceptibility, 328
Raman effect, 443
    stimulated, 443f
Raman process, 343
Rationalized MKS system of units, 15
Rayleigh criterion, 417
RC circuit, 208f
    time constant, 209
Red shifts
    law of, 163
Reflectance, 290, 382
Reflection, 4, 267f
    multiple from a thin plate, 381f
    phase change, 270
    and polarization, 285
    total internal, 269
Reflection coefficient, 381
Refraction, 4, 267f
    double, 286f
    of electric field lines, 182
    Snell's law, 268
Refractive index (see Index of refraction)
Remanence, 255
Resistance, 118, 195
    internal, 207
    parallel connection of resistors, 207
    and the separation of living cells, 196f
    series connection of resistors, 206
Resistivity, 195
    of some pure metals, 196
Resolving power, 385
    of diffraction grating, 406
    Fabry-Perot interferometer, 385
Resonance
    cyclotron, 82, 230
    electron paramagnetic (see Electron
        paramagnetic resonance)
    electron spin, 342
    magnetic, 5, 317f
    nuclear magnetic, 317
    nuclear quadrupole (see Nuclear
        quadrupole resonance)
Resonant cavity, 430
Ribonucleic acid (RNA), 308

Ribosome, 314
Rock magnetism, 257
Rotary dispersion, 283
    dextrorotatory, 283
    laevorotatory, 283
Rotating coordinate axes, 457f
Rowland ring, 258
Rutherford, Ernest, 10, 57

Saturation broadening, 326
Saturation factor, 325
Savart, Felix, 92
Scalar potential, 169
Scattering, 297
Scattering angle, 11
Scattering cross section, 304
Schawlow, Arthur L., 430
Schrieffer, Robert, 234
Schroedinger equation
    solution for a gas of free electrons, 199f
Schroedinger, Erwin, 5
Schwinger, Julian, 4
Screening constant, 334
Second moment, 337
Seismic waves, 56
Seismogram, 56
Self-focusing, 445
Self inductance, 127
Semiconductor, 224f
    energy gap, 224
    extrinsic, 228
    impurity, 227f
        Fermi level, 228
        n-type, 227
        p-type, 227
    intrinsic, 224f
        conductivity of, 226
        Fermi level, 226
Semiconductor laser, 437f
Shockley, William B., 232
Skin depth, 272
Snell's law, 268
Solenoid, 95
Solid angle, 49
Solid lasers, 435f
Solvated electron, 277
Source
    coherent, 357
Spatial coherence, 365f
Spatial coherence
    lateral, 365
Spatial filtering, 413f
    high-pass, 414
    low-pass, 414
Spatial frequencies, 413
Spatial quantization, 112

Spectrograph
  beta-ray, 134
Spectrum
  angular, 277
  electromagnetic, 150
Speed
  drift, 194
  of electromagnetic waves, 265
  Fermi, 203
  of light, 147
Speed of light cylinder, 165
Spin-coupling constant, 334
Spin-lattice coupling, 322
Spin-lattice relaxation
  electron, 343f
  nuclear, 338f
Spin-lattice relaxation time, 324
Spin magnetic moment, 109
Spin-rotation interaction, 340
Spin-spin relaxation time, 325
Spontaneous emission process, 310
Spontaneous lifetime, 312
Stark effect, 190
Statcoulomb, 15
Stellarator, 105
Steradian, 50
Stern-Gerlach experiment, 112
Stern, Otto, 112
Stimulated emission process, 310
Stimulated Raman emission, 443f
Stokes' theorem, 138, 454f
Structure factor, 409
Sum frequencies
  generation of, 444
Superconductor, 234f
Superconductivity
  BCS theory, 234, 236
    Cooper pairs, 236
  critical field, 235
  energy gap, 235
  isotope effect, 235
  Meissner effect, 235
  penetration depth, 240
  transition temperatures, 235
  trapped magnetic flux, 237
  weakly coupled
    Josephson junctions, 237
Superposition
  principle, 357
Susceptibility
  diamagnetic, 247
  electric, 175
  magnetic, 242
  of a metal, 251
  of a nuclear paramagnetic system, 252
  paramagnetic, 248f
  radio-frequency, 328
Synchrotron radiation, 162

Synchrotron radiation (cont.)
  from quasars, 166f
Systems of units
  CGS electromagnetic, 15
  CGS electrostatic, 15
  rationalized MKS, 15

Tamm, Igor, 277
Temperature
  Curie, 187, 248, 254
  superconducting transition, 235
Temporal coherence, 360f
Thompson, Joseph J., 57
Three-dimensional diffraction grating, 409
Thymine, 29
Thymine-thymine dimer, 31, 309
Time constant
  LR circuit, 211
  RC circuit, 209
Tokamak, 105
Tomonaga, Shin-Ichiro, 4
Toroid, 96
Torque
  on electric dipole moment, 27
  on electric quadrupole, 39
  on a magnetic dipole moment, 107
Torrey, Henry, 317
Total internal reflection, 269
Townes, Charles, 5, 430
Transfer matrix
  for multilayer film, 388
Transistor
  base, 233
  collector, 233
  collector current vs collector voltage, 234
  emitter, 233
Transistor, bipolar junction, 232f
Transmission coefficient, 381
Transmittance, 290, 382
Transfer matrix, 388
Twenty-one cm radiation, 162f
Twiss, R.Q., 371

Uniaxial crystal, 286
Units
  CGS electromagnetic system, 15
  CGS electrostatic system, 15
  rationalized MKS system, 15
Unpolarized radiation, 284
Unpolarized wave, 148
Upatnieks, Juris, 418
Urey, Harold, 307

Vacuum permittivity, 14
Valence band, 225

Van Allen, James A., 84
Van Allen radiation zones, 84*f*
   inner and outer radiation zones, 84
   slot region, 84
Vector potential, 169
Velocity filter, 88
Visibility
   fringe, 361
Volt, 32
Voltage drop, 206

Watson, James, 29
Watt, James, 205
Wave
   electromagnetic (*see* Electromagnetic wave)
   evanescent, 269, 277
   harmonic polarization, 427
   object, 417
   plane (*see* Plane wave)
   primary diffracted, 419
   reconstruction, 418
   reference, 417
   secondary diffracted, 419
   twin, 418
   unpolarized, 148

Wave equation, 145*f*
Wave vector, 264
Wavelength
   Compton, 304
Weber, 76, 121
Weber per square meter, 76
Weiss, Pierre, 260
Whistler, 289
Wigner, Eugene, 5
Wilkins, Maurice, 29
Work function, 301

X-ray diffraction, 409f

Young, Thomas, 3, 357, 377
Young's double slit interference experiment, 377

Zavoisky, E., 317
Z-axis modulation, 62